Recent Advances in Polyphenol Research

Recent Advances in Polyphenol Research

A series for researchers and graduate students whose work is related to plant phenolics and polyphenols, as well as for individuals representing governments and industries with interest in this field. Each volume in this biennial series focuses on several important research topics in plant phenols and polyphenols, including chemistry, biosynthesis, metabolic engineering, ecology, physiology, food, nutrition, and health.

Recent Advances in Polyphenol Research

Volume 5

Edited by

Kumi Yoshida

Professor, Natural Product and Bioorganic Chemistry
Graduate School of Information Science
Nagoya University, Japan

Véronique Cheynier

Research Director, Plant and Food chemistry
Institut National de la Recherche Agronomique
UMR1083 Sciences pour l'Œnologie
Montpellier, France

Stéphane Quideau

Professor, Organic and Bioorganic Chemistry
Institut des Sciences Moléculaires, CNRS-UMR 5255
University of Bordeaux, France

WILEY Blackwell

Registered Office
John Wiley & Sons, Ltd, The Atrium, Southern Gate, Chichester, West Sussex, PO19 8SQ, UK

Editorial Offices
9600 Garsington Road, Oxford, OX4 2DQ, UK
The Atrium, Southern Gate, Chichester, West Sussex, PO19 8SQ, UK
111 River Street, Hoboken, NJ 07030-5774, USA

For details of our global editorial offices, for customer services and for information about how to apply for permission to reuse the copyright material in this book please see our website at www.wiley.com/wiley-blackwell.

Library of Congress Cataloging-in-Publication Data

ISBN: 9781118883266

Recent advances in polyphenol research

ISSN 2474-7696

A catalogue record for this book is available from the British Library.

Wiley also publishes its books in a variety of electronic formats. Some content that appears in print may not be available in electronic books.

Cover image: The ICP2014 Organizing Committee

Set in 10/13pt Times by SPi Global, Pondicherry, India
Printed and bound in Malaysia by Vivar Printing Sdn Bhd

10 9 8 7 6 5 4 3 2 1

Dedication

To **Michel Bourzeix**—one of the founders of Groupe Polyphénols and its secretary from 1972 to 1995—who devoted his career to promoting research on polyphenols and supported GP activities and conferences with dedication and enthusiasm

To **Dieter Treutter**—a faithful member of the Groupe Polyphénols board for many years and the organiser of ICP2000

in memoriam

The editors wish to thank all of the members of the "Groupe Polyphénols" Board Committee (2012–2014) for their guidance and assistance throughout this project.

Contents

Contributors

Yoshinori Asakawa, Faculty of Pharmaceutical Sciences, Tokushima Bunri University, Yamashiro-cho, Tokushima, Japan.

Vincent L. Chiang, Forest Biotechnology Group, Department of Forestry and Environmental Resources, North Carolina State University, Raleigh, NC, USA.

Ling Chuang, Forest Biotechnology Group, Department of Forestry and Environmental Resources, North Carolina State University, Raleigh, NC, USA.

Katherine H. M. Cox, Centre for Human Psychopharmacology, School of Health Sciences, Swinburne University, Melbourne, Victoria, Australia.

Olivier Dangles, UMR 408 INRA, Sécurité et Qualité des Produits d'Origine Végétale, University of Avignon, Avignon Cedex 9, France.

Joel J. Ducoste, Civil, Construction and Environmental Engineering, North Carolina State University, Raleigh, NC, USA.

Claire Dufour, UMR 408 INRA, Sécurité et Qualité des Produits d'Origine Végétale, Centre de Recherche PACA, University of Avignon, Avignon Cedex 9, France.

Sheena R. Gayomba, Department of Biology and Center for Molecular Signaling, Wake Forest University, Winston-Salem, NC, USA.

Hideaki Hara, Molecular Pharmacology, Department of Biofunctional Evaluation, Gifu Pharmaceutical University, Gifu, Japan.

Yoshihiro Hase, Ion Beam Mutagenesis Research Group, Quantum Beam Science Directorate, Japan Atomic Energy Agency, Takasaki, Gunma, Japan.

Fikret Isik, NCSU Cooperative Tree Improvement Program, Department of Forestry and Environmental Resources, North Carolina State University, Raleigh, NC, USA.

Yong Jiang, State Key Laboratory of Natural and Biomimetic Drugs, School of Pharmaceutical Sciences, Peking University, Beijing, China.

Herbert Kolodziej, Institute of Pharmacy, Pharmaceutical Biology, Freie Universität Berlin, Berlin, Germany.

Tadao Kondo, Graduate School of Information Science, Nagoya University, Chikusa, Nagoya, Japan.

Laigeng Li, National Key Laboratory of Plant Molecular Genetics and CAS Center for Excellence in Molecular Plant Sciences, Institute of Plant Physiology and Ecology, Chinese Academy of Sciences, Shanghai, China.

Quanzi Li, State Key Laboratory of Tree Genetics and Breeding, Chinese Academy of Forestry, Beijing, China.

Jie Liu, Forest Biotechnology Group, Department of Forestry and Environmental Resources, North Carolina State University, Raleigh, NC, USA.

Philip L. Loziuk, W.M. Keck Fourier Transform Mass Spectrometry Laboratory, Department of Chemistry, North Carolina State University, Raleigh, NC, USA.

Hai-Ning Lv, State Key Laboratory of Natural and Biomimetic Drugs, School of Pharmaceutical Sciences, Peking University, Beijing, China.

Yosuke Matsuo, Department of Natural Product Chemistry, Graduate School of Biomedical Sciences, Nagasaki University, Nagasaki, Japan.

Gloria K. Muday, Department of Biology and Center for Molecular Signaling, Wake Forest University, Winston-Salem, NC, USA.

David C. Muddiman, W.M. Keck Fourier Transform Mass Spectrometry Laboratory, Department of Chemistry, North Carolina State University, Raleigh, NC, USA.

Punith P. Naik, Civil, Construction and Environmental Engineering, North Carolina State University, Raleigh, NC, USA.

Kenjirou Ogawa, Molecular Pharmacology, Department of Biofunctional Evaluation, Gifu Pharmaceutical University, Gifu, Japan.

Kin-ichi Oyama, Research Center for Materials Science, Nagoya University, Chikusa, Nagoya, Japan.

Ikuo Saiki, Division of Pathogenic Biochemistry, Institute of Natural Medicine (INM), University of Toyama, Toyama, Japan.

Andrew Scholey, Centre for Human Psychopharmacology, School of Health Sciences, Swinburne University, Melbourne, Victoria, Australia.

Ronald R. Sederoff, Forest Biotechnology Group, Department of Forestry and Environmental Resources, North Carolina State University, Raleigh, NC, USA.

Rui Shi, Forest Biotechnology Group, Department of Forestry and Environmental Resources, North Carolina State University, Raleigh, NC, USA.

Yue-Lin Song, Modern Research Center for Traditional Chinese Medicine, Beijing University of Chinese Medicine, Beijing, China.

Takashi Tanaka, Department of Natural Product Chemistry, Graduate School of Biomedical Sciences, Nagasaki University, Nagasaki, Japan.

Claire Tonnelé, Chimie des Matériaux Nouveaux, University of Mons, Mons, Belgium.

Patrick Trouillas, UMR 850 INSERM, Faculté de Pharmacie, University of Limoges, Limoges Cedex, France.

Takanori Tsuda, College of Bioscience and Biotechnology, Chubu University, Kasugai, Aichi, Japan.

Peng-Fei Tu, State Key Laboratory of Natural and Biomimetic Drugs, School of Pharmaceutical Sciences, Peking University, Beijing, China.

Sermsawat Tunlaya-Anukit, Forest Biotechnology Group, Department of Forestry and Environmental Resources, North Carolina State University, Raleigh, NC, USA.

Jack P. Wang, Forest Biotechnology Group, Department of Forestry and Environmental Resources, North Carolina State University, Raleigh, NC, USA.

Justin M. Watkins, Department of Biology and Center for Molecular Signaling, Wake Forest University, Winston-Salem, NC, USA.

Cranos M. Williams, Electrical and Computer Engineering, North Carolina State University, Raleigh, NC, USA.

Peng Xu, National Key Laboratory of Plant Molecular Genetics and Key Laboratory of Synthetic Biology, Institute of Plant Physiology and Ecology, Chinese Academy of Sciences, Shanghai, China.

Chenmin Yang, Forest Biotechnology Group, Department of Forestry and Environmental Resources, North Carolina State University, Raleigh, NC, USA.

Ting-Feng Yeh, Forestry and Resource Conservation, National Taiwan University, Taipei, Taiwan.

Kumi Yoshida, Natural Product and Bioorganic Chemistry, Graduate School of Information Science, Nagoya University, Chikusa, Nagoya, Japan.

Ke-Wu Zeng, State Key Laboratory of Natural and Biomimetic Drugs, School of Pharmaceutical Sciences, Peking University, Beijing, China.

Preface

Polyphenols are secondary metabolites that are variously distributed in the plant kingdom and characterized by a wide diversity of chemical structures. On behalf of the international scholarly society "*Groupe Polyphénols*," which organizes the biennial conference, "*International Conference on Polyphenols*" (ICP), we define the term "polyphenol" as related to plant products exclusively derived from the shikimate/phenylpropanoid and/or the polyketide pathway, featuring more than one phenolic unit and deprived of nitrogen-based functions (http://www.groupepolyphenols.com/the-society/why-bother-with-polyphenols/). The number of known plant polyphenols is quite large, from structurally simple compounds such as the stilbenoid resveratrol or the flavonoid quercetin to complex macromolecules such as the proanthocyanidin oligomers or the lignin polymer. It is thus not surprising that their functions in plant and physicochemical properties are also quite varied. In the early 20th century, investigations on polyphenols were mainly dedicated to the determination of their structures and their roles in traditional medicines, as well as in vegetable tanning. Nowadays, research on plant polyphenols concerns a much wider area of science with novel and multidisciplinary efforts made toward the understanding of their properties and exploitation thereof in *inter alia* the development of new materials, the innovation in agriculture and food products, including the development of new crops and flowers, the higher fixation of carbon dioxide, and the formulation of functional foods with human health benefits, as well as the discovery of new pharmaceutical medicines.

This book series "*Recent Advances in Polyphenol Research*" began its publication in 2008 on the occasion of the 24th ICP in Salamanca, Spain. The content of this first volume was already mostly based on review articles written by plenary lecturers of the previous ICP, which had taken place in Winnipeg, Canada. Since then, this flagship publication of the *Groupe Polyphénols* has been released without any discontinuity every 2 years to provide the reader with authoritative updates on various topics of polyphenol research written by ICP plenary lecturers and by invited expert contributors.

This book, the fifth volume of the series, is concerned with the topics that were covered during the 27th ICP, which was organized jointly with the 8th edition of the *Tannin Conference* in September 2014 in Nagoya, Japan. In more than 40 years of the history of the *Groupe Polyphénols*, it was the first time that the *International Conference on*

Polyphenols took place in Asia. Six different main topics of the polyphenol science were selected for the scientific program of this memorable ICP2014 edition:

1) **Chemistry, Physicochemistry, and Materials Science**, covering structures, reactivity, organic synthesis, molecular modeling, fundamental aspects, chemical analysis, spectroscopy, molecular associations, and interactions of polyphenols.
2) **Biosynthesis, Genetics, and Metabolic Engineering**, covering molecular biology, genetics, enzymology, gene expression and regulation, trafficking, biotechnology, horticultural science, and molecular breeding related to polyphenols.
3) **Plants and Ecosystems, Lignocellulose Biomass**, covering plant growth and development, biotic and abiotic stress, resistance, ecophysiology, sustainable development, valorization, plant environmental system, forest chemistry, and lignin and lignan.
4) **Food, Nutrition, and Health**, covering food ingredients, nutrient components, functional food, mode of action, bioavailability and metabolism, food processing, influence on food and beverages properties, cosmetics, and antioxidant activity of polphenols.
5) **Natural Medicine and Kampo**, a special session for this first conference held in Asia covering oriental traditional medicine, herbal medicine, Chinese herbal medicine, folklore, mode of action, metabolism, natural products chemistry, and drug discovery.
6) **Tannins and Their Functions**, another special session on the occasion of this joint meeting with the *Tannin Conference* covering research topics related to condensed tannins, hydrolyzable tannins, tea, wine, persimmon, seed-coat color, mode of action, and enzymatic reactions.

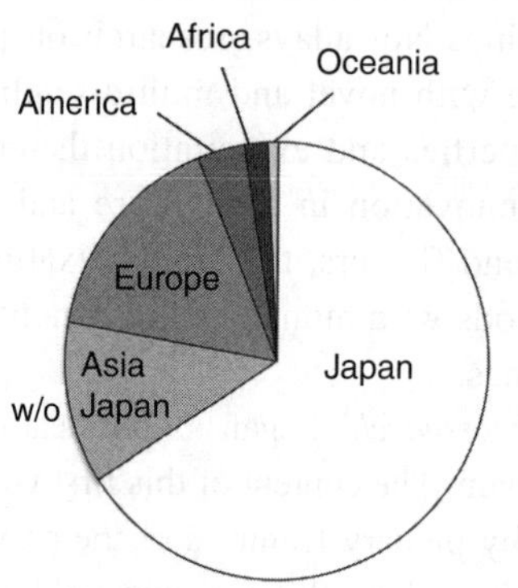

Asia	Europe	America
China	Austria	Canada
Hong Kong	Belgium	Chile
India	Finland	French Guiana
Indonesia	France	Mexico
Korea	Germany	United States
Malaysia	Italy	**Africa**
Oman	Latvia	Algeria
Saudi Arabia	Netherlands	Botswana
Taiwan	Poland	Morocco
Turkey	Portugal	South Africa
	Spain	
	Switzerland	
	United Kingdom	**Oceania**
		Australia
		New Caledoniaa

More than 500 scientists from 35 countries attended the conference, with 321 paper contributions that comprised 61 oral communications and 260 poster presentations. The fifth volume of "*Recent Advances in Polyphenol Research*" contains chapters from 14 guest speakers of the conference. The support and assistance of the *Groupe Polyphénols*, the *Tannin Conference* Group, several Japanese academic associations and foundations, notably the Nagoya University, the City of Nagoya and the Nagoya Convention & Visitors Bureau, and numerous private sponsors are gratefully acknowledged, as the great success of these joint editions of the *International Conference on Polyphenols* and the *Tannin*

Conference would not have been possible without their contributions. As a final note, we would also like to deeply thank all of the plenary, communication, and poster presenters for the quality of their contributions, from basic science to more applied fields, and all of the attendees, with a special thank to the numerous Asian researchers for their first participation in the ICP and for expressing their eagerness to attend the next ICP meetings.

Kumi Yoshida
Véronique Cheynier
Stéphane Quideau

Chapter 1

The Physical Chemistry of Polyphenols: Insights into the Activity of Polyphenols in Humans at the Molecular Level

Olivier Dangles, Claire Dufour, Claire Tonnelé and Patrick Trouillas

Abstract: This chapter reviews the following versatile physicochemical properties of polyphenols in relation with their potential activity in humans:

1) Interactions with proteins and lipid–water interfaces. These interactions must be qualified with respect to the current knowledge on polyphenol bioavailability and metabolism. They are expected to mediate most of the cell signaling activity of polyphenols.
2) A general reducing capacity that may be expressed in the gastrointestinal tract submitted to postprandial oxidative stress and also in cells, for example, by direct scavenging of reactive oxygen species, especially if preliminary deconjugation of metabolites takes place
3) The complex relationships with transition metal ions involving binding and/or electron transfer in close connection with the antioxidant versus pro-oxidant activity of polyphenols

Keywords: polyphenol, flavonoid, Health effectsbiological activity, mechanism, antioxidant, protein, membrane, metal ion, gastrointestinal tract, DFT methods.

1.1 Introduction

The activity, functions, and structural diversity of polyphenols in plants, food, and humans reflect the remarkable diversity of their physicochemical properties: UV–visible absorption, electron donation, affinity for metal ions, propensity to develop molecular interactions (van der Waals, hydrogen bonding) with proteins and lipid–water interfaces, and

Recent Advances in Polyphenol Research, First Edition. Edited by Kumi Yoshida, Véronique Cheynier and Stéphane Quideau.

nucleophilicity. This chapter aims to exemplify how polyphenols act to promote health in humans at the molecular level. It rests on two common assumptions based on epidemiological evidence and food analysis (Manach *et al.*, 2005; Crozier *et al.*, 2010; Del Rio *et al.*, 2013):

- The consumption of fruit and vegetables helps prevent chronic diseases and, in particular, favors cardiovascular health.
- Phenolic compounds, from the simple hydroxybenzoic and hydroxycinnamic acids to the complex condensed and hydrolyzable tannins, constitute the most abundant class of plant secondary metabolites in our diet and take part in this protection.

By contributing to the sensorial properties of food, for example, color and astringency, native polyphenols and their derivatives obtained after technological and domestic processing can directly influence the consumer's choice. Moreover, polyphenols undergo only minimal enzymatic conversion in the oral cavity and in the gastric compartment although their release from the food matrix (bioaccessibility) is an important issue. Thus, intact food polyphenols may directly promote health benefits in the upper digestive tract, in particular by fighting postprandial oxidative stress resulting from an unbalanced diet (Sies *et al.*, 2005; Kanner *et al.*, 2012). Beyond the gastric compartment, polyphenol bioavailability[1] (Fig. 1.1) must be considered as a priority to tackle any biological effects (Manach *et al.*, 2005; Crozier *et al.*, 2010; Del Rio *et al.*, 2013). Indeed, even for polyphenols that can be partially absorbed in the upper intestinal tract (aglycones, glucosides), most of the dietary intake reaches the colon where extensive catabolism by the microbiota takes place: hydrolysis of glycosidic and ester bonds, release of flavanol monomers from proanthocyanidins, hydrogenation of the C=C double bond of hydroxycinnamic acids, deoxygenation of aromatic rings, cleavage of the central heterocycle of flavonoids, and so on. Conjugation of polyphenols and their bacterial metabolites in intestinal and liver cells eventually results in a complex mixture of circulating polyphenol *O*-β-D-glucuronides and *O*-sulfo forms (less rigorously called sulfates). When present, catechol groups are also partially methylated.

The concentration of circulating polyphenols is usually evaluated after treatment by a mixture of glucuronidases and sulfatases that release the aglycones and their *O*-methyl ethers. This concentration is usually quite low (barely higher than 0.1 μM) and much lower than that of typical plasma antioxidants such as ascorbate (>30 μM). At first sight, this does not argue in favor of nonspecific biological effects, such as the antioxidant activity by radical scavenging or chelation of transition metal ions to form inert complexes. This seems all the more true that the catechol group, displayed by many common dietary polyphenols and which is a critical determinant of the electron-donating and metal-binding capacities, is generally either absent in the circulating metabolites (bacterial deoxygenation) or at least

[1]Bioavailability: the fraction of ingested polyphenol (native form + metabolites) that enters the general blood circulation and is thus potentially available for health effects.

Bioaccessibility: the first step of bioavailability, the fraction of ingested polyphenol (native form + metabolites) that is released from the food matrix and is thus potentially available for intestinal absorption.

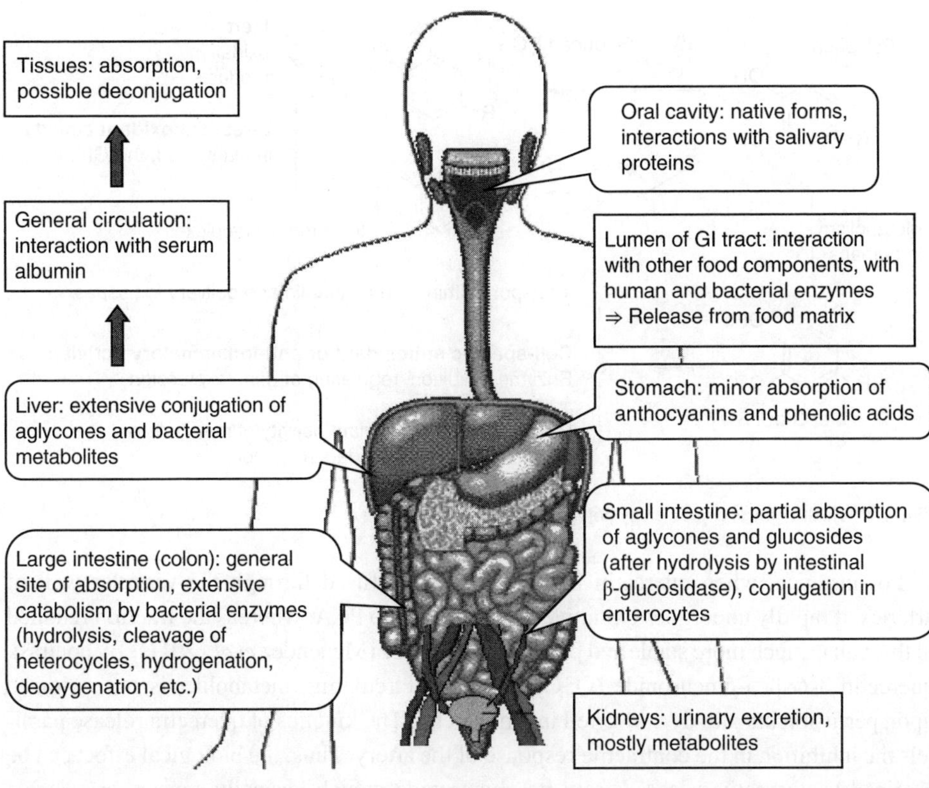

Fig. 1.1 A simplified view of polyphenol bioavailability. (*See insert for color representation of the figure*)

partially conjugated. However, the claim that *in vivo* polyphenol concentrations are low should be nuanced for the following reasons:

1) The complete assessment of polyphenol bioavailability must include the bacterial catabolites and their conjugates, some being much more abundant in the circulation than the parent phenol. A spectacular example can be found in the case of anthocyanins. Indeed, after consumption of blood orange juice, the total amount of native cyanidin 3-*O*-β-D-glucoside (C3G) in plasma is 0.02% of the ingested dose versus 44% for (unconjugated) protocatechuic acid (PCA), its main catabolite (Vitaglione *et al.*, 2007). When the fecal content is also taken into account, PCA eventually represents ca. 73% of the metabolic fate of ingested C3G. Its absence in urine (unlike C3G) also suggests that it takes part in the antioxidant protection and is thus oxidized in tissues.
2) The circulating concentration and its time dependence say nothing concerning either the possibility of polyphenol metabolites accumulation at a much higher local concentration at specific sites of inflammation and oxidative stress or their deconjugation into more active forms.

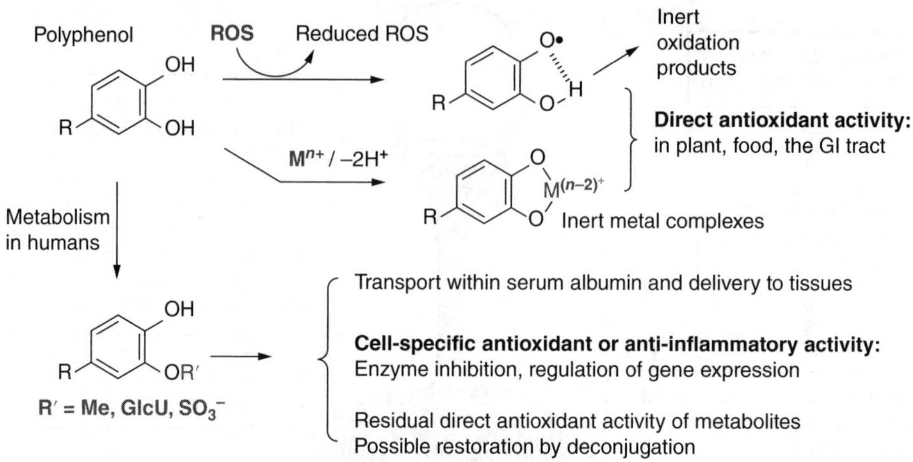

Fig. 1.2 Health effects expressed by polyphenols.

For instance, when quercetin is continuously perfused through the vascular wall of arteries, it rapidly undergoes oxidative degradation into PCA, whereas the fraction retained in the wall is much more stable and partially methylated (Menendez *et al.*, 2011). By contrast, quercetin 3-*O*-β-D-glucuronide (Q3G), the main circulating metabolite, is not oxidized upon perfusion but slowly converted into quercetin. The kinetics of quercetin release parallels the inhibition in the contractile response of the artery. Thus, the biological effect can be ascribed to quercetin released from its glucuronide, which basically appears as a stable storage form. A schematic view for the bioactivity of polyphenols is summed up in Fig. 1.2.

1.2 Molecular complexation of polyphenols

The phenolic nucleus can be regarded as a benchmark chemical group for molecular interactions as it combines an acidic OH group liable to develop hydrogen bonds (both as a donor and as an acceptor) and an aromatic nucleus for dispersion interactions (the stabilizing component of van der Waals interactions).

1.2.1 Polyphenol–protein binding

Polyphenol–protein binding of nutritional relevance can be classified as follows:

- Binding processes within the gastrointestinal (GI) tract, that is, with food proteins, mucins, and the digestive enzymes, with an impact on the bioaccessibility of polyphenols and the digestibility of macronutrients
  - Interactions with plasma proteins, with an impact on transport and the rate of clearance from the general circulation
  - Interactions with specific cell proteins (enzymes, receptors, transcription factors, etc.) that would mediate the nonredox health effects of polyphenols

As the last two situations lie downstream the intestinal absorption and passage through the liver, they concern the circulating polyphenol metabolites. However, some exceptions may be found. For instance, epigallocatechin 3-*O*-gallate (EGCG), the major green tea flavanol, is a rare example of a polyphenol entering the blood circulation mostly in its initial (nonconjugated) form (Manach *et al.*, 2005). No less remarkable, EGCG is also one of the rare polyphenols for which a specific receptor has been identified, namely the 67-kDa laminin receptor (67LR) that is expressed on the surface of various tumor cells (Umeda *et al.*, 2008). EGCG-67LR binding leads to myosin phosphatase activation and actin cytoskeleton rearrangement, thus inhibiting cell growth. It provides a strong basis for interpreting the *in vivo* anticancer activity of EGCG and its anti-inflammatory activity in endothelial cells (Byun *et al.*, 2014).

It is not the authors' purpose to provide the reader with an exhaustive updated report on polyphenol–protein binding processes (see Dangles and Dufour (2008) for a specific review on this topic). Only a few recent important examples will be discussed with an emphasis on works dealing with polyphenol metabolites.

1.2.1.1 Interactions in the digestive tract

In the postprandial phase, black tea drinking leads to vasorelaxation as evidenced by flow-mediated dilation experiments in humans and a strong increase in the activity of endothelial nitric oxide synthase (eNOS) (Lorenz *et al.*, 2007). However, these effects are completely abolished when 10% milk is added to black tea. Experiments with isolated fractions of milk proteins show that caseins are actually responsible for this inhibition. It can thus be proposed that caseins bind and probably precipitate black tea polyphenols in the GI tract, thereby preventing their intestinal absorption. This is a spectacular example of how food proteins may sequester oligomeric polyphenols and cancel their bioaccessibility and downstream biological effects.

The binding between dietary polyphenols and the digestive enzymes is best evidenced with large polyphenols such as oligomeric proanthocyanidins (OPAs). For instance, OPAs inhibit pancreatic elastase, a serine protease, proportionally to their mean degree of polymerization (Bras *et al.*, 2010). A K_i value of ca. 0.5 mM was estimated for a catechin tetramer. However, a mixture of *n*-mers ($n = 2$–6) rich in 3-*O*-galloyl flavanol units binds much more tightly ($K_i \approx 14\,\mu M$). Similar data were obtained with trypsin (Goncalves *et al.*, 2007). By slowing down the digestion, such interactions could prolong the sensation of satiety and help fight weight gain and obesity. By contrast, simple phenols were shown to mildly enhance pepsin activity at pH 2 in the following order: resveratrol ≥ quercetin > EGCG > catechin (Tagliazucchi *et al.*, 2005). Tannins are known to inhibit pancreatic lipase (McDougall *et al.*, 2009), thereby possibly contributing to lowering fat intake. Polyphenol-rich berry extracts also inhibit pancreatic α-amylase (thus decreasing starch digestibility) and intestinal α-glucosidase, with tannins and anthocyanins being, respectively, the main contributors to the observed inhibition (McDougall *et al.*, 2005). These mild inhibitory effects could help regulate the circulating D-glucose concentration.

1.2.1.2 Interactions beyond intestinal absorption

In the circulating blood, polyphenol metabolites likely travel in association with serum albumin, the most abundant plasma protein, which displays several binding sites for the transport of drugs, free fatty acids, and other nutrients. Our recent work (Khan *et al.*, 2011) has shown that flavanone glucuronides (conjugation at the A- or B-ring) are moderate serum albumin ligands ($K_b = 3–6 \times 10^4\ M^{-1}$) that bind site 2 (subdomain IIIA), in contrast to the more planar flavones and flavonols, which bind site 1 (subdomain IIA).

Once delivered to tissues, polyphenol metabolites are expected to bind specific cell proteins to express their biological effects, in particular their well-documented anti-inflammatory activity (Pan *et al.*, 2010; Spencer *et al.*, 2012; Wu & Schauss, 2012). Inflammation is an adaptive response to deleterious stimuli, activating the immune system. What is at stake with dietary polyphenols is the inhibition of chronic low-grade inflammation (in contrast to acute inflammation following microbial infection) associated with the development of degenerative diseases, such as type 2 diabetes and cardiovascular disease. Indeed, this pathological state is deeply influenced by lifestyle and environmental factors, especially dietary habits.

At the cell level, inflammation involves complex signaling pathways and cascades (Fig. 1.3). In particular, mitogen-activated protein kinases (MAPKs, e.g., ERK, JNK, and

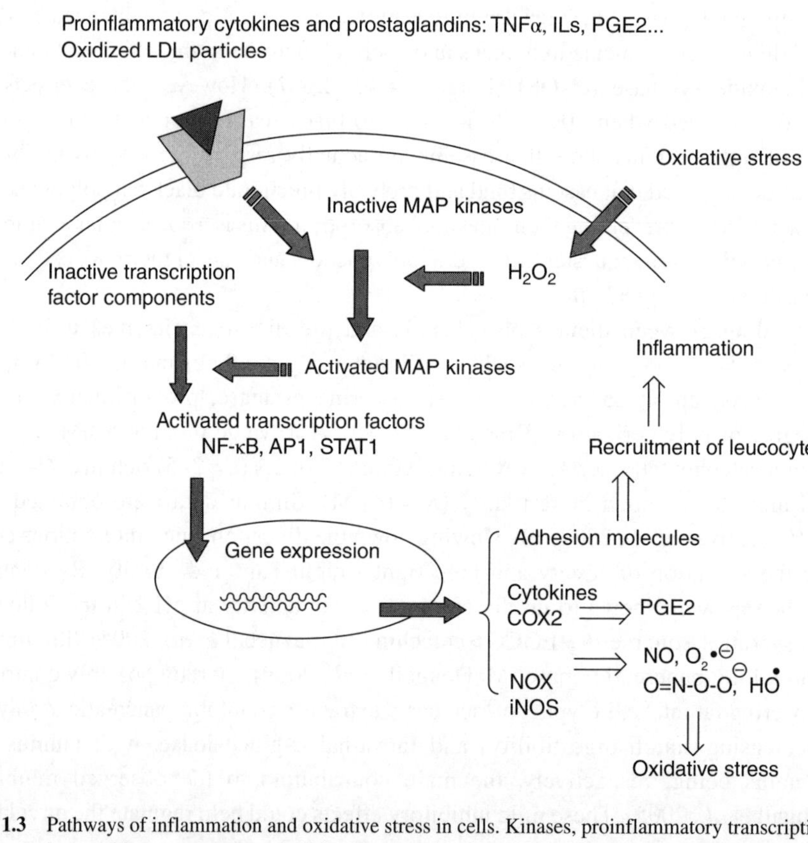

Fig. 1.3 Pathways of inflammation and oxidative stress in cells. Kinases, proinflammatory transcription factors, and pro-oxidant enzymes are possible target proteins for polyphenols and their metabolites.

p38) are important in the transduction of extracellular signals into cellular responses. When activated by oxidative stress or proinflammatory eicosanoids (prostaglandins, leukotrienes) and cytokines (e.g., TNFα, interleukins, and C-reactive protein), MAPKs phosphorylate both cytosolic and nuclear target proteins resulting in the assembly and translocation of transcription factors such as NF-κB, STAT1, and AP1. By upregulating the expression of inducible NO synthase (iNOS), cycloxygenase-2 (COX2), NADPH oxidase (NOX), cell adhesion molecules, cytokines, and cytokine receptors, these transcription factors trigger cell damage, inflammation, or apoptosis. MAPKs and the subsequently activated transcription factors (or their cytosolic components) are all potential targets of polyphenols and their metabolites, which rationalize their anti-inflammatory action. However, such mechanisms are subtle and not easy to track down to the highest level of resolution, that is, polyphenols interacting with specific proteins.

An additional difficulty also stems from the complex interplay between inflammation and oxidative stress. For instance, activated leucocytes (macrophages) produce reactive oxygen species (ROS) via the activity of NOX and iNOS. Conversely, NF-κB can be directly activated by ROS (Gloire *et al.*, 2006). Indeed, H_2O_2 is known to inhibit Tyr phosphatases via oxidation of Cys residues in the catalytic domain, thereby triggering Tyr kinase activity and downstream signaling. Thus, the overall biological effects of polyphenols in cells may be a complex combination of anti-inflammatory and antioxidant activities.

The anti-inflammatory activity of polyphenols can develop through the following:

- The inhibition of the cycloxygenase (COX) and lipoxygenase (LOX) enzymes responsible for the production of the inflammatory mediators prostaglandins and leukotrienes from arachidonic acid, respectively
- The downregulation of proinflammatory genes

A few recent examples are reported as follows with an emphasis on the possible activity of polyphenol metabolites.

Among the two main circulating quercetin metabolites, namely quercetin 3-*O*-β-D-glucuronide (Q3G) and 3′-*O*-sulfoquercetin, only the latter is a potent 5-LOX inhibitor in activated monocytes reducing accumulation of LTB_4 by ca. 50% at 2 μM (Loke *et al.*, 2008a). Unlike quercetin and 3′-*O*-methylquercetin, both metabolites were ineffective at inhibiting PGE_2 production. By contrast, with its free electron-rich catechol nucleus, Q3G is a much better inhibitor of LDL (low-density lipoprotein) peroxidation than 3′-*O*-sulfoquercetin.

NF-κB and STAT1 are important transcription factors for iNOS expression in macrophages. A structure–activity relationship with a series of flavonoid aglycones (Hamalainen *et al.*, 2007) has shown that the inhibition of iNOS expression and NO production in activated macrophages is due to the inhibition of the nuclear translocation of either the sole transcription factor NF-κB (flavone, the flavanone naringenin, 3′-*O*-methylquercetin) or both NF-κB and STAT1 (the flavonols kaempferol and quercetin, the isoflavones genistein and daidzein). However, the inhibition is modest at low flavonoid concentration (10 μM) and abolished with the corresponding flavonoid glycosides. It is thus doubtful that the main flavonoid circulating metabolites, that is, glucuronides, could exert a substantial anti-inflammatory activity via this mechanism, unless preliminary deconjugation takes place. A similar study in mouse microglia cells failed to demonstrate the anti-inflammatory activity of 3′-*O*-sulfoquercetin (Chen *et al.*, 2005). More encouraging is a recent investigation dealing with the porcine isolated coronary artery instead of cultured cells (Al-Shalmani *et al.*, 2011). In this study, it was shown that the lipopolysaccharide-induced alteration of the contractile response was significantly inhibited by low quercetin concentrations (0.1 μM) and higher concentrations (10 μM) of 3′-*O*-sulfoquercetin and Q3G. Moreover, NO production and iNOS expression were reduced. As the protection of the contractile response was abolished by an NF-κB inhibitor and persisted in endothelium-denuded segments, it can be proposed that quercetin and its metabolites act by inhibiting the NF-κB pathway in the vasculature, possibly by stabilizing the complex combining NF-κB and its cytosolic repressor IkB.

A direct binding between polyphenols and NF-κB proteins was suggested from experiments showing that procyanidin dimers B1 (epicatechin-β-4,8-catechin) and B2 (epicatechin-β-4,8-epicatechin), but not the more rigid A1 and A2, actually inhibit NF-κB-DNA binding (Mackenzie *et al.*, 2009; Fraga *et al.*, 2010). Docking experiments support a binding mode involving H-bonding between three phenolic OH groups of the dimers (the C3′-OH and C4′-OH of the terminal unit + the C7-OH of the extension unit) and two NF-κB Arg residues.

The anti-inflammatory activity of flavonoid metabolites in endothelial cells could also be mediated by their ability to inhibit the MAPK pathway. For instance, high D-glucose concentration is known to induce oxidative stress (evidenced by elevated H_2O_2 concentration) and subsequent activation of NOX and c-JUN N-terminal protein kinase (JNK) and caspase-3, which ultimately leads to apoptosis. Interestingly, D-glucose-induced JNK and caspase-3 activation and oxidative stress in endothelial cells are efficiently inhibited by physiological concentration (0.3 μM) of 3′-*O*-sulfoquercetin and Q3G (Chao *et al.*, 2009).

Finally, the flavanone metabolites showing conjugation at the B-ring, namely 3′-*O*-sulfohesperetin, hesperetin 3′-*O*-β-D-glucuronide, and naringenin 4′-*O*-β-D-glucuronide, were also demonstrated to inhibit the adhesion of monocytes to TNFα-activated endothelial cells (ca. −20% at 2 μM) (Chanet *et al.*, 2013). Gene expression analysis suggests that the protection involves the downregulation of genes coding for NF-κB, cell adhesion molecules, and cytoskeleton proteins.

Inhibition of pro-oxidant enzymes is also a mechanism for polyphenol metabolites to fight oxidative stress in cells. As an example, Q3G is a potent inhibitor of myeloperoxidase, which is secreted by neutrophils and macrophages at a site of inflammation and may be involved in LDL oxidation (Loke *et al.*, 2008b; Shiba *et al.*, 2008). Docking experiments suggest binding to a hydrophobic region of the enzyme with the B-ring pointing to the heme pocket.

Epicatechin glucuronides are even more potent than epicatechin at inhibiting NOX activity in stimulated endothelial cells (Steffen *et al.*, 2008). Experiments with disintegrated cells showed that unlike epicatechin (which simply scavenges superoxide), the glucuronides are true NOX inhibitors (IC50 ≈ 5 μM). Similar observations were made with quercetin and its glucuronides.

1.2.2 Interactions with membranes

There is growing evidence that interaction of phenolic compounds with biomembranes is important to rationalize their beneficial effects and toxicity. Nowadays, experimental techniques tackling this issue (fluorescence spectroscopy and microscopy, solid-state NMR, surface plasmon resonance, atomic force microscopy, Langmuir–Blodgett trough) are elegantly supported by molecular dynamics simulations for a detailed description of the different aspects of polyphenol–membrane interaction (penetration, partitioning, positioning, crossing).

As a first approach, the partition coefficient (log*P*) of flavonoid aglycones, which measures the relative lipophilicity (e.g., flavones are more lipophilic than the corresponding flavanones), was shown to correlate with their antioxidant capacity to protect membranes (Saija *et al.*, 1995). Nevertheless, log*P* does not reliably describe the amphiphilic character of polyphenols, a property that is of crucial importance to rationalize their membrane penetration and location.

Experimental (Hendrich *et al.*, 2002; Ollila *et al.*, 2002; Oteiza *et al.*, 2005) and theoretical (Sinha *et al.*, 2011; Kosinova *et al.*, 2012) works have shown that many flavonoids (flavonols, flavones, flavanones, flavan-3-ols, isoflavonoids) can penetrate lipid bilayers and preferentially lie in the polar head-group region rather than being deeply buried within the lipid chains. The driving forces of interaction and penetration arise from the amphiphilic character of polyphenols. Aromatic rings provide the hydrophobic character for interactions with lipid chains while the phenolic OH groups mainly act as hydrogen bond donors to the polar head groups of phospholipids. Such intermolecular hydrogen bonds tend to maintain polyphenols just below membrane surface, thus slowing down membrane crossing (passive diffusion).

The importance of planarity has been suggested by comparing the capacity of various phenolic compounds to penetrate lipid bilayers (Areias *et al.*, 2001; Lopez *et al.*, 2014). However, this must be nuanced with flavonoids, as the torsion between the C- and B-rings is rather flexible. Indeed, catechin derivatives, which are nonplanar, penetrate membranes and lie at a similar location as quercetin derivatives. By favoring multiple H-bonding, 3-*O*-galloylation of catechins enhances membrane affinity but favors a more superficial contact (Sirk *et al.*, 2008). Indeed, EGCG strongly binds through its B- and galloyl rings to the phosphodiester O-atoms and remains adsorbed on the bilayer surface. By contrast, EC mainly binds through its A-ring to the acyl O-atoms and is thus absorbed more deeply in the membrane.

The flavonolignan silybin locates at the interface of microsomal bilayers (Parasassi *et al.*, 1984) as well as genistein and daidzein (Raghunathan *et al.*, 2012), the former isoflavonoid being slightly more buried than the latter in agreement with its slightly higher lipophilicity (log*P* = 3.04 and 2.51, respectively). Interestingly, the lipophilic stilbenoid resveratrol (at physiological concentrations) appears more buried than most flavonoids and was shown to intercalate between phospholipid chains (Brittes *et al.*, 2010; Olas & Holmsen, 2012). However, the resveratrol–membrane interactions depend on lipids (length of acyl chains, degree of unsaturation, nature of the head group). As another example, the relatively hydrophobic gallotannin 1,2,3,4,5-penta-*O*-galloyl-β-D-glucopyranose (log*P* = 2.0) inserts more deeply into a lipid bilayer than the similar sized but much more hydrophilic catechin-α-4,8-catechin-α-4,8-catechin (log$P$ = −0.92) (Yu *et al.*, 2011).

Polyphenol penetration into membranes appears pH-dependent. For instance, quercetin displays pK_a values of 5.7, 7.1, 8.0 in water, corresponding to the three most acidic groups, namely C7-OH, C4′-OH, and C3-OH, respectively. At low pH, quercetin has a better capacity to penetrate lipid bilayers, whereas at neutral or basic pH, it locates closer to the polar domains, because of the repulsion between negative charges at the interface. Here, the experimental evidence (Movileanu *et al.*, 2000) agrees with molecular dynamics simulations (Kosinova *et al.*, 2012).

Phenolic compounds are known to aggregate by π-stacking and H-bonding interactions. The aggregation of flavanols at the membrane surface slows down penetration, especially when the 3-*O*-galloyl moiety is present (Sirk *et al.*, 2009). As a consequence, the partition coefficient of EGCG decreases with increasing concentration. The role of molecular size has also been suggested from molecular dynamics simulations. However, within the microsecond timescale, the difference in size between catechin and EGCG might weakly influence the penetration. Aggregation inside the lipid bilayer has also been indirectly evidenced with quercetin, due to segregation of the flavonol and clustering within microdomains (Movileanu *et al.*, 2000). At relatively high concentration, curcumin aggregates within the lipid chains as well, which consequently decreases lipid ordering (Loverde, 2014). Quercetin and other flavonoids (rutin, naringenin, genistein) were also shown to stabilize membranes through a decrease in lipid fluidity (Arora *et al.*, 2000). The authors suggested that this decrease in membrane fluidity might slow down free radical reactions. By contrast, resveratrol increases membrane fluidity. Its permeation of the membrane even in the gel phase confirms its high affinity to biomembranes.

In liposomes, flavonoids (e.g., quercetin) were proposed to inhibit lipid peroxidation by reducing the propagating lipid peroxyl radicals (Ioku *et al.*, 1995). The location of flavonoids just below the polar head surface could be critical: if the compound is slightly more buried, it can inhibit the propagation stage. If it is closer to the polar head groups, its access to the lipid peroxyl radicals may be lost. Such slight changes may be driven by lipid composition, lipid phase, pH of the aqueous phase, and the polyphenol's pK_a and log$P$ values. The specific location of flavonoids, that is, slightly less buried than vitamin E, is also ideal to enable regeneration of vitamin E.

As polyphenols are more likely to bind membranes as conjugates, it is quite relevant to compare aglycones with *O*-β-D-glucuronides and *O*-sulfo forms. In the case of quercetin, molecular dynamic simulations (Kosinova *et al.*, 2012) clearly show that the polar conjugates bind in a more superficial manner than the aglycone (Fig. 1.4). Thus, whereas quercetin lies below the interface, mostly parallel to the surface and with its 5-OH groups at 1.5 (±0.2) nm from the center of the DOPC membrane, Q3G is pulled to the surface by the glucuronyl moiety protruding in the aqueous phase, so that the conjugate lies in average at 1.8 (±0.2) nm from the center of the membrane. It can thus be anticipated that the quercetin conjugates are less efficient than quercetin at scavenging lipid peroxyl radicals, as suggested by the decrease in their ability to protect LDL (Loke *et al.*, 2008b). Similarly, the capacity of the metabolites at regenerating vitamin E in membranes is predicted to be lower than for quercetin. Again, deconjugation is

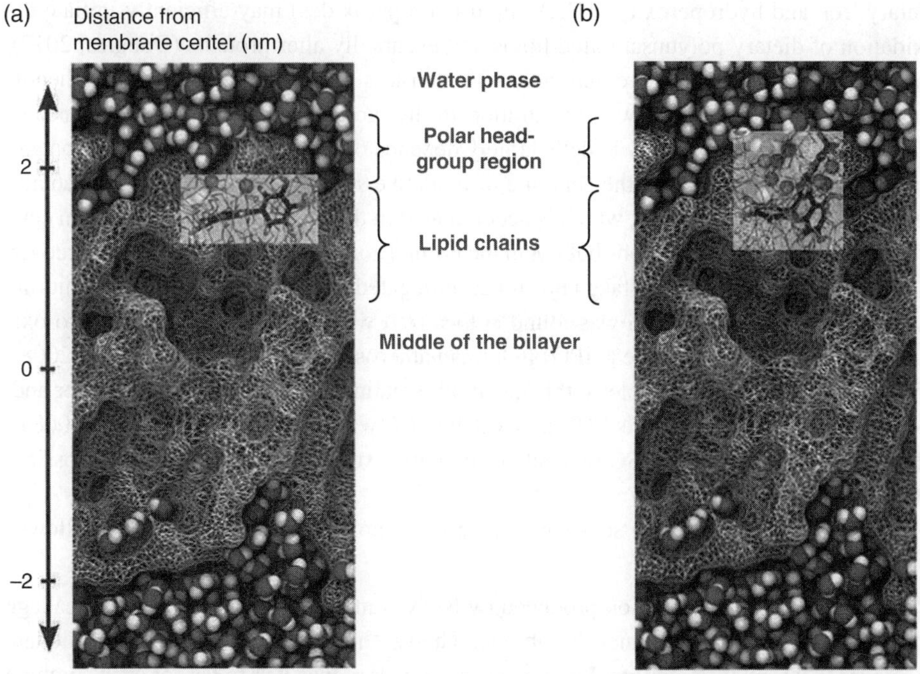

Fig. 1.4 Snapshots of the location of quercetin (a) and quercetin 3-*O*-β-D-glucuronide (b) in a 1,2-dioleoyl-*sn*-glycero-3-phosphatidylcholine lipid bilayer. Spheres represent the phosphatidyl groups (Source: Adapted from Kosinova *et al.* 2012). (*See insert for color representation of the figure*)

expected to markedly increase the ability of polyphenols to protect membranes against oxidation, not only by restoring the redox activity but also by favoring their penetration into lipid bilayers.

Nonbioavailable oligomeric proanthocyanidins (OPAs) may also exert their bioactivity via direct interactions with the membrane of intestinal cells. Quite importantly, hexameric PAs were shown to specifically bind the lipid rafts of the Caco-2 cell membrane (Da Silva *et al.*, 2012; Verstraeten *et al.*, 2013), that is, more rigid domains rich in cholesterol, glycosphingolipids, and sphingomyelin and incorporating proteins involved in major cellular events. The interaction is cholesterol-dependent, results in a superficial decrease in membrane fluidity, and inhibits deoxycholate-induced cell permeabilization. Consequently, OPAs also inhibit the activation of MAP kinases and NOX in intestinal cells and could thus help fight chronic colonic inflammation and oncogenesis.

1.3 Polyphenols as electron donors

The catechol nucleus of many common polyphenols is a potent electron/H-atom donor for the reduction of the ROS involved in oxidative stress (Fig. 1.2). Given the limited intestinal absorption and extensive metabolism of dietary polyphenols in humans, this classic mechanism of antioxidant activity now seems especially relevant in the digestive tract where dietary iron and hydroperoxides (H_2O_2, lipid hydroperoxides) may efficiently initiate the oxidation of dietary polyunsaturated lipids and eventually alter proteins (Dangles, 2012). As polyphenol metabolites are generally (i) much less reducing than native polyphenols and (ii) recovered in only low concentration in the blood circulation, the importance of their ROS-scavenging activity in cells largely depends on their possible accumulation and substantial deconjugation on the very site of oxidative stress. This possibility has actually been demonstrated with Q3G, which is accumulated in the macrophage-derived foam cells of human atherosclerotic lesions but not in the normal aorta (Kawai *et al.*, 2008). Moreover, the metabolite is significantly taken up and deconjugated into quercetin in activated murine macrophages. Similarly, Q3G was found colocalized with macrophages and the pro-oxidant enzyme myeloperoxidase (MPO) in human atherosclerotic aorta (Shiba *et al.*, 2008). Owing to its free catechol nucleus, this metabolite retains a strong reducing character and, for instance, efficiently inhibits LDL peroxidation (Kawai *et al.*, 2008; Loke *et al.*, 2008a). Thus, the accumulation of Q3G on a site of oxidative stress strongly suggests its possible ROS-scavenging activity *in vivo*.

The most relevant ROS for scavenging by polyphenols are (Dangles, 2012) as follows:

- The superoxide radical anion produced by NOX, xanthine oxidase, or electron leakage from the mitochondrial inner membrane. The $O_2^{\bullet-}$ radical may then disproportionate (under SOD catalysis) to form hydrogen peroxide, or reduce Fe^{III} to Fe^{II}, or even combine with NO (produced by iNOS at inflammation sites) to form peroxynitrite.
- The hydroxyl radical produced by one-electron reduction of H_2O_2 by Fe^{II} (Fenton reaction) or by decomposition of peroxynitrite in acidic conditions
- The hypervalent Fe^{IV}=O species formed upon activation by hydroperoxides of heme proteins such as (met)myoglobin, myeloperoxidase, and COX.

1.3.1 The physicochemical bases of polyphenol-to-ROS electron transfer

1.3.1.1 Thermodynamics descriptors

The catechol but also the pyrogallol ring (e.g., the B-ring of several common flavonoids and phenolic acids, the galloyl residues of hydrolyzable tannins, and green tea flavanols) are particularly efficient electron/H-atom donors to scavenge free radicals. Their activity is enhanced if they are part of a long conjugation path (e.g., in hydroxycinnamic acids, flavones, and flavonols). The C3-OH group of flavonols and the guaiacol ring are also two moieties having efficient H-atom abstraction capacity (Goupy *et al.*, 2003; Trouillas *et al.*, 2006). These structure–activity relationships (SARs) related to free radical scavenging by antioxidants are well interpreted by the O—H bond dissociation enthalpy (BDE). Its evaluation is a complex experimental issue for polyphenol derivatives, while density functional theory (DFT) calculations are a powerful alternative to evaluate it, as the difference in standard enthalpy between the polyphenol (ArOH) and the aryloxyl radical ($ArO^{\bullet}$) obtained after H-atom abstraction.

$$\mathrm{BDE} = H^{\circ}_{(298\mathrm{K})}\left(\mathrm{ArO}^{\bullet}\right) + H^{\circ}_{(298\mathrm{K})}\left(\mathrm{H}^{\bullet}\right) - H^{\circ}_{(298\mathrm{K})}\left(\mathrm{ArOH}\right)$$

The computed O—H BDEs are particularly predictive. As a characteristic example, the relative H-donating capacity of the C3-, C3′-, C4′-, C5-, and C7-OH groups of quercetin is clearly confirmed by calculations. Indeed, the BDE values (in kcal mol^{-1}) are as follows: below 80 for the C3- and C4′-OH groups (very active); in the range 80–85 for C3′-OH (active); in the range 84–89 for C7-OH (poorly active)[2]; and higher than 90 for C5-OH (inactive). The SAR of $DPPH^{\bullet}$ scavenging is perfectly predicted by the sole BDE descriptor. When compared to other antioxidant assays (e.g., $ABTS^{+\bullet}$, ORAC, electrochemistry), required for a comprehensive antioxidant evaluation, BDE might not be sufficient and other minor descriptors are required for rationalization, which can be evaluated by quantum calculations as well (e.g., spin density distribution, electron transfer and deprotonation energies, number of "active" OH groups and H-bonds, frontier orbital energies and distribution).

Acting as an antioxidant, the polyphenol (ArOH) transfers a H-atom to the free radical ($R^{\bullet}$). The standard enthalpy of H-atom transfer (HAT) reaction from ArOH to $R^{\bullet}$ is given by

$$\Delta H^{\circ}_{298\mathrm{K}} = \mathrm{BDE}(\mathrm{ArOH}) - \mathrm{BDE}(\mathrm{RH})$$

The reaction is exothermic if BDE (ArOH) < BDE (RH). The DPPH-H BDE is ca. 80 kcal mol^{-1} (Pratt *et al.*, 2004)[3]; therefore, the $DPPH^{\bullet}$ assay adequately discriminates active from nonactive phenolic compounds.

Concerning peroxyl radicals, BDE(ROO—H) is in the range 84–88 kcal mol^{-1} (Blanksby *et al.*, 2001; Ramond *et al.*, 2002), depending on the R substituent. This makes peroxyl radicals easier to scavenge than $DPPH^{\bullet}$. However, the SAR of $DPPH^{\bullet}$ and peroxyl scavenging are usually similar (Trouillas *et al.*, 2008). Of course, when peroxyl scavenging is measured in liposomes (R = hydrocarbon chain of fatty acid residues), the H-donating capacity of the phenol groups is not the only factor for consideration and its ability to penetrate the membrane must be considered as well (see Section 1.2.2).

The robustness of DFT methodologies has been repeatedly tested to accurately predict thermodynamic antioxidant descriptors on various phenolic compounds. A review of the huge amount of publications on BDE (DFT-based) calculations is far beyond the scope of this chapter (see, e.g., the review by Leopoldini *et al.*, 2011). The reader must be aware that care should be given to the choice of functionals, basis sets, and solvent models.[4]

[2]The C7-OH group is poorly active as pure H-atom donor. However, it is the most acidic group of many flavonoid derivatives, therefore being in some cases involved in the antioxidant action.

[3]The DPHH-H BDE is ca. 80.4 kcal mol^{-1} as obtained by IEFPCM-B3P86/6-31 + G(d,p).

[4]Solvent effects have to be accurately taken into account. The ideal treatment would be to include large numbers of solvent molecules around the antioxidant (solute) and simulate the entire system at the quantum mechanical level. Unfortunately, this is not possible due to constraints in computing power. Two different approaches can be adopted to alleviate this problem. Either a small number of solvent molecules are included and quantum mechanics calculations are performed using a continuum dielectric (hybrid approach), or a large number of solvent molecules are included and Monte Carlo or molecular dynamics simulations are performed. As an alternative to these sophisticated methods, the solvent can simply be treated as a continuum, for example, the polarizable continuum model (PCM). In this case, the antioxidant molecule sits in a shape-adapted cavity surrounded by the continuum solvent.

However, we believe that the 15-year background in this domain is now sufficient to calculate these major descriptors of the antioxidant action on large series of compounds.

1.3.1.2 Kinetics of hydrogen atom transfer

BDE and other thermodynamic descriptors correlate with static parameters (stoichiometry and EC50) obtained in antioxidant assays. However, under chemical and biological environments when several types of ROS are present, the impact of an antioxidant can be a matter of kinetics as well. Indeed, rate constants of ROS-scavenging may be very different from one ROS to another.

The hydroxyl radical is so reactive that it can react with any biomolecule with rate constants as high as 10^9–$10^{10}\,M^{-1}\,s^{-1}$. $HO^\bullet$ can initiate lipid peroxidation either by HAT or by addition on the PUFA (polyunsaturated fatty acid) carbon–carbon double bonds. In both cases, a carbon-centered radical is formed. This step exhibits high rate constants (e.g., $5\times10^8\,M^{-1}\,s^{-1}$ in lecithin bilayers) (Antunes *et al.*, 1996). The following step (O_2 addition to form peroxyl radicals) is also very fast, and the apparent rate constant for these two steps is ca. $10^9\,M^{-1}\,s^{-1}$ (Kamal-Eldin & Appelqvist, 1996). The rate constants of the propagation step (chain reaction in lipid bilayers) are much lower (10^1–$10^5\,M^{-1}\,s^{-1}$) (Tang *et al.*, 2000), and depend on the type of PUFA.

To inhibit the lipid peroxidation process, an antioxidant may scavenge radical initiators as well as peroxyl radicals formed in lipid bilayers. An effective antioxidant such as vitamin E can scavenge the peroxyl radicals with rate constants ranging from 10^3 to $10^6\,M^{-1}\,s^{-1}$, thus in competition against the propagation stage and acting by chain breaking in lipid bilayers (Antunes *et al.*, 1996). Interestingly, the scavenging of the *t*-butoxyl radical by α-tocopherol exhibits much higher rate constants of ca. $10^9\,M^{-1}\,s^{-1}$, when measured in solution (Evans *et al.*, 1992).

The rate constants for $DPPH^\bullet$ scavenging were measured for a series of phenolic acids and flavonoids (Goupy *et al.*, 2003; Roche *et al.*, 2005). Using a simple second-order kinetic scheme, the rate constants for the first (most labile) H-atom abstraction from the antioxidant were estimated as ca. 10^3, 10^3, and $2\times10^3\,M^{-1}\,s^{-1}$ for caffeic acid, catechin, and quercetin (catechol moiety), respectively, and $4\times10^3\,M^{-1}\,s^{-1}$ for epigallocatechin (pyrogallol moiety). Even if phenolic compounds react much faster with peroxyl radicals than with $DPPH^\bullet$, the rate constants to scavenge both radicals were correlated (Foti *et al.*, 2010). However, whereas α-tocopherol acts both as inhibitor of the initiation and propagation steps, phenolic compounds seem to inhibit mainly initiation. As described earlier, this strongly depends on the polyphenol chemical structure and the lipid composition of the membrane. Namely, slight changes of one of these parameters may bury more or less the compound in the bilayer, therefore modulating this conclusion.

The scavenging of the superoxide radical (in acid–base equilibrium with the perhydroxyl radical $HO_2^\bullet$) by active polyphenols is in the range 10^4–$10^7\,M^{-1}\,s^{-1}$ at pH 7. Again, the catechol and pyrogallol moieties provide fast superoxide scavenging. As an interesting example, theaflavin appears particularly efficient with a rate constant of $10^7\,M^{-1}\,s^{-1}$ compared to 6.4×10^4 and $7.3\times10^5\,M^{-1}\,s^{-1}$ for catechin and epigallocatechin gallate, respectively (Jovanovic & Simic, 2000). At pH 7, the following hierarchy was obtained: catechin ~ epicatechin < epigallocatechin ~ epicatechin gallate < epigallocatechin gallate < theaflavin. At pH 10, one has galangin < kaempferol < catechin < quercetin ~ rutin, with variation in the rate constants of less than two orders of magnitude in this series. Phenolic compounds can also efficiently scavenge oxyl radicals (De Heer *et al.*, 2000) and singlet oxygen (Jovanovic & Simic, 2000; He *et al.*, 2009). The rate constants of $^1O_2$ scavenging by active flavonoids (Jovanovic & Simic, 2000; Mukai *et al.*, 2005) are in the range 10^6–$10^8\,M^{-1}\,s^{-1}$, that is, comparable to vitamin E (10^7–$10^8\,M^{-1}\,s^{-1}$) (Kamal-Eldin & Appelqvist, 1996) but lower than carotenoids ($10^{10}\,M^{-1}\,s^{-1}$) (Ouchi *et al.*, 2010).

1.3.1.3 Kinetics and mechanisms

Rate constants are related to BDEs (Foti *et al.*, 2010; Mayer, 2011) but not strictly correlated. Small variations in BDE may be associated with huge variations in rate constants. Therefore, BDE is only a primary descriptor that provides basic understanding and rough estimate of kinetics. To properly estimate kinetic parameters, a thorough knowledge of the scavenging mechanism is required. Four mechanisms are usually proposed for free radical ($R^•$) scavenging by phenolic compounds (ArOH):

HAT (H-atom transfer) and/or PCET (proton-coupled electron transfer)

$$ArO-\mathbf{H}+R^{\bullet} \rightarrow ArO^{\bullet}+R-\mathbf{H} \tag{1}$$

SET–PT (sequential electron transfer–proton transfer)

$$ArO-\mathbf{H}+R^{\bullet} \rightarrow ArOH^{\bullet+}+R^{-} \rightarrow ArO^{\bullet}+R-\mathbf{H} \tag{2}$$

SPLET (sequential proton loss–electron transfer)

$$ArO-\mathbf{H} \rightarrow ArO^{-}+H^{+};\ ArO^{-}+R^{\bullet} \rightarrow ArO^{\bullet}+R^{-};\ R^{-}+H^{+} \rightarrow R-\mathbf{H} \tag{3}$$

AF (adduct formation)

$$ArO-\mathbf{H}+R^{\bullet} \rightarrow \left[ArOH-R\right]^{\bullet} \rightarrow ArO^{\bullet}+R-\mathbf{H} \text{ and/or other oxidation products} \tag{4}$$

Reaction (1) is the direct HAT, which corresponds to the homolytic dissociation of the O—H bond of any active (low BDE) phenolic OH group. The HAT mechanism should refer to all processes in which electron and proton are transferred in one kinetic step. The PCET (Huynh & Meyer, 2007) terminology is often used to distinguish a specific HAT, according to the number of molecular orbitals or electrons that are involved in the process (Fig. 1.5). In PCET, even if being transferred in the same kinetic step, electron and proton follow different routes. Namely, in the case of a phenolic compound scavenging an oxygen-centered free radical, the proton is transferred from the OH group of the former to the oxygen lone pair of the latter, in principle across the hydrogen bond pre-established between both reactants (Fig. 1.5a). This proton transfer occurs in the plane defined by the aromatic ring.

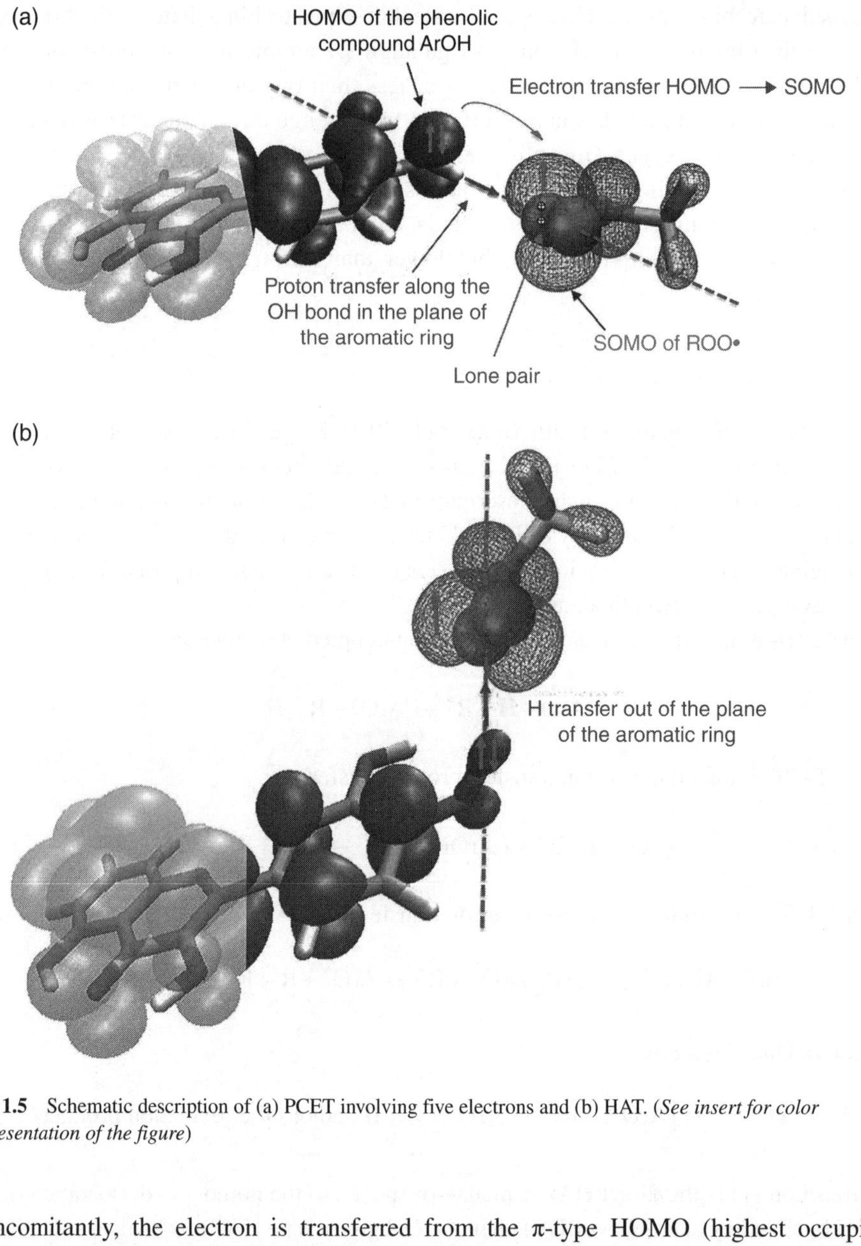

Fig. 1.5 Schematic description of (a) PCET involving five electrons and (b) HAT. (*See insert for color representation of the figure*)

Concomitantly, the electron is transferred from the π-type HOMO (highest occupied molecular orbital) of ArOH to the π-type SOMO (singly occupied molecular orbital) of the free radical, both MOs being perpendicular to the plane in which the proton transfer occurs. In HAT (Fig. 1.5b), both proton and electron are transferred through the same σ-type MO.

The prereaction complexes are of crucial importance to determine which mechanism of action proceeds and its effectiveness (Di Meo *et al.*, 2013). Depending on the phenolic antioxidant and the free radical, various noncovalent arrangements are possible according to the type of interactions (e.g., H-bonding, XH–π interaction, lone pair–π interaction, or π–π stacking, see Fig. 1.6) (DiLabio & Johnson, 2007; Foti *et al.*, 2010; Inagaki

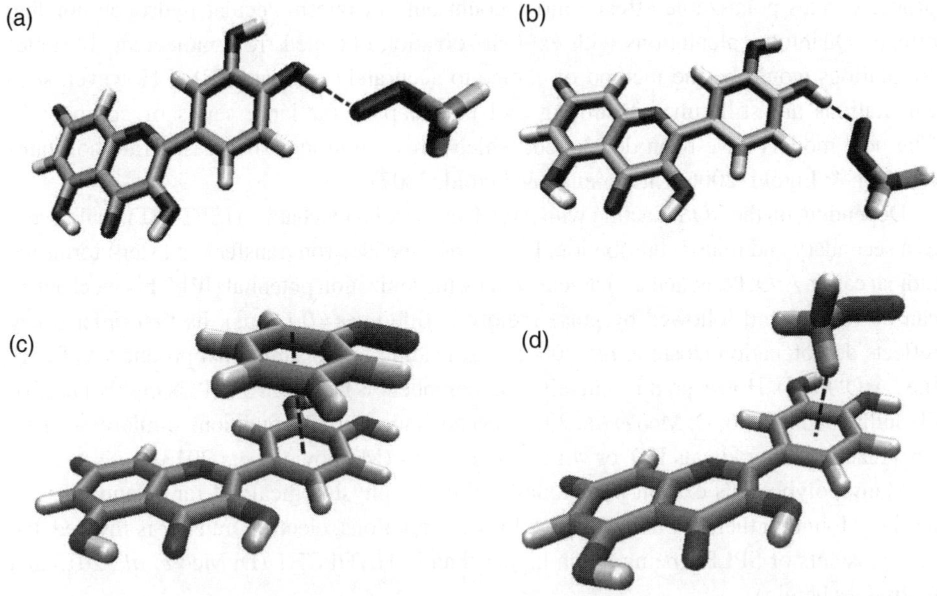

Fig. 1.6 Various geometries of noncovalent ArOH---ROO$^{\bullet}$ arrangements. (a) *transoid* H-bonding, (b) *cisoid* H-bonding, (c) π–π stacking, and (d) XH–π stacking complexes. (*See insert for color representation of the figure*)

et al., 2011). According to the molecular arrangements in the prereaction complexes, there are many cases with phenols in which reaction (1) is neither a pure PCET nor a pure HAT, therefore being a mixed quantum process.

Theoretical studies have evaluated the rate constants of free radical scavenging by phenolic compounds with a reasonable agreement with experiment (Lingwood *et al.*, 2006; Tejero *et al.*, 2007; Chiodo *et al.*, 2010; Galano *et al.*, 2012; Di Meo *et al.*, 2013; Garzon *et al.*, 2014). These studies were based on the transition state theory, requiring calculation of the HAT/PCET transition states. Again, DFT provides a relevant compromise between accuracy and reasonable computational time. However, a careful choice of the functional is even more critical for kinetics than for thermodynamics. The HAT/PCET mechanism may be better described by MPWB1K (Zhao & Truhlar, 2004; Lingwood *et al.*, 2006), knowing that classical hybrid functionals usually underestimate Gibbs energy of activation of HAT processes. However, in the case of peroxyl radical scavenging by polyphenols, B3LYP also provided reasonable agreement with experimental values (DiLabio & Johnson, 2007). Other functionals such as MO5-2X also seem adapted to describe those transition states (Galano *et al.*, 2012). In any event, even if we believe that methodological refinements are still needed, DFT-based rate constants are particularly accurate at predicting (i) SAR and (ii) competition between PCET and electron transfer processes (Di Meo *et al.*, 2013). Interestingly, most of these studies have highlighted the necessity to account for tunneling effect, which, depending on the system, may contribute to rate constant values by up to five orders of magnitude.

The PCET mechanism is known to exhibit kinetic solvent effects (KSEs). PCM provides a perfect description of nonpolar solvents. Concerning polar solvents, PCM-type solvation properly takes polarizable effects into account but not intermolecular hydrogen bonding effects. Quantum calculations with explicit solvation obtained from molecular dynamics simulations would be the method of choice to accurately evaluate KSEs. However, such calculations are still time-consuming and not adapted for large series of compounds. Classical models have been developed, which stress significant decreased rate constants (Nielsen & Ingold, 2006; Litwinienko & Ingold, 2007).

Depending on the ROS reacting with ArOH, the second mechanism (SET–PT) could occur as a secondary and minor contribution. In this case, the electron transfer (first step) forms the radical cation $ArOH^{\bullet+}$, which is characterized by the ionization potential (IP). This mechanism can be formed and followed by pulse radiolysis (lifetimes = 0.1–1 μs). Its first-order decay reflects deprotonation (Joshi *et al.*, 2004), thus leading to the same final products as PCET (i.e., $ArO^{\bullet}$ + RH). However, it is unlikely with polyphenols because $ArOH^{\bullet+}$ is highly unstable (Trouillas *et al.*, 2006; Di Meo *et al.*, 2013), except under certain conditions similarly with the very reactive free radicals $HO^{\bullet}$ or with various bases (Markovic *et al.*, 2013).

Many polyphenols exhibit pK_a values within the physiological pH range and thus are partially found in their phenolate forms. In such situations, electron transfer is favored, the rate constants of SPLET being much higher than in HAT/PCET (Di Meo *et al.*, 2013 and references herein).

From a kinetics point of view, the process is advantageously analyzed within the Marcus theory. In its original formulation, the semiclassical Marcus equation defines the activation barrier as the sum of the Gibbs energy $\Delta G°$ for the ET reaction and the reorganization energy λ of the system along the reaction coordinate corresponding to ET (Huynh & Meyer, 2007; Borrelli & Peluso, 2013).

$$k_{\mathrm{ET}} = k_{\mathrm{A}} \exp\left(\frac{\lambda + \Delta G°}{4\lambda k_{\mathrm{B}} T}\right)$$

The multiplicative coefficient k_A corresponds to preassociation.[5] Near the diffusion-controlled limit, 10^9–$10^{11}\ M^{-1}\,s^{-1}$, both rate constants for diffusion (k_D) and electron transfer (k_{ET}) are comparable and kinetically coupled according to (Huynh & Meyer, 2007) the following:

$$\frac{1}{k_{\mathrm{obs}}} = \frac{1}{k_{\mathrm{D}}} + \frac{1}{k_{\mathrm{ET}}}$$

The reorganization energy λ can be separated into two components corresponding to the energy barriers for the geometrical relaxation along the ET reaction coordinate of solute (λ_i) and solvent (λ_S) molecules, respectively (Fig. 1.7a). The internal contribution λ_i can also be described as the average of the two terms λ_{i1} and λ_{i2}[6] (Fig. 1.7b). The classical

[5]It depends on the internuclear separation distance, Avogadro's number, and electrostatic energy of association.

[6]λ_{i1} is the difference between the energy of the reactants in the geometry characteristic of the products and that in their equilibrium geometry; λ_{i2} corresponds to the difference between the energy of the products in the geometry characteristic of the reactants.

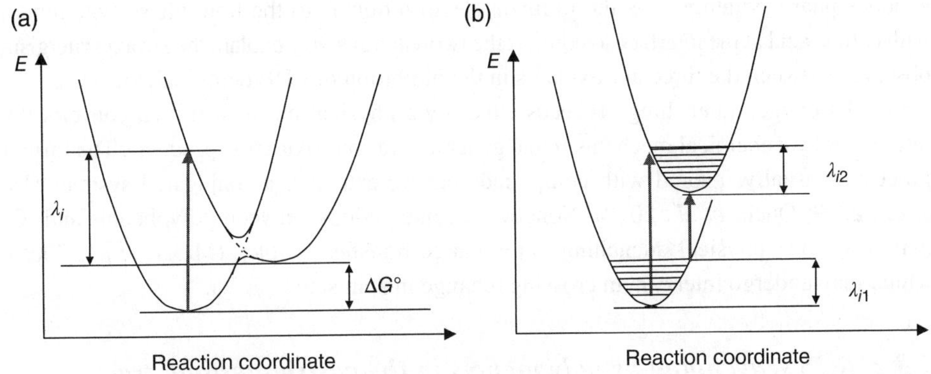

Fig. 1.7 Schematic potential energy surfaces with respect to the reaction coordinate (a) corresponding to the electron transfer and (b) defining the internal reorganization energies.

dielectric continuum model, as initially developed by Marcus for ET between spherical ions in solution, can be used to estimate λ_S (Huynh & Meyer, 2007). The kinetic description of electron transfer is described as nonadiabatic, that is, geometries do not have time to evolve on the ET timescale. For a comprehensive description, the ET probability should be treated within a full quantum approach, including at least the vibrational modes of importance along the reaction coordinate (Borrelli & Peluso, 2013). The Marcus–Levich–Jortner formulation of the ET process introduces electronic coupling between the antioxidant HOMO and the free radical SOMO in the geometry of the prereaction complexes. Within this formalism, the tunneling effect has been added. The type and stability of noncovalent prereaction complexes determine MO overlap and thus k_{ET} (Di Meo *et al.*, 2013). A comprehensive evaluation of all noncovalent prereaction associations is mandatory.

Both HAT/PCET and SPLET mechanisms are in competition to scavenge DPPH$^•$, peroxyl, hydroxyl, oxyl, and superoxide radicals. The latter mechanism is faster but is only favored in polar solvents and under neutral to basic conditions, according to the pK_a of the phenolic antioxidant. As said earlier, SET–PT is unlikely with polyphenols and the AF mechanism more specific to HO$^•$. ROS scavenging of polyphenols should actually be seen as a complex process following various regimes. The four mechanisms exhibit the same thermodynamic balance since reactants and products are the same (see reactions 1–4). Consequently, the global process is well described by BDE values.

It is also noteworthy that polyphenols have been described to react by HAT with the aryloxyl radical of vitamin E, thereby enabling its regeneration.

$$\text{VitE}-\text{OH}+\text{R}^{\bullet} \rightarrow \text{VitE}-\text{O}^{\bullet}+\text{RH};\ \text{ArOH}+\text{VitE}-\text{O}^{\bullet} \rightarrow \text{ArO}^{\bullet}+\text{VitE}-\text{OH} \quad (5)$$

Among other examples, catechin and epicatechin are as effective as vitamin C at regenerating vitamin E within lipid bilayers (Mukai *et al.*, 2005). Electron spin resonance measurements in micelles also showed that caffeic acid can reduce the aryloxyl radical of vitamin E, while vitamin C in turn can reduce the semiquinone radical derived from caffeic acid (Laranjinha & Cadenas, 1999). This sequence of electron/H-atom transfers from the aqueous phase (vitamin C as the ultimate electron donor) to the lipid phase (vitamin E) with caffeic acid at the interface to connect the two vitamins may explain the strong synergism observed between the three antioxidants in the inhibition of LDL peroxidation.

Singlet oxygen quenching proceeds either by a physical mechanism that converts $^1O_2$ into $^3O_2$ or by a chemical mechanism that generates various oxidation products. The former process is usually favored with compounds having extended π-conjugated systems (He *et al.*, 2009; Ouchi *et al.*, 2010). Noncovalent interactions between polyphenols and $^1O_2$ may also favor physical quenching via a charge transfer exciplex (Mukai *et al.*, 2005), which may undergo intersystem crossing (change in spin state).

1.3.2 ROS scavenging by polyphenols in the gastrointestinal tract

From both *in vitro* and *in vivo* studies, it is becoming clear that advanced lipid oxidation end products (ALEs) can accumulate in the stomach under postprandial conditions (Kanner *et al.*, 2012). The most efficient route to ALEs is probably the heme-induced autoxidation of PUFA residues in dietary triacylglycerols, phospholipids, and cholesterol esters. The process is especially fast in the gastric compartment due to the presence of O_2 in nonlimiting concentration and to acidic conditions making heme iron (the main dietary iron form, typically brought by muscle meat) particularly pro-oxidant. It is triggered by the heme-induced homolytic cleavage of hydroperoxide traces inevitably contaminating lipid-rich food. This reaction is the first step of a Heme-Fe^{III}/Heme-Fe^{IV} redox cycle yielding propagating lipid peroxyl radicals ($LOO^•$) through which lipid hydroperoxides (LOOHs) are formed. Thus, lipid hydroperoxides are both consumed during initiation and formed during propagation, so that their net accumulation is highly variable. For instance, when minipigs were fed with a meal based on sunflower oil (major PUFA = linoleic acid, C18 : 2) and red meat, LOOHs detected by their conjugated diene (CD) moiety were only moderately accumulated in the stomach (a maximum of +35% over ca. 3 hours), whereas malondialdehyde increased by a factor 5 over 4 hours (Gobert *et al.*, 2014). Even ALEs such as malondialdehyde and 4-hydroxyalkenals may be underestimated because of the fast reaction of these strong electrophiles with nucleophilic protein residues (Lys, His, Cys). After digestion of the protein-bound fraction, ALEs may be absorbed in the intestinal cells and express cytotoxic effects. For instance, there is strong evidence suggesting that dietary 4-hydroxyalkenals are central in the relationship between excessive red meat consumption and the risk of colon cancer (Pierre *et al.*, 2007). Moreover, modification of LDL by dietary ALEs is a likely mechanism leading to modified LDL particles with increased susceptibility to oxidation and atherogenic character (Kanner *et al.*, 2012). Hence, fighting dietary PUFA autoxidation in the stomach is one of the most important antioxidant mechanisms for polyphenols. Indeed, native forms possessing a free catechol group (caffeic acid, (epi)catechin, quercetin, and their derivatives) are abundant in the diet and, with the possible exception of flavanol polymers (highly polymerized proanthocyanidins), are readily released from the food matrix upon gastric digestion (Bouayed *et al.*, 2011). For instance, fruit and vegetables were found as efficient as their polyphenol-rich extracts at inhibiting malondialdehyde accumulation in minipigs (Gobert *et al.*, 2014).

Based on *in vitro* experiments, the mechanism of heme-induced lipid peroxidation in the stomach is expected to be sensitive to pH variations and the action of the digestive enzymes. *In vivo* measurements showed that the gastric pH quickly increases from a basal value of 2 to ca. 5 after food ingestion, then slowly decays back to its basal value along the gastric emptying (Gobert *et al.*, 2014). Model experiments in micelles and emulsions showed that lipid peroxidation induced by metmyoglobin (CD monitoring) at pH 5–6 is fast and accompanied by a gradual degradation of the heme (Goupy *et al.* 2007). At pH 4, the heme cofactor (hematin) is quickly dissociated from the protein moiety (globin) and its degradation with release of free iron is faster than at higher pH. Although those changes do not markedly affect the overall rate of CD accumulation in emulsions, they strongly alter the performance of the phenolic antioxidants (Lorrain *et al.*, 2012). Indeed, common dietary flavonoids and hydroxycinnamic acids are potent inhibitors of metmyoglobin-induced lipid peroxidation in emulsions of sunflower oil stabilized by proteins or phospholipids at pH 5–6, whereas they are much less active at pH 4. On the other hand, polyphenols can efficiently reduce ferrylmyoglobin ($MbFe^{IV}{=}O$), prepared by reacting metmyoglobin ($MbFe^{III}$—OH) with H_2O_2. It is thus proposed that polyphenols inhibit metmyoglobin-induced lipid peroxidation by reducing hypervalent heme species formed by reaction of $MbFe^{III}$—OH with lipid hydroperoxides (Lorrain *et al.*, 2010). This mechanism (inhibition of initiation) requires an intact metmyoglobin particle and is much less efficient at pH 4 when hematin is released followed by free iron. Indeed, at pH 4, polyphenols are unlikely to bind free iron and quench its pro-oxidant activity. Unlike polyphenols, α-tocopherol does not reduce $MbFe^{IV}{=}O$ and acts by reducing the lipid peroxyl radicals (inhibition of propagation). However, both polyphenols and α-tocopherol can protect the heme from degradation (Goupy *et al.*, 2007). The mechanism by which phenolic antioxidants can fight heme-induced lipid peroxidation in the stomach is summarized in Fig. 1.8.

The gastric protease pepsin and the gastric lipase could influence both lipid peroxidation in the stomach and its inhibition by polyphenols. For instance, pepsin could modify the structure of the interface by partial hydrolysis of adsorbed proteins. Moreover, mild pepsin-catalyzed proteolysis is known to convert metmyoglobin into more efficient inducers of lipid peroxidation (Carlsen & Skibsted, 2004). Gastric lipase can release free fatty acids and diacylglycerols within emulsions of triacylglycerols, thereby contribution to 5–30% of lipid digestion. However, addition of gastric fluid did not alter the rate of lipid peroxidation in sunflower emulsions but strongly attenuated the inhibitory capacity of quercetin (Lorrain *et al.*, 2012).

1.4 Polyphenols as ligands for metal ions

Polyphenols bearing a catechol group can bind a variety of metal ions, including aluminum, iron, and copper ions. Flavonoids may also display additional binding moieties, such as 3-hydroxy-4-oxo and 5-hydroxy-4-oxo substitutions (Engelmann *et al.*, 2005). The pH-dependent displacement of protons from those groups by metal ions yields chelates, and this binding can have important consequences for the polyphenols' redox activity (stability, pro- and antioxidant activity) (Perron & Brumaghim, 2009) and bioavailability.

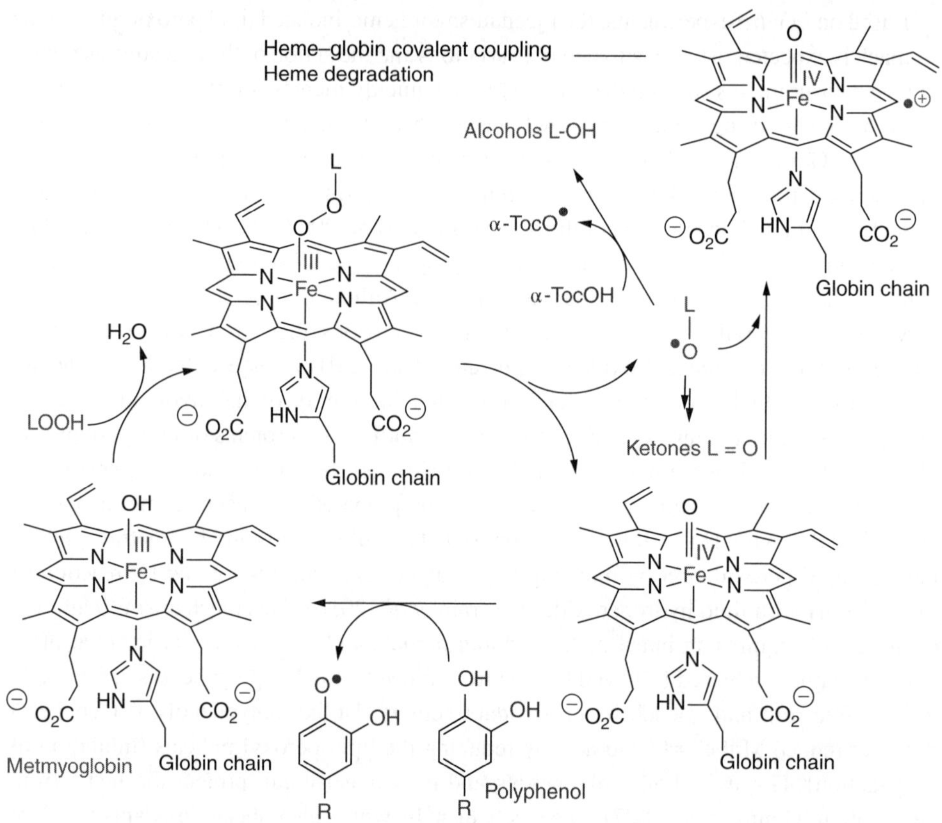

Fig. 1.8 Activation of heme iron (metmyoglobin) by lipid hydroperoxides and the role of phenolic antioxidants.

For instance, iron–flavonoid complexes present in colored beans are not absorbed through the intestinal barrier (Hu *et al.*, 2006).

1.4.1 Interactions of polyphenols with iron and copper ions

As iron is the most abundant transition metal ion in human diet and in tissues, iron–polyphenol interactions have been thoroughly investigated. Depending on pH and the presence of other competing ligands (e.g., buffer components), such interactions may be coupled with different redox phenomena (Fig. 1.9):

- In strongly acidic aqueous solution or water–MeOH mixture (pH 1–3), despite the strong competition between protons and iron ions for the binding site, fast Fe^{III} binding can be observed with flavonols at the 3-hydroxy-4-oxo group (as for quercetin, in which additional binding at the catechol group is possible) and with flavanols and hydroxycinnamic acids at the catechol group. The latter binding mode is evidenced by the weak phenol-to-Fe^{III} charge transfer absorption appearing at 700 nm. In such acidic conditions, Fe^{III} sufficiently oxidizes, so phenol-to-iron electron transfers take place, thereby resulting

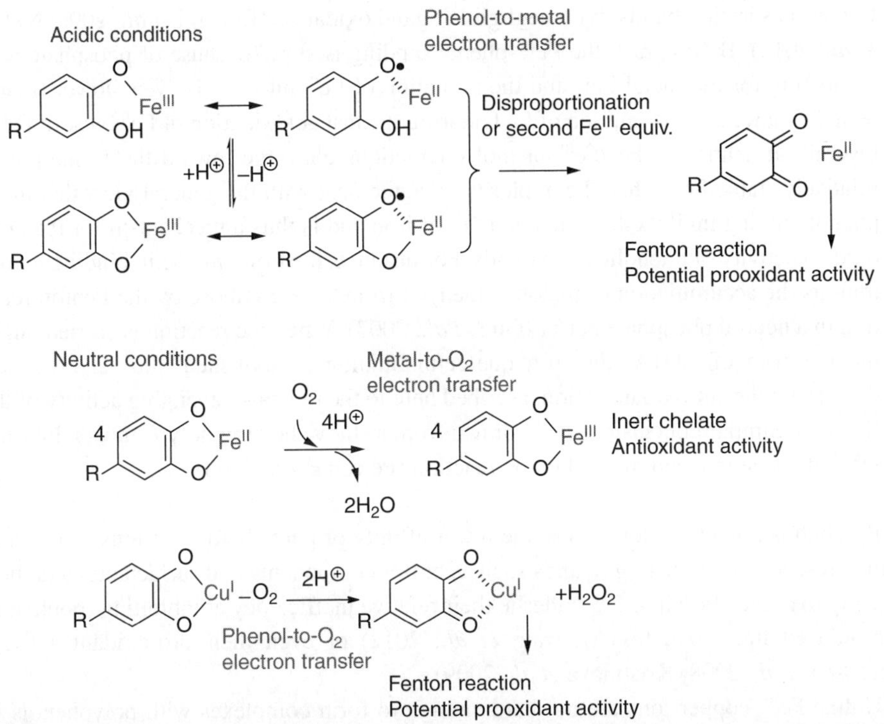

Fig. 1.9 Polyphenols and transition metal ions. Acidic conditions: iron binding and subsequent electron transfer. Neutral conditions: binding and autoxidation.

in the reduction of several Fe^{III} equivalents. By contrast, with the flavanone naringin, binding at the 5-hydroxy-4-oxo group is not followed by electron transfer (Hynes & O'Coinceanainn, 2004; El Hajji *et al.*, 2006; Ryan & Hynes, 2007, 2008). The pH dependence of the observed rate constant of electron transfer suggests that the reaction preferentially involves a monoprotonated catecholate ligand. With ferulic and sinapic acids, electron transfer takes place with no evidence of an intermediate Fe^{III}–phenol complex (outer sphere mechanism).

- In mildly acidic to neutral solution, Fe^{III} becomes less oxidizing and Fe^{II} more prone to autoxidation. In a (1 : 1) MeOH–water mixture, high-spin Fe^{III}–quercetin complexes are observed by ESR as well as its partial conversion into a high-spin (ESR-silent) Fe^{II} complex. The formation of Fe^{II}–quercetin complexes is also confirmed by MS (Mira *et al.*, 2002; Guo *et al.*, 2007). Similar observations were reported for phenolic acids bearing a catechol group in neutral N-containing buffers susceptible to stabilize Fe^{II} (Chvatalova *et al.*, 2008). However, the Fe^{III}-to-Fe^{II} conversion is slow and remains limited (typically, <10% after 1 hour). By contrast, catechol-bearing phenolic acids and flavonoids efficiently accelerate the reverse process, that is, the autoxidation of Fe^{II} into Fe^{III} (Chvatalova *et al.*, 2008; Perron *et al.*, 2010). Consistently, in a neutral phosphate buffer, Fe^{II}–catechol binding is fast and accompanied by a rapid autoxidation of Fe^{II} into Fe^{III} (confirmed by the ferrozine test and the development of catechol-to-Fe^{III} charge transfer

UV-vis absorption bands) with negligible ligand oxidation (El Hajji *et al.*, 2006; Nkhili *et al.*, 2014). By contrast, the Fe^{III}–phenol binding is slow (because of phosphate ions competing for the metal ion) and the reduction of Fe^{III} into Fe^{II} is even slower (manifested by a weak consumption of O_2 featuring the fast autoxidation of Fe^{II} back to Fe^{III}). Overall, at neutrality, Fe^{II}/Fe^{III}–phenol interactions converge toward the formation of relatively stable Fe^{III}–phenol complexes, in agreement with the general view that iron–phenol binding inhibits the redox activity of iron and is thus a mechanism for the antioxidant activity of phenolic compounds. For instance, the iron–quercetin binding totally inhibits the accumulation of malondialdehyde from 2-deoxyribose by the Fenton reaction in a neutral phosphate buffer (Guo *et al.*, 2007). When the reaction is carried out in the presence of EDTA, the iron–quercetin binding is abolished, thus significantly decreasing the antioxidant action, ascribed here to the radical-scavenging activity of the flavonol. Similar conclusions were drawn from a study showing polyphenols inhibiting DNA damage induced by the Fenton reaction (Perron *et al.*, 2008).

By contrast, in acidic conditions, the lower affinity of phenols for iron ions (especially in the presence of competing ligands such as buffer components and surfactants) and their capacity to reduce Fe^{III} into Fe^{II} underlie their relative inefficiency at inhibiting (nonheme) iron-induced lipid oxidation (Lorrain *et al.*, 2012) or even their pro-oxidant activity (Sorensen *et al.*, 2008; Kristinova *et al.*, 2009).

Unlike Fe^{III}, copper ions are not hard enough to form complexes with polyphenols in acidic conditions. However, the binding is fast and quasi-irreversible in neutral conditions even in phosphate buffer (El Hajji *et al.*, 2006; Nkhili *et al.*, 2014). Remarkably, copper–polyphenol complexes, especially involving Cu^{I}, are quite labile and quickly evolve toward oxidation products with concomitant O_2 consumption, H_2O_2 production, and maintenance of copper at the low-valence level + I. Thus, Cu^{I}–phenol complexes transfer electrons from the ligand (instead of the metal center) to the Cu^{I}-bound O_2. The persistence of Cu^{I} (even when starting from the Cu^{II}–phenol complexes) and the accumulation of H_2O_2 favor the onset of the Fenton reaction. It is thus no surprise that polyphenols are less efficient at inhibiting copper-induced than iron-induced biologically relevant oxidation processes and may even behave as pro-oxidants in the presence of copper (Perron *et al.*, 2011). For instance, green tea flavanols display contrasted influences on DNA damage induced by Cu^{I}–H_2O_2: while EC and EGC are pro-oxidant throughout the entire concentration range (0.1–3000 μM), EGCG behaves as a weak antioxidant and ECG as both pro-oxidant and antioxidant at low and high concentrations, respectively.

Interactions of polyphenols with transition metal ions, even as unidentified metal traces, can mediate polyphenol autoxidation and the concomitant production of H_2O_2 and electrophilic *o*-quinones or *p*-quinone methides. Interestingly, this phenomenon can occur in cell culture with subsequent activation of genes of the electrophile-responsive element (EpRE) that governs the expression of several phase II detoxifying enzymes such as NAD(P)H–quinone oxidoreductase and enzymes involved in glutathione biosynthesis and transfer (Erlank *et al.*, 2011). Indeed, H_2O_2 and electrophiles derived from polyphenol oxidation could modify critical Cys residues of Keap1, a repressor protein bound to EpRE transcription factor Nrf2 in the cytosol. The subsequent dissociation of Keap1 from Nrf2 triggers the translocation of Nrf2 into the nucleus and the onset of gene expression. Thus, the mild oxidative stress caused by polyphenol autoxidation stimulates the cellular antioxidant defense. The ability of flavonoids to upregulate EpRE genes is correlated to their electron-donating capacity expressed by the HOMO energy (Lee-Hilz *et al.*, 2006). The enol group of flavonols seems more important that the catechol group as shown by the variations of induction factor: quercetin (IF = 10) > 3′-*O*-methylquercetin (IF = 7.8) > kaempferol (IF = 7.1) > luteolin (IF = 3). Thus, if flavonoid metabolites (glucuronides and sulfates) undergo partial hydrolysis in cells, the concomitant restoration of the reducing capacity could be beneficial, not only because of the enhanced antioxidant (ROS-scavenging) activity but also because of the enhanced pro-oxidant activity leading to increased detoxifying capacity.

1.4.2 A preliminary theoretical study of iron–polyphenol binding

Various theoretical studies have rationalized metal–phenol binding in active sites of non-heme iron-containing enzymes such as dioxygenases (Nakatani *et al.*, 2009), whereas less theoretical studies directly deal in details with metal–polyphenol binding in aqueous solution (Furia *et al.*, 2014 and references herein), especially to rationalize oxidative processes. In this case, the influence of pH is of crucial importance and must be investigated. This section provides some highlights based on theoretical calculations performed with DFT[7] on catechol, chosen as an adequate polyphenol prototype (C. Tonnelé & P. Trouillas, unpublished results). Calculations performed on 3-*O*-methylquercetin (as a model of the common 3-*O*-glycosides of quercetin) revealed similar electronic features (data not shown), thus confirming that catechol (mono- and bi-deprotonated) is a relevant model to tackle metal–polyphenol binding and the further redox processes related to their antioxidant/pro-oxidant effects.

1.4.2.1 Charge states, spin states, and geometries

Metal–polyphenol binding requires a thorough analysis of charge and spin states. Complexation of Fe^{III} with the catecholate monoanion leads to the $[Fe^{III}(Cat^{1-})(H_2O)_5]^{2+}$ (coordination of the catechol ligand through the sole deprotonated O-atom, i.e., monodentate mode) and $[Fe^{III}(Cat^{1-})$

[7] A proper and reliable description of iron complex spin states constitutes a very challenging issue that can hardly be tackled by the most accurate quantum methods but rather by the less computational costly DFT approach. However, the results highly depend on the functional chosen. Among all the investigated DFT functionals, OPBE, which combines Handy's and Cohen's optimized exchange OPTX with PBE correlation, was shown to be one of the most performing. Thus, in the present work, all geometries were optimized at the OPBE/6-31 + g(d,p) level, and the LANL2DZ pseudopotential/basis set was used to describe the iron ion. Solvent effects were accounted using a polarizable continuum model within its IEFPCM formalism; explicit solvent molecules were also included in the coordination sphere of the iron ion. The Gibbs free energies were obtained by summing (i) thermal correction to free energies as obtained on OPBE/6-31 + g(d,p)-LANL2DZ geometries to (ii) electronic energy, as obtained by a single-point calculations performed with OPBE/6-311 g(3df,3pd)-LANL2DZ. All calculations were performed with the Gaussian09 package.

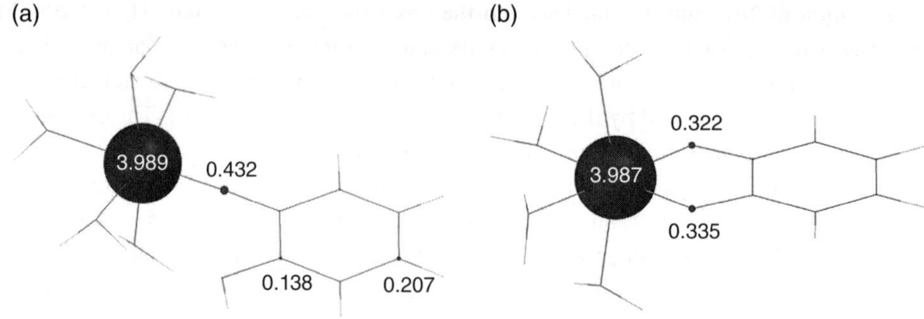

Fig. 1.10 Spin density distribution: (a) $[Fe^{III}(Cat^{1-})(H_2O)_5]^{+2}$ and (b) $[Fe^{III}(Cat^{2-})(H_2O)_4]^{+1}$.

$(H_2O)_4]^{2+}$ (coordination of the catechol ligand through both O atoms, i.e., bidentate mode) complexes.[8] A clear high-spin octahedral coordination is observed with both complexes. In $[Fe^{III}(Cat^{1-})(H_2O)_5]^{2+}$, the spin propagation occurs via the iron–oxygen bond and further develops into the aromatic ring (Fig. 1.10a). The charge and spin features support the hypothesis of a Fe^{II}–semiquinone radical character, in agreement with ligand-to-Fe^{III} ET (Fig. 1.11). Interestingly, $[Fe^{III}(Cat^{1-})(H_2O)_4]^{2+}$ did not exhibit any significant spin delocalization on the catechol ligand. This complex is most probably an intermediate toward a second deprotonation of the catechol moiety, thus yielding the corresponding bideprotonated complex $[Fe^{III}(Cat^{2-})(H_2O)_4]^{+1}$. This complex is more stable in its high-spin state (sextet) by 40.0 and 36.5 kcal mol^{-1} compared to the lower-spin states (quartet and doublet), respectively. Interestingly, it only exhibits a slight spin density propagation over the ligand (Fig. 1.10b), which suggests that ligand-to-Fe^{III} ET within this complex is less favorable than within $[Fe^{III}(Cat^{1-})(H_2O)_5]^{2+}$, in agreement with the pH dependence of the experimental rate constant of electron transfer (Hynes & O'Coinceanainn, 2004). The bideprotonated catechol can also bind to ferrous ions, forming the $[Fe^{II}(Cat^{2-})(H_2O)_4]^0$ complex also stabilized in its high-spin state, that is, quintet.

According to the metal/polyphenol molar ratio, one can also expect the formation of multicatechol complexes. The 1 : 3 stoichiometry provides the most stable complex (binding energy of −202.0 kcal mol^{-1}), while the two 1 : 2 complexes (equatorial–equatorial and axial–equatorial arrangements) exhibit similar binding energies of −180.5 and −181.6 kcal mol^{-1}, respectively. For comparison, the binding energy for 1 : 1 stoichiometry is expectedly the highest (less stable complex) and amounts to −129.1 kcal mol^{-1}.[9]

1.4.2.2 Oxidation of the bideprotonated catechol

The charge distribution in $[Fe^{III}(Cat^{2-})(H_2O)_4]^{+1}$ and its weak Fe^{II}–semiquinonate radical character in solution suggest the possibility of ligand-to-Fe^{III} ET (Fig. 1.11). On the basis of this assumption, we propose a mechanism involving metal-induced disproportionation

[8] In this case, the conformational rearrangement suggested an incomplete coordination of the iron ion, leading to less stable complex.

[9] It is worth noting that the absolute values are hardly comparable to related experimental data; however, the relative energies are considered reliable.

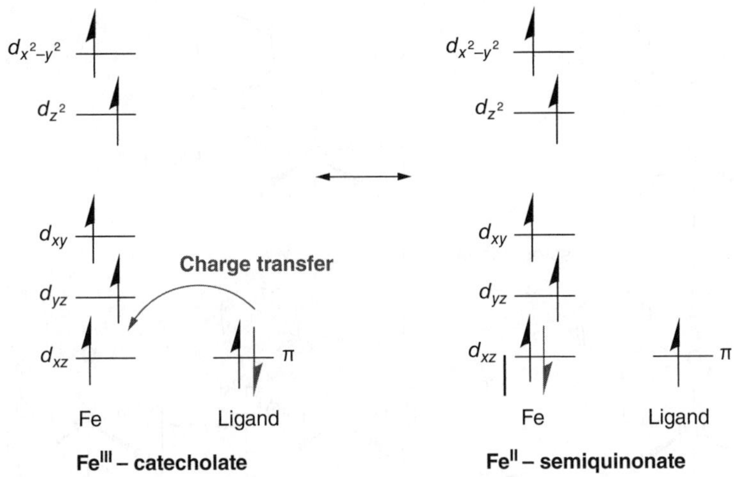

Fig. 1.11 Electronic structure of the Fe^{III}–Catecholate complex in its high spin (sextet) state, showing the two possible electronic configurations.

of the complex giving products **1** and **2** (Fig. 1.12). Compound **1** is the reduced product and corresponds to the $[Fe^{II}(Cat^{2-})(H_2O)_4]^0$ complex. Compound **2** is the oxidized product, in which the catechol moiety has been oxidized to the *o*-quinone form ($[Fe^{II}(Quinone)(H_2O)_4]^{+2}$).

In complex **2**, the ligand exhibits features of a quinone form with localized π-bonds, with reduced C–O bond lengths (1.26Å), and a pattern of alternating single and double bonds, together with a total charge close to 0.

1.5 Conclusions

Whereas clinical studies have established causal relationships between diets rich in polyphenols and health, biochemical investigations in cells including the true circulating metabolites are a critical step toward understanding the molecular mechanisms at work. Among polyphenol metabolites, the simplified structures delivered by the colonic microbiota and their conjugates by the intestinal and liver enzymes probably deserve special attention as the large intestine is the major site of absorption for dietary polyphenols. Considering the global pool of metabolites, polyphenols are more bioavailable than we once believed. For instance, it was estimated that ca. 70% of green tea flavanol intake was recovered in the urine of human volunteers (Del Rio *et al.*, 2013). As polyphenols may exert a direct antioxidant action by ROS scavenging in the digestive tract, the possible accumulation in this site of polyphenol oxidation products also deserves specific investigation as nothing is known about the bioavailability and potential bioactivity of such compounds. It must also be kept in mind that the low concentrations of individual circulating metabolites typically observed after a meal rich in polyphenols do not exclude accumulation of

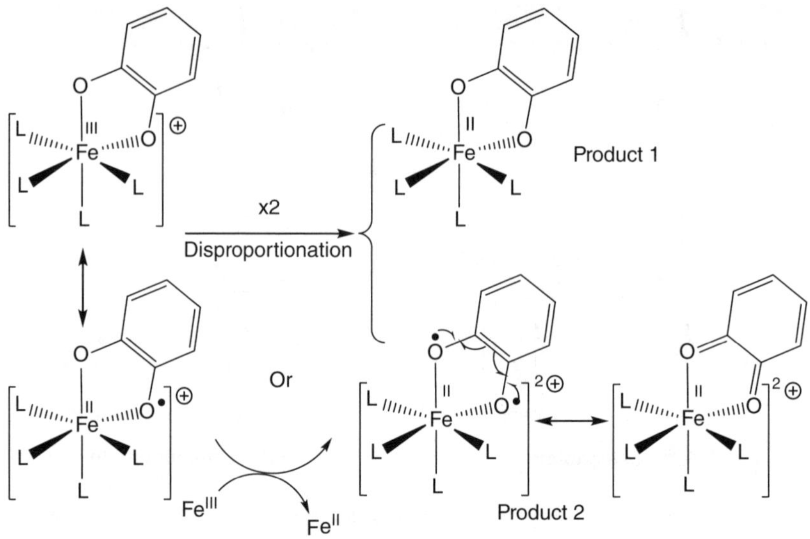

Fig. 1.12 Proposed mechanism for the Fe^{III}-induced oxidation of catechol ($L=H_2O$).

much higher concentration at specific inflammation sites and subsequent deconjugation into more active forms. Finally, besides their activity in human cells, dietary polyphenols may also contribute to human health by modulating the activity of the colonic microbiota (Anhe *et al.*, 2015).

Experimental and computational physical chemistry may be of great help in understanding the health effects of polyphenols at the molecular level. For investigating the activity in the GI tract, approaches combining simple chemical models with experiments in dynamic digesters including physiological parameters deduced from studies in humans and animals could be quite rewarding. After intestinal absorption, the ongoing identification of potential targets, including cytosolic and membrane proteins but also specific membrane lipid domains, will foster further binding and molecular modeling investigations to bring the knowledge to the highest resolution of intra- and intermolecular hydrogen bonding, van der Waals and ionic interactions, and electron transfer within defined supramolecular assemblies. It is thought that these multidisciplinary approaches will efficiently contribute to establishing ever more firmly the important role of dietary polyphenols in maintaining human health.

References

Al-Shalmani, S., Suri, S., Hughes, D.A. *et al.* (2011) Quercetin and its principal metabolites, but not myricetin, oppose lipopolysaccharide-induced hyporesponsiveness of the porcine isolated coronary artery. *British Journal of Pharmacology*, **162**, 1485–1497.

Anhe, F.F., Roy, D., Pilon, G. *et al.* (2015) A polyphenol-rich cranberry extract protects from diet-induced obesity, insulin resistance and intestinal inflammation in association with increased *Akkermansia* spp. population in the gut microbiota of mice. *Gut*, **64**, 872–883.

Antunes, F., Salvador, A., Marinho, H.S., Alves, R. and Pinto, R.E. (1996) Lipid peroxidation in mitochondrial inner membranes. I. An integrative kinetic model. *Free Radical Biology and Medicine*, **21**, 917–943.

Areias, F.M., Rego, A.C., Oliveira, C.R. and Seabra, R.M. (2001) Antioxidant effect of flavonoids after ascorbate/Fe^{2+}-induced oxidative stress in cultured retinal cells. *Biochemical Pharmacology*, **62**, 111–118.

Arora, A., Byrem, T.M., Nair, M.G. and Strasburg, G.M. (2000) Modulation of liposomal membrane fluidity by flavonoids and isoflavonoids. *Archives of Biochemistry and Biophysics*, **373**, 102–109.

Blanksby, S.J., Ramond, T.M., Davico, G.E. *et al.* (2001) Negative-ion photoelectron spectroscopy, gas-phase acidity, and thermochemistry of the peroxyl radicals CH3OO and CH3CH2OO. *Journal of the American Chemical Society*, **123**, 9585–9596.

Borrelli, R. and Peluso, A. (2013) Elementary electron transfer reactions: from basic concepts to recent computational advances. *Wiley Interdisciplinary Reviews: Computational Molecular Science*, **3**, 542–559.

Bouayed, J., Hoffmann, L. and Bohn, T. (2011) Total phenolics, flavonoids, anthocyanins and antioxidant activity following simulated gastro-intestinal digestion and dialysis of apple varieties: bioaccessibility and potential uptake. *Food Chemistry*, **128**, 14–21.

Bras, N.F., Goncalves, R., Mateus, N., Fernandes, P.A., Ramos, M.J. and De Freitas, V. (2010) Inhibition of pancreatic elastase by polyphenolic compounds. *Journal of Agricultural and Food Chemistry*, **58**, 10668–10676.

Brittes, J., Lucio, M., Nunes, C., Lima, J.L.F.C. and Reis, S. (2010) Effects of resveratrol on membrane biophysical properties: relevance for its pharmacological effects. *Chemistry and Physics of Lipids*, **163**, 747–754.

Byun, E.-B., Yang, M.-S., Kim, J.-H. *et al.* (2014) Epigallocatechin-3-gallate-mediated Tollip induction through the 67-kDa laminin receptor negatively regulating TLR4 signaling in endothelial cells. *Immunobiology*, **219**, 866–872.

Carlsen, C.U. and Skibsted, L.H. (2004) Myoglobin species with enhanced prooxidative activity is formed during mild proteolysis by pepsin. *Journal of Agricultural and Food Chemistry*, **52**, 1675–1681.

Chanet, A., Milenkovic, D., Claude, S. *et al.* (2013) Flavanone metabolites decrease monocyte adhesion to TNF-a-activated endothelial cells by modulating expression of atherosclerosis-related genes. *British Journal of Nutrition*, **110**, 587–598.

Chao, C.-L., Hou, Y.-C., Chao, P.-D.L., Weng, C.-S. and Ho, F.-M. (2009) The antioxidant effects of quercetin metabolites on the prevention of high glucose-induced apoptosis of human umbilical vein endothelial cells. *British Journal of Nutrition*, **101**, 1165–1170.

Chen, J.-C., Ho, F.-M., Chao, P.-D.L. *et al.* (2005) Inhibition of iNOS gene expression by quercetin is mediated by the inhibition of IkB kinase, nuclear factor-kappa B and STAT1, and depends on heme oxygenase-1 induction in mouse BV-2 microglia. *European Journal of Pharmacology*, **521**, 9–20.

Chiodo, S.G., Leopoldini, M., Russo, N. and Toscano, M. (2010) The inactivation of lipid peroxide radical by quercetin. A theoretical insight. *Physical Chemistry Chemical Physics*, **12**, 7662–7670.

Chvatalova, K., Slaninova, I., Brezinova, L. and Slanina, J. (2008) Influence of dietary phenolic acids on redox status of iron: ferrous iron autoxidation and ferric iron reduction. *Food Chemistry*, **106**, 650–660.

Crozier, A., Del Rio, D. and Clifford, M.N. (2010) Bioavailability of dietary flavonoids and phenolic compounds. *Molecular Aspects of Medicine*, **31**, 446–467.

Da Silva, M., Jaggers, G.K., Verstraeten, S.V., Erlejman, A.G., Fraga, C.G. and Oteiza, P.I. (2012) Large procyanidins prevent bile-acid-induced oxidant production and membrane-initiated ERK1/2, p38, and Akt activation in Caco-2 cells. *Free Radical Biology & Medicine*, **52**, 151–159.

Dangles, O. (2012) Antioxidant activity of plant phenols: chemical mechanisms and biological significance. *Current Organic Chemistry*, **16**, 692–714.

Dangles, O. and Dufour, C. (2008) Flavonoid–protein binding processes and their potential impact on human health, in *Recent Advances in Polyphenol Research*, vol. 2 (eds F. Daayf and V. Lattanzio), Blackwell Publishing, Oxford, pp. 67–87.

De Heer, M.I., Mulder, P., Korth, H.-G., Ingold, K.U. and Lusztyk, J. (2000) Hydrogen atom abstraction kinetics from intramolecularly hydrogen bonded ubiquinol-0 and other (poly)methoxy phenols. *Journal of the American Chemical Society*, **122**, 2355–2360.

Del Rio, D., Rodriguez-Mateos, A., Spencer, J.P.E., Tognolini, M., Borges, G. and Crozier, A. (2013) Dietary (poly)phenolics in human health: structures, bioavailability, and evidence of protective effects against chronic diseases. *Antioxidants & Redox Signaling*, **18**, 1818–1892.

Di Meo, F., Lemaur, V., Cornil, J. *et al.* (2013) Free radical scavenging by natural polyphenols: atom versus electron transfer. *Journal of Physical Chemistry A*, **117**, 2082–2092.

DiLabio, G.A. and Johnson, E.R. (2007) Lone pair $-\pi$ and $\pi-\pi$ interactions play an important role in proton-coupled electron transfer reactions. *Journal of the American Chemical Society*, **129**, 6199–6203.

El Hajji, H., Nkhili, E., Tomao, V. and Dangles, O. (2006) Interactions of quercetin with iron and copper ions: complexation and autoxidation. *Free Radical Research*, **40**, 303–320.

Engelmann, M.D., Hutcheson, R. and Cheng, I.F. (2005) Stability of ferric complexes with 3-hydroxyflavone (flavonol), 5,7-dihydroxyflavone (chrysin), and 3′,4′-dihydroxyflavone. *Journal of Agricultural and Food Chemistry*, **53**, 2953–2960.

Erlank, H., Elmann, A., Kohen, R. and Kanner, J. (2011) Polyphenols activate Nrf2 in astrocytes via H_2O_2, semiquinones, and quinones. *Free Radical Biology & Medicine*, **51**, 2319–2327.

Evans, C., Scaiano, J.C. and Ingold, K.U. (1992) Absolute kinetics of hydrogen abstraction from alpha-tocopherol by several reactive species including an alkyl radical. *Journal of the American Chemical Society*, **114**, 4589–4593.

Foti, M.C., Amorati, R., Pedulli, G.F., Daquino, C., Pratt, D.A. and Ingold, K.U. (2010) Influence of "remote" intramolecular hydrogen bonds on the stabilities of phenoxyl radicals and benzyl cations. *Journal of Organic Chemistry*, **75**, 4434–4440.

Fraga, C.G., Galleano, M., Verstraeten, S.V. and Oteiza, P.I. (2010) Basic biochemical mechanisms behind the health benefits of polyphenols. *Molecular Aspects of Medicine*, **31**, 435–445.

Furia, E., Marino, T. and Russo, N. (2014) Insights into the coordination mode of quercetin with the Al(III) ion from a combined experimental and theoretical study. *Dalton Transactions*, **43**, 7269–7274.

Galano, A., Leon-Carmona, J.R. and Alvarez-Idaboy, J.R. (2012) Influence of the environment on the protective effects of guaiacol derivatives against oxidative stress: mechanisms, kinetics, and relative antioxidant activity. *Journal of Physical Chemistry B*, **116**, 7129–7137.

Garzon, A., Bravo, I., Barbero, A.J. and Albaladejo, J. (2014) Mechanistic and kinetic study on the reactions of coumaric acids with reactive oxygen species: a DFT approach. *Journal of Agricultural and Food Chemistry*, **62**, 9705–9710.

Gloire, G., Legrand-Poels, S. and Piette, J. (2006) NF-kappaB activation by reactive oxygen species: fifteen years later. *Biochemical Pharmacology*, **72**, 1493–1505.

Gobert, M., Remond, D., Loonis, M., Buffiere, C., Sante-Lhoutellier, V. and Dufour, C. (2014) Fruits, vegetables and their polyphenols protect dietary lipids from oxidation during gastric digestion. *Food & Function*, **5**, 2166–2174.

Goncalves, R., Soares, S., Mateus, N. and de Freitas, V. (2007) Inhibition of trypsin by condensed tannins and wine. *Journal of Agricultural and Food Chemistry*, **55**, 7596–7601.

Goupy, P., Dufour, C., Loonis, M. and Dangles, O. (2003) Quantitative kinetic analysis of hydrogen transfer reactions from dietary polyphenols to the DPPH radical. *Journal of Agricultural and Food Chemistry*, **51**, 615–622.

Goupy, P., Vulcain, E., Caris-Veyrat, C. and Dangles, O. (2007) Dietary antioxidants as inhibitors of the heme-induced peroxidation of linoleic acid: mechanism of action and synergism. *Free Radical Biology & Medicine*, **43**, 933–946.

Guo, M., Perez, C., Wei, Y. *et al.* (2007) Iron-binding properties of plant phenolics and cranberry's bio-effects. *Dalton Transactions*, 4951–4961.

Hamalainen, M., Nieminen, R., Vuorela, P., Heinonen, M. and Mollanen, E. (2007) Anti-inflammatory effects of flavonoids: genistein, kaempferol, quercetin and daidzein inhibit STAT-1 and NF-kB activations, whereas flavone isorhamnetin, naringenin, and pelargonidin inhibit only NF-kB activation along with their inhibitory effects on iNOS expression and NO production in activated macrophages. *Mediators of Inflammation*, **2007**, 45673–45682.

He, S., Jiang, L., Wu, B., Pan, Y. and Sun, C. (2009) Pallidol, a resveratrol dimer from red wine, is a selective singlet oxygen quencher. *Biochemical and Biophysical Research Communications*, **379**, 283–287.

Hendrich, A.B., Malon, R., Pola, A., Shirataki, Y., Motohashi, N. and Michalak, K. (2002) Differential interaction of Sophora isoflavonoids with lipid bilayers. *European Journal of Pharmaceutical Sciences*, **16**, 201–208.

Hu, Y., Cheng, Z., Heller, L.I., Krasnoff, S.B., Glahn, R.P. and Welch, R.M. (2006) Kaempferol in red and pinto bean seed (*Phaseolus vulgaris* L.) coats inhibits iron bioavailability using an *in vitro* digestion/human Caco-2 cell model. *Journal of Agricultural and Food Chemistry*, **54**, 9254–9261.

Huynh, M.H.V. and Meyer, T.J. (2007) Proton-coupled electron transfer. *Chemical Reviews*, **107**, 5004–5064.

Hynes, M.J. and O'Coinceanainn, M. (2004) The kinetics and mechanisms of reactions of iron(III) with caffeic acid, chlorogenic acid, sinapic acid, ferulic acid and naringin. *Journal of Inorganic Biochemistry*, **98**, 1457–1464.

Inagaki, T., Yamamoto, T. and Kato, S. (2011) Proton-coupled electron transfer of the phenoxyl/phenol couple: effect of Hartree–Fock exchange on transition structures. *Journal of Computational Chemistry*, **32**, 3081–3091.

Ioku, K., Tsushida, T., Takei, Y., Nakatani, N. and Terao, J. (1995) Antioxidative activity of quercetin and quercetin monoglucosides in solution and phospholipid bilayers. *Biochimica et Biophysica Acta - Biomembranes*, **1234**, 99–104.

Joshi, R., Naumov, S., Kapoor, S., Mukherjee, T., Hermann, R. and Brede, O. (2004) Phenol radical cations and phenoxyl radicals in electron transfer from the natural phenols sesamol, curcumin and trolox to the parent radical cations of 1-chlorobutane. *Journal of Physical Organic Chemistry*, **17**, 665–674.

Jovanovic, S.V. and Simic, M.G. (2000) Antioxidants in nutrition. *Annals of the New York Academy of Sciences*, **899**, 326–334.

Kamal-Eldin, A. and Appelqvist, L.-A. (1996) The chemistry and antioxidant properties of tocopherols and tocotrienols. *Lipids*, **31**, 671–701.

Kanner, J., Gorelik, S., Roman, S. and Kohen, R. (2012) Protection by polyphenols of postprandial human plasma and low-density lipoprotein modification: the stomach as a bioreactor. *Journal of Agricultural and Food Chemistry*, **60**, 8790–8796.

Kawai, Y., Nishikawa, T., Shiba, Y. *et al.* (2008) Macrophage as a target of quercetin glucuronides in human atherosclerotic arteries. Implication in the anti-atherosclerotic mechanism of dietary flavonoids. *Journal of Biological Chemistry*, **283**, 9424–9434.

Khan, M.K., Rakotomanomana, N., Dufour, C. and Dangles, O. (2011) Binding of citrus flavanones and their glucuronides and chalcones to human serum albumin. *Food & Function*, **2**, 617–626.

Kosinova, P., Berka, K., Wykes, M., Otyepka, M. and Trouillas, P. (2012) Positioning of antioxidant quercetin and its metabolites in lipid bilayer membranes: implication for their lipid–peroxidation inhibition. *Journal of Physical Chemistry B*, **116**, 1309–1318.

Kristinova, V., Mozuraityte, R., Storro, I. and Rustad, T. (2009) Antioxidant activity of phenolic acids in lipid oxidation catalyzed by different prooxidants. *Journal of Agricultural and Food Chemistry*, **57**, 10377–10385.

Laranjinha, J. and Cadenas, E. (1999) Redox cycles of caffeic acid, alpha-tocopherol, and ascorbate: implications for protection of low-density lipoproteins against oxidation. *IUBMB Life*, **48**, 57–65.

Lee-Hilz, Y.Y., Boerboom, A.M., Westphal, A.H., Berkel, W.J., Aarts, J.M. and Rietjens, I.M. (2006) Pro-oxidant activity of flavonoids induces EpRE-mediated gene expression. *Chemical Research in Toxicology*, **19**, 1499–1505.

Leopoldini, M., Russo, N. and Toscano, M. (2011) The molecular basis of working mechanism of natural polyphenolic antioxidants. *Food Chemistry*, **125**, 288–306.

Lingwood, M., Hammond, J.R., Hrovat, D.A., Mayer, J.M. and Borden, W.T. (2006) MPW1K performs much better than B3LYP in DFT calculations on reactions that proceed by proton-coupled electron transfer (PCET). *Journal of Chemical Theory and Computation*, **2**, 740–745.

Litwinienko, G. and Ingold, K.U. (2007) Solvent effects on the rates and mechanisms of reaction of phenols with free radicals. *Accounts of Chemical Research*, **40**, 222–230.

Loke, W.M., Proudfoot, J.M., Stewart, S. *et al.* (2008a) Metabolic transformation has a profound effect on anti-inflammatory activity of flavonoids such as quercetin: lack of association between antioxidant and lipoxygenase inhibitory activity. *Biochemical Pharmacology*, **75**, 1045–1053.

Loke, W.M., Proudfoot, J.M., Mckinley, A.J. *et al.* (2008b) Quercetin and its *in vivo* metabolites inhibit neutrophil-mediated low-density lipoprotein oxidation. *Journal of Agricultural and Food Chemistry*, **56**, 3609–3615.

Lopez, S., Bermudez, B., Montserrat-de la Paz, S. *et al.* (2014) Membrane composition and dynamics: a target of bioactive virgin olive oil constituents. *Biochimica et Biophysica Acta - Biomembranes*, **1838**, 1638–1656.

Lorenz, M., Jochmann, N., von Krosigk, A. *et al.* (2007) Addition of milk prevents vascular protective effects of tea. *European Heart Journal*, **28**, 219–223.

Lorrain, B., Dufour, C. and Dangles, O. (2010) Influence of serum albumin and the flavonol quercetin on the peroxidase activity of metmyoglobin. *Free Radical Biology & Medicine*, **48**, 1162–1172.

Lorrain, B., Dangles, O., Loonis, M., Armand, M. and Dufour, C. (2012) Dietary iron-initiated lipid oxidation and its inhibition by polyphenols in gastric conditions. *Journal of Agricultural and Food Chemistry*, **60**, 9074–9081.

Loverde, S.M. (2014) Molecular simulation of the transport of drugs across model membranes. *Journal of Physical Chemistry Letters*, **5**, 1659–1665.

Mackenzie, G.G., Delfino, J.M., Keen, C.L., Fraga, C.G. and Oteiza, P.I. (2009) Dimeric procyanidins are inhibitors of NF-kB–DNA binding. *Biochemical Pharmacology*, **78**, 1252–1262.

Manach, C., Williamson, G., Morand, C., Scalbert, A. and Remesy, C. (2005) Bioavailability and bioefficacy of polyphenols in humans. I. Review of 97 bioavailability studies. *American Journal of Clinical Nutrition*, **81** (suppl), 230S–242S.

Markovic, Z., Amic, D., Milenkovic, D., Dimitric-Markovic, J.M. and Markovic, S. (2013) Examination of the chemical behavior of the quercetin radical cation towards some bases. *Physical Chemistry Chemical Physics*, **15**, 7370–7378.

Mayer, J.M. (2011) Understanding hydrogen atom transfer: from bond strengths to Marcus theory. *Accounts of Chemical Research*, **44**, 36–46.

McDougall, G.J., Shpiro, F., Dobson, P., Smith, P., Blake, A. and Stewart, D. (2005) Different polyphenolic components of soft fruits inhibit α-amylase and α-glucosidase. *Journal of Agricultural and Food Chemistry*, **53**, 2760–2766.

McDougall, G.J., Kulkarni, N.N. and Stewart, D. (2009) Berry polyphenols inhibit pancreatic lipase activity *in vitro*. *Food Chemistry*, **115**, 193–199.

Menendez, C., Duenas, M., Galindo, P. *et al.* (2011) Vascular deconjugation of quercetin glucuronide. The flavonoid paradox revealed? *Molecular Nutrition & Food Research*, **55**, 1780–1790.

Mira, L., Fernandez, M.T., Santos, M., Rocha, R., Florencio, M.H. and Jennings, K.R. (2002) Interactions of flavonoids with iron and copper ions: a mechanism for their antioxidant activity. *Free Radical Research*, **36**, 1199–1208.

Movileanu, L., Neagoe, I. and Flonta, M.L. (2000) Interaction of the antioxidant flavonoid quercetin with planar lipid bilayers. *International Journal of Pharmaceutics*, **205**, 135–146.

Mukai, K., Nagai, S. and Ohara, K. (2005) Kinetic study of the quenching reaction of singlet oxygen by tea catechins in ethanol solution. *Free Radical Biology and Medicine*, **39**, 752–761.

Nakatani, N., Nakao, Y., Sato, H. and Sakaki, S. (2009) Theoretical study of dioxygen binding process in iron(III) catechol dioxygenase: "oxygen activation" vs "substrate activation". *Journal of Physical Chemistry B*, **113**, 4826–4836.

Nielsen, M.F. and Ingold, K.U. (2006) Kinetic solvent effects on proton and hydrogen atom transfers from phenols. Similarities and differences. *Journal of the American Chemical Society*, **128**, 1172–1182.

Nkhili, E., Loonis, M., Mihai, S., El Hajji, H. and Dangles, O. (2014) Reactivity of food phenols with iron and copper ions: binding, dioxygen activation and oxidation mechanisms. *Food & Function*, **5**, 1186–1202.

Olas, B. and Holmsen, H. (2012) Interaction of resveratrol with membrane glycerophospholipids in model system *in vitro*. *Food and Chemical Toxicology*, **50**, 4028–4034.

Ollila, F., Halling, K., Vuorela, P., Vuorela, H. and Slotte, J.P. (2002) Characterization of flavonoid–biomembrane interactions. *Archives of Biochemistry and Biophysics*, **399**, 103–108.

Oteiza, P.I., Erlejman, A.G., Verstraeten, S.V., Keen, C.L. and Fraga, C.G. (2005) Flavonoid–membrane interactions: a protective role of flavonoids at the membrane surface? *Clinical & Developmental Immunology*, **12**, 19–25.

Ouchi, A., Aizawa, K., Iwasaki, Y. *et al.* (2010) Kinetic study of the quenching reaction of singlet oxygen by carotenoids and food extracts in solution. Development of a singlet oxygen absorption capacity (SOAC) assay method. *Journal of Agricultural and Food Chemistry*, **58**, 9967–9978.

Pan, M.-H., Laia, C.-S. and Ho, C.-T. (2010) Anti-inflammatory activity of natural dietary flavonoids. *Food & Function*, **1**, 15–31.

Parasassi, T., Martellucci, A., Conti, F. and Messina, B. (1984) Drug–membrane interactions: silymarin, silybin and microsomal membranes. *Cell Biochemistry and Function*, **2**, 85–88.

Perron, N.R. and Brumaghim, J.L. (2009) A review of the antioxidant mechanisms of polyphenol compounds related to iron binding. *Cellular Biochemistry and Biophysics*, **53**, 75–100.

Perron, N.R., Hodges, J.N., Jenkins, M. and Brumaghim, J.L. (2008) Predicting how polyphenol antioxidants prevent DNA damage by binding to iron. *Inorganic Chemistry*, **47**, 6153–6161.

Perron, N.R., Wang, H.C., DeGuire, S.N., Jenkins, M., Lawson, M. and Brumaghim, J.L. (2010) Kinetics of iron oxidation upon polyphenol binding. *Dalton Transactions*, **39**, 9982–9987.

Perron, N.R., Garcia, C.R., Pinzon, J.R., Chaur, M.N. and Brumaghim, J.L. (2011) Antioxidant and prooxidant effects of polyphenol compounds on copper-mediated DNA damage. *Journal of Inorganic Biochemistry*, **105**, 745–753.

Pierre, F., Tache, S., Gueraud, F., Rerole, A.L., Jourdan, M.-L. and Petit, C. (2007) Apc mutation induces resistance of colonic cells to lipoperoxide-triggered apoptosis induced by faecal water from haem-fed rats. *Carcinogenesis*, **28**, 321–327.

Pratt, D.A., Blake, J.A., Mulder, P., Walton, J.C., Korth, H.-G. and Ingold, K.U. (2004) O–H bond dissociation enthalpies in oximes: order restored. *Journal of the American Chemical Society*, **126**, 10667–10675.

Raghunathan, M., Zubovski, Y., Venable, R.M., Pastor, R.W., Nagle, J.F. and Tristram-Nagle, S. (2012) Structure and elasticity of lipid membranes with genistein and daidzein bioflavonoids using X-ray scattering and MD simulations. *Journal of Physical Chemistry B*, **116**, 3918–3927.

Ramond, T.M., Blanksby, S.J., Kato, S. *et al.* (2002) Heat of formation of the hydroperoxyl radical HOO via negative ion studies. *Journal of Physical Chemistry A*, **106**, 9641–9647.

Roche, M., Dufour, C., Mora, N. and Dangles, O. (2005) Antioxidant activity of olive phenols: mechanistic investigation and characterization of oxidation products by mass spectrometry. *Organic & Biomolecular Chemistry*, **3**, 423–430.

Ryan, P. and Hynes, M.J. (2007) The kinetics and mechanisms of the complex formation and antioxidant behaviour of the polyphenols EGCg and ECG with iron(III). *Journal of Inorganic Biochemistry*, **101**, 585–593.

Ryan, P. and Hynes, M.J. (2008) The kinetics and mechanisms of the reaction of iron(III) with quercetin. *Journal of Inorganic Biochemistry*, **102**, 127–136.

Saija, A., Scalese, M., Lanza, M., Marzullo, D., Bonina, F. and Castelli, F. (1995) Flavonoids as antioxidant agents: importance of their interaction with biomembranes. *Free Radical Biology & Medicine*, **19**, 481–486.

Shiba, Y., Kinoshita, T., Chuman, H. *et al.* (2008) Flavonoids as substrates and inhibitors of myeloperoxidase: molecular actions of aglycone and metabolites. *Chemical Research in Toxicology*, **21**, 1600–1609.

Sies, H., Stahl, W. and Sevanian, A. (2005) Nutritional, dietary and postprandial oxidative stress. *Journal of Nutrition*, **135**, 969–972.

Sinha, R., Gadhwal, M.K., Joshi, U.J., Srivastava, S. and Govil, G. (2011) Interaction of quercetin with DPPC model membrane: molecular dynamic simulation, DSC and multinuclear NMR studies. *Journal of the Indian Chemical Society*, **88**, 1203–1210.

Sirk, T.W., Brown, E.F., Sum, A.K. and Friedman, M. (2008) Molecular dynamics study on the biophysical interactions of seven green tea catechins with lipid bilayers of cell membranes. *Journal of Agricultural and Food Chemistry*, **56**, 7750–7758.

Sirk, T.W., Brown, E.F., Friedman, M. and Sum, A.K. (2009) Molecular binding of catechins to biomembranes: relationship to biological activity. *Journal of Agricultural and Food Chemistry*, **57**, 6720–6728.

Sorensen, A.D.M., Haahr, A.M., Becker, E.M. *et al.* (2008) Interactions between iron, phenolic compounds, emulsifiers, and pH in omega-3-enriched oil-in-water emulsions. *Journal of Agricultural and Food Chemistry*, **56**, 1740–1750.

Spencer, J.P.E., Vafeiadou, K., Williams, R.J. and Vauzour, D. (2012) Neuroinflammation: modulation by flavonoids and mechanisms of action. *Molecular Aspects of Medicine*, **33**, 83–97.

Steffen, Y., Gruber, C., Schewe, T. and Sies, H. (2008) Mono-O-methylated flavanols and other flavonoids as inhibitors of endothelial NADPH oxidase. *Archives of Biochemistry and Biophysics*, **469**, 209–219.

Tagliazucchi, D., Verzelloni, E. and Conte, A. (2005) Effect of some phenolic compounds and beverages on pepsin activity during simulated gastric digestion. *Journal of Agricultural and Food Chemistry*, **53**, 8706–8713.

Tang, L., Zhang, Y., Qian, Z. and Shen, X. (2000) The mechanism of Fe2+-initiated lipid peroxidation in liposomes: the dual function of ferrous ions, the roles of the pre-existing lipid peroxides and the lipid peroxyl radical. *Biochemical Journal*, **352**, 27–36.

Tejero, I., Gonzalez-Garcia, N., Gonzalez-Lafont, A. and Lluch, J.M. (2007) Tunneling in green tea: understanding the antioxidant activity of catechol-containing compounds. A variational transition-state theory study. *Journal of the American Chemical Society*, **129**, 5846–5854.

Trouillas, P., Marsal, P., Siri, D., Lazzaroni, R. and Duroux, J.-L. (2006) A DFT study of the reactivity of OH groups in quercetin and taxifolin antioxidants: the specificity of the 3-OH site. *Food Chemistry*, **97**, 679–688.

Trouillas, P., Marsal, P., Svobodova, A. *et al.* (2008) Mechanism of the antioxidant action of silybin and 2,3-dehydrosilybin flavonolignans: a joint experimental and theoretical study. *Journal of Physical Chemistry A*, **112**, 1054–1063.

Umeda, D., Yano, S., Yamada, K. and Tachibana, H. (2008) Green tea polyphenol epigallocatechin-3-gallate signaling pathway through 67-kDa laminin receptor. *Journal of Biological Chemistry*, **283**, 3050–3058.

Verstraeten, S.V., Jaggers, G.K., Fraga, C.G. and Oteiza, P.I. (2013) Procyanidins can interact with Caco-2 cell membrane lipid rafts: involvement of cholesterol. *Biochimica et Biophysica Acta*, **1828**, 2646–2653.

Vitaglione, P., Donnarumma, G., Napolitano, A. *et al.* (2007) Protocatechuic acid is the major human metabolite of cyanidin-glucosides. *Journal of Nutrition*, **137**, 2043–2048.

Wu, X. and Schauss, A.G. (2012) Mitigation of inflammation with foods. *Journal of Agricultural and Food Chemistry*, **60**, 6703–6717.

Yu, X., Chu, S., Hagerman, A.E. and Lorigan, G.A. (2011) Probing the interaction of polyphenols with lipid bilayers by solid-state NMR spectroscopy. *Journal of Agricultural and Food Chemistry*, **59**, 6783–6789.

Zhao, Y. and Truhlar, D.G. (2004) Hybrid meta density functional theory methods for thermochemistry, thermochemical kinetics, and noncovalent interactions: the MPW1B95 and MPWB1K models and comparative assessments for hydrogen bonding and van der Waals interactions. *Journal of Physical Chemistry A*, **108**, 6908–6918.

Chapter 2
Polyphenols in Bryophytes: Structures, Biological Activities, and Bio- and Total Syntheses

Yoshinori Asakawa

Abstract: The Marchantiophyta (liverworts) are rich sources of polyphenols, especially, cyclic and acyclic bis-bibenzyls, which are rare natural products in the plant kingdom. At present, 103 bis-bibenzyls have been found in liverworts. The structurally unique cyclic and acyclic bis-bibenzyls show, among others, various biological activities such as antimicrobial, antifungal, cytotoxic, muscle relaxation, antioxidant, tubulin polymerization inhibitory, antitrypanosomal activities. They are biosynthesized from dimerization of lunularic acid via dihydrocoumaric acid and prelunularin. This chapter deals with the distribution of bis-bibenzyls in liverworts and their structures, bio- and total syntheses, and biological activities.

Keywords: liverworts, bis-bibenzyls, bio- and total syntheses, cytotoxicity, antimicrobial, antifungal, anticancer, muscle relaxing activity.

2.1 Introduction

The bryophytes, the Marchantiophyta (liverworts), Bryophyta (mosses), and Anthocerotophyta (hornworts), are distributed everywhere in the world except in the oceans. They are taxonomically placed between the algae and the pteridophytes and 25,000 species are now known worldwide. The most charming countries of the bryophytes are New Zealand, Argentina, and Japan. There are many endemic genera in Southern hemispheric countries. On the contrary, in North America, Europe, and Africa, the distribution of the endemic genera of bryophytes is very limited. The bryophytes have not been studied chemically over the past century because they are generally very tiny, and their identification and collection in large amounts are difficult. Among the three classes, only liverworts contain beautiful oil bodies in the cells. The bryophytes have been considered useless for human

Recent Advances in Polyphenol Research, First Edition. Edited by Kumi Yoshida, Véronique Cheynier and Stéphane Quideau.

Fig. 2.1 Marchantia polymorpha. (*See insert for color representation of the figure*)

diets; however, 24 species of the bryophytes have been used in China as medicinal plants (Ding, 1982). For example, one of the popular liverworts, *Marchantia polymorpha* (Fig. 2.1) shows antipyretic, antidotal, antihepatic, and diuretic activities. *Conocephalum conicum* (Fig. 2.2) also shows antimicrobial and antifungal activities as well as elimination of gall stones. *Frullania tamarisci* and *Reboulia hemisphaerica* (Fig. 2.3) show antiseptic and hemostasis activities, respectively. Some mosses are also used to cure burns, bruises, external wounds, snake bite, pulmonary tuberculosis, neurasthenia, fractures, convulsions, scald, uropathy, pneumonia, neurasthenia, and so on. Although the bryophytes possess such pharmacologically interesting substances, their isolation and structural elucidation were neglected over the past century.

In this chapter, the distribution and chemical structures of polyphenols, especially cyclic and acyclic bis-bibenzyls, in liverworts, their bio- and total syntheses, and activities are surveyed.

2.2 Distribution of cyclic and acyclic bis-bibenzyls in Marchantiophyta (liverworts)

The occurrence, conformation, biosynthesis, biological activity, and synthesis of bis-bibenzyls have been reviewed by Asakawa (1982, 1990a, 1990b, 1993, 1994, 1995, 1999, 2001, 2008a, 2008b, 2008c) and his group (Asakawa & Ludwiczuk, 2008, 2009; Asakawa *et al.*, 2009, 2013a), Keseru and Nogradi (1995a, 1995b, 1996), Zinsmeister *et al.* (1991), and Lou (2013). These compounds are basically oligomers of polyhydroxylated dihydrostilbenes, most of which are cyclic dimers with biaryl (C—C) and/or diarylether (C—O) linkage, respectively, with noted C- and O-ring connections for simplicity as shown in Table 2.1.

Fig. 2.2 Marchantia paleacea var. diptera. (*See insert for color representation of the figure*)

Fig. 2.3 Pellia epiphylla. (*See insert for color representation of the figure*)

Such macrocyclic and acyclic bis-bibenzyls are rare plant metabolites possessing structures that occur exclusively in the liverworts. However, only riccardin H (**92a**) (Oiso *et al.*, 1999) and riccardin C (**3**) (Kosenkova *et al.*, 2007) were found in the fern *Hymenophyllum barbatum* and the higher plant *Primula macrocalyx*, respectively.

The characteristic bis-bibenzyls are not only significant chemical markers of several Marchantiophyta families but also important for considering the phylogeny of the bryophytes and the evolutionary processes of the lower terrestrial spore-forming plants (Asakawa, 1995, 2004; Oiso *et al.*, 1999; Asakawa *et al.*, 2008, 2013a).

In Table 2.1, the distribution of cyclic and acyclic bis-bibenzyls has been shown (Asakawa & Ludwiczuk, 2012). Since the first two cyclic bis-bibenzyls, riccardin A (**1**) and marchantin A (**14**), were isolated from the thalloid liverworts, *Riccardia multifida* and *M. polymorpha* (Asakawa, 1982), 103 bis-bibenzyls have been isolated from Jungermanniales, Metzgeriales, and Marchantiales species as shown in Figs. 2.4, 2.5, 2.6, 2.7, 2.8, and 2.9 (Asakawa, 1995; Asakawa *et al.*, 2000, 2013b; Asakawa & Ludwiczuk, 2012; Fang *et al.*, 2007).

The methanol extract (105 g) of a Japanese specimen of *M. polymorpha* was chromatographed over silica gel and Sephadex LH-20 to give the cyclic bis-bibenzyls, marchantin A (**14**) (30 g), and its derivatives, marchantins B (**15**), C (**16**), D (**17**), E (**18**), G (**19**), and J (**22**). The yield of **14** is dependent upon the particular *Marchantia* species being investigated. Pure **14** (80–120 g) has been isolated from 6.67 kg of the dried *Marchantia paleacea* var. *diptera*.

The isolated bis-bibenzyls are divided into two groups: cyclic (**1–84**) and acyclic (**85–103**). The former group is further classified into two bibenzyls that are connected with (i) one ether (C_1—C_2') and one biphenyl linkages (C_{12}—C_{10}', C_{14}—C_{10}', C_{14}—C_{12}'), (ii) one ether (C_{14}—C_{11}') and one biphenyl (C_6—C_2') linkages, (iii) one ether (C_6—C_2') and one biphenyl (C_{13}—C_{10}') linkages, (iv) two ether (C_1—C_2' and C_{13}—C_{12}', C_{14}—C_{11}', C_{13}—C_{10}', C_{12}—C_{11}') linkages, (v) two biphenyl (C_6—C_2' and C_{14}—C_{12}') linkages, and (vi) two biphenyl (C_6—C_3' and C_{14}—C_{11}) linkages among which type 4 (two ether linkages between C_1—C_2' and C_{14}—C_{11}') and type 5 (two ether linkages between C_6—C_2' and C_{14}—C_{12}') are the most abundant. The former type is distributed in the Marchantiales and the latter in the Jungermanniales.

Among riccardin-type bis-bibenzyls, riccardin C (**3**) is distributed in the Marchantiaceae and *Plagiochasma* species belonging to the Aytoniaceae. The type 1 bis-biphenyls–bibenzyl with biphenyl units connected between C_6—C_2' and C_{14}—C_{12}' is distributed in the stem-leafy liverwort, *Plagiochila fruticosa*, collected in the southern Shikoku Island, Japan.

Plagiochin and isoplagiochin types (type 1) of bis-bibenzyls have been distributed in the *Plagiochila*, *Heteroscyphus*, and *Herbertus* species belonging to the Jungermanniales. Type 1 is restricted to the Aneuraceae and Blasiaceae, which belong to the Metzgeriales, and the Aytoniaceae, Marchantiaceae, Monocleaceae, and Ricciaceae belonging to the Marchantiales.

Since 1995, 14 acyclic bis-bibenzyls (**85–103**) have been isolated from the Marchantiophyta and their structures elucidated. These are mainly distributed in the *Radula, Plagiochila, Nardia,* and *Jungermannia* belonging to the Jungermanniales, *Pellia* to the Metzgeriales, and *Marchantia* and *Lunularia* to the Marchantiales.

2.3 Biosynthesis of bis-bibenzyls

The naturally occurring bis-bibenzyls are categorized into three structure types, with the macrocyclic rings constructed via (i) two biphenyl ether C—O bonds, (ii) a diphenyl ether C—O and a biaryl C—C bond, and (iii) two biphenyl C—C bonds.

Table 2.1 Bis-bibenzyls found in the Marchantiophyta (liverworts) (Asakawa & Ludwiczuk, 2012).

Structure number	Name of compounds	A–C and B–D ring connection pattern	A–C and B–D ring connection position	Plant source
1	Riccardin A	O–C	C_1–C_2'/C_{14}–C_{12}'	*Riccardia multifida* *R. multifida* subsp. *decrescens* *Riccardia nagasakiensis*
2	Riccardin B	O–O	C_1–C_2'/C_{13}–C_{12}'	*Asterella angusta* *Preissia quadrata* *R. multifida*
3	Riccardin C	O–C	C_1–C_2'/C_{14}–C_{12}'	*Blasia pusilla* *Dumortiera hirsuta* *Jungermannia infusca* *Mastigophora diclados* *Plagiochasma pterospermum* *Plagiochasma rupestre* *Plagiochasma intermedium* *Plagiochila* sp.
4	Riccardin D (=Plagiochin E)	O–C	C_1–C_2'/C_{14}–C_{12}'	*A. angusta* *D. hirsuta* *Marchantia polymorpha* *Plagiochasma japonica* *Plagiochila cristata*
5	Riccardin F	O–C	C_1–C_2'/C_{14}–C_{12}'	*B. pusilla* *P. intermedium*
6	Riccardin G	O–C	C_1–C_2'/C_{14}–C_{12}'	*Marchantia chenopoda*
7	Riccardin H	O–C	C_1–C_2'/C_{14}–C_{12}'	*M. polymorpha*
8	Isoriccardin C	O–C	C_1–C_2'/C_{14}–C_{12}'	*Marchantia paleacea* *P. intermedium* *P. rupestre*
9	Isoriccardin D	O–C	C_1–C_2'/C_{14}–C_{12}'	*M. polymorpha*
10	IsoriccardinquinoneA	O–C	C_1–C_2'/C_{14}–C_{12}'	*M. paleacea*
11	Isoriccardinquinone B	O–C	C_1–C_2'/C_{14}–C_{12}'	*M. paleacea*
12	13,13′-*O*-isopropylidenericcardin D	O–C	C_1–C_2'/C_{14}–C_{12}'	*M. polymorpha*
13	Polymorphatin A	O–C	C_1–C_2'/C_{14}–C_{12}'	*M. polymorpha*
14	Marchantin A	O–O	C_1–C_2'/C_{14}–C_{12}'	*Marchantia emarginata* subsp. *tosana* *M. paleacea* var. *diptera* *M. polymorpha* *Plagiochasma appendiculatum*
15	Marchantin B	O–O	C_1–C_2'/C_{14}–C_{12}'	*M. paleacea* var. *diptera* *M. polymorpha* *P. appendiculatum* *P. rupestre*

Table 2.1 (Continued)

Structure number	Name of compounds	A–C and B–D ring connection pattern	A–C and B–D ring connection position	Plant source
16	Marchantin C	O–O	C_1–C_2'/C_{14}–C_{12}'	*D. hirsuta* *J. infusca* *Marchantia foliacea* *M. paleacea* var. *diptera* *P. appendiculatum* *Plagiochila barteri* *Schistochila glaucescens* *Reboulia hemisphaerica* *R. nagasakiensis*
17	Marchantin D	O–O	C_1–C_2'/C_{14}–C_{12}'	*M. polymorpha*
18	Marchantin E	O–O	C_1–C_2'/C_{14}–C_{12}'	*M. paleacea* var. *diptera* *M. polymorpha*
19	Marchantin G	O–O	C_1–C_2'/C_{14}–C_{12}'	*M. polymorpha*
20	Marchantin H	O–O	C_1–C_2'/C_{14}–C_{12}'	*A. angusta* *P. intermedium* *P. barteri*
21	Marchantin I	O–O	C_1–C_2'/C_{14}–C_{12}'	*R. multifida* subsp. *decrescens*
22	Marchantin J	O–O	C_1–C_2'/C_{14}–C_{12}'	*M. polymorpha*
23	Marchantin K	O–O	C_1–C_2'/C_{14}–C_{12}'	*P. rupestre*
24	Marchantin M	O–O	C_1–C_2'/C_{14}–C_{12}'	*A. angusta* *R. hemisphaerica*
25	Marchantin N	O–O	C_1–C_2'/C_{14}–C_{12}'	*R. hemisphaerica*
26	Marchantiaquinone	O–O	C_1–C_2'/C_{14}–C_{12}'	*R. hemisphaerica* *Mannia subpilosa*
27	Marchantin O	O–O	C_1–C_2'/C_{14}–C_{12}'	*R. hemisphaerica*
20	Marchantin P	O–O	C_1–C_2'/C_{14}–C_{12}'	*A. angusta* *M. chenopoda* *M. foliacea*
29	Isomarchantin C	O–O	C_1–C_2'/C_{14}–C_{12}'	*Bryopteris filicina* *D. hirsuta* *M. paleacea* var. *diptera* *M. foliacea*
30	Isomarchantin C-1′-methyl ether	O–O	C_1–C_2'/C_{14}–C_{12}'	*M. foliacea*
31	Neoisomarchantin C	O–O	C_1–C_2'/C_{14}–C_{12}'	*M. foliacea*
32	Neomarchantin A	O–O	C_1–C_2'/C_{14}–C_{12}'	*M. polymorpha* *P. intermedium* *P. quadrata* *S. glaucescens*
33	Neomarchantin B	O–O	C_1–C_2'/C_{14}–C_{12}'	*S. glaucescens*
34	Asterellin A	O–C–O	C_1–C_2'/C_{14}–C_{12}'	*A. angusta*
35	Asterellin B	O–C–O	C_1–C_2'/C_{14}–C_{12}'	*A. angusta*
36	11-*O*-demethylmarchantin I	O–O	C_1–C_2'/C_{14}–C_{12}'	*A. angusta*
37	Dihydroptychantol A	O–O	C_1–C_2'/C_{14}–C_{12}'	*A. angusta*
38	Pakyonol	O–O	C_1–C_2'/C_{14}–C_{12}'	*P. intermedium* *P. pterospermum*
39	Planusin A	O–C	C_1–C_2'/C_{14}–C_{12}'	*Heteroscyphus planus*
40	Plagiochin D	O–C	C_1–C_2'/C_{14}–C_{12}'	*Plagiochila sciophila*
41	Plagiochin A	O–C	C_1–C_2'/C_{14}–C_{12}'	*P. sciophila*
42	Plagiochin C	O–C	C_1–C_2'/C_{14}–C_{12}'	*P. sciophila*
43	Plagiochin B	O–C	C_1–C_2'/C_{14}–C_{12}'	*P. sciophila*

(*Continued*)

Table 2.1 (Continued)

Structure number	Name of compounds	A–C and B–D ring connection pattern	A–C and B–D ring connection position	Plant source
44	Isoplagiochin A	C–O	$C_6-C_2'/C_{14}-C_{11}'$	*H. planus* *Plagiochila diversifolia* *Plagiochila fruticosa* *Plagiochila permista* var. *integerrima*
45	Isoplagiochin B	C–O	$C_6-C_2'/C_{14}-C_{11}'$	*P. fruticosa*
46	Isoplagiochin D	C–C	$C_6-C_2'/C_{14}-C_{12}'$	*Herbertus sakuraii* *Lepidozia incurvata* *P. fruticosa* *Plagiochila* sp.
47	Isoplagiochin C	C–C	$C_6-C_2'/C_{14}-C_{12}'$	*Bazzania trilobata* *H. sakuraii* *Lepidozia fauriana* *P. fruticosa* *Plagiochila* sp.
48	Isoplagiochin E	C–O	$C_6-C_2'/C_{14}-C_{11}'$	*P. permista* var. *integerrima*
49	Isoplagiochin F	C–O	$C_6-C_2'/C_{14}-C_{11}'$	*P. permista* var. *integerrima* *Plagiochila* sp.
50	Isoplagiochin G	C–O	$C_6-C_2'/C_{14}-C_{11}'$	*Plagiochila* sp.
51	2′,10,10′,12,14′-Pentachloro-7′,8′-dehydroisoplagiochin D	C–O	$C_6-C_2'/C_{14}-C_{12}'$	*Bazzania tricrenata*
52	12-Chloroisoplagiochin D	C–C	$C_6-C_2'/C_{14}-C_{12}'$	*M. diclados* *P. permista* var. *integerrima*
53	2,12-Dichloroisoplagiochin D	C–C	$C_6-C_2'/C_{14}-C_{12}'$	*H. sakuraii* *M. diclados*
54	12,7′-Dichloroisoplagiochin D	C–C	$C_6-C_2'/C_{14}-C_{12}'$	*H. sakuraii*
55	12,10′-Dichloroisoplagiochin C	C–C	$C_6-C_2'/C_{14}-C_{12}'$	*H. sakuraii*
56	Bazzanin A	C–C	$C_6-C_2'/C_{14}-C_{12}'$	*B. trilobata*
57	Bazzanin B	C–C	$C_6-C_2'/C_{14}-C_{12}'$	*B. trilobata*
58	Bazzanin C	C–C	$C_6-C_2'/C_{14}-C_{12}'$	*B. trilobata*
59	Bazzanin D	C–C	$C_6-C_2'/C_{14}-C_{12}'$	*B. trilobata*
60	Bazzanin E	C–C	$C_6-C_2'/C_{14}-C_{12}'$	*B. trilobata*
61	Bazzanin F	C–C	$C_6-C_2'/C_{14}-C_{12}'$	*B. trilobata*
62	Bazzanin G	C–C	$C_6-C_2'/C_{14}-C_{12}'$	*B. trilobata*
63	Bazzanin H	C–C	$C_6-C_2'/C_{14}-C_{12}'$	*B. trilobata*
64	Bazzanin I	C–C	$C_6-C_2'/C_{14}-C_{12}'$	*B. trilobata*
65	Bazzanin J	C–C	$C_6-C_2'/C_{14}-C_{12}'$	*B. trilobata*
66	Bazzanin K	C–C	$C_6-C_2'/C_{14}-C_{12}'$	*B. trilobata*
67	Bazzanin L	C–C	$C_6-C_2'/C_{14}-C_{12}'$	*L. incurvata*
67a	Cavicularin	O–C	$C_6-C_2'/C_{14}-C_{12}'$	*Cavicularia densa*
68	Bazzanin M	C–C	$C_6-C_2'/C_{14}-C_{12}'$	*L. incurvata*
69	Bazzanin N	C–C	$C_6-C_2'/C_{14}-C_{12}'$	*L. incurvata*
70	Bazzanin O	C–C	$C_6-C_2'/C_{14}-C_{12}'$	*L. incurvata*
71	Bazzanin P	C–C	$C_6-C_2'/C_{14}-C_{12}'$	*L. incurvata*
72	Bazzanin Q	C–C	$C_6-C_2'/C_{14}-C_{12}'$	*L. incurvata*
73	Bazzanin R	C–C	$C_6-C_2'/C_{14}-C_{12}'$	*L. incurvata*
74	Bazzanin S	C–C	$C_6-C_2'/C_{14}-C_{12}'$	*B. trilobata*
75	Ptychantol A	O–O	$C_1-C_2'/C_{13}-C_{10}'$	*Ptychanthus striatus*
76	Ptychantol B	O–O	$C_1-C_2'/C_{13}-C_{10}'$	*P. striatus*
77	Ptychantol C	O–O	$C_1-C_2'/C_{13}-C_{10}'/$	*P. striatus*

Table 2.1 (Continued)

Structure number	Name of compounds	A–C and B–D ring connection pattern	A–C and B–D ring connection position	Plant source
78	Pusilatin A	O—C/O—C	$C_1{-}C_2'/C_{14}{-}C_{12}'/$ $C_1''{-}C_2'''C_{14}''{-}C_{12}'''/C_{12}{-}C_{12}''$	*B. pusilla*
79	Pusilatin B (=6′,6‴-bisriccardin C)	O—C/O—C	$C_1{-}C_2'/C_{14}{-}C_{12}'/C_1''{-}C_2'''/$ $C_{14}''{-}C_{12}'''/C_6'{-}C_6'''$	*B. pusilla* *Ricciocarpos natans*
80	Pusilatin C	O—C/C—O	$C_1{-}C_2'/C_{14}{-}C_{12}'/C_1''{-}C_2'''/$ $C_{14}''{-}C_{12}'''/C_{12}{-}C_6'''$	*B. pusilla*
81	Pusilatin D	O—C/C—O	$C_1{-}C_2'/C_{14}{-}C_{12}'/C_1''{-}C_2'''/$ $C_{14}''{-}C_{12}'''/C_{14}'{-}C_{11}''$ $C_{11}''{-}C_{12}'''/C_{11}''{-}C_{13}'''$	*B. pusilla*
82	Pusilatin E	O—C/O—C	$C_1{-}C_2'/C_{14}{-}C_{12}'/C_1''{-}C_2'''/$ $C_{14}''{-}C_{12}'''/C_6'{-}C_6'''$	*R. multifida* subsp. *decrescens*
85	12′,12‴-bis(10′-hydroxyperrottetin E)	O—C—O	$C_1{-}C_2'/C_1''{-}C_2'''/C_{12}'{-}C_{12}'''$	*Pellia epiphylla*
83	GBB A (glaucescens bis-bibenzyl A)	O—O	$C_1{-}C_2'/C_{12}{-}C_{11}'$	*S. glaucescens*
84	GBB B (glaucescens bis-bibenzyls B)	O—O	$C_1{-}C_2'/C_{12}{-}C_{11}'$	*S. glaucescens*
86	Glossophyllin	C—C—C	$C_{6\text{-}(7)}{-}C_2'$	*Lethocolea glossophylla*
87	Plagilin	C—C	$C_5{-}C_5'$	*Plagiochila* sp.
88	Isoplagilin	C—C	$C_5{-}C_6'$	*Plagiochila* sp.
89	Plagiolin	C—C	$C_5{-}C_7'$	*Plagiochila* sp.
90	Perrottetin E	O	$C_1{-}C_2'$	*A. angusta* *Frullania convoluta* *Jungermannia comata* *J. infusca* *M. polymorpha* *Monoclea forsteri* *Nardia subclavata* *P. epiphylla* *Radula laxiramea*
91	Perrottetin F	O	$C_1{-}C_2'$	*F. convoluta*
92	Perrottetin G	O	$C_1{-}C_2'$	*F. convoluta*
93	7′,8′-Dehydroperrottetin F	O	$C_1{-}C_2'$	*F. convoluta*
94	Perrottetin E 11′-methyl ether	O	$C_1{-}C_2'$	*P. epiphylla*
95	Paleatin B	O	$C_{14}{-}C_{11}'$	*N. subclavata*
96	2,2′,10,10′,12,12′-Hexachloroisoperrottetin A	O	$C_6{-}C_6'$	*Jamesoniella colorata*
97	Perrottetin E-11-methyl ether	O	$C_6{-}C_6'$	*P. epiphylla*
98	14′-Hydroxyperrottetin E	O	$C_6{-}C_6'$	*P. epiphylla*
99	10′-Hydroxyperrottetin E	O	$C_6{-}C_6'$	*P. epiphylla*
100	10′-Hydroxyperrottetin E-11-methyl ether	O	$C_6{-}C_6'$	*P. epiphylla*
101	10,10′-Dihydroxyperrottetin E	O	$C_6{-}C_6'$	*P. epiphylla*
102	Isoperrottetin A	C	$C_6{-}C_6'$	*Radula perrottetii*
103	Cruciatin	O	$C_1{-}C_2'/C_1''{-}C_2'''$	*Lunularia cruciata*

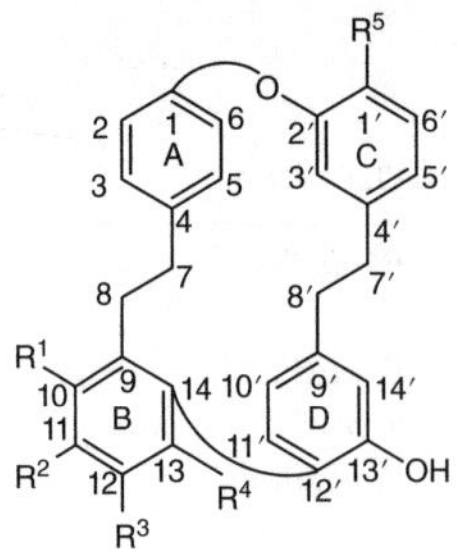

1 Riccardin A: R^1=R^3=R^4=R^5=H, R^2=OMe
3 Riccardin C: R^1=R^3=R^4=R^5=H, R^2=OH
4 Riccardin D: R^1=R^2=R^3=R^5=H, R^2=OH
5 Riccardin F: R^1=R^2=R^4=H, R^2=OH, R^5=Me
6 Riccardin G: R^1=R^2=R^3=H, R^4=OH, R^5=Me
7 Riccardin H: R^1=R^2=R^3=OH, R^4=R^5=H

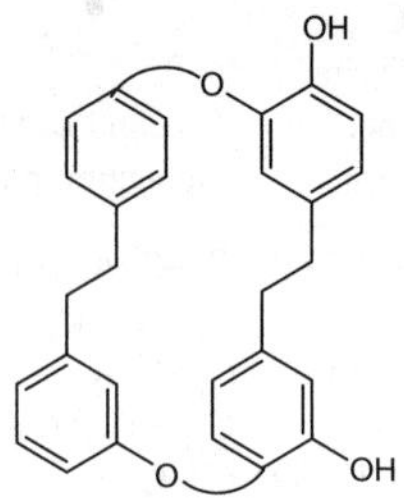

2 Riccardin B

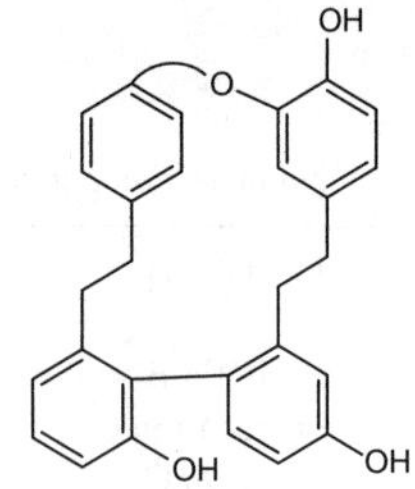

4a Plagiochin E:
(tentative structure)

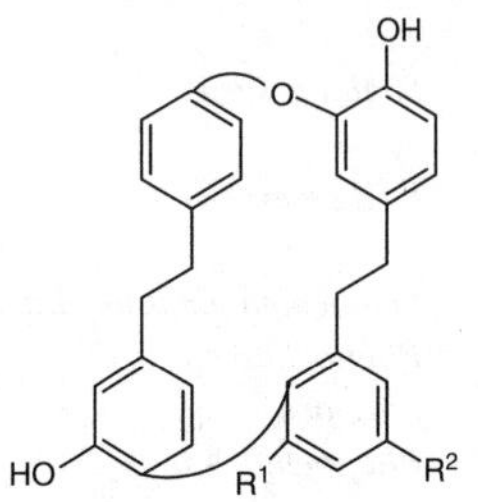

8 Isoriccardin C: R^1=OH, R^2=H
9 Isoriccardin D: R^1=H, R^2=OH

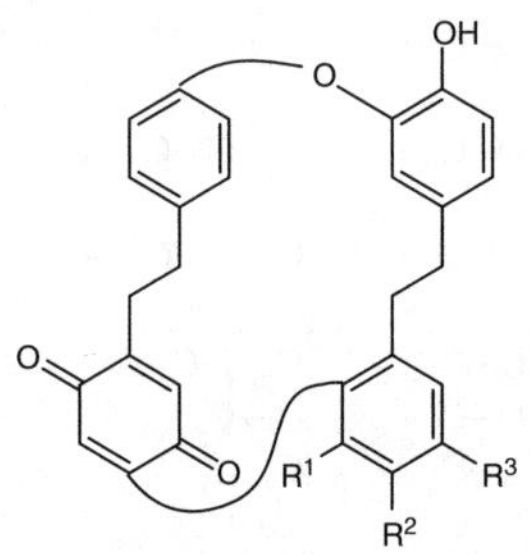

10 Isoriccardinquinone A:
R^1=OMe, R^2=H, R^3=OH

11 Isoriccardinquinone B:
R^1=R^3=H, R^2=OMe

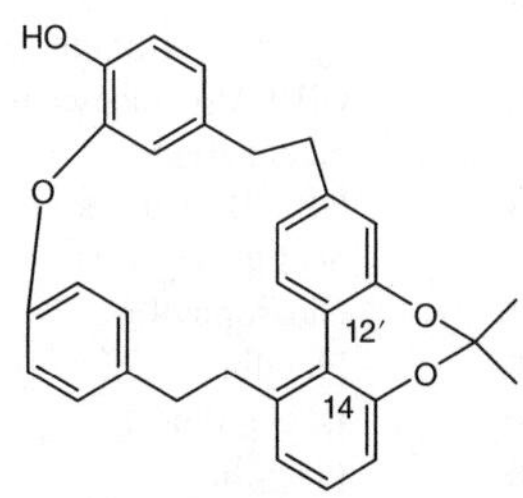

12 13,13′-*O*-Isopropylidenericcardin D

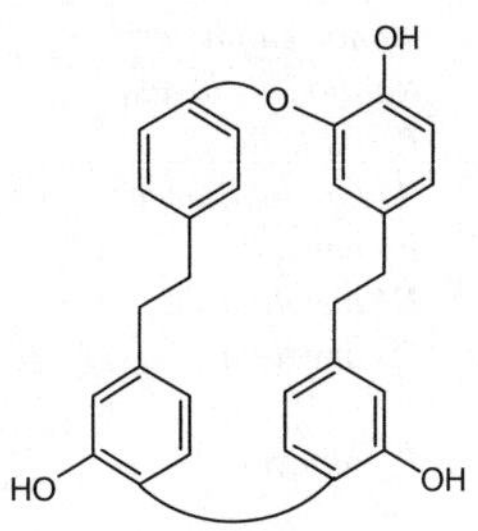

13 Polymorphation A

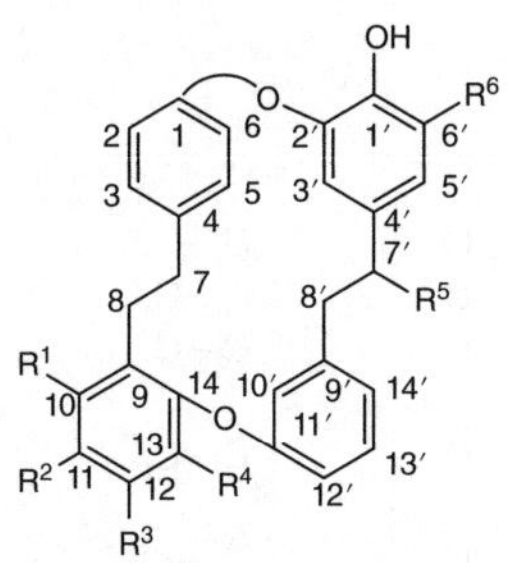

14 Marchantin A: R^1=R^2=R^3=R^5=H, R^4=R^6=OH
15 Marchantin B: R^1=R^2=R^5=H, R^3=R^4=R^6=OH
16 Marchantin C: R^1=R^2=R^3=R^5=R^6=H, R^4=OH
17 Marchantin D: R^1=R^2=R^3=H, R^4=R^5=R^6=OH
18 Marchantin E: R^1=R^2=R^3=H, R^4=R^6=OH, R^5=OMe
20 Marchantin H: R^1=R^2=R^3=R^5=H, R^4=R^6=OH
21 Marchantin I: R^1=R^3=R^4=R^5=R^6=H, R^2=OMe
22 Marchantin J: R^1=R^2=R^3=H, R^4=R^6=OH, R^5=OH
23 Marchantin K: R^1=R^2=H, R^3=R^4=R^6=OH, R^5=OMe
24 Marchantin M: R^1=OH, R^2= R^3=R^5=R^6=H, R^4=OMe
27 Marchantin O: R^1=R^2= R^3=R^5=R^6=H, R^4=OMe

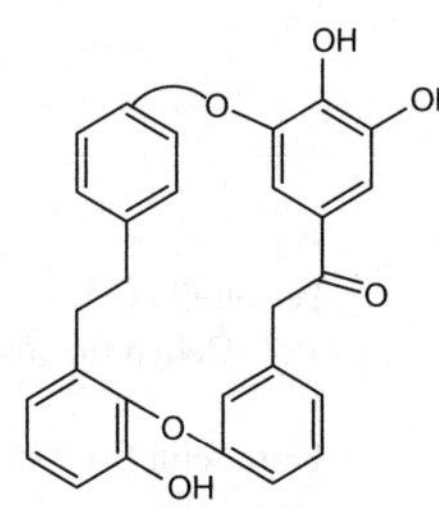

19 Marchantin G

Fig. 2.4 Structures of bis-bibenzyls (**1–24, 27**) found in the Marchantiophyta (liverworts).

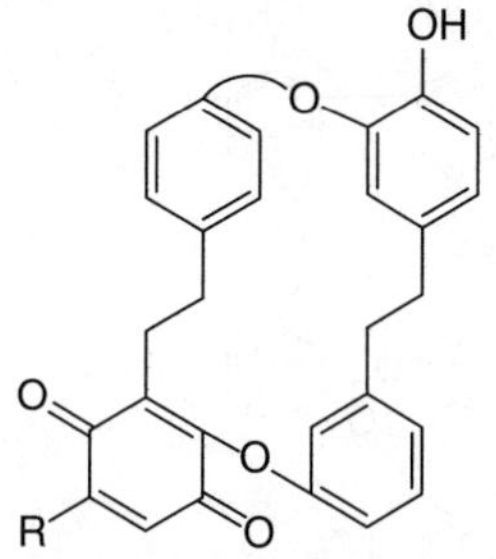

25 Marchantin N: R=OMe
26 Marchantiaquinone: R=H

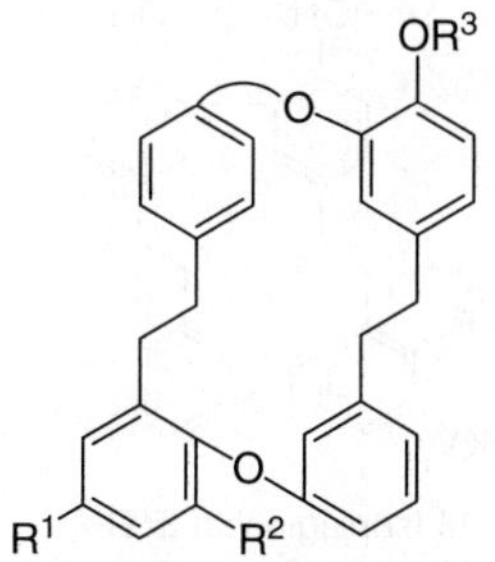

28 Marchantin P: R^1=H, R^2=OH, R^3=Me
36 11-*O*-Demethylmarchantin I: R^1=OH, R^2= R^3=Me

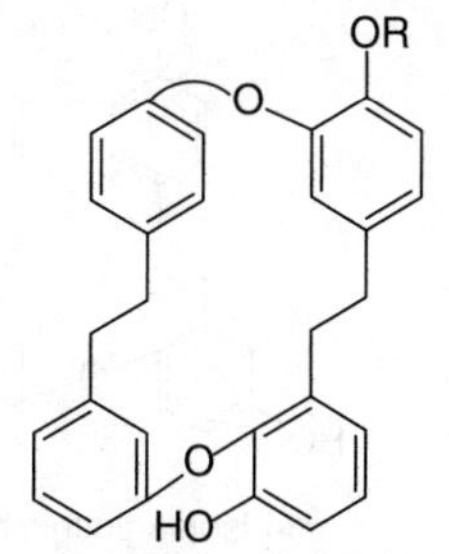

29 Isomarchantin C: R=H
30 Isomarchantin C: C1′-methyl ether: R=Me

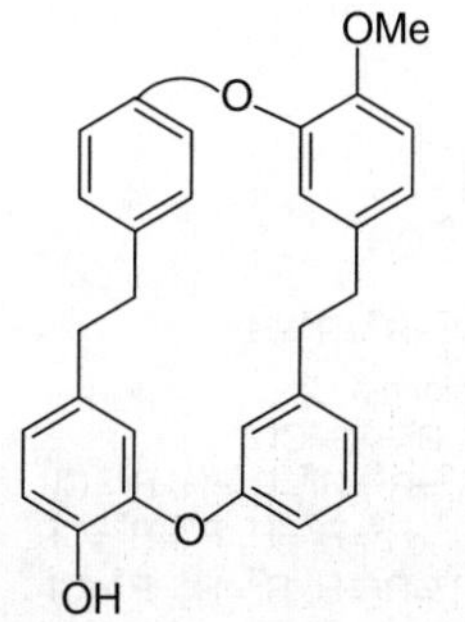

31 Neoisomarchantin C

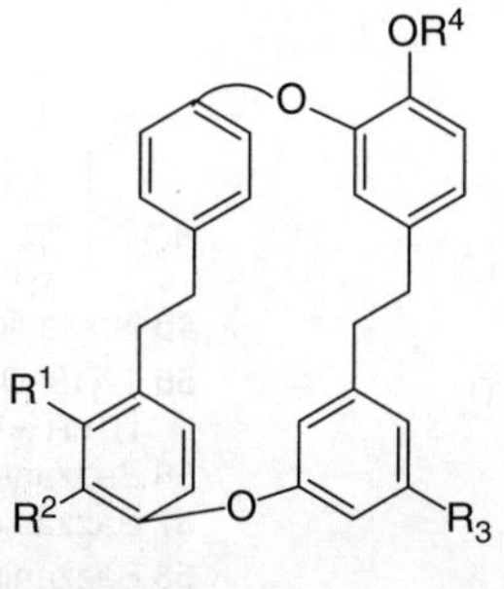

32 Neomarchantin A: R^1=R^2=R^4=H, R^2=OH
33 Neomarchantin B: R^1=R^2=OH, R^3=R^4=H
37 Dihydropthychantol : R^1=R^2=R^4=H, R^2=OH
38 Pakyonol:R^1=R^3=H, R^2=OH, R^4=Me

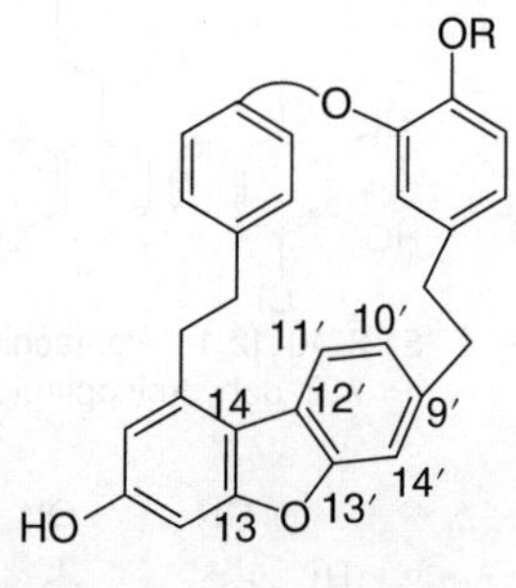

34 Asterelin A: R=H
35 Asterelin B: R=Me

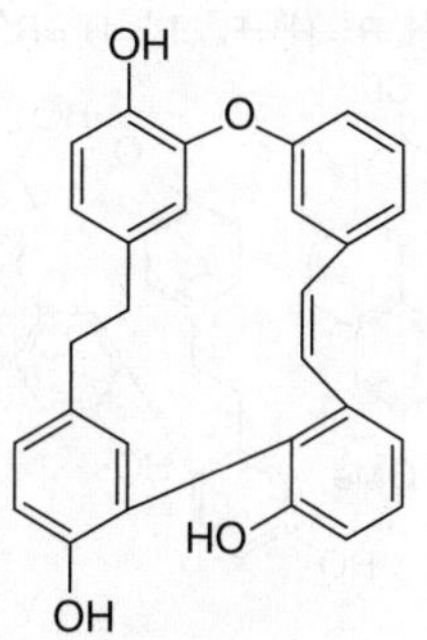

39 Planusin A

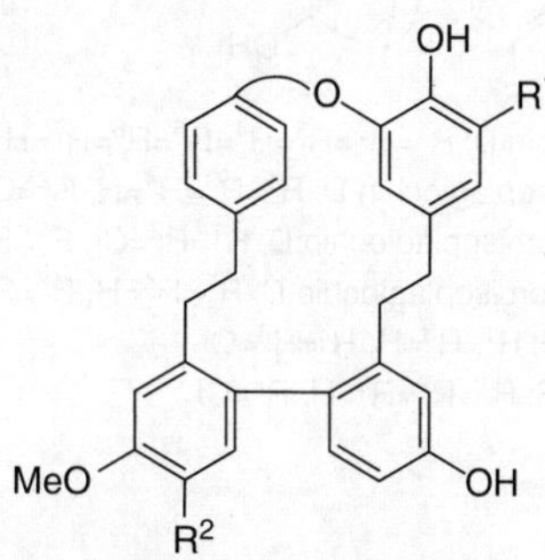

40 Plagiochin D: R^1=R^2=H
41 Plagiochin A: R^1=R^2=OH
42 Plagiochin C: R^1=H, R^2=OH
43 Plagiochin B: R^1=OH, R^2=H

Fig. 2.5 Structures of bis-bibenzyls (**25, 26–43**) found in the Marchantiophyta (liverworts).

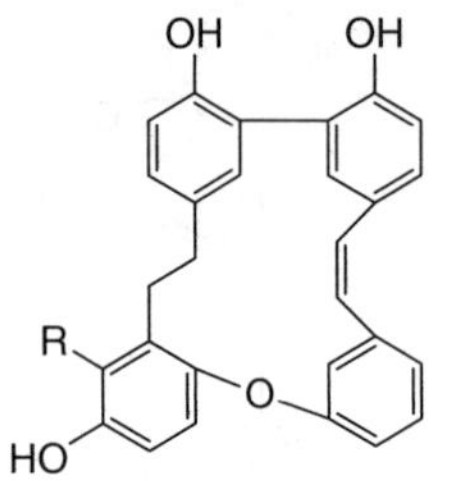

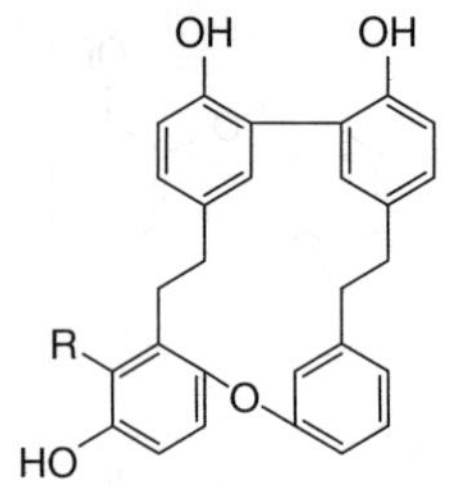

44 Isoplagiochin A: R=H
45 Isoplagiochin B: R=OH

48 Isoplagiochin E: R=H
49 Isoplagiochin F: R=OH

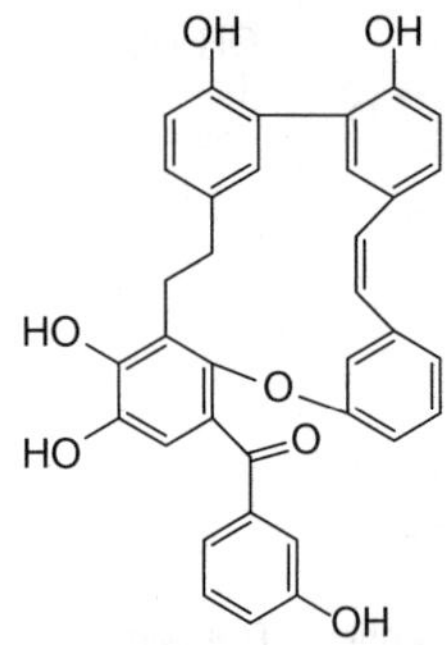

50 Isoplagiochin G

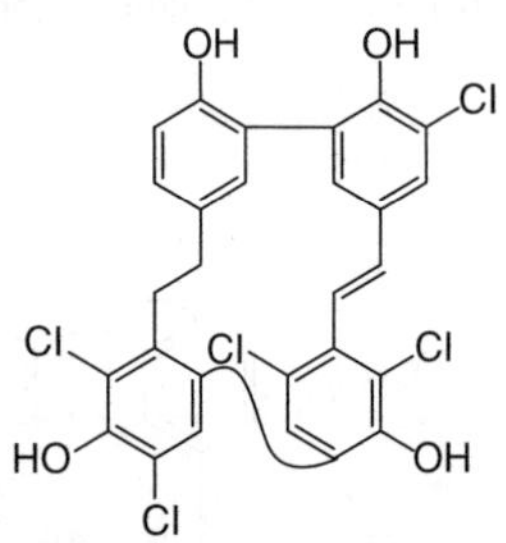

51 2′,10′,12,14′-Pentachloro-7′,8′-dehydroisoplagiochin D

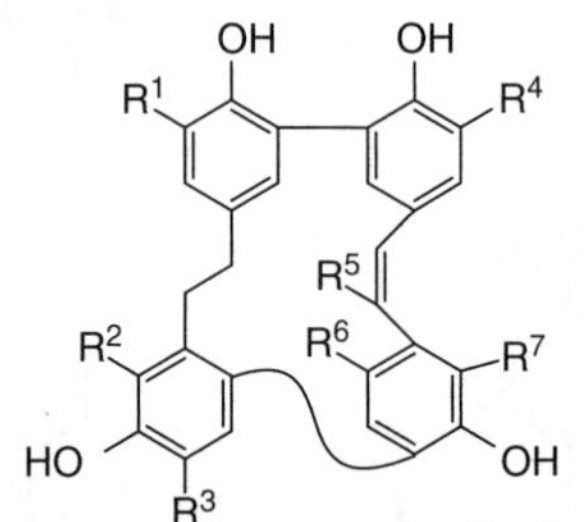

46 Isoplagiochin C: $R^1=R^2=R^3=R^4=H$
55 12,10′-Dichloroisoplagiochin C: $R^1=R^2=R^4=R^5=R^7=H$, $R^3=R^6=Cl$
56 Bazzanin A: $R^1=R^2=R^3=R^5=R^6=R^7=H$, $R^4=Cl$
57 Bazzanin B: $R^1=R^2=R^3=R^6=R^7=H$, $R^4=R^5=Cl$
58 Bazzanin C: $R^1=R^2=R^6=R^7=H$, $R^3=R^4=R^5=Cl$
59 Bazzanin D: $R^1=R^2=R^3=R^7=H$, $R^4=R^5=R^6=Cl$
60 Bazzanin E: $R^1=R^2=R^7=H$, $R^3=R^4=R^5=R^6=Cl$
61 Bazzanin F: $R^1=R^3=R^6=H$, $R^2=R^4=R^5=R^7=Cl$
62 Bazzanin G: $R^3=R^7=H$, $R^1=R^2=R^4=R^5=R^6=Cl$
63 Bazzanin H: $R^1=R^3=H$, $R^2=R^4=R^5=R^6=R^7=Cl$
64 Bazzanin I: $R^2=H$, $R^1=R^3=R^4=R^5=R^6=R^7=Cl$

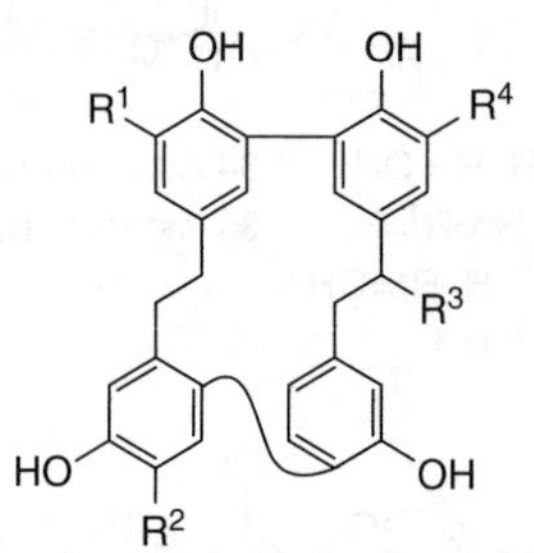

47 Isoplagiochin D: $R^1=R^2=R^3=R^4=R^5=R^6=R^7=H$
52 12-Chloroisoplagiochin D: $R^1=R^3=R^4=H$, $R^2=Cl$
53 2,12-Dichloroisoplagiochin D: $R^1=R^2=Cl$, $R^3=R^4=H$
54 12,7′-Dichloroisoplagiochin D: $R^1=R^4=H$, $R^2=R^3=Cl$
65 Bazzanin J: $R^1=R^3=H$, $R^2=R^4=Cl$
66 Bazzanin S: $R^1=R^2=R^3=H$, $R^4=Cl$

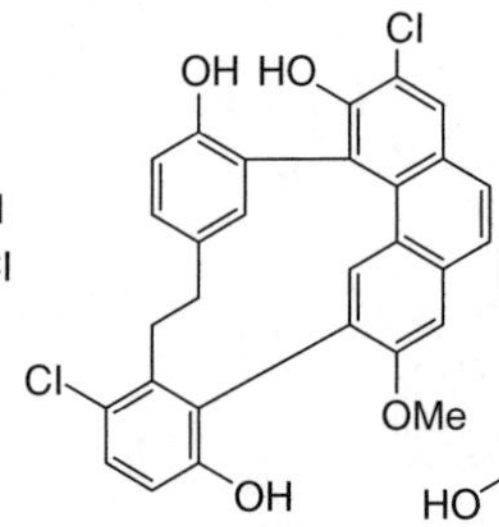

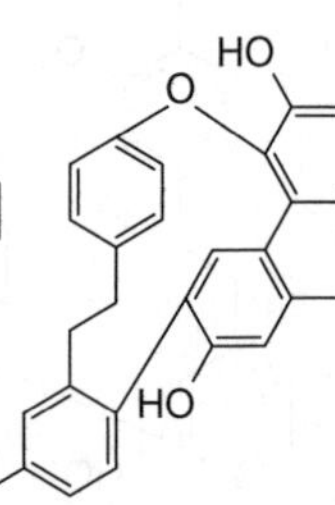

67 Bazzanin K **67a** Cavicularin

Fig. 2.6 Structures of bis-bibenzyls (**44–67a**) found in the Marchantiophyta (liverworts).

Asakawa and Matsuda (1982) proposed that cyclic bis-bibenzyls, such as riccardin C (**3**) and marchantin A (**14**), might be biosynthesized from phenolic bibenzyls that correspond chemically to phenolic dihydrostilbenes. This assumption was proved by feeding experiments of radioactive and ^{13}C-labelled precursors, such as L-[U-^{14}C]

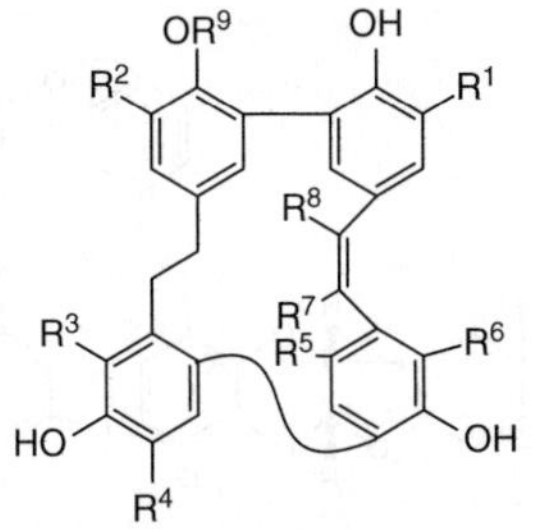

68 Bazzanin L: $R^1=R^2=R^6=R^7=R^8=H$, $R^3=R^4=R^5=Cl$, $R^9=Me$
69 Bazzanin M: $R^1=R^2=R^6=R^7=R^8=R^9=H$, $R^3=R^4=R^5=Cl$
70 Bazzanin N: $R^1=R^6=R^7=R^8=R^9=H$, $R^2=R^3=R^4=R^5=Cl$
71 Bazzanin O: $R^2=R^7=R^8=H$, $R^1=R^3=R^4=R^5=R^6=Cl$, $R^9=Me$
72 Bazzanin P: $R^1=R^2=R^3=R^4=R^5=Cl$, $R^6=R^7=R^8=R^9=H$
73 Bazzanin Q: $R^1=R^2=R^3=R^4=R^5=R^6=Cl$, $R^7=R^8=R^9=H$
74 Bazzanin R: $R^1=R^2=R^3=R^4=R^5=R^6=R^7=R^8=Cl$, $R^9=H$

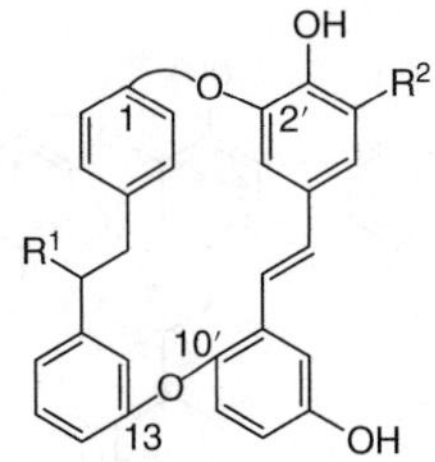

75 Ptychantol A: $R^1=R^2=H$
76 Ptychantol B: $R^1=H$, $R^2=OH$
77 Ptychantol C: $R^1=OH$, $R^2=H$

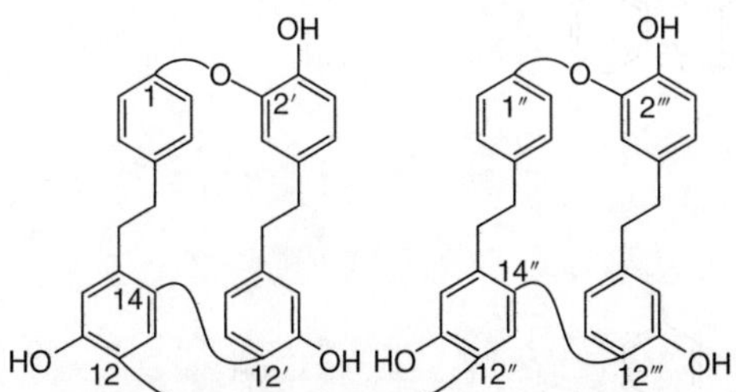

78 Pusilatin A

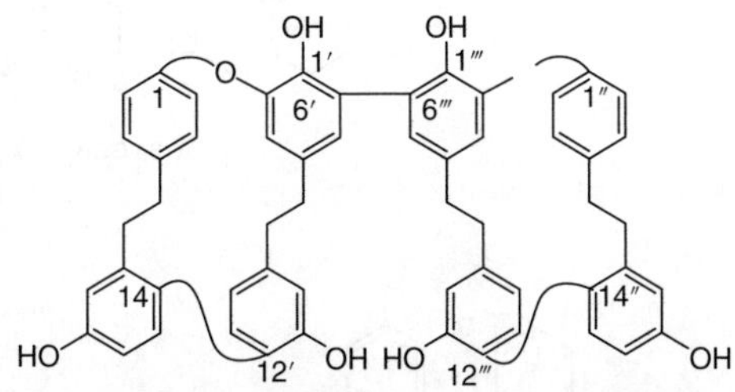

79 Pusilatin B

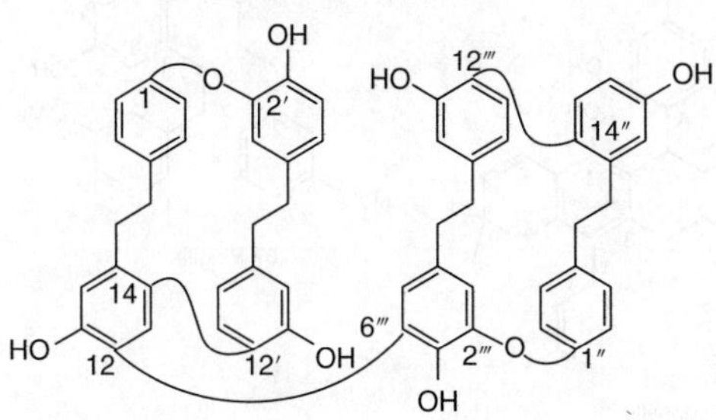

80 Pusilatin C

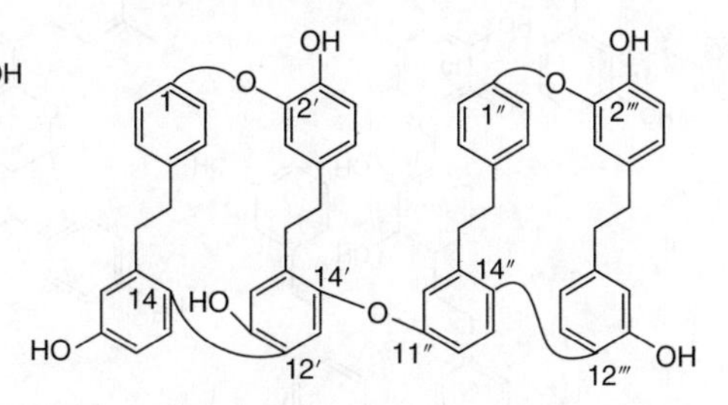

81 Pusilatin D

Fig. 2.7 Structures of bis-bibenzyls (**68–77**) and their dimers (**78–81**) found in the Marchantiophyta (liverworts).

phenylalanine, [U-^{14}C]dihydro-*p*-coumaric acid, [2-^{13}C]acetate, and L-[^{13}COOH]-phenylalanine, as shown in Fig. 2.10 (Friederich *et al.*, 1999a, 1999b). The aqueous precursor solutions were applied to 0.5 cm^2 samples of aseptic thallus tissue of *M. polymorpha* and incubated for 24 hours. The A- and B-rings of the marchantin molecule are derived from the benzene ring of L-phenylalanine via *trans*-cinnamic acid and *p*-coumaric acid. Application of the ^{13}C-labeled precursor with subsequent ^{13}C NMR spectroscopy established that dihydrocoumaric acid is an intermediate in

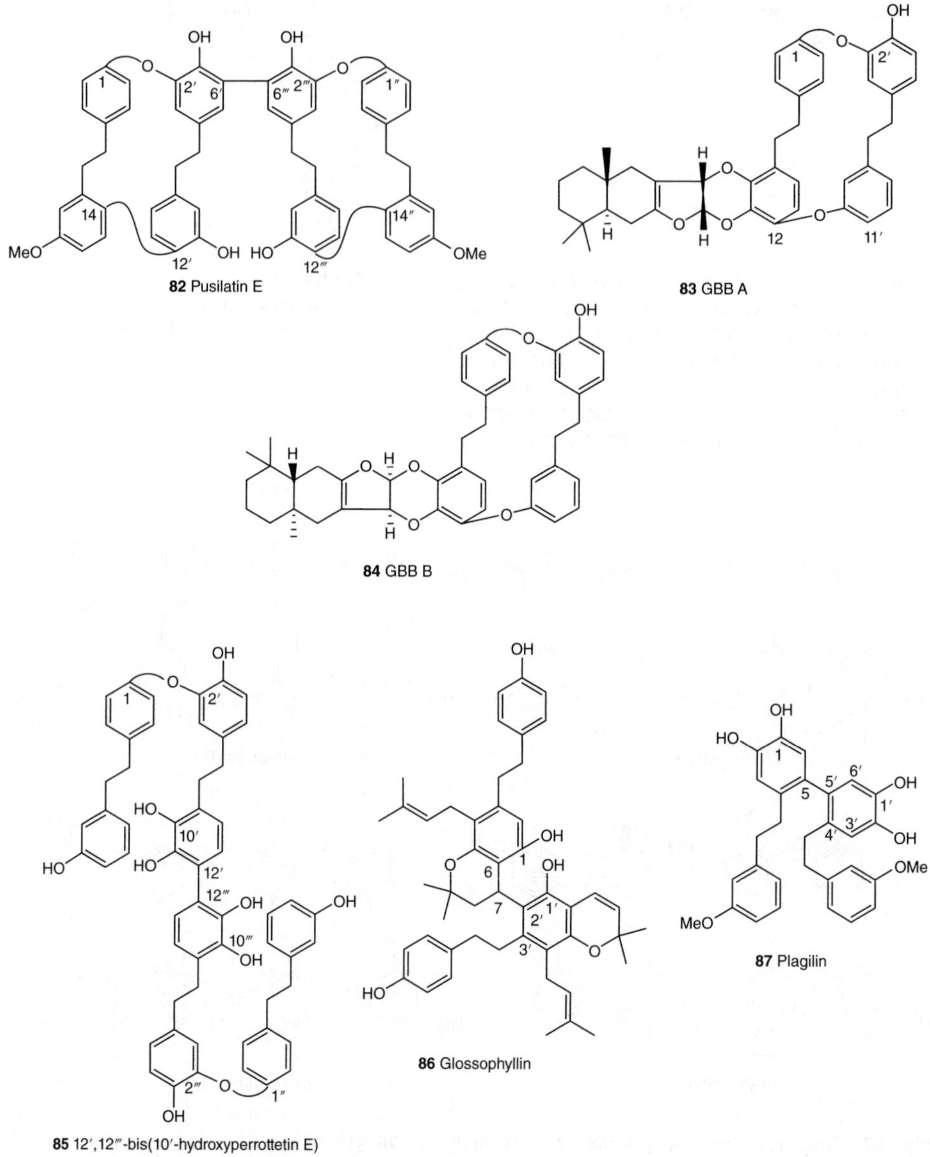

Fig. 2.8 Structures of bis-bibenzyls (**82–85**) found in the Marchantiophyta (liverworts).

marchantin biosynthesis. Dihydrocoumaric acid and then coumaric acid condense with three molecules of malonyl-CoA to form prelunularic acid (**105**). The latter is aromatized to yield lunularic acid (**106**) and possibly lunularin (**104**), which is followed by condensation of lunularin or lunularic acid to afford marchantin A (**14**). The mechanism of this final coupling step is still unknown.

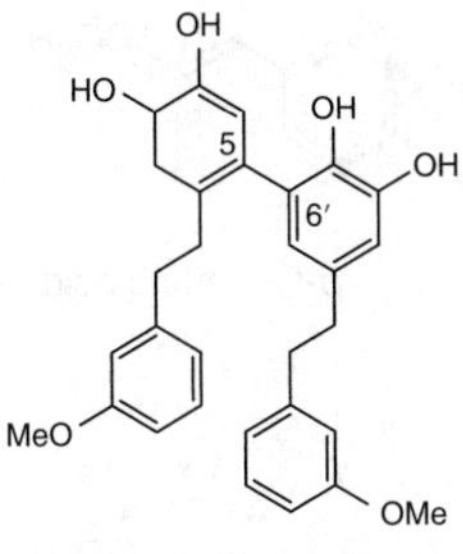

88 Isoplagilin

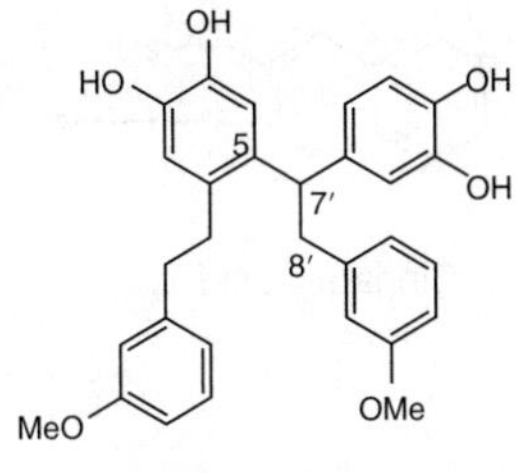

89 Plagiolin

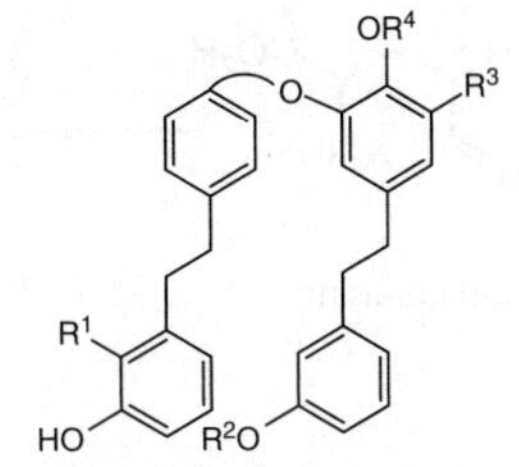

90 Perrottetin E: $R^1=R^2=R^3=R^4=H$
91 Perrottetin F: $R^1=R^2=R^4=H$, $R^3=OH$
92 Perrottetin G: $R^1=R^2=H$, $R^3=OH$, $R^4=Me$
92a Perrottetin H: $R^1=R^3=OH$, $R^2=R^4=H$
94 Perrottetin E–11′-methyl ether: $R^1=R^3=R^4=H$, $R^2=Me$,

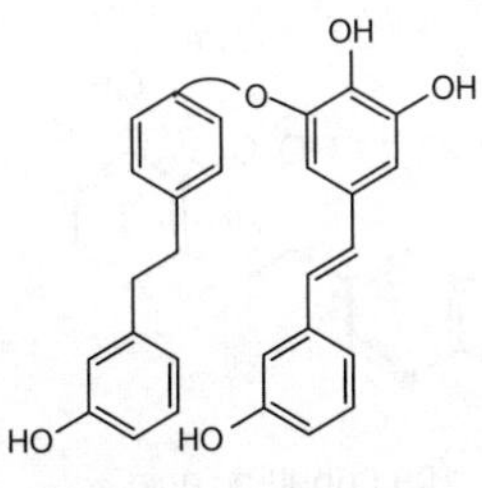

93 7′,8-Dehydroperrottetin F

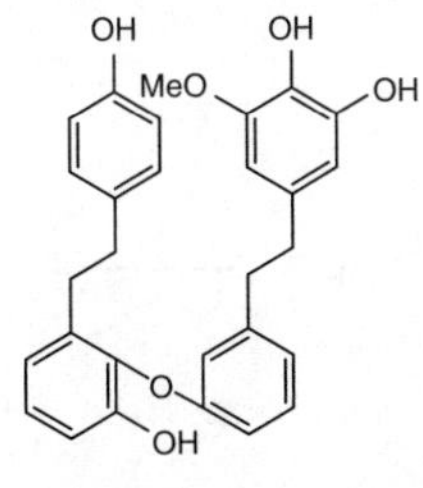

95 Paleatin B

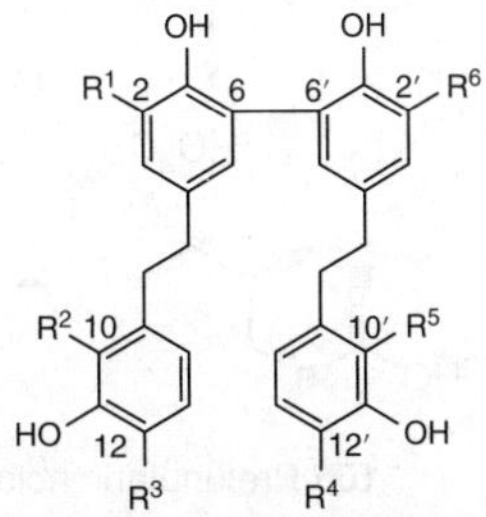

96 2,2′,10,10′,12,12′-Hexachloroisoperrottetin A: $R^1=R^2=R^3=R^4=R^5=R^6=Cl$
102 Isoperrottetin A: $R^1=R^2=R^3=R^4=R^5=R^6=H$

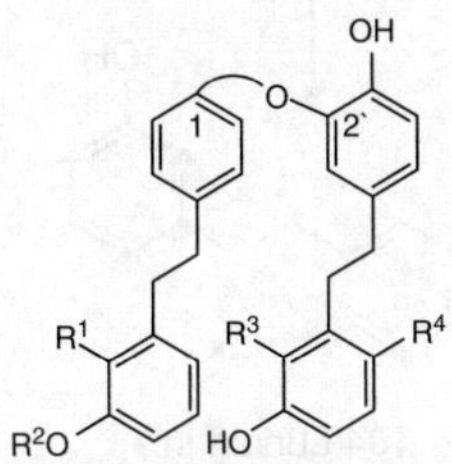

97 Perrottetin E-11-methyl ether: $R^2=Me$, $R^1=R^3=R^4=H$
98 14′-Hydroxyperrottetin E: $R^1=R^2=R^3=H$, $R^4=OH$
99 10′-Hydroxyperrottetin E: $R^1=R^2=R^4=H$, $R^3=OH$
100 10′-hydroxyperrottetin 11-methyl ether: $R^1=R^4=H$, $R^2=Me$, $R^4=OH$
101 10,10′-dihydroxyperrottetin E: $R^1=R^3=OH$, $R^2=R^4=H$

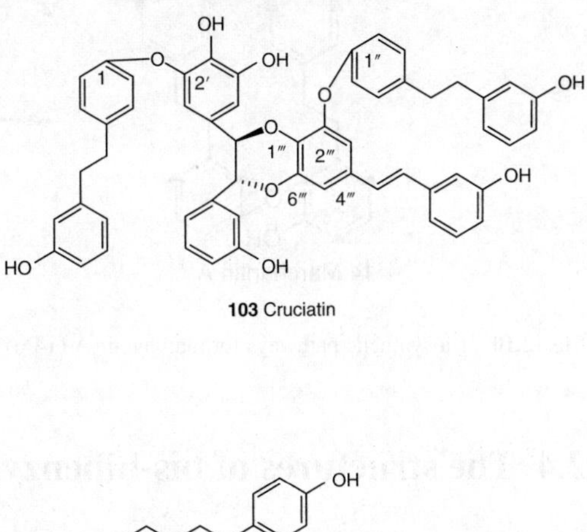

103 Cruciatin

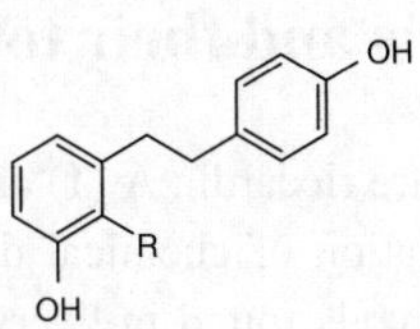

104 Lunularin R=H
106 Lunularic acid $R=CO_2H$

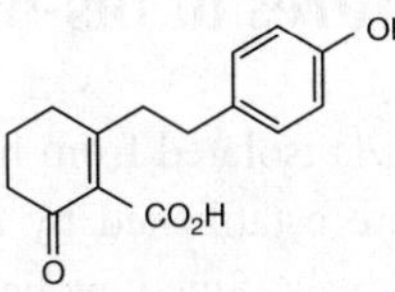

105 Prelunularin

Fig. 2.9 Structures of acyclic bis-bibenzyl dimers (**82, 85**), bis-bibenzyls (**83, 84**), and acyclic bis-bibenzyls (**86, 87**) found in the Marchantiophyta (liverworts).

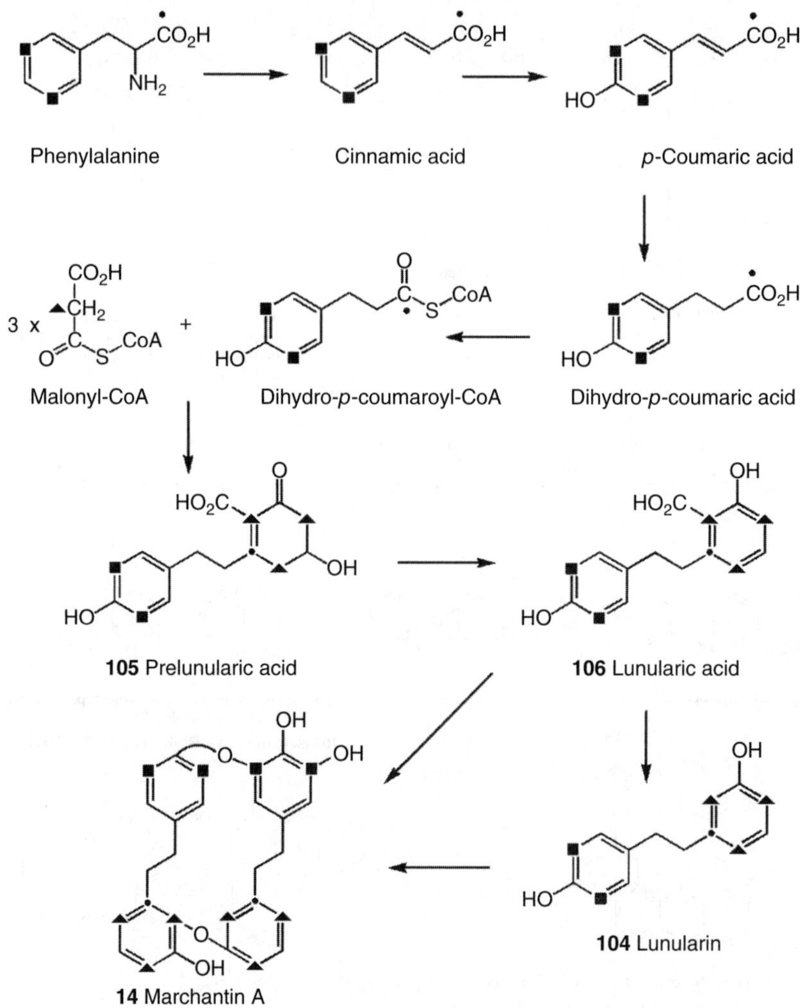

Fig. 2.10 Biosynthetic pathways for marchantin A (**14**) (Friederich *et al.*, 1999a, 1999b).

2.4 The structures of bis-bibenzyls and their total synthesis

The first bis-bibenzyls isolated from liverwort are riccardin A (**1**) and marchantin A (**14**). Their structures were established by a combination of chemical degradation and X-ray crystallographic analysis. Since cyclic bis-bibenzyls found in liverworts possess unusual structures and various interesting biological activities, several organic chemists have focused on their total synthesis (Speicher *et al.,* 2005, 2010; Harrowven & Kostiuk, 2012).

The first total syntheses of riccardin B (**2**) and marchantin A (**14**) were accomplished by Japanese groups (Kodama *et al.*, 1985, 1988; Shiobara *et al.*, 1985). Gottsegen *et al.* (1990) synthesized riccardins A–C (**1–3**) using a combination of Ullmann, Wittig, and Wurtz reactions and Ni(0)-assisted intermolecular coupling reaction. The total synthesis of riccardin

C (**3**), which possesses nuclear receptor LXRα selective agonist activity, was also accomplished by Harrowven *et al.* (2005), using isovanillin as the starting material in the course of the total synthesis of cavicularin (**67a**). The total synthesis of riccardin C (**3**) and its seven *O*-methylated derivatives were achieved by Hioki *et al.* (2009) via intramolecular Suzuki–Miyaura coupling to construct the requisite 18-membered biaryl linkage. Riccardin C (**3**) has been synthesized by Takiguchi *et al.* (2011) using an intramolecular S_N reaction of a sulfinylfluorobenzene (A-ring) by an internal phenolate (C-ring moiety), providing the key macrocyclic ring closure in high yield.

Plagiochin E (**4a**) was isolated from the Chinese *M. polymorpha* (Niu *et al.*, 2006) and *Asterella angusta* (Qu *et al.*, 2007). A short and efficient total synthesis of plagiochin E (**4a**) was accomplished by Speicher *et al.* (2009b). However, its spectroscopic data were not the same as those of the natural product **4a**, but identical with those of riccardin D (**4**), which was isolated from *P. fruticosa* (Asakawa, 1995). Thus, the structure of **4a** should be revised as riccardin D (**4**). Riccardin G (**6**), the methyl ether of **4**, has been found in *Marchantia chenopoda* (Asakawa, 1995).

Xie *et al.* (2010) accomplished the total synthesis of marchantin C (**16**) over 12 steps from 2-hydroxy-3-methoxybenzaldehyde in 23% overall yield. The total synthesis of marchantiaquinone (**26**) was accomplished by Lopez *et al.* (2000) by macrocyclization of a dichloride precursor using an active nickel complex under high dilution conditions. Subsequent demethylation of marchantin M (**24**) trimethyl ether with boron tribromide and oxidation using silver oxide afforded marchantiaquinone (**26**).

The New Zealand *Marchantia foliacea* was reinvestigated chemically to give the two new bis-bibenzyls isomarchantin C-1′-methyl ether (**30**) and neoisomarchantin C (**31**), together with the known marchantin C (**16**), isomarchantin C (**29**), and marchantin P (**28**) (Hosoda, 2003). The structure of **29** was confirmed by a combination of X-ray crystallographic analysis. The ^{1}H and ^{13}C NMR spectra of **30** resembled those of isomarchantin C (**29**), except for evidence for the presence of a methoxy group in place of a phenolic hydroxy group, indicating **30** to be methoxyisomarchantin C-1′-methyl ether. This was confirmed by the formation of isomarchantin C-1′,13′-dimethyl ether from **30** and isomarchantin C (**29**) by methylation with methyl iodide. The positions of the methoxy group of **29** and the two methoxy groups of **30** were confirmed by the NOESY experiments and Molecular Operating Environment (MOE) version 2002.03 (Chemical Computing Group KK) calculations carried out on both compounds. Thus, the structure of **30** was established as isomarchantin C-1′-methyl ether. A conclusive structure for neoisomarchantin C (**31**) possesses the same molecular formula as determined by HRMS. Its ^{1}H and ^{13}C NMR data were obtained by careful analysis of the 2D NMR (COSY, HMBC, and NOESY) data of the B- and D-ring regions of the molecule (Hosoda, 2003).

Tori *et al.* (1985) reported the isolation and structure elucidation of marchantin H (**20**) from *Plagiochasma intermedium.* The same Chinese species elaborates pakyonol (**38**) (Wang & Lou, 2000; Xie *et al.*, 2010), along with riccardin C (**3**), riccardin F (**5**), isoriccardin C (**8**), and neomarchantin A (**32**) (Xie *et al.*, 2010).

Plagiochasma pterospermum produces riccardin C (**3**) and pakyonol (**38**) (Hashimoto *et al.*, 1998). Lahlou *et al.* (2000) reported that the methanol extract of *Plagiochasma japonica* contains riccardin D (**4**). Nógrádi *et al.* (1990) accomplished the total synthesis of pakyonol (**38**) by Ullmann coupling of 3-benzoyloxy-4-(3-formylphenoxy)benzoic acid benzyl ester with Wittig and modified Wittig reactions.

Qu *et al.* (2007) reported the presence of four new bis-bibenzyls, asterellin A (**34**), asterellin B (**35**), 11-*O*-demethylmarchantin I (**36**), and dihydroptychantol A (**37**), together with the known macrocyclic bis-bibenzyls, riccardin B (**2**), riccardin D (**4**), marchantin H (**20**), marchantin M (**24**), marchantin P (**28**), and perrottetin E (**90**) in *A. angusta*. The structures of the new compounds were established by a combination of 2D NMR and/or X-ray crystallographic data analysis, as well as by comparison with reference values mainly established by Asakawa and his group (Asakawa, 1995; Asakawa *et al.*, 2000).

Plagiochila sciophila (=*P. acanthophylla* subsp. *japonica*) produced the new plagiochins, plagiochins A (**41**), B (**43**), C (**42**), and D (**40**), which possess an additional *ortho* biphenyl linkage between the two benzyl groups (Hashimoto *et al.*, 1987). The stereochemistry of **41** was established by chemical reactions (acetylation and methylation) and analysis of ^{1}H NMR spectrum and NOE of the pentamethyl ether as well as by an X-ray crystallographic analysis of the latter. This showed that ring A is perpendicular to ring C and parallel with ring C. The proton at C_3' is strongly shielded by both rings A and C, causing high field shift to δ 4.83 ppm in plagiochin A (**41**). This phenomenon has also been seen in the marchantin and riccardin series. The structures of the remaining plagiochins were also deduced from double resonance and NOE difference spectra of their peracetylated derivatives.

Bioactivity-guided fractionation of the ether extract of *Plagiochila ovalifolia* using a DHHP radical-scavenging assay resulted in the isolation of plagiochin D (**40**) (Sadamori, 2009). The total synthesis of plagiochin C (**42**) and plagiochin D (**40**) was accomplished by Keseru *et al.* (1992). Fukuyama *et al.* (1999) carried out the total synthesis of plagiochin D (**40**) using *m*-anisaldehyde as the starting material, followed by a Wadsworth–Emmons condensation reaction, hydrogenation, and a Stille–Kelly reaction using hexamethylditin and tetrakis(triphenylphosphine)palladium, to yield plagiochin D (**40**), after removal of the MOM group by HBr in methanol. Molales *et al.* (2011) also synthesized plagiochin D (**40**) by intramolecular S_NAr reaction.

The new bis-bibenzyl planusin (**39**) and a known isoplagiochin A (**44**) were isolated from *Heteroscyphus planus*. The positions of the substituents on the four benzene rings of planusin (**39**) were established by a combination of HMBC and NOE experiments. The ether linkage between C_6 and C_2' was assigned as the sole ether linkage by taking into account the steric distortion of the molecule (Nabeta *et al.* 1998).

The Japanese *P. fruticosa* produced the known isoplagiochins A (**44**) and B (**45**) (Asakawa, 1995) and two novel macrocyclic bis-bibenzyls named isoplagiochin C (**46**) and isoplagiochin D (**47**), the spectroscopic data of which were similar to those of isoplagiochin A (**44**) except for the presence of a (*Z*)-bond and a signal pattern of the D-ring indicating that compound **46** possesses two C—C biphenyl linkages with one stilbene moiety and four phenolic hydroxy groups. This was confirmed by the preparation of the tetraacetate by acetylation. The structure was defined on the basis of COSY, HMBC, HMQC, and NOESY experiments. Isoplagiochin D (**47**) has spectroscopic data similar to those of **46** except for the presence of four instead of two methylene signals. Acetylation and methylation of **47** gave the tetraacetate and tetramethyl ether, indicating the presence of four phenolic hydroxy groups. Hydrogenation of the tetraacetate of **46** afforded a dihydroderivative with spectroscopic data identical to those of the tetraacetate of **47**. Thus, the structure of **47** was established as dihydroisoplagiochin C. Isoplagiochins A (**44**) and B (**45**) possess a C_6–C_2' biphenyl bond and a C_{14}–C_{11}' ether linkage, whereas isoplagiochins C (**46**) and D (**47**) have two C_6–C_2' and C_{14}–C_{12}' biphenyl bonds. Isoplagiochins A–D (**44–47**) may be biosynthesized from isoperrottetin A (**102**), isolated from the liverwort *Radula perrottetii* (Asakawa, 1995). The latter was formed by dimerization of lunularin (**104**) found in most of the liverworts. This is the first report of the isolation of macrocyclic bis-bibenzyls possessing two biphenyl linkages between rings A and C/rings B and D (Hashimoto *et al.*, 1996).

The total synthesis of isoplagiochin A (**44**) was achieved by Gerencser *et al.* (1997) in 23 steps by coupling of methyl 3-(2-methoxy-5-formylphenyl)-4-methoxybenzoate with methyl [(2-(hydroxymethyl)-4-methoxyphenoxy)]benzoate, in an overall yield of 15%.

It is suspected that the halogenated bis-bibenzyls (**53–55**) isolated from *Herbertus sakuraii* (Hashimoto *et al.*, 2000) might be artifacts originating from isoplagiochin C (**46**) and isoplagiochin D (**47**), coexisting in *H. sakuraii*, during chromatographic workup. However, when **46** and **47** were dissolved in a mixture containing chloroform, methanol, and Sephadex LH-20 and refluxed for 4 hours, only the starting materials were recovered and no halogenated compounds were detected. Thus, the new chlorinated compounds are true natural products. Speicher *et al.* (2003) found that the chlorinated bis-bibenzyls did not constitute artifacts of extraction because they were directly detected by NAKDI–TOF mass spectrometry. It was confirmed that chlorinated bis-bibenzyls are biosynthesized by a chloroperoxidase from *Caldariomyces fumago* in KCl or KBr and hydrogen peroxide at pH 3.0, and an enzyme of this type was detected in the liverwort, *Bazzania trilobata*, which produces a number of chlorinated bis-bibenzyls (**52, 56–74**) (Martini *et al.*, 1998; Anton *et al.*, 1999; Scher *et al.*, 2003). This is the first report of a haloperoxidase from the liverworts (Speicher *et al.*, 2003). However, compounds **46** and **47** both showed a positive optical rotation ($[\alpha]_D$ +74.8° for **46** and +47.5° for **47**) and ECD effects (($\lambda(\Delta\varepsilon)$=231(+24.4), 213(−63.2) for **46** and 252 (+2.4), 225 (+22.5) nm(mol^{-1}dm^{3}cm^{-1})] for **47**) (Bringmann *et al.*, 2004). The separation of the enantiomers of isoplagiochin C (**46**) and isoplagiochin D (**47**) on a chiral-phase HPLC provided enantiomers of almost all bis-bibenzyls isolated from *B. trilobata.* A direct stereochemical online analysis of the peaks by LC-ECD coupling exhibited opposite ECD curves, clearly revealing that the separated products are enantiomers, but do not occur in a pure enantiomeric form in Nature. The assignment of these two peaks to the respective enantiomers was achieved by quantum chemical ECD calculations using a molecular dynamics (MD)-based approach. Isoplagiochin C (**46**) and isoplagiochin D (**47**) were not obtained enantiomerically pure, but occurred in a 85:15 and 48:52 ratio, in favor of the enantiomer possessing the (*P*)-configuration [(*P*)-**46** and (*P*)-**47**] at the stereochemically most stable axis as shown in Fig. 2.11. Racemization rates of isoplagiochin C (**46**) were determined at different temperatures from 85 to 145°C. From the corresponding Arrhenius plot, activation energy of 101.6 kJ/mol for the racemization was calculated, in good accordance with the theoretically predicted value of 115.0 kJ/mol. This result led to the expectation of a molecule configurationally stable at room temperature. However, the specific rotation value varied even for the same compound depending on the plant source and the isolation procedure (Scher *et al.*, 2004a, 2004b). A decrease of the enantiomeric purity may occur during standard isolation methods at a temperature higher than 50°C.

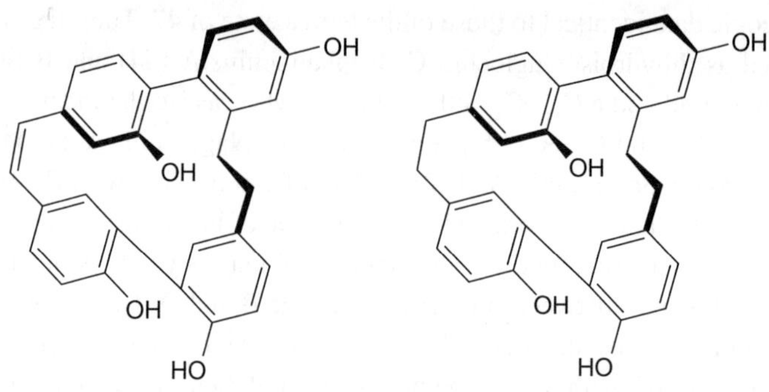

(*P*)-46
[(*P*)-isoplagiochin C]

(*P*)-47
[(*P*)-isoplagiochin D]

Fig. 2.11 Stereochemistry of isoplagiochins C (**46**) and D (**47**).

B. trilobata contained isoplagiochin D (**47**) and 6′-chloroisoplagiochin D (named bazzanin S; **66**) (Scher *et al.*, 2004a, 2004b). The latter compound was reported for the first time from a liverwort; however, earlier it was obtained as a derivative in the total synthesis of bazzanin A (**56**) (Speicher *et al.*, 2002).

The Taiwanese *Lepidozia fauriana* produced isoplagiochin D (**47**). This was the first isolation of a bis-bibenzyl from this species, although members of the *Bazzania* genus belonging to the same family Lepidoziaceae produce a number of bis-bibenzyls, as mentioned earlier (Martini *et al.*, 1998; Shy & Wu, 2006).

Eicher *et al.* (1998) reported the total synthesis of isoplagiochin C (**46**). The total synthesis of isoplagiochin D (**47**), a highly strained macrocyclic bis-bibenzyl with two biaryl units was achieved by Esumi *et al.* (2004) in 1.6% overall yield over 11 steps, via the formation of tetramethoxy isolplagiochin D by a palladium(0)-catalyzed Suzuki–Miyaura cross-coupling reaction, followed by demethylation with boron tribromide in dichloromethane. The tetramethyl ether of isoplagiochin D (**47**) and related compounds possessing (*Z*)- and (*E*)-olefinic moieties with more rigid biaryl bridges were synthesized using the Sonogashira and McMurry protocols (Speicher *et al.*, 2009a).

The Costa Rican *Plagiochila permista* var. *integerrima* afforded the three new bis-bibenzyls, isoplagiochin E (**48**), isoplagiochin F (**49**), and 12-chloroisoplagiochin D (**52**), together with the two known bis-bibenzyls, isoplagiochin A (**44**) and perrottetin E (**90**) (Anton *et al.*, 1999). The substitution patterns proposed in rings C and D of the new compounds **48** and **49** were based on a comparison of their 2D NMR spectra with those of isoplagiochin A (**44**). The presence of a chlorine atom was also confirmed by comparison of a combination of the mass spectrometric and NMR spectroscopic data with those of isoplagiochin D (**47**) isolated from *P. fruticosa* (Asakawa, 1995).

The Japanese primitive stem-leafy liverwort *H. sakuraii* (Herbertaceae) was found to contain the two new optically active chlorinated compounds, (−)-12,7′-dichloroisoplagiochin D (**54**) ($[\alpha]_D^{20}$ −2.7°; ECD effects $\lambda(\Delta\varepsilon)$ = 300(+0.74), 281(−0.39), 234(+5.34) nm($mol^{-1}dm^3cm^{-1}$)) and (−)-12,10′-dichloroisoplagiochin C (**55**) ($[\alpha]_D^{21}$ −19.1°; $\lambda(\Delta\varepsilon)$= 322(+7.80), 282(−7.47), 244(+48.08), 214(−157.72) nm($mol^{-1}dm^3cm^{-1}$)), along with an optically inactive compound, 2,12-dichloroisoplagiochin D (**53**) (Hashimoto *et al.*, 2000).

These were obtained together with the two known optically active nonhalogenated substances, isoplagiochin C (**46**) and isoplagiochin D (**47**), isolated from *P. fruticosa* (Hashimoto *et al.*, 1996). The structures of **53–55** were established by comparison of analysis of their spectroscopic data with those of isoplagiochins C and D, and/or X-ray crystallographic analysis. Compound **53** was present in the racemic form since neither an optical rotation, nor an ECD effect could be observed. The bis-bibenzyls **46** and **47** showed a positive specific rotation and a positive first Cotton effect in their ECD spectra, indicating that both compounds are optically active. The chlorinated bis-bibenzyls have been found in the liverworts *B. trilobata* (Martini *et al.*, 1998) and *Plagiochila oresitropha* (Anton *et al.*, 1997), as mentioned earlier. This was the third report of the isolation of chlorinated bis-bibenzyls from liverworts. Compound **54** possesses planar chirality since it shows a negative specific rotation and positive, negative, and positive ECD effects. Compound **55** showed the same chiroptical data as mentioned earlier. However, it proved to be difficult to establish the absolute configuration of both **54** and **55** and the nonhalogenated **46** and **47** from their ECD spectra, because of the possibility of the presence of four atropisomers for each isolated compound (Hashimoto *et al.*, 2000).

Two optically inactive chlorinated bis-bibenzyls, 12-chloroisoplagochin D (**52**) and 2,12-dichloroisoplagiochin D (**53**), were obtained from *Mastigophora diclados* (Hashimoto *et al.*, 2000). This was the first isolation and structure elucidation of halogenated and nonhalogenated bis-bibenzyls from a species in the Mastigophoroideae. The former compound has been obtained from *Plagiochila deflexa* (Anton *et al.*, 1997). Matrix-assisted laser desorption/ionization time-of-flight (MALDI–TOF) mass measurement confirmed that the chlorinated compounds are not artifacts (Speicher *et al.*, 2001).

Guo *et al.* (2007) reported the rapid identification of bis-bibenzyls using electron ionization (EI)-TOF and electrospray ionization triple-quadruple (ESI-TQ) mass spectrometry. Xing *et al.* (2007) developed a simple and rapid LC-DAD/MS/MS method using full-scan MS, MS/MS precursor ion scan, and MS/MS product-ion scan modes for the identification of bis-bibenzyls including when present in the crude ether extracts of liverworts. The European *B. trilobata* afforded 11 cyclic bis-bibenzyls named bazzanins A–K (**56–67**) (Martini *et al.*, 1998). Compounds A–I are isoplagiochin C (**46**) derivatives with one to six chlorine atoms in their molecules, with compound J (**65**) being the 16,6′-dichloro derivative of isoplagiochin D (**47**) (Asakawa, 1995). Bazzanin A (**56**) was assigned as 6′-chloroisoplagiochin C by DCI mass spectrometry and from its COSY, HMBC, and NOESY NMR spectroscopic data. Bazzanin B (**57**) possesses two chlorine atoms, the positions of the chlorine atoms for compound **57** being based on a comparison of its spectroscopic data with those of **56** and by analyses of the COSY, HMBC, and NOESY spectra. The proton H-7′, when compared with that of **56**, showed an alternation of multiplicity in the ^1H NMR spectrum from a doublet ($J = 12\,Hz$) to a singlet, indicating the position of the second chlorine atom at C_8'. Bazzanins C–I (**58–65**) were assigned with a chlorine atom at C_8' by spectroscopic data comparison with isoplagiochin C and bazzanin A (**56**). All other structural assignments for these new compounds were supported by analyses of their COSY, HMBC, and NOESY spectra. Bazzanin J (**65**) is a 7′,8′ dihydro compound, the structure of which was again determined by detailed analyses of its COSY, HMBC, and NOESY spectra. Bazzanin K (**67**) possesses a bibenzyl and phenanthrene moiety, and its structure also was based on the analysis of 2D NMR data (Martini *et al.*, 1998). The enantiomeric purities of bazzanins A–I (**56–64**), K (**67**), and S (**66**) were determined by means of enantioselective chromatography with online measurement of

ECD spectra (Martini *et al.*, 1998; Scher *et al.*, 2004b). *Bazzania tricrenata* also produces three chlorinated bis-bibenzyls, of which the structure of **51** was characterized as 2′,10,10′,12,14′-pentachloro-7′,8′-dehydroisopalagiochin D (Asakawa, 2008c).

The Costa Rican *Lepidozia incurvata* elaborated isoplagiochin C-type chlorinated bis-bibenzyls, named bazzanins L–R (**68–74**). The structures of all of these newly compounds were established using a combination of their CI- and DCI/MS and 2D NMR spectroscopic data (Scher *et al.*, 2003). These chlorinated bis-bibenzyls were the first such compounds reported from a *Lepidozia* species, although the genus *Bazzania*, belonging to the same family, Lepidoziaceae, produces a number of chlorinated bis-bibenzyls, that is, bazzanins A–K and some nonchlorinated bis-bibenzyls (Martini *et al.*, 1998).

Speicher *et al.* (2002) accomplished the total synthesis of 12-chloroisopalgiochin D (**52**) isolated from *Plagiochila* species, and 6′-chloroisopalgiochin C (=bazzanin A) (**56**) and 6,12′-dichloroisoplagiochin D (=bazzanin J) (**65**) by construction of the biaryl moiety using regioselective *Suzuki* protocols and coupling to acyclic bibenzyl and cyclic bibenzyls by Wittig and MacMurry procedures, followed by hydrogenation and deprotection. The presence of atropisomers among the bis-bibenzyls has been discussed by Hashimoto *et al.* (1996) and Asakawa *et al.* (2000). Bazzanins L–R (**68–74**) displayed specific rotations between $[\alpha]_D = +54$ and +225° (Scher *et al.*, 2003).

Three new bis-bibenzyls named ptychantols A–C (**75–77**), possessing an (*E*)-stilbene moiety, were isolated from the Japanese *Ptychanthus striatus* (Hashimoto *et al.*, 1999). The gross structure of ptychantol A (**75**) was suggested by the analysis of the 2D NMR data and of its permethylated derivatives. The structure proposed was confirmed by X-ray crystallographic analysis. The structures of compounds **76** and **77** were also determined by a combination of their 2D NMR spectroscopic data and those of their permethylated derivatives. Ptychantol C (**77**) is an optically active compound with a specific rotation $[\alpha]_D = -14.8°$ and a series of Cotton effects $\lambda(\Delta\varepsilon) = 289(+10.6)$, 270(−3.0), 253(+21.2), 235(−30.8) nm ($mol^{-1}dm^3cm^{-1}$) in its ECD spectrum. The absolute configuration of **77** remains to be clarified. Ptychantols are the first macrocyclic bis-bibenzyls with an (*E*)-stilbene moiety found in a species in the Marchantiophyta; however, an acyclic bis-bibenzyl with an (*E*)-stilbene moiety, 7′,8′-dehydroperrottetin F (**93**) was identified in *Lunularia cruciata* (Asakawa, 1995). Ptychantol B (**76**) might be formed from **93** by phenolic oxidation (Hashimoto *et al.*, 1999).

The Japanese *Blasia pusilla* produces the bis-bibenzyl dimers (**78–81**). The structure of pusilatin A (**78**) was established by X-ray crystallographic analysis of its peracetate (Yoshida *et al.*, 1996). *R. multifida ssp. decrescens* elaborated a new bis-bibenzyl dimer pusilatin E (**82**), together with riccardin A (**1**) and marchantin I (**21**). The structure of **82** was elucidated as the C_6'/C_6'''-coupled symmetrical dimer of riccardin A (**1**) using a combination of its FABMS and 2D NMR spectroscopic data (Yoshida *et al.*, 1997). Further evidence of the structure of **82** was obtained by its partial synthesis from riccardin A (**1**). Oxidative coupling of **1** in ethanol with manganese triacetate gave a C_6'/C_6'''-coupled symmetrical dimer having spectroscopic data identical to those of the natural product (**82**). Furthermore, the demethylation of **82** with boron tribromide gave the naturally occurring pusilatin B (**79**) (Yoshida *et al.*, 1997).

Anton *et al.* (1997) studied a methanol extract of the Costa Rican *P. deflexa* and isolated four bis-bibenzyls, named isoplagiochin G (**50**), plagilin (**87**), isoplagilin (**88**), and plagiolin

(**89**), together with the known riccardin C (**3**), and isoplagiochins C (**46**), D (**47**), and F (**49**). Their structures were proven by 2D NMR spectroscopic methods. The biphenyl linkages at C_5–C_5' in **87**, C_5–C_6' in **88**, and C_5–C_7' in **89**, between two 3,4-dihydroxy-3′-methoxybibenzyl moieties, were established by the analysis of NOESY correlations. Plagiolin (**89**) is the first report of the presence of bibenzyls with a C–C linkage of a phenyl C-atom to a C-atom of the ethano bridge. The NMR spectrum of isoplagiochin G (**50**) is similar to that of isoplagiochin B (**45**), indicating that **50** contains a benzophenone moiety. The position of a phenyl ether linkage at C_{13} was confirmed by means of a NOESY experiment. Isoplagiochin F (**49**) has been found in *P. permista* var. *integerrima* (Anton *et al.*, 1997).

Monoclea forsteri produces perrottetin E (**90**) and marchantin-type bis-bibenzyls (Asakawa, 1995). Reinvestigation of the same species gave perrottetin E (**90**) in a good yield (22.2% of the crude extract) (Omatsu, 2004). *Jungermannia comata* (Nagashima *et al.*, 1996a) and *Jungermannia infusca* (Nagashima *et al.*, 1996b) gave **90** as a major component.

The Ecuadorian *Frullania convoluta* was found to contain four known bis-bibenzyls, perrottetins E–G (**90–92**) and 7′,8′-dehydroperrottetin F (**93**). Their spectroscopic data were identical with those reported for these bis-bibenzyls (Flegel *et al.*, 1999). While the New Zealand *L. cruciata* produced perrottetin F (**91**) isolated from the same species, when collected in Japan, it was not detected (Nakaishi, 1996). *Nardia subclavata* and *Pellia epiphylla* produced perrottetin E (**90**) (Asakawa, 1995), perrottetin E and its 11′-*O*-methyl ether (**94**), respectively (Asakawa *et al.*, 2008). Both compounds were isolated from *P. epiphylla* and *Pellia endiviifolia* (Asakawa, 1995).

Jamesoniella colorata afforded the chlorinated bis-bibenzyl (**96**). Direct CI mass spectrometry of **96** indicated the presence of six chlorine atoms in the molecule. The structure of this compound was established as 2,2′,10,10′,12,12′-hexachloroisoperrottetin A by means of its HMBC and NOESY spectra. This was the first isolation of a bis-bibenzyl from the genus *Jamesoniella* (Hertewich *et al.*, 2003).

From the European *P. epiphylla*, the five bis-bibenzyls **97–101** and the bis-bibenzyl dimer **85** were obtained, together with perrottetin E (**90**) (Cullmann *et al.*, 1997; Cullmann & Becker, 1998). Compounds **99** and **100** were also isolated from the sporophytes of this liverwort (Cullmann & Becker, 1998). Compound **97** is an *O*-methyl ether of perrottetin E according to the molecular formula, its location at C_{11} being confirmed via the NOESY spectrum. Thus, the structure of **97** was established as perrottetin E-11-*O*-methyl ether. Its isomer, perrottetin E-11′-*O*-methyl ether (**94**), has been isolated from *P. endiviifolia* (Asakawa, 1995). The structure of compound **98** could be assigned as 14′-hydroxyperrottetin E based on the similarity of the NMR spectra of **98** and 14-hydroxyperrottetin E isolated from *P. endiviifolia* (Asakawa, 1995) and as a result of its total synthesis (Seoane *et al.*, 1996). Previously, compound **99** was reported as 14′-hydroxyperrottetin E. However, this structure was revised as 10′-hydroxyperrottetin E (**99**) by the re-evaluation of the NMR spectroscopic data and by total synthesis (Seoane *et al.*, 1996; Cullmann *et al.*, 1997).

The structure of **100** was assigned as 10′-hydroxyperrottetin E-11-*O*-methyl ether, according to the similarity of ^1H and ^{13}C NMR spectra with those of **99** and by analysis of its NOESY spectrum. Previously, 14,14′-dihydroxyperrottetin E was reported as one of the perrottetin E series from *P. epiphylla*. However, the structure was revised as 10,10′-dihydroxyperotttetin E (**101**) by total synthesis, using a twofold *Wittig* reaction (Cullmann *et al.*, 1997).

The ^{1}H and ^{13}C NMR spectra of compound **85**, $C_{56}H_{50}O_{10}$, showed signals reminiscent of a bis-bibenzyl moiety, but according to its mass spectrum, it resulted from the oxidative combination of two perrottetin units. Analysis of the entire ^{1}H and ^{13}C NMR data led to the structure 13′,13‴-bis(10′-hydroxyperrottetin E) (**85**). The proposed structure was confirmed by its total synthesis (Cullmann *et al.*, 1997).

Schistochila glaucescens has been known to produce the cyclic bis-bibenzyls, marchantin C (**16**), neomarchantin A (**32**), and neomarchantin B (**33**) (Asakawa, 1995). Furthermore, two additional cyclic bis-bibenzyls, glaucescens bis-bibenzyl (GBB) A (**83**) and GBB B (**84**), were isolated together with neomarchantins A (**32**) and B (**33**) (Scher *et al.*, 2002). The new bibenzyls were shown to be isomers by ESIMS. These bibenzyls showed NMR spectra similar to those of neomarchantin B (**33**), indicating that these compounds possess a neomarchantin B (**33**) moiety with a sesquiterpenoid unit. Further analysis of the HMBC and NOESY spectra of **83** and **84** led to the full structures of these bibenzyls. Sher *et al.* (2002) proposed the biosynthesis of both compounds. From farnesyl pyrophosphate, a furanosesquiterpene might be formed, followed by cyclization with bis-bibenzyl ortho-quinones to afford **83** and **84**. Reinvestigation of the ether extract of *S. glaucescens* collected in various locations in New Zealand led to the isolation of the same bis-bibenzyl sesquiterpene dimers (**83** and **84**), along with marchantin C (**16**) and neomarchantin A (**32**) (Nagashima *et al.*, 2004; Asakawa, 2008c).

2.5 Biological activity of bis-bibenzyls

In earlier reviews (Asakawa, 1982, 1990a, 1990b, 1995), chemical constituents isolated from liverworts were discussed and demonstrated to possess characteristic fragrances, bitterness, pungency, and sweetness, as well as allergenic contact dermatitis, cytotoxic, antimicrobial, antifungal, calmodulin inhibitory, cardiotonic, insect antifeedant, 5-lipoxygenase inhibitory, molluscicidal, muscle relaxant, neurotrophic, plant growth regulatory, superoxide release inhibitory, thromboxane synthase inhibitory, and vasopressin antagonist activities. Several reviews dealing with biologically active compounds from liverworts have been published more recently (Asakawa, 2008a, 2008b; Asakawa & Ludwiczuk, 2008, 2009; Asakawa *et al.*, 2009, 2013a, 2013b; Harinantenaina & Asakawa, 2007). Some of the most interesting biologically active compounds are macrocyclic bis-bibenzyls, such as those of the marchantin, riccardin, and isoplagiochin series. Especially, marchantin-type compounds show nitric oxide production inhibitory, α-glucosidase inhibitory, apoptotic, liver X-receptor agonist (LXRα) agonist receptor, farnesoid X-receptor (FXR) activation, anti-influenza, antimicrobial and antifungal, cytotoxicity, anticancer, and muscle relaxing activity (Asakawa, 1995, Asakawa *et al.*, 2013b; Harinantenaina *et al.*, 2005; Morita *et al.*, 2009; Sun *et al.*, 2009). Marchantin A (**14**) shows antitrypanosomal (Otoguro *et al.*, 2012), anticancer (Asakawa, 1995; Shi *et al.*, 2008a, 2008b, 2009; Huang *et al.*, 2010; Asakawa *et al.*, 2013b), muscle relaxing (Taira *et al.*, 1994), and antiinfluenza activities (Iwai *et al.*, 2011). Marchantin C (**16**) also shows anticancer activity (Shi *et al.*, 2009). Riccardin C (**31**) shows antiobesity activity (Tamehiro *et al.*, 2005). In Table 2.2, biological activity, bioactive compounds and the names of the bioactive liverworts are shown.

Table 2.2 Biologically active bis-bibenzyls from liverworts (Asakawa, 1982, 1995, 2008a; Asakawa *et al.*, 2013a, 2013b).

Biological activity	Bioactive compounds	Liverworts
Muscle relaxant activity	Marchantin A (**14**)	*Marchantia polymorpha*
Calcium inhibitory activity	Riccardin A (**1**)	*Riccardia multifida*
	Marchantin A (**14**)	*M. polymorpha*
Farnesoid X-receptor (FXR) activation	Marchantin A (**14**)	*M. polymorpha*
	Marchantin E (**18**)	*M. polymorpha*
Liver X-receptor alpha (LXRα) agonist activity	Riccardin C (**3**)	*Reboulia hemisphaerica*
Cytotoxic and apoptotic activity	Marchantin A (**14**)	*M. polymorpha*
	Marchantin B (**15**)	*Schistochila glaucescens*
	Marchantin C (**15**)	*S. glaucescens*
	Marchantin D (**16**)	*S. glaucescens*
	Neomarchantin A (**32**)	*S. glaucescens*
	Neomarchantin B (**33**)	*S. glaucescens*
	Pusilatin B (**79**)	*Blasia pusilla*
	Pusilatin C (**80**)	*B. pusilla*
	Bis-bibenzyl dimer GBB A (**83**)	*S. glaucescens*
	Bis-bibenzyl dimer GBB B (**84**)	*S. glaucescens*
	Perrottetin F (**91**)	*Radula perrottetii*
	Paleatin B (**95**)	*Marchantia paleacea* var. *diptera*
Tubulin polymerization inhibition	Marchantin C (**16**)	*M. polymorpha*
	Isoplagiochin A (**44**)	
	Isoplagoichin B (**45**)	*Plagiochila fruticosa*
		P. fruticosa
Antimicrobial activity (*Acinetobacter calcoaceticus, Bacillus cereus, Bacillus megaterium, Bacillus subtilis, Cryptococcus neoformans*, and *Staphylococcus aureus*)	Marchantin A (**14**)	*M. polymorpha*
		M. palearea var. diptera
		M. tosana
	Riccardin C (**3**)	*Plagiochasma intermedium*
	Riccardin D (=Plagiochin E) (**4**)	*M. polymorpha*
	Riccardin F (**5**)	*P. intermedium*
	Riccardin H (**7**)	*M. polymorpha*
	Isoriccardin C (**8**)	*Bazzania trilobata*
	13,13′-*O*-isopropylidene riccardin D (12)	*Marchantia polygmorpha*
	Marchantin A (**14**)	*M. polygmorpha*
	Marchantin B (**15**)	*M. polymorpha*
	Marchantin C (**16**)	*S. glaucescens*
	Marchantin E (**18**)	*M. polymorpha*
	Marchantin H (**20**)	*P. intermedium*
	Marchantin M (**24**)	*P. intermedium*
	Marchantin P (**28**)	*P. intermedium*
	Neomarchantin A (**32**)	*S. glaucescens*
	Asterellin A (**34**)	*Asterella angusta*
	Asterellin B (**35**)	*A. angusta*
	11-*O*-demethylmarchantin I (**36**)	*A. angusta*
	Dihydroptychantol A (**37**)	*A. angusta*
	Pakyonol (**38**)	*A. angusta*
	Isoplagiochin D (**46**)	*B. trilobata*
	6′,8′-Dichloroisoplagiochin C (=bazzanin B) (**57**)	*B. trilobata*
	6′-Chloroisoplagiochuin D (=bazzanin S) (**66**)	*B. trilobata*
	Perrottetin E (**90**)	*A. angusta*

(*Continued*)

Table 2.2 (Continued)

Biological activity	Bioactive compounds	Liverworts
Antiviral activity (Influenza)	Marchantin A (**14**)	*M. polymorpha*
Antiviral activity (HIV-1)	Marchantin B (**15**)	*M. polymorpha*
	Marchantin E (**18**)	*M. polymorpha*
	Plagiochin A (**41**)	*M. polymorpha*
	Perrottetin F (**91**)	*M. polymorpha*
	Marchantin A (**14**)	*M. polymorpha*
	Marchantin B (**15**)	*M. polymorpha*
	Marchantin D (**17**)	*M. polymorpha*
	Perrottetin F (**91**)	*R. perrottetii*
	Paleatin B (**95**)	*M. paleacea* var. *diptera*
Cathepsin L and cathepsin B inhibitory activity	Isomarchantin C (**29**)	*Bryopteris filicina*
Antiplatelet activity	Marchantiaquinone (**26**)	*Plagiochila elegans*
Neurotrophic activity	Plagiochin A (**41**)	*P. fruticosa*
Antioxidant activity	Marchantin A (**14**)	*M. polymorpha*
	Marchantin H (**20**)	*M. polymorpha*
	Plagiochin D (**40**)	*P. fruticosa*
Antitrypanosomal activity	Marchantin A (**14**)	*M. polymorpha*
	Plagiochin A (**41**)	*P. fruticosa*
Cardiotonic activity	Marchantin A (**14**)	*M. polymorpha*
α-Glucosidase inhibitory activity	Marchantin C (**16**)	*M. polymorpha*
Antithrombin activity	Perrottetin E (**90**)	*Jungermannia* species
Nitric oxide production inhibition	Riccardin A (**1**)	*R. multifida*
	Riccardin C (**3**)	*M. polymorpha*
	Riccardin F (**5**)	*M. polymorpha*
	Marchantin A (**14**)	*M. polymorpha*
	Marchantin B (**15**)	*M. polymorpha*
	Marchantin D (**17**)	*M. polymorpha*
	Marchantin E (**18**)	*M. polymorpha*
	Perrottetin F (**91**)	*R. perrottetii*

2.6 Conclusions

The Marchantiophyta (liverworts) are rich sources of cyclic and acyclic bis-bibenzyls possessing two or more phenolic hydroxy groups, bibenzyls, phenolic sesquiterpenoids, several of which show interesting biological activity such as antimicrobial, antifungal and antiviral, cytotoxic, antioxidant, muscle relaxing, and NO production inhibitory activities. Some cyclic bis-bibenzyls have optical activity and the structures of two conformers have been established by ECD spectroscopic analysis. It is noteworthy that the marchantin and isoplagiochin series obtained from the liverworts, *M. polymorpha* and *P. fruticosa,* showed similar microtubule polymerization inhibitory activity as the clinical anticancer drug taxol (paclitaxel). A number of flavonoids including bi- and triflavones have been isolated from and detected in mosses; however, their biological activity is unknown (Asakawa, 1995; Asakawa *et al.*, 2013b). The distribution of polyphenols in hornworts is very poor. Only caffeic acid derivatives have been found in a few species (Asakawa *et al.*, 2013a).

Acknowledgments

The author thanks the late Prof. Guy Ourisson (Univ. Louis Pasteur, France); the late Prof. Meinhart H. Zenk (University of Munich, Germany; Danforth Institute of Plant Science, USA); Profs. M. Toyota, M. Tori, T. Hashimoto, A. Fukuyama, T. Kuzuhara, and Dr. Nagahima (Tokushima Bunri University, Japan); and the former postdoctoral fellows Dr. A. Ludwiczuk (Medical University of Lublin, Poland), Dr. L. Harinantenaina (Ohio University, USA), Dr. L. Harisson and Dr. M. Buchanan (University of Glasgow, Scotland), and Dr. A. Bardón and Dr. C. Socolsky (Tucuman National University of Argentina, Argentina) for their valuable discussion. Thanks are indebted to Prof. R. Gradstein (National Museum of Natural History, Paris); Dr. M. Mizutani (Hattori Botanical Laboratory, Nichinan, Japan); Dr. T. Furuki (Natural History Museum and Institute, Chiba, Japan); and P. Raharivelomanana, J.-P. Bianchini, and Dr. A. Pham (University of French Polynesia) for their collection and identification of liverworts. A part of this work was supported by a Grant-in-Aid for the Scientific Research (A) (no. 11309012) and the Open and Senryaku Research from the Ministry of Education, Culture, Sports, Science and Technology.

References

Anton, H., Kraut, L., Mues, R. and Moralez, I. (1997) Phenanthrenes and Bisbibenzyls from *Plagiochila* species. *Phytochemistry*, **46**, 1069–1075.

Anton, H., Schoeneborn, R. and Mues, R. (1999) Bibenzyls and bisbibenzyls from a Neotropical *Plagiochila* species. *Phytochemistry*, **52**, 1639–1645.

Asakawa, Y. (1982) Chemical constituents of the Hepaticae, in *Progress in the Chemistry of Organic Natural Products*, vol. **42** (eds W. Herz, H. Grisebach and G.W. Kirby), Springer, Vienna, pp. 1–285.

Asakawa, Y. (1990a) Biologically active substances from bryophytes, in *Bryophyte Development: Physiology and Biochemistry* (eds R.N. Chopra and S.C. Bhatla), CRC Press, Boca Raton, FL, pp. 259–287.

Asakawa, Y. (1990b) Terpenoids and aromatic compounds with pharmacological activity from bryophytes, in *Bryophytes: Their Chemistry and Chemical Taxonomy* (eds D.H. Zinsmeister and R. Mues), Oxford University Press, Oxford, pp. 369–410.

Asakawa, Y. (1993) Biologically active terpenoids and aromatic compounds from liverworts and inedible mushroom *Cryptoporus volvatus*, in *Bioactive Natural Products: Detection, Isolation, and Structural Determination* (eds S.M. Colegate and R.J. Molyneux), CRC Press, Boca Raton, FL, pp. 319–347.

Asakawa, Y. (1994) Highlights in phytochemistry of Hepaticae—biologically active terpenoids and aromatic compounds. *Pure and Applied Chemistry*, **66**, 2193–2196.

Asakawa, Y. (1995) Chemical constituents of the bryophytes, in *Progress in the Chemistry of Organic Natural Products*, vol. **65** (eds W. Herz, W.B. Kirby, R.E. Moore, W. Steglich and C. Tamm), Springer, Vienna, pp. 1–618.

Asakawa, Y. (1999) Phytochemistry of bryophytes: biologically active terpenoids and aromatic compounds from liverworts, in *Phytochemicals in Human Health Protection, Nutrition, and Plant Defense* (ed J. Romeo), Kluwer Academic/Plenum Publishers, New York, pp. 319–342.

Asakawa, Y. (2001) Recent advances in phytochemistry of bryophytes—acetogenins, terpenoids and bis(bibenzyl)s from selected Japanese, Taiwanese, New Zealand, Argentinean and European liverworts. *Phytochemistry*, **56**, 297–312.

Asakawa, Y. (2004) Chemosystematics of the Hepaticae. *Phytochemistry*, **65**, 623–669.

Asakawa, Y. (2008a) Liverworts—potential source of medicinal compounds. *Current Pharmaceutical Design*, **14**, 3067–3088.

Asakawa, Y. (2008b) Recent advances of biologically active substances from Marchantiophyta. *Natural Product Communications*, **3**, 77–92.

Asakawa, Y. (2008c) Marchantiophyta: a good source of biologically active compounds, in *Bryology in the New Millennium* (eds H. Mohamed, B.B. Baki, A. Nasrulhaq-Boyce and P.K.Y. Lee), University of Malaya, Kuala Lumpur, pp. 367–394.

Asakawa, Y. and Ludwiczuk, A. (2008) Bryophytes—chemical diversity, bioactivity and chemosystematics. Part I. Chemical diversity and bioactivity. *Rośliny Lecznicze w Polsce i na Świecie (Medical Plants in Poland and in the World)*, **2**, 33–53.

Asakawa, Y. and Ludwiczuk, A. (2009) Marchantiophyta: bio- and chemical diversity and bioactivity. *Malaysian Journal of Science*, **28**, 229–257.

Asakawa, Y. and Ludwiczuk, A. (2012) Distribution of cyclic and acyclic bis-bibenzyls in the Marchantiophyta (liverworts), ferns and higher plants and their biological activities, biosynthesis, and total synthesis. *Heterocycles*, **86**, 891–917.

Asakawa, Y. and Matsuda, R. (1982) Riccardin C, a novel cyclic bibenzyl derivative from *Reboulia hemisphaerica*. *Phytochemistry*, **21**, 2143–2144.

Asakawa, Y., Toyota, M., Tori, M. and Hashimoto, T. (2000) Chemical structures of macrocyclic bis(bibenzyls) isolated from liverworts (Hepaticae). *Spectroscopy*, **14**, 149–175.

Asakawa, Y., Toyota, M., Nagashima, F. and Hashimoto, T. (2008) Chemical constituents of selected Japanese and New Zealand liverworts. *Natural Product Communications*, **3**, 289–300.

Asakawa, Y., Ludwiczuk, A., Nagashima, F. *et al.* (2009) Bryophytes: bio- and chemical diversity, bioactivity and chemosystematics. *Heterocycles*, **7**, 99–150.

Asakawa, Y., Ludwiczuk, A. and Nagashima, F. (2013a) Phytochemical and biological studies of bryophytes. *Phytochemistry*, **55**, 55–88.

Asakawa, Y., Ludwiczuk, A. and Nagashima, F. (2013b) Chemical constituents of bryophytes: bio- and chemical diversity, biological activity and chemosystematics, in *Progress in the Chemistry of Organic Natural Products*, vol. **42** (eds D.A. Kinghorn, H. Falk and J. Kobayashi), Springer, Vienna, pp. 1–796.

Bringmann, G., Mühlbacher, J., Reichert, M., Dreyer, M., Kolz, J. and Speicher, A. (2004) Stereochemistry of isoplagiochin C, a macrocyclic bisbibenzyl from liverworts. *Journal of the American Chemical Society*, **126**, 9283–9133.

Cullmann, F. and Becker, H. (1998) Terpenoid and phenolic constituents of sporophytes and spores from the liverwort *Pellia epiphylla*. *Journal of the Hattori Botanical Laboratory*, **8**, 285–295.

Cullmann, F., Becker, H., Pandolfi, E., Roeckner, E. and Eicher, T. (1997) Bibenzyl from *Pellia epiphylla*. *Phytochemistry*, **45**, 1235–1247.

Ding, H. (1982) *Zhong guo Yao Yun Bao zi Zhi Wu*, Kexue Jishu Chuban She, Shanghai, pp. 1–409.

Eicher, T., Fey, S., Puhl, W., Büchel, E. and Speicher, A. (1998) Bryophyte constituents: part 9. Syntheses of cyclic bisbibenzyls systems. *European Journal of Organic Chemistry*, 877–888.

Esumi, T., Wada, M., Mizushima, E. *et al.* (2004) Efficient synthesis of isoplagiochin D, a macrocyclic bis(bibenzyls), by utilizing an intramolecular Suzuki–Miyaura reaction. *Tetrahedron Letters*, **45**, 6941–6945.

Fang, L., Guo, H.-F. and Lou, H.-X. (2007) Three new bisbibenzyl derivatives from the Chinese liverwort *Marchantia polymorpha*. *Helvetica Chimica Acta*, **90**, 748–752.

Flegel, M., Adam, K.-P. and Becker, H. (1999) Sesquiterpene lactones and bisbibenzyl derivatives from the neotropical liverwort *Frullania convoluta*. *Phytochemistry*, **52**, 1633–1638.

Friederich, S., Maier, U.H., Deus-Meumann, B., Asakawa, Y. and Zenk, M.H. (1999a) Biosynthesis of cyclic bis(bibenzyls) in *Marchantia polymorpha*. *Phytochemistry*, **50**, 589–598.

Friederich, S., Rueffer, M., Asakawa, Y. and Zenk, M.H. (1999b) Biosynthesis of cyclic bis(bibenzyls) in *Marchantia polymorpha*. *Phytochemistry*, **52**, 1192–1202.

Fukuyama, Y., Yaso, H., Nakamura, K. and Kodama, M. (1999) Total synthesis of plagiochin D, a macrocyclic bis(bibenzyl) from liverworts by intramolecular Still-Kelly reaction. *Tetrahedron Letters*, **40**, 105–108.

Gerencser, J., Keseru, G.M., Macsari, I., Nogradi, M., Kajtar-Peredy, M. and Szollosy, A. (1997) Synthesis of isoplagiochin A. *The Journal of Organic Chemistry*, **62**, 3666–3670.

Gottsegen, A., Nógrádi, M., Vermes, B., Kajtár-Peredy, M. and Bihátsi-Karsai, E. (1990) Total syntheses of riccardins A, B, and C, cytotoxic macrocyclic bis(bibenzyls) from liverworts. *Journal of the Chemical Society, Perkin Transactions 1*, 315–320.

Guo, H., Xing, J., Xie, C., Qu, J., Gao, Y. and Lou, H. (2007) Study of bis(bibenzyls) in bryophytes using electron ionization time-of-flight and ionization triple-quadruple mass spectrometry. *Rapid Communications in Mass Spectrometry*, **21**, 1367–1374.

Harinantenaina, L. and Asakawa, Y. (2007) Malagasy liverworts, source of new and biologically active compounds. *Natural Product Communications*, **2**, 701–709.

Harinantenaina, L., Luang, D.N., Nishizawa, T. *et al.* (2005) Bis(bibenzyls) from liverworts inhibit lipopolysaccharide-induced inducible NOS in RAW 264.7 cells: a study of structure–activity relationships and molecular mechanism. *Journal of Natural Products*, **68**, 1779–1781.

Harrowven, D. and Kostiuk, S.L. (2012) Macrocyclic bisbibenzyl natural products and their chemical synthesis. *Natural Product Reports*, **29**, 121–328.

Harrowven, D., Woodcock, T. and Howes, P.D. (2005) Total synthesis of cavicularin and riccardin C: addressing the synthesis of an arene that adopts a boat configuration. *Angewandte Chemie International Edition*, **44**, 3899–3901.

Hashimoto, T., Tori, M., Asakawa, Y. and Fukazawa, Y. (1987) Plagiochins A, B, C, and D, new type of macrocyclic bis(bibenzyls) having a biphenyl linkage between the *ortho* positions to the benzyl methylenes, from the liverwort *Plagiochila acantophylla* subsp. *japonica*. *Tetrahedron Letters*, **28**, 6295–6298.

Hashimoto, T., Kanayama, S., Kan, Y., Tori, M. and Asakawa, Y. (1996) Isoplagiochins C and D, new type of macrocyclic bis(bibenzyls), having two biphenyl linkages from the liverwort *Plagiochila fruticosa*. *Chemistry Letters*, **25**, 741–742.

Hashimoto, T., Irita, H. and Asakawa, Y. (1998) Chemical constituents of the Japanese liverwort *Plagiochasma pterospermum*. *Journal of Hattori Botanical Laboratory*, **85**, 239–244.

Hashimoto, T., Ikeda, H., Takaoka, S., Tanaka, M. and Asakawa, Y. (1999) Ptychantols A–C, macrocyclic bis(Bibenzyls), possessing a *trans*-stilbene structure from the liverwort *Ptychanthus striatus*. *Phytochemistry*, **52**, 501–509.

Hashimoto, T., Irita, H., Takaoka, S., Tanaka, M. and Asakawa, Y. (2000) New chlorinated cyclic bis(bibenzyls) from the liverworts *Herbertus sakuraii* and *Mastigophora diclados*. *Tetrahedron*, **56**, 3153–3159.

Hertewich, U.M., Zapp, J. and Becker, H. (2003) Secondary metabolites from the liverwort *Jamesoniella colorata*. *Phytochemistry*, **63**, 227–233.

Hioki, H., Shima, N., Kawaguchi, K. *et al.* (2009) Synthesis of riccardin C and its seven analogues. Part 1: the role of their phenolic hydroxy groups as LXRα agonists. *Bioorganic and Medicinal Chemistry Letters*, **19**, 738–741.

Hosoda, A. (2003) Phytochemical study on the chemical constituents of some New Zealand and Japanese liverworts. Master thesis. Tokushima Bunri University, Tokushima.

Huang, W.-J., Wu, C.-L., Lin, C.-W. *et al.* (2010) Marchantin A, a cyclic bis(bibenzyl ether), isolated from the liverwort *Marchantia emarginata* subsp. *tosana* induces apoptosis in human MCF-7 breast cancer cells. *Cancer Letters*, **291**, 108–119.

Iwai, Y., Murakami, K., Gomi, Y., Hashimoto, T. *et al.* (2011) Anti-influenza activity of marchantins, macrocyclic bisbibenzyls contained in liverworts. *PLoS One*, **6**, e19825.

Keseru, G.M. and Nogradi, M. (1995a) The biological activity of cyclic bis(bibenzyls): a rotational approach. *Bioorganic and Medicinal Chemistry*, **3**, 1511–1517.

Keseru, G.M. and Nogradi, M. (1995b) The chemistry of macrocyclic bis(bibenzyls). *Natural Product Reports*, **12**, 69–75.

Keseru, G.M. and Nogradi, M. (1996) Molecular similarity analysis on biologically active macrocyclic bis(bibenzyls). *Journal of Molecular Recognition*, **9**, 133–138.

Keseru, G.M., Mezeyvendor, G., Nogradi, M. *et al.* (1992) Total synthesis of plagiochin-C and plagiochin D, macrocyclic bis(bibenzyl) constituents of *Plagiochila acanthophylla*. *Tetrahedron*, **48**, 912–922.

Kodama, M., Shiobara, Y., Matsumura, K. and Sumitomo, H. (1985) Total synthesis of marchantin A, a novel cytotoxic bis(bibenzyl) isolated from liverworts. *Tetrahedron Letters*, **26**, 877–880.

Kodama, M., Shiobara, Y., Sumitomo, H., Matsumura, K., Tsukamoto, M. and Harada, C. (1988) Total syntheses of marchantin A and riccardin B, cytotoxic bis(bibenzyls) from liverworts. *Journal of Organic Chemistry*, **53**, 72–77.

Kosenkova, Y.S., Polovinka, M.P., Komarova, N.I. *et al.* (2007) Riccardin C, a bisbibenzyl compound from *Primula macrocalyx*. *Chemisty of Natural Compounds*, **43**, 712–713.

Lahlou, E.H., Hashimoto, T. and Asakawa, Y. (2000) Chemical constituents of liverworts *Plagiochasma japonica* and *Marchantia tosana*. *Journal of Hattori Botanical Laboratory*, **88**, 271–275.

Lopez, G.V., Pandolfi, E.M. and Seoane, G.A. (2000) Total synthesis of Marchantiaquinone. *Synthesis*, **10**, 1403–1408.

Lou, H.-X. (2013) *Chemistry and Biology of Bryophytes*, Beijing Science Publisher, Beijing, pp. 1–672.

Martini, U., Zapp, J. and Becker, H. (1998) Chlorinated macrocyclic bisbibenzyls from the liverwort *Bazzania trilobata*. *Phytochemistry*, **47**, 89–96.

Molales, J.C.C., Torres, A.G. and Gonzalez-Zaamora, E. (2011) Total synthesis of plagiochin D by intramolecular S(N) Ar reaction. *European Journal of Organic Chemistry*, **17**, 3165–3170.

Morita, H., Tomizawa, Y., Tsuchiya, T., Hirasawa, Y., Hashimoto, T. and Asakawa, Y. (2009) Antimitotic activity of two macrocyclic bis(bibenzyls), isoplagiochins A and B from the liverwort *Plagiochila fruticosa*. *Bioorganic and Medicinal Chemistry Letters*, **19**, 493–496.

Nabeta, K., Ohkubo, S., Hozumi, R. and Katoh, K. (1998) Macrocyclic bisbibenzyls in cultured cells of the liverwort *Heteroscyphus planus*. *Phytochemistry*, **49**, 1941–1943.

Nagashima, F., Momosaki, S., Watanabe, Y., Toyota, M., Huneck, S. and Asakawa, Y. (1996a) Terpenoids and aromatic compounds from six liverworts. *Phytochemistry*, **41**, 207–211.

Nagashima, F., Tamada, A. and Asakawa, Y. (1996b) A new monocyclic diterpene from the liverwort *Jungermannia infusca* (Mitt.) Steph. *Chemical and Pharmaceutical Bulletin*, **44**, 1628–1630.

Nagashima, F., Sekiguchi, T., Takaoka, S. and Asakawa, Y. (2004) Terpenoids and aromatic compounds from the New Zealand liverwort *Plagiochila*, *Schistochila* and *Heteroscyphus* species. *Chemical and Pharmaceutical Bulletin*, **52**, 556–560.

Nakaishi, E. (1996) Phytochemical studies on the New Zealand liverworts. Master thesis. Tokushima Bunri University, Tokushima.

Niu, C., Qu, J.-B. and Lou, H.-X. (2006) Antifungal bis(bibenzyls) from the Chinese liverwort *Marchantia polymorpha*. *Chemistry and Biodiversity*, **3**, 34–40.

Nógrádi, M., Vermes, B., Kajtár-Peredy, M. and Novikov, V.P. (1990) Synthese von Pakyonol, einem macrocyclischen bis(bibenzylether) aus *Mannia fragrans*. *Liebigs Annalen der Chemie*, 299–301.

Oiso, Y., Toyota, Y. and Asakawa, Y. (1999) Occurrence of a bisbibenzyl derivative in the Japanese fern *Hymenophyllum barbatum*: first isolation and identification of perrottetin H from the Pteridophytes. *Chemical and Pharmaceutical Bulletin*, **47**, 297–298.

Omatsu, I. (2004) Phytochemical studies on 14 New Zealand liverworts. Master thesis. Tokushima Bunri University, Tokushima.

Otoguro, K., Ishiyama, A., Iwatsuki, M. *et al.* (2012) *In vitro* antitrypanosomal activity of bis(bibenzyls) and bibenzyls from the liverworts against *Trypanosoma brucei*. *Journal of Natural Medicines*, **66**, 377–382.

Qu, J., Xie, C., Guo, H., Yo, W. and Lou, H. (2007) Antifungal dibenzofuran bis(bibenzyl)s from the liverwort *Asterella angusta*. *Phytochemistry*, **66**, 1767–1774.

Sadamori, M. (2009) Studies on the new biologically active substances of Tahitian and Tokushima's *Plagiochila* genus. Master thesis. Tokushima Bunri University, Tokushima.

Scher, J.M., Burgess, E.J., Lorimer, S.D. and Perry, N.B. (2002) A cytotoxic sesquiterpene and unprecedented sesquiterpene–bisbibenzyl compounds from the liverwort *Schistochila glaucescens*. *Tetrahedron*, **58**, 7875–7882.

Scher, J.M., Zapp, J., Schmidt, A. and Becker, H. (2003) Bazzanins L–R, chlorinated macrocyclic bisbibenzyls from the liverwort *Lepidozia incurvata*. *Phytochemistry*, **64**, 791–796.

Scher, J.M., Speakman, J.-B., Zapp, J. and Becker, H. (2004a) Bioactivity guided isolation of antifungal compounds from the liverwort *Bazzania trilobata* (L.) S.F. Gray. *Phytochemistry*, **65**, 2583–2588.

Scher, J.M., Zapp, J., Becker, H. *et al.* (2004b) Optically active bisbibenzyls from *Bazzania trilobata*: isolation and stereochemical analysis by chromatographic, chiroptical, and computational methods. *Tetrahedron*, **60**, 9877–9881.

Seoane, G., Pandolfi, E., Gamenara, D. and Eicher, T. (1996) Total synthesis of bisbibenzylic compounds isolated from bryophytes. *Journal of the Brazilian Chemical Society*, **7**, 307–317.

Shi, Y.-Q., Liao, Y.-X., Qu, X.-J. *et al.* (2008a) Marchantin C, a macrocyclic bisbibenzyl, induces apoptosis of human glioma A172 cells. *Cancer Letters*, **262**, 173–182.

Shi, Y.-Q., Qu, X.-J. and Liao, Y.-X. (2008b) Reversal effect of a macrocyclic bisbibenzyl plagiochin E on multidrug resistance in adriamycin-resistant K562/A02 cells. *European Journal of Pharmacology*, **584**, 66–71.

Shi, Y.-Q., Zhu, C.-J. and Yuan, H.-Q. (2009) Marchantin C, a novel microtubule inhibitor from liverwort with anti-tumor activity both *in vivo* and *in vitro*. *Cancer Letters*, **276**, 160–170.

Shiobara, Y., Sumitomo, H., Tsukamoto, M., Harada, C. and Kodama, M. (1985) Synthesis and structure confirmation of riccardin B, a macrocyclic bis(bibenzyl) from the liverwort, *Riccardia multifida*. *Chemistry Letters*, **14**, 1587–1588.

Shy, H.-S. and Wu, C.-L. (2006) Eudesmane-type sesquiterpenoids from the liverwort *Lepidozia fauriana*. *Journal of Asian Natural Products Research*, **8**, 723–731.

Speicher, A., Hollemeyer, K. and Heinzle, E. (2001) Rapid detection of chlorinated bisbibenzyls in *Bazzania trilobata* using MALDI–TOF mass spectrometry. *Phytochemistry*, **57**, 303–306.

Speicher, A., Kolz, J. and Sambanje, R.P. (2002) Syntheses of chlorinated bisbibenzyls from bryophytes. *Synthesis*, **17**, 2503–2512.

Speicher, A., Heisel, R. and Kolz, J. (2003) First detection of a chloroperoxidase in bryophytes. *Phytochemistry*, **62**, 679–682.

Speicher, A., Backes, T. and Grosse, S. (2005) Syntheses of macrocyclic bibenzyls on solid support. *Tetrahedron*, **61**, 11692–11696.

Speicher, A., Backes, T., Hesidens, K. and Kolz, J. (2009a) Ring strain and total syntheses of modified macrocycles of the Isoplagiochin type. *Beilstein Journal of Organic Chemistry*, **5**. doi: 10.3762/bjoc.5.71.

Speicher, A., Groh, M., Zapp, J., Schaumloffel, A., Knauer, M. and Bringmann, G. (2009b) Synthesis-driven structure revision of 'plagiochin E', a highly bioactive bisbibenzyl. *Synlett*, 1852–1858.

Speicher, A., Groh, M., Hennnrich, M. *et al.* (2010) Syntheses of macrocyclic bis(bibenzyl) compounds from perrottetin E. *European Journal of Organic Chemistry*, **35**, 6760–6778.

Sun, B., Yuan, H.Q., Xi, G.M. *et al.* (2009) Synthesis and multidrug resistance reversal activity of dihydroptychantol A and its novel derivatives. *Bioorganic and Medicinal Chemistry*, **17**, 4981–4989.

Taira, Z., Takei, M., Endo, K., Hashimoto, T., Sakiya, Y. and Asakawa, Y. (1994) Marchantin A trimethyl ether: its molecular structure and tubocurarine-like skeletal muscle relaxation activity. *Chemical and Pharmaceutical Bulletin*, **42**, 52–56.

Takiguchi, H., Ohmori, K. and Suzuki, K. (2011) Concise synthesis of riccardin C, macrocyclic bisbibenzyl natural product. *Chemistry Letters*, 1069–1071.

Tamehiro, N., Sato, Y., Suzuki, T. *et al.* (2005) Riccardin C: a natural product that functions as a liver X receptor (LXR)α agonist and an LXRβ antagonist. *FEBS Letters*, **579**, 5299–5304.

Tori, M., Toyota, M., Harrison, L.J., Takikawa, K. and Asakawa, Y. (1985) Total assignment of ^{1}H and ^{13}C NMR spectra of marchantins isolated from liverworts and its application to structure determination of two new macrocyclic bis(bibenzyls) from *Plagiochasma intermedium* and *Riccardia multifida*. *Tetrahedron Letters*, **26**, 4735–4738.

Wang, F.-Q. and Lou, H.-X. (2000) Chemical studies on the constituents of *Plagiochasma intermedium* L. *Acta Pharmaceutica Sinica*, **35**, 587–591.

Xie, C.-F., Qu, J.-B., Wu, X.-Z., Liu, N., Ji, M. and Lou, H.-X. (2010) Antifungal macrocyclic bis(bibenzyls) from the Chinese liverwort *Plagiochasma intermedium* L. *Natural Product Research*, **24**, 515–520.

Xing, J., Xie, C., Qu, J., Guo, H., Lv, B. and Lou, H.-X. (2007) Rapid screening for bisbibenzyls in bryophyte crude extracts using liquid chromatography/tandem mass spectrometry. *Rapid Communications in Mass Spectrometry*, **21**, 2467–2476.

Yoshida, T., Hashimoto, T., Takaoka, S., Kan, Y., Tori, M. and Asakawa, Y. (1996) Phenolic constituents of the liverwort: four novel cyclic bisbibenzyl dimers from *Blasia pusilla* L. *Tetrahedron*, **52**, 14487–14500.

Yoshida, T., Toyota, M. and Asakawa, Y. (1997) Isolation, structure elucidation, and chemical derivatization of new cyclic bisbibenzyl dimer, pusilatin E, from the liverwort *Riccardia multifida* subsp. *decrescens*. *Journal of Natural Products*, **60**, 145–147.

Zinsmeister, H.D., Becker, H. and Eicher, T. (1991) Bryophytes, a source of biologically active naturally occurring material? *Angewandte Chemie International Edition*, **30**, 130–147.

Chapter 3
Oxidation Mechanism of Polyphenols and Chemistry of Black Tea

Yosuke Matsuo and Takashi Tanaka

Abstract: The tea plant (*Camellia sinensis*) is unique in its ability to accumulate monomeric catechins with both pyrogallol-type and catechol-type B-rings. Black tea is produced by mechanical crushing of partially withered leaves, and enzymes in the leaves catalyze oxidation of catechins with air oxygen to generate B-ring *o*-quinones. The resulting electron-deficient quinones react with electron-rich tea catechins to produce oxidation products characteristic to black tea. Heating of the fermented leaves at the final drying stage is also important because some black tea polyphenols are produced by degradation of heat-sensitive quinone dimers. The composition of the oxidation products is extremely complex, and only part of the total black tea polyphenols have been chemically characterized to date. This review summarizes recent results of studies on the catechin oxidation mechanism, including the production mechanisms for theaflavins, theasinensins, and theacitrins. The spectroscopic properties of chemically uncharacterized black tea polyphenols, the so-called thearubigins, are also introduced.

Keywords: black tea, catechin, oxidation, theaflavin, theasinensin, thearubigins.

3.1 Introduction

Tea plant, *Camellia sinensis* (L.) O. Kuntze (Theaceae), is an evergreen tree originally distributed around the border of South China and Myanmar. The plant was selected for use as a medicine thousands of years ago, probably because it contains a high concentration of caffeine, similar to the way in which coffee beans were selected in Africa. Since tea was first exported from China to Europe in the 17th century, it has remained an important international trade commodity. Today, the tea plant is cultivated worldwide, with the main crop areas in China, India, Kenya, Sri Lanka, Turkey, Vietnam, Indonesia, Argentina, and Japan. Tea is now recognized as the most important polyphenol-rich beverage, because a growing

---

Recent Advances in Polyphenol Research, First Edition. Edited by Kumi Yoshida,
Véronique Cheynier and Stéphane Quideau.

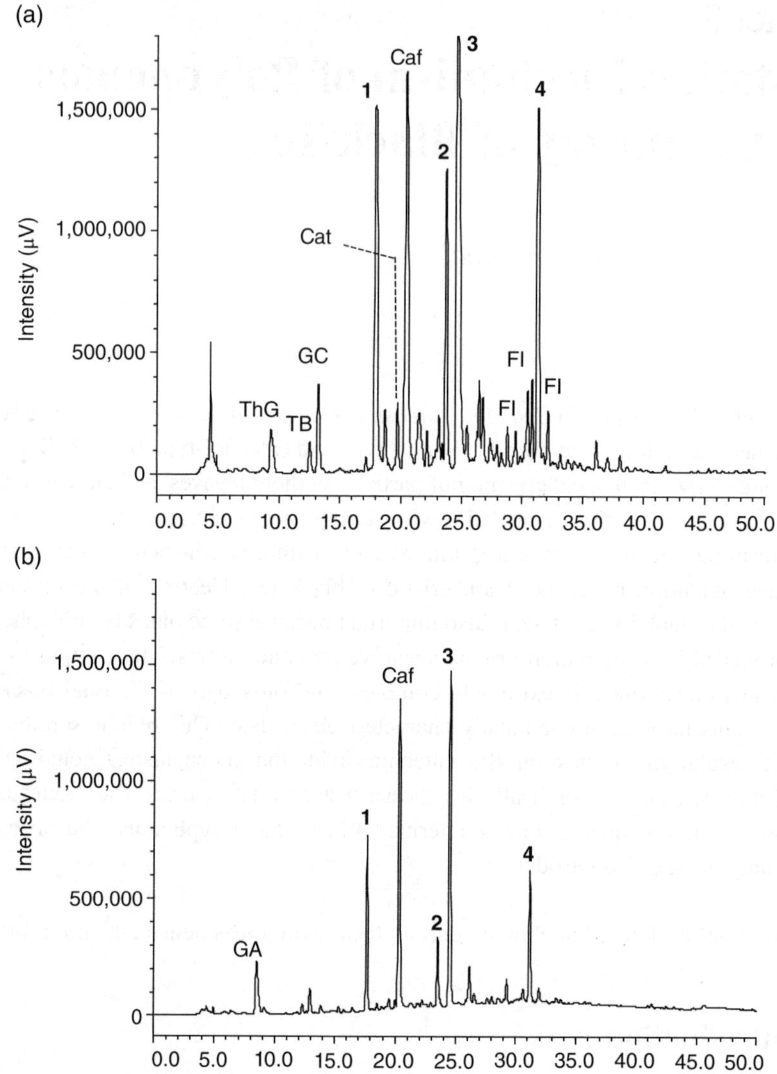

Fig. 3.1 Reversed-phase high-performance liquid chromatography (HPLC) profiles of major tea products (maximum absorbance): (a) green tea (Japan), (b) postfermented tea produced under aerobic conditions (Hunan, China), (c) oolong tea (Fujian, China), and (d) black tea (Assam, India). Caf, caffeine; Cat, catechin; Fl, unidentified flavonol glycoside; GA, gallic acid; GC, gallocatechin; GCg, gallocatechin-3-*O*-gallate; TB, theobromine; TCA, theacitrin A; ThG, theogallin; epigallocatechin (**1**), epicatechin (**2**), epigallocatechin-3-*O*-gallate (**3**), epicatechin-3-*O*-gallate (**4**), theaflavin (**5**), theaflavin-3-*O*-gallate (**6**), theaflavin-3′-*O*-gallate (**7**), theaflavin-3,3′-di-*O*-gallate (**8**), theasinensin A (**9**), theasinensin B (**10**), and theasinensin C (**11**). HPLC conditions: Cosmosil 5C_{18}-AR II (Nacalai Tesque, Japan) column (4.6 × 250 mm i.d.) with gradient elution from 4 to 30% (39 min) and from 30 to 75% (15 minutes) of CH_3CN in 50-mM H_3PO_4, flow rate, 0.8 ml/minute; column temperature, 35°C; detection, JASCO photodiode array detector MD-910.

body of evidence reveals that plant polyphenols in foods and medicinal plants contribute to human health and have done so for many years. Among plant polyphenols, catechins (flavan-3-ols) and proanthocyanidins are the most important groups because of their wide

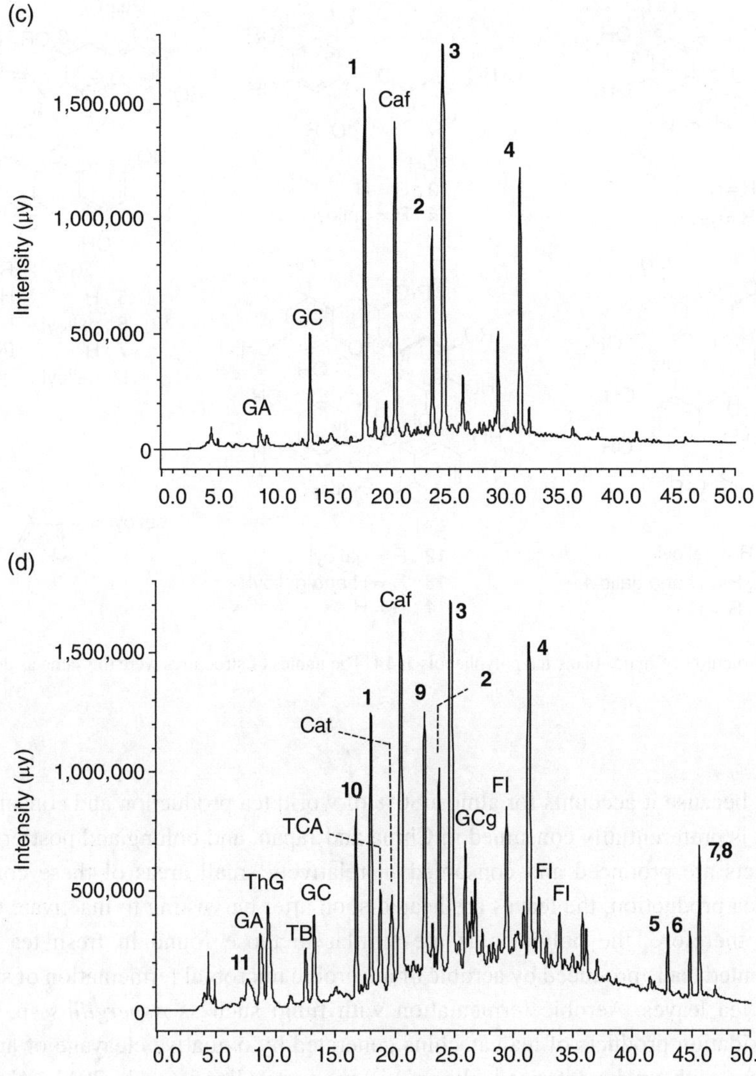

Fig. 3.1 (Continued)

distribution in the plant kingdom. In many polyphenol-rich plants, such as grape (Ricardo da Silva *et al.*, 1991), apple (Guyot *et al.*, 2003), and blueberry (Matsuo *et al.*, 2010), catechins coexist with proanthocyanidins. However, the tea plant is unique in accumulating four monomeric catechins, namely epigallocatechin (**1**) and epicatechin (**2**) and their galloyl esters (**3** and **4**) (Figs. 3.1a and 3.2), and the concentration of proanthocyanidins is relatively low (Hashimoto *et al.*, 1989, 1992).

Commercial tea products derived from the tea leaves are usually classified into four major categories: unfermented (green), fermented (black), semifermented (oolong), and postfermented (Pu-erh tea and related products) (Ho *et al.*, 2008). Although each category includes a wide variety of products, black tea is generally considered the most

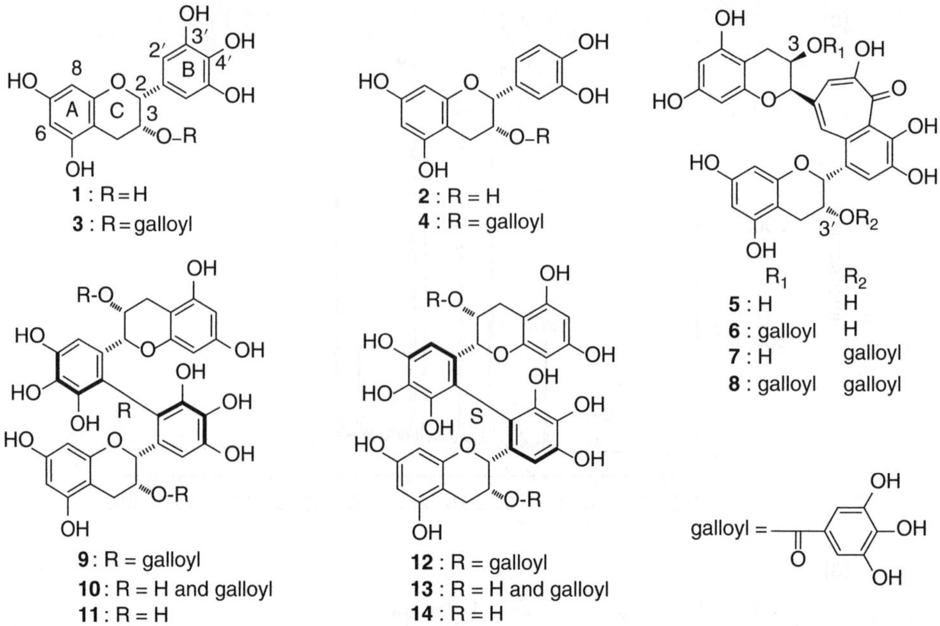

Fig. 3.2 Structures of major black tea polyphenols **1–14**. The names of structures were the same as described in Fig. 3.1.

important because it accounts for almost 80% of world tea production and consumption. Green tea is preferentially consumed in China and Japan, and oolong and postfermented tea products are produced and consumed in relatively small areas of these countries. In green tea production, the leaves are heated soon after harvesting to inactivate the leaf enzymes; therefore, the polyphenols are similar to those found in fresh tea leaves. Postfermented tea is produced by aerobic or anaerobic microbial fermentation of steamed or boiled tea leaves. Aerobic fermentation with fungi such as *Aspergillus* sp. mainly affords oxidation products of tea catechins generated by oxidative cleavage of aromatic rings, along with uncharacterized oligomeric products (Jiang *et al.*, 2011; Wulandari *et al.*, 2011). In contrast, the products of anaerobic fermentation with bacteria such as *Lactobacillus* sp. contain reduction products that are similar to the tea catechin metabolites produced by mammalian intestinal bacteria (Tanaka *et al.*, 2011, 2012). In the production of oolong tea, the fresh leaves are first subjected to drying stress and are then partially rolled and roasted to produce a characteristic flavor. The chromatographic profile of oolong tea obtained by high-performance liquid chromatography (HPLC) is similar to that of green tea but different from that of black tea to varying degrees among commercial products (Fig. 3.1). In oolong tea, the presence of characteristic oligomeric polyphenols with strong lipase inhibition activity has been reported (Nakai *et al.*, 2005). Black tea is usually categorized as a fermented tea, although the production process does not include microbial fermentation. Therefore, the process is sometimes called "tea fermentation." The key chemical process of the tea fermentation is catechin oxidation, which is conducted by enzymes originally contained in the tea leaves with oxygen molecules as oxidant. At the initial step of black tea production, the withered leaves are curled and crushed, and the catechins are mixed with oxidation enzymes, which are accumulated in different cells in the living leaf. In the subsequent tea-fermentation process, the leaves are mixed with air oxygen to oxidize catechins. This chapter summarizes our recent results of the study of the reaction mechanism of the enzymatic oxidation of tea catechins.

3.2 Catechin oxidation and production of theaflavins

Isolation of catechin oxidation products from commercial black tea and subsequent determination of the chemical structures have so far revealed the presence of characteristic polyphenols that are only found in black tea. In particular, the chemical characterization of the reddish-yellow pigmented theaflavins (**5–8**) featuring a benzotropolone chromophore was one of the most successful and notable results in food chemistry (Takino *et al.*, 1964, 1965; Bryce *et al.*, 1970; Collier *et al.*, 1973). These pigments can only be produced by oxidative coupling between pyrogallol-type and catechol-type B-rings via a bicyclo[3.2.1]octane intermediate (Yanase *et al.*, 2005) followed by decarboxylation (Fig. 3.3). These pigments are found only in black tea because the tea plant has the unique ability to accumulate monomeric catechins with both pyrogallol-type and catechol-type B-rings as the major polyphenols. Interestingly, theaflavins can be prepared by treatment of **1** and **2** with homogenates prepared from various fruits and vegetables (Tanaka *et al.*, 2002a). In our previous study, plant homogenates of Japanese pear, loquat, and banana each showed strong ability to produce theaflavin if only **1** and **2** were added. These fruits do not contain catechins; however, the enzymes of these plants preferentially use catechins as electron donors to reduce molecular oxygen. Japanese pear homogenate is well suited for use in catechin oxidation experiments because the homogenate shows no substantial background peaks by HPLC and can be easily performed on a large scale. Although the reactions observed in the model oxidation experiments are not identical to those observed in tea leaves, essentially the same products, such as theaflavins, theasinensins, and theacitrins, are obtained, even in large-scale experiments. The remarkable difference between the model experiments and oxidation in crushed tea leaves is the production of quinone dimer **1b** (Fig. 3.6) and its galloyl esters (Valcic *et al.*, 1999, 2000), which were major products in model experiments and were not detected in the crushed tea leaves.

Some proanthocyanidins, such as that found in persimmon (Tanaka *et al.*, 1994), are composed of catechin units with pyrogallol-type and catechol-type B-rings, and reddish-brown pigments are generated when the fruits are physically damaged. However, evidence for the production of benzotropolone units in proanthocyanidin molecules has not been observed (Tanaka *et al.*, 2005a). Oxidation of the B-rings in proanthocyanidin molecules leads to the formation of an A-type linkage by intramolecular addition of the A-ring hydroxy group to the C-2 of the upper unit (Morimoto *et al.*, 1987; Tanaka *et al.*, 2000a) (Fig. 3.3).

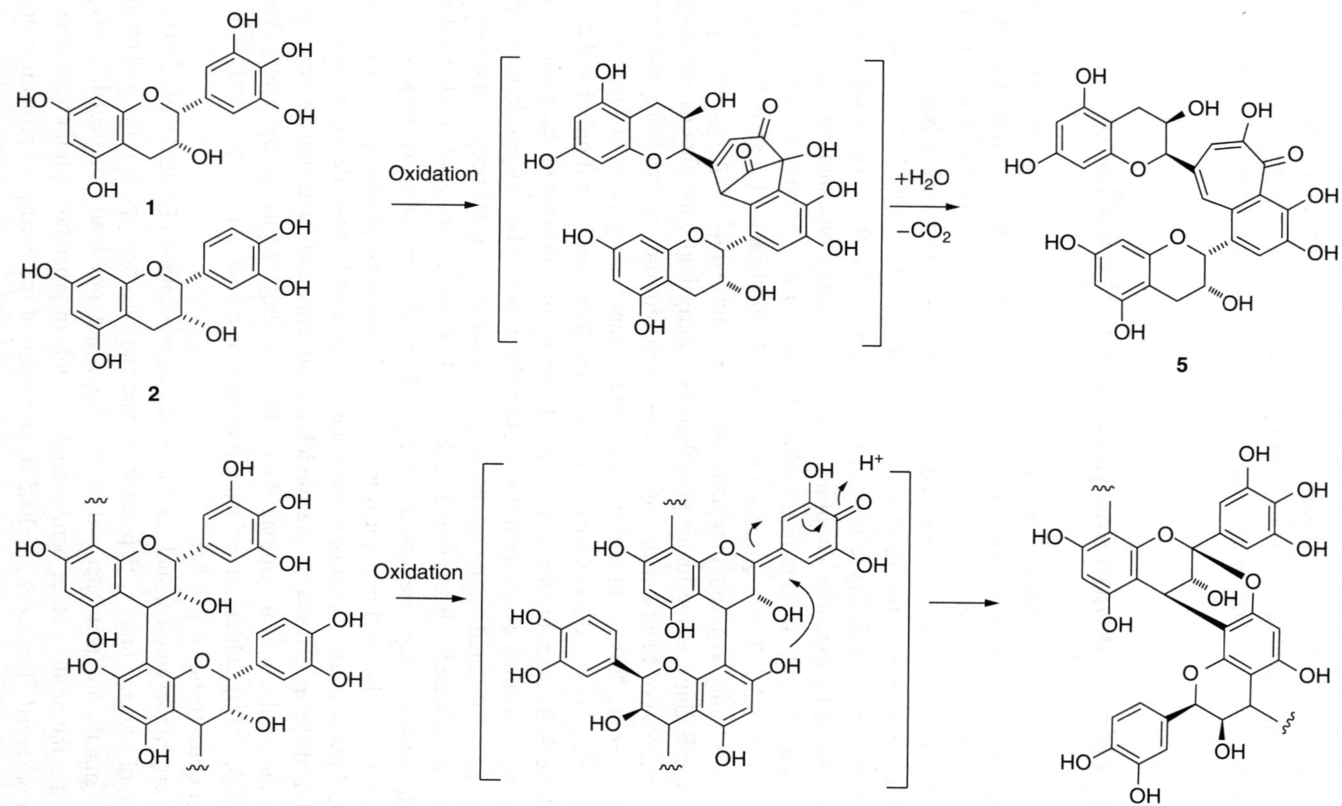

Fig. 3.3 Production of theaflavin and oxidation of proanthocyanidins.

3.3 Theasinensins

Theasinensins (**9–14**) are another major group of catechin dimers found in black tea at concentrations comparable with those of theaflavins. Production of these catechin dimers during tea fermentation was originally presumed by Roberts and Myers (1959), and their presence in black tea was confirmed chemically almost 10 years later (Ferretti *et al.*, 1968). The complete structures of theasinensins A–E and H (**9–14**), including the absolute configuration of biphenyl bonds, were established by Hashimoto *et al.* (Hashimoto *et al.*, 1988; Shii *et al.*, 2011b). Apparently, the chemical structures suggested that these products are produced by simple intermolecular phenol coupling between two pyrogallol-type B-rings of **1** and **3**. However, theasinensins are not detected in crushed tea leaves despite the appearance of prominent peaks in HPLC chromatograms attributable to theaflavins (Fig. 3.4a). Theasinensins were only detected in crushed leaves after heating at 80°C (Fig. 3.4c). Appearance of broad peaks before heating and disappearance of the peaks after heating indicated that theasinensins are produced by degradation of heat-sensitive precursors. Based on the assumption that the precursors have *o*-quinone structures (Fig. 3.5), the crushed tea leaves were directly treated with *o*-phenylenediamine and production of phenazine derivatives (**15a–c**) of theasinensin quinones was confirmed (Fig. 3.4b) (Tanaka *et al.*, 2002c).

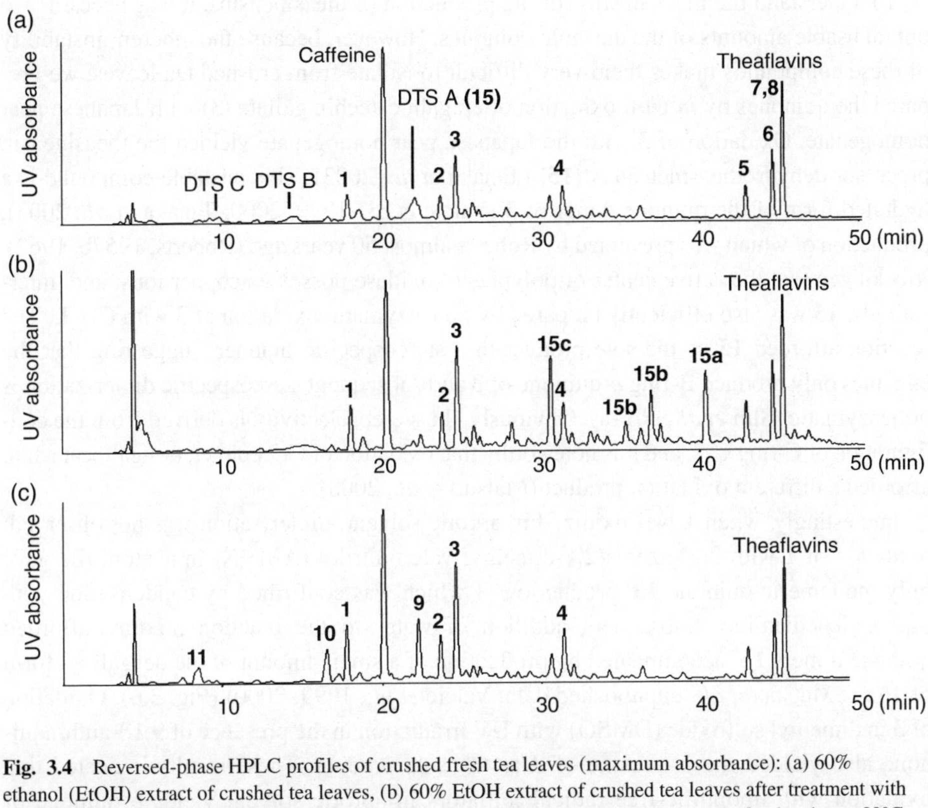

Fig. 3.4 Reversed-phase HPLC profiles of crushed fresh tea leaves (maximum absorbance): (a) 60% ethanol (EtOH) extract of crushed tea leaves, (b) 60% EtOH extract of crushed tea leaves after treatment with *o*-phenylenediamine–acetic acid/EtOH, and (c) 60% EtOH extract of crushed tea leaves after heating at 80°C. HPLC conditions were the same as described in Fig. 3.1.

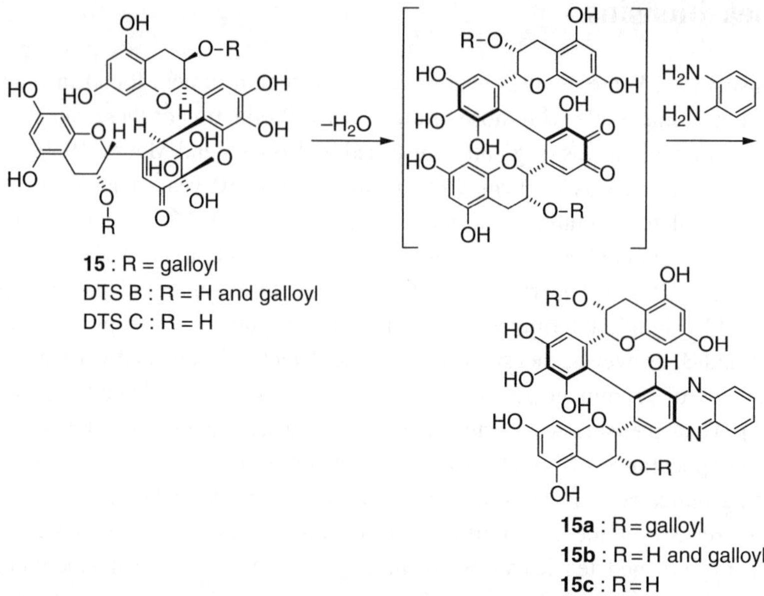

Fig. 3.5 Structures of dehydrotheasinensin A (**15**) and phenazine derivatives.

To understand the mechanisms for the production of theasinensins, it was necessary to obtain usable amounts of the unstable quinones. However, because the inherent instability of these compounds makes them very difficult to isolate from crushed tea leaves, we prepared the quinones by *in vitro* oxidation of epigallocatechin gallate (**3**) with Japanese pear homogenate. Oxidation of **3** with the Japanese pear homogenate yielded the theasinensin precursor dehydrotheasinensin A (**15**) (Tanaka *et al.*, 2003). This unstable compound is a hydrated form of the quinone dimer of **3** (Valcic *et al.*, 1999, 2000; Tanaka *et al.*, 2003), production of which was presumed by Roberts almost 50 years ago (Roberts, 1957b, 1962). It is known that the active center of polyphenol oxidase possesses copper ions, and, interestingly, **15** was also efficiently prepared by nonenzymatic oxidation of **3** with $CuCl_2$. The reaction afforded **15** as the sole product in a stereospecific manner, suggesting that the enzymes only produce B-ring *o*-quinone of **3** and subsequent stereospecific dimerization is nonenzymatic (Shii *et al.*, 2011a). Obviously, the stereoselectivity is derived from the configuration of C-ring C-2, and it is noteworthy that oxidation of C-2 epimer, (−)-gallocatechin, afforded a different oxidation product (Matsuo *et al.*, 2008).

Interestingly, when **1** was oxidized in aprotic solvent, dimerization was not observed: oxidation of **1** with 2,2′-azobis(2,4-dimethyl-valeronitrile) (AMVN) in acetonitrile gave only monomeric quinone **1a**, production of which was confirmed by condensation with *o*-phenylenediamine. Subsequent addition of water to the reaction mixture afforded quinone dimers **1b**, accompanied by production of a small amount of the desgalloyl form of **15** (Y. Matsuo *et al.*, unpublished data; Valcic *et al.*, 1999, 2000) (Fig. 3.6). Oxidation of **3** in dimethyl sulfoxide (DMSO) with UV irradiation in the presence of 9,10-anthraquinone also yielded the *o*-quinone of **3** as the sole product. These results indicated that oxidation with lipophilic-free radical initiators in aprotic solvent yields *o*-quinone of

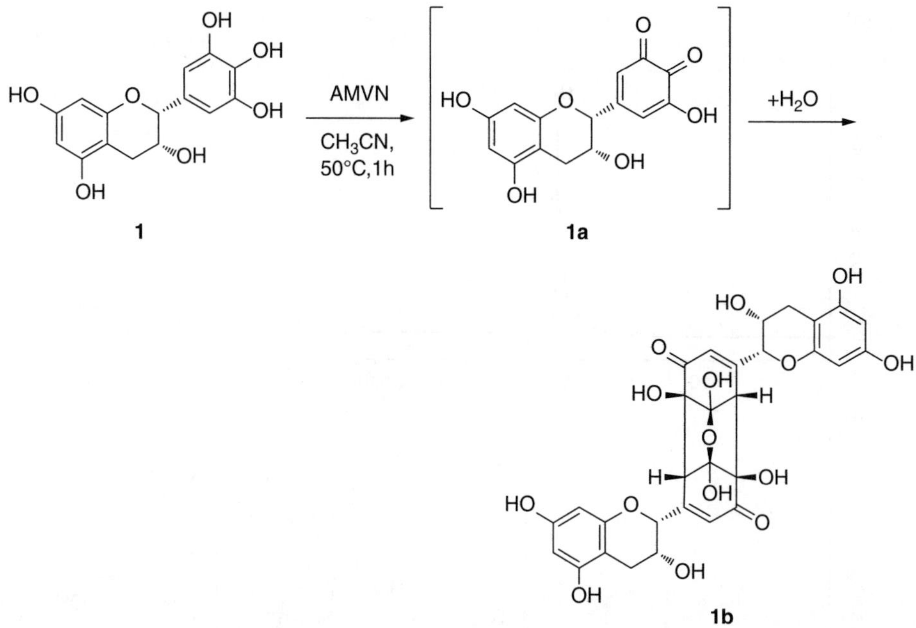

Fig. 3.6 Oxidation of **1** with radical initiator. AMVN: 2,2′-azobis(2,4-dimethyl-valeronitrile).

B-ring as the major product, and hydration of the quinone carbonyl group is necessary to form quinone dimers, such as **15** and **1b**.

Reduction of **15** with ascorbic acid yielded theasinensin A (**9**) in quantitative yield. Conversely, heating of **15** in aqueous solution, as a mimic of the final drying stage in commercial black tea production, afforded **9** and its atropisomer theasinensin D (**12**). The reaction was accompanied by isomerization of **15** to give a black tea pigment dehydrotheasinensin AQ (**16**) (Fig. 3.7). Theasinensins **9** and **12** are reduction products, and therefore, the occurrence of reduction–oxidation disproportionation on heating was suggested. Spontaneous degradation of **15** in aqueous phosphate buffer at pH 7.0 similarly yielded **9** and **12** together with oxidation products, such as galloyl oolongtheanin (**17**) (Matsuo *et al.*, 2015) and **18** (Tanaka *et al.*, 2003) (Fig. 3.8). However, the yield of the oxidation products isolated in the experiment was much lower than the yield of reduction products (Tanaka *et al.*, 2003). Production of many minor degradation products, including oligomeric substances observed as a broad hump on the HPLC baseline, was presumed (Fig. 3.7).

3.4 Coupled oxidation mechanism

The redox potential of the pyrogallol-type B-ring of **1** is much lower than that of catechol-type B-ring of **2** (Roberts, 1957a), making **1** more susceptible to oxidation than **2**. This was corroborated by the fact that only **1** and **3** were oxidized when tea catechin mixture was treated with $CuCl_2$ (Shii *et al.*, 2011a). However, on enzymatic oxidation with tea leaf enzyme or Japanese pear homogenate, **2** and **4** were oxidized to their quinones and

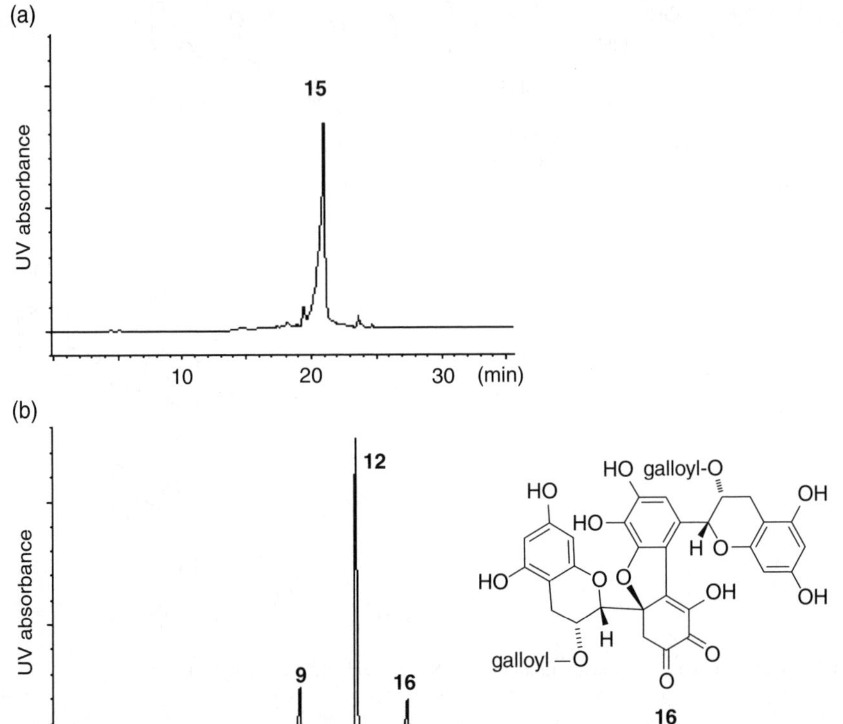

Fig. 3.7 Degradation of dehydrotheasinensin A (**15**) on heating: (a) before heating and (b) after heating of aqueous solution of **15** at 90°C for 30 minutes. HPLC column: Develosil C30-UG-5 (250 × 4.6 mm, Nomura Chemical) with gradient elution 4–30% (39 minutes) and 30–75% (15 minutes) of CH_3CN in 50-mM H_3PO_4; flow rate, 0.8 ml/minute; column temperature, 35°C; detection, JASCO photodiode array detector MD-910.

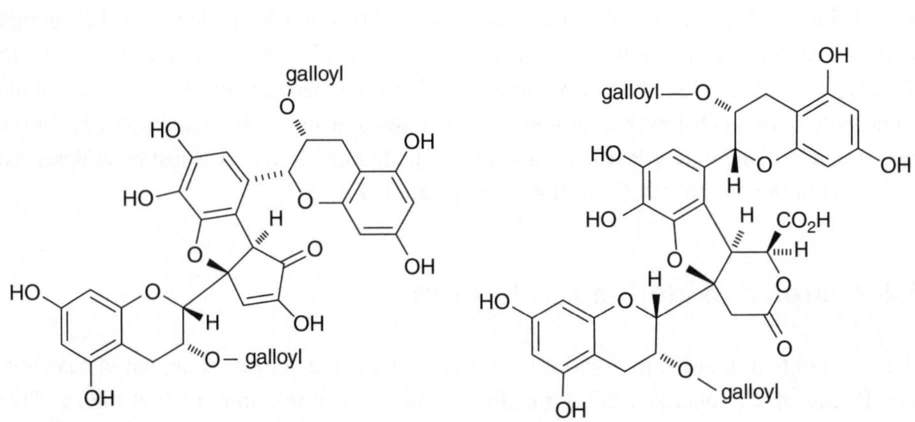

17 **18**

Fig. 3.8 Structures of oxidation products of **15**.

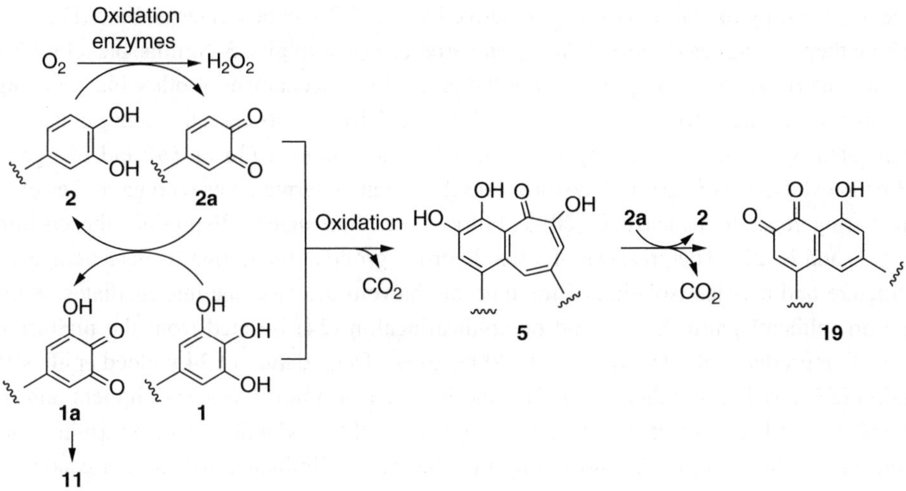

Fig. 3.9 Production and oxidation of theaflavin by coupled oxidation of **1** and **2**.

subsequently coupled with **1** and **3** to produce theaflavins (Fig. 3.9). This phenomenon was attributed to the so-called coupled oxidation mechanism based on higher enzyme specificity toward catechol ring types (Roberts, 1957a; Robertson, 1983). As the enzyme oxidizes the catechol-type B-ring, the resulting *o*-quinone **2a** is reduced to catechol ring (**2**) by electron transfer from pyrogallol ring (**1**), which is in turn oxidized to the *o*-quinone (**1a**). The apparent decrease of **2** in the reaction mixture is much slower than the decrease of **1**; however, the addition of glutathione to the mixture to trap the quinones yielded the glutathione adduct of **2a** as a major product, confirming the rapid oxidation of **2**. This mechanism was further evidenced by oxidation of myricitrin (3,5,7,3′,4′,6′-hexahydroxyflavone 3-*O*-rhamnoside): the flavonol was not oxidized by enzymes directly; however, addition of (+)-catechin to the mixture dramatically accelerated the oxidation to give dimeric product structurally related to theasinensins (Roberts, 1957a; Takino & Imagawa, 1963). When a mixture of **1** and **2** was stirred with Japanese pear homogenate, the quinone **2a** oxidized **1**, and the concentration of **1** decreased much faster than **2**. The *o*-quinone **1a** was consumed to produce dehydrotheasinensin C (Matsuo *et al.*, 2006). When **1** was completely consumed, **2a** began to oxidize **5**, and the major oxidation product was theanaphthoquinone (**19**) (Tanaka *et al.*, 2000b, 2002a).

3.5 Bicyclo[3.2.1]octane intermediates

Theaflavin production starts from the formation of an electron-transfer complex between electron-deficient epicatechin quinone (**2a**) and electron-rich epigallocatechin (**1**) (Haslam, 1998; Drynan *et al.*, 2010). Subsequent carbon–carbon bond formation in the complex accompanied by oxidation affords a putative bicyclo[3.2.1]octane intermediate (Fig. 3.3), which then undergoes decarboxylation and aromatization to give **5**. Similar bicyclo[3.2.1] octane intermediates were presumed in the production mechanism of other black tea pigments, such as theacitrins (Davis *et al.*, 1997) (Fig. 3.10). In this case, the intermediate was generated by oxidative coupling between pyrogallol quinone (**1a** or **3a**) and pyrogallol (**1** or **3**) (Matsuo *et al.*, 2011). The bicyclo[3.2.1]octane intermediates with galloyl esters at the C-ring generate theacitrin C (**20**) and further rearrangement of **20** produce theacitrinins A (**21**) and B (**22**). The presence of a free hydroxy group at the C-ring forms a hemiacetal structure and enables isolation of the unstable bicyclo[3.2.1]octane intermediates, which are proepitheaflagallin B (**23**) and proepitheaflagallin (**24**) isolated from the mixture of oxidation products of **1** (Matsuo *et al.*, 2006, 2009). Degradation of **24** yielded epitheaflagallin (**25**) and hydroxytheaflavin (**26**): the former is a minor black tea pigment and the latter has not been detected in black tea to date. In the production of theacitrinin A and epitheaflagallin, a set of A- and C-ring moieties were eliminated and the corresponding hydroxychroman analog (**27**) was isolated from black tea (Tanaka *et al.*, 2003).

3.6 Structures of catechin oxidation products

As shown in the HPLC profile of black tea (Fig. 3.1d), the composition of black tea polyphenols is extremely complex and numerous small peaks attributed to phenolic substances showing similar UV absorptions are observed. Separation of each peak component from the black tea extract is difficult and impractical because of the presence of a large amount of oligomeric substance, which is detected as a broad hump on the HPLC baseline. Therefore, *in vitro* enzymatic oxidation of purified catechins was effective in understanding the catechin oxidation mechanism. By applying this method under various conditions, many oxidation products were characterized and each production mechanism was proposed (Fig. 3.11). Dehydrotheasinensin E (**28**) is a yellow pigment produced by oxidation of **1** with crude enzyme prepared from fresh tea leaf homogenate (Tanaka *et al.*, 2002a; Shii *et al.*, 2011b). Compounds **29** (Tanaka *et al.*, 2005b) and **30** (Tanaka *et al.*, 2002b) are dimers of **3** with aliphatic cage-like structures, and compounds **31–33** were minor oxidation products of **1** obtained by oxidation with Japanese pear homogenate (Matsuo *et al.*, 2013). Compound **34** is a minor oxidation product of **3**, indicating that the galloyl group also participates in oxidation (Tanaka *et al.*, 2005b; Li *et al.*, 2007). However, our *in vitro* experiments strongly suggested that the contribution of the galloyl group to oxidative coupling is limited, and dimeric oxidation products are generated by oxidative coupling between two B-rings of four tea catechins.

Theaflavin (**5**) is not a final oxidation product and is further oxidized to give theanaphthoquinone (**19**) together with **35** and **36**, which are produced by oxidative coupling between two molecules of **5** and condensation between **5** and **19**, respectively (Tanaka *et al.*, 2001). When the *in vitro* enzymatic oxidation of theaflavin was terminated by addition of ethanol to the reaction mixture, some products possessing ethoxy groups were obtained. The products indicated oxidation of the benzotropolone ring and nucleophilic 1,4-addition of the benzotropolone to catechol *o*-quinones (Li *et al.*, 2010).

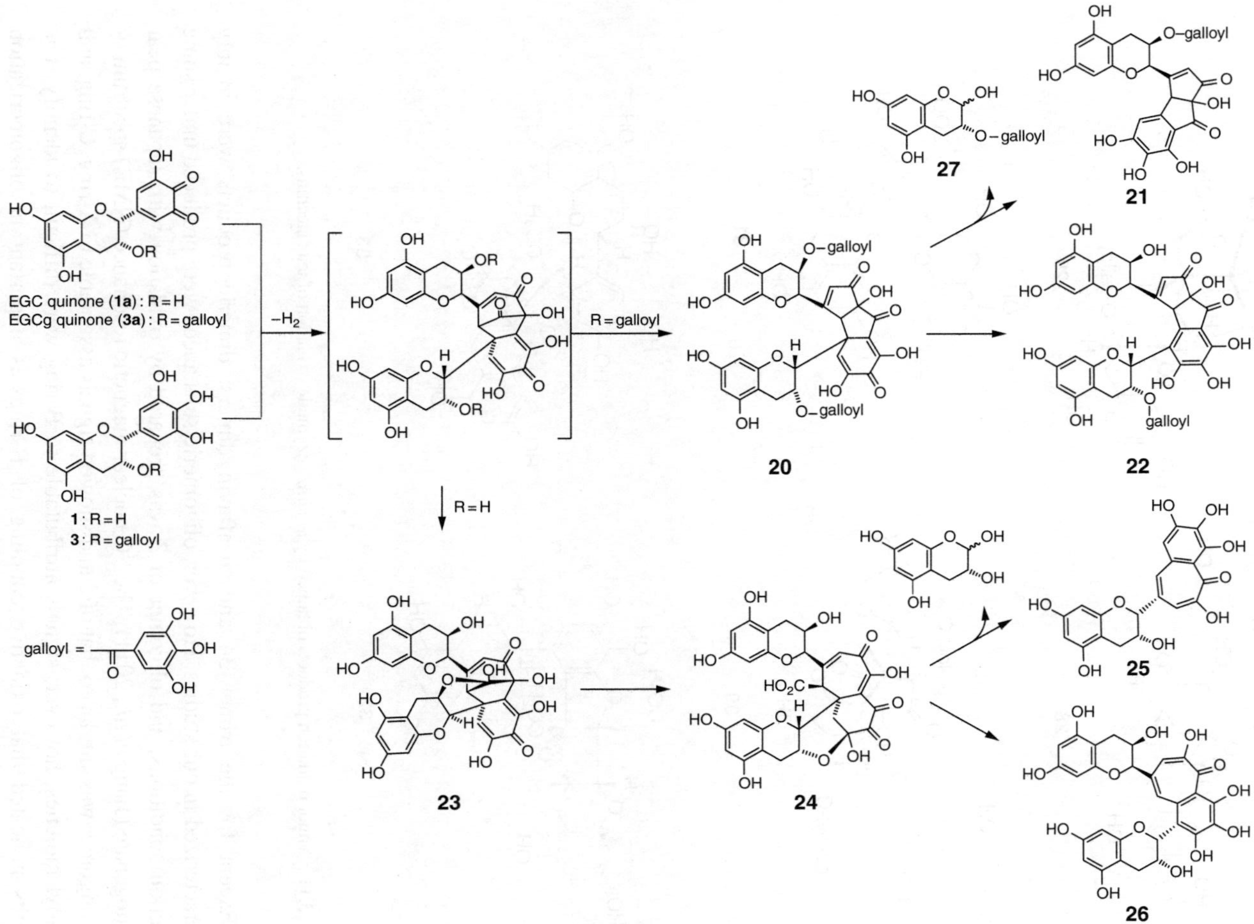

Fig. 3.10 Production of black tea polyphenols from bicyclo[3.2.1]octane intermediates. EGC, epigallocatechin.

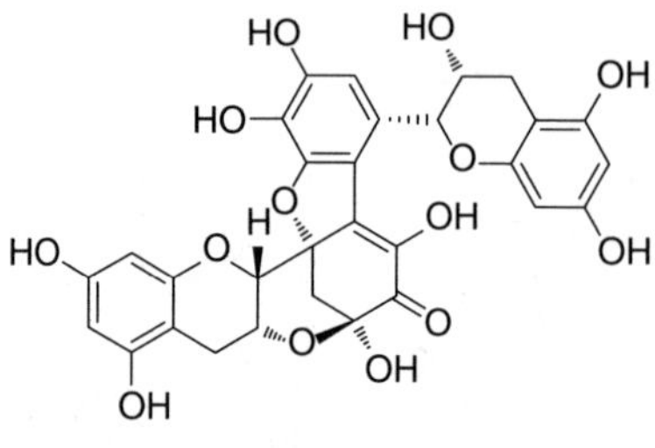

28

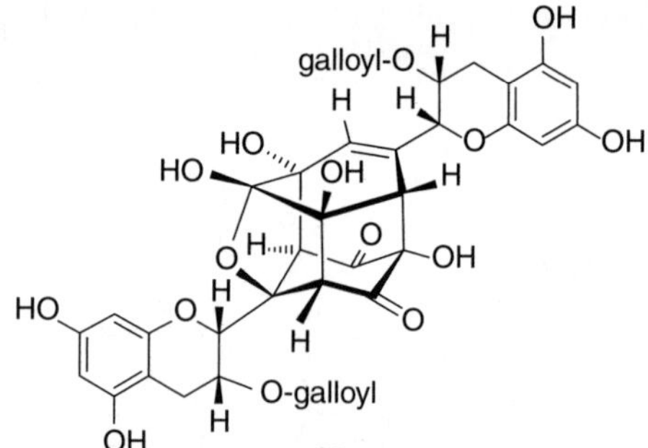

29

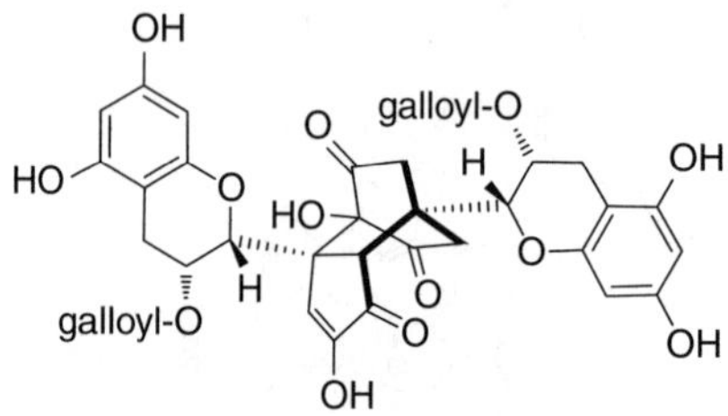

30

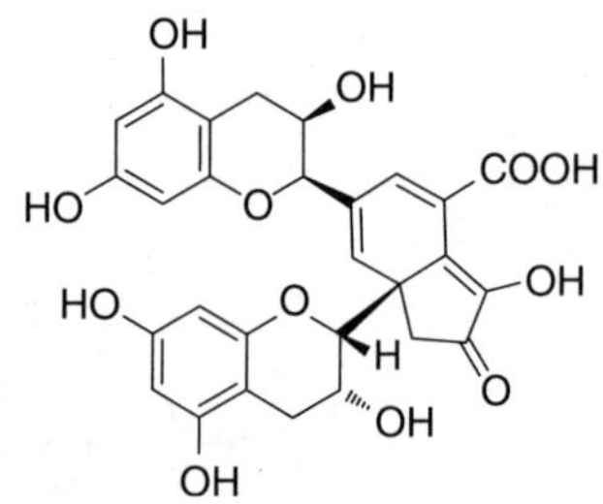

31

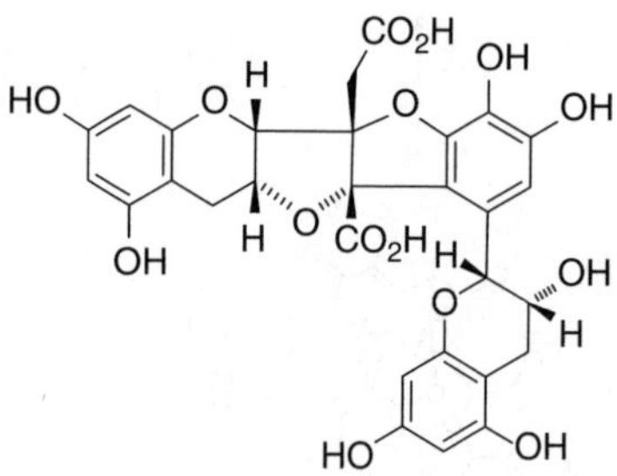

32

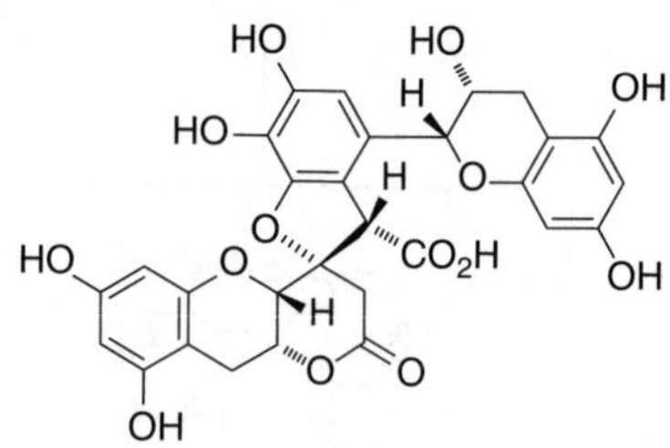

33

Fig. 3.11 Minor oxidation products obtained by *in vitro* enzymatic oxidation of tea catechins.

Except for the trimer **34** and theaflavin dimers, dimeric products were mainly characterized in our studies. However, oligomeric substances were produced under some reaction conditions, and oligomer of **3** was prepared by oxidation with Japanese pear homogenate (Jiang *et al.*, 2011). The ^{13}C nuclear magnetic resonance (NMR) spectrum of the oligomer was similar to that of **3** and showed signals assignable to A-ring, C-ring, and galloyl moieties; however, signals attributable to B-ring were difficult to identify. The results indicated that oxidative coupling of B-rings is important in oligomerization mechanisms.

Because **1** and **3** account for about 70% of total tea catechins, the oxidation of pyrogallol is a very important step in the production of black tea polyphenols. However, it should be noted that enzymatic oxidation of epicatechin (**2**) alone produces dimers, trimers (**37**), and

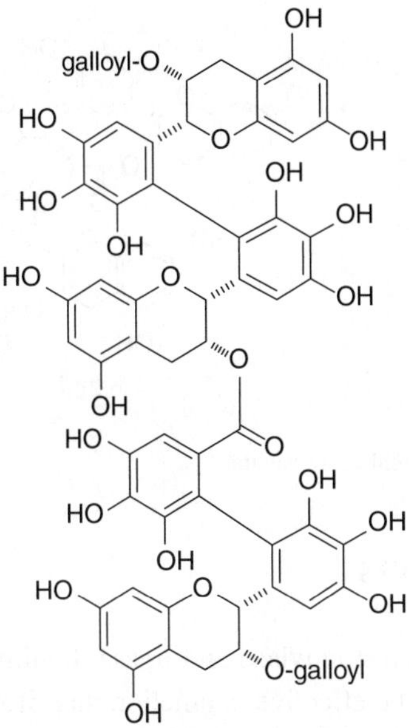

34

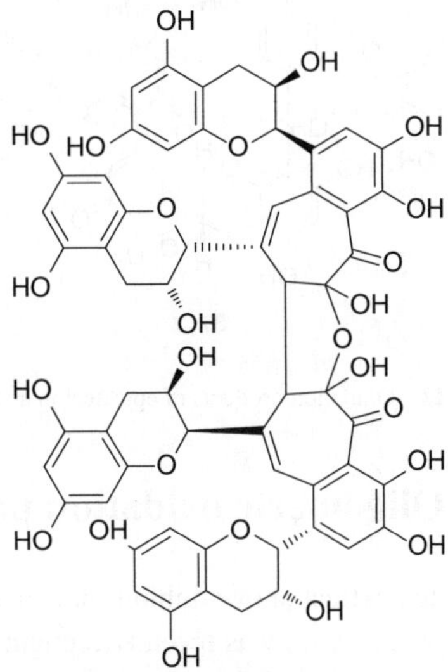

35

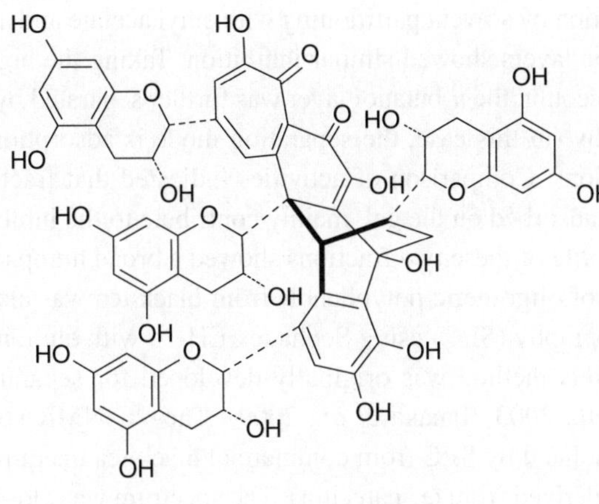

36

Fig. 3.11 (Continued)

oligomers by oxidative condensation, mainly between A- and B-rings (Guyot *et al.*, 1996; Tanaka *et al.*, 2002a) (Fig. 3.12). Furthermore, enzymatic oxidation of **4** alone affords condensation products, such as **38**, with a benzotropolone ring produced by oxidative coupling between galloyl groups and the catechol B-rings (Wan *et al.*, 1997; Lewis *et al.*, 1998; Sang *et al.*, 2002, 2004; Kusano *et al.*, 2007). In the oxidation of **4**, coupling products between A- and B-rings were not isolated. In tea fermentation, **4** is less reactive than other catechins.

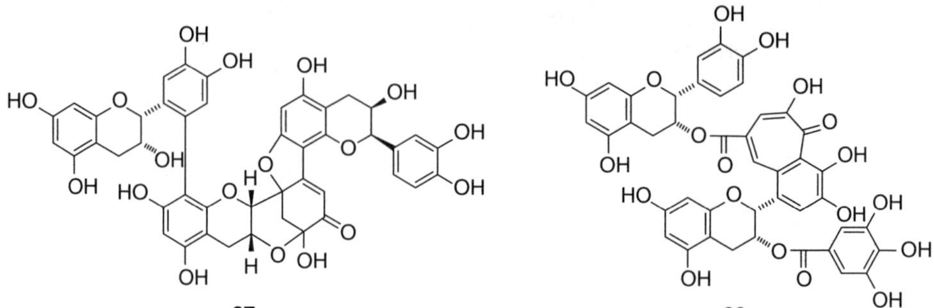

37 **38**

Fig. 3.12 Oxidation products of epicatechin and epicatechin-3-*O*-gallate.

3.7 Oligomeric oxidation products

Black tea extract shows inhibition activity against amylase and lipase. Inhibition of these digestive enzymes was recently recognized to be effective in ameliorating lifestyle-related diseases, such as diabetes and hyperlipidemia. Activity-guided fractionation of black tea extract has indicated that oligomeric oxidation products of tea catechins show strong inhibition activities (Kusano *et al.*, 2008; Shii *et al.*, 2011b). In comparison of inhibition activities after fractionation by solvent partitioning with ethyl acetate and *n*-butanol, it was found that the two organic layers showed similar inhibition. Taking the higher yield of the *n*-butanol fraction into account, the *n*-butanol layer was further separated by Sephadex LH-20 column chromatography (in this case, the separation mode is adsorption) (Yanagida et al., 2003) into five fractions. Comparison of activities indicated that fractions 4 and 5, which were most strongly adsorbed on the gel, mainly contribute to the inhibition activity of black tea. The HPLC profile of these two fractions showed a broad hump on the baseline (Fig. 3.13).

Separation of oligomeric polyphenols from black tea was also achieved by size-exclusion chromatography (SEC) using Sephadex LH-20 with elution by aqueous acetone containing urea; this method was originally developed for separation of proanthocyanidins (Yanagida *et al.*, 2003; Tanaka *et al.*, 2009). The ^{13}C NMR spectrum of the oligomeric polyphenols isolated by SEC from commercial black tea unequivocally confirmed that the oligomer was derived from tea catechins. The spectrum was closely related to the spectrum of **3** (Fig. 3.14) except for the absence of signals for the pyrogallol-type B-ring C-2,6 carbons. Signals attributable to catechol-type B-rings indicated that the oligomer contains units derived from **2** and **4**. Elongation of the molecules by intermolecular formation of a benzotropolone ring as observed in oxidation of **4** was suggested (Menet *et al.*, 2004); however, signals attributable to benzotropolone moieties were not observed in the spectrum. Elemental analysis indicated that the presence of nitrogen atom was negligible, suggesting that proteins and amino acids do not participate in the oligomerization. These results strongly suggest that oxidative coupling of pyrogallol-type B-rings plays a key role in the formation of oligomeric polyphenols, which is probably the same as, or similar to, the thearubigins, a chemically undefined group of catechin oxidation products in black tea (Roberts, 1962; Haslam, 2003; Stodt & Engelhardt, 2013).

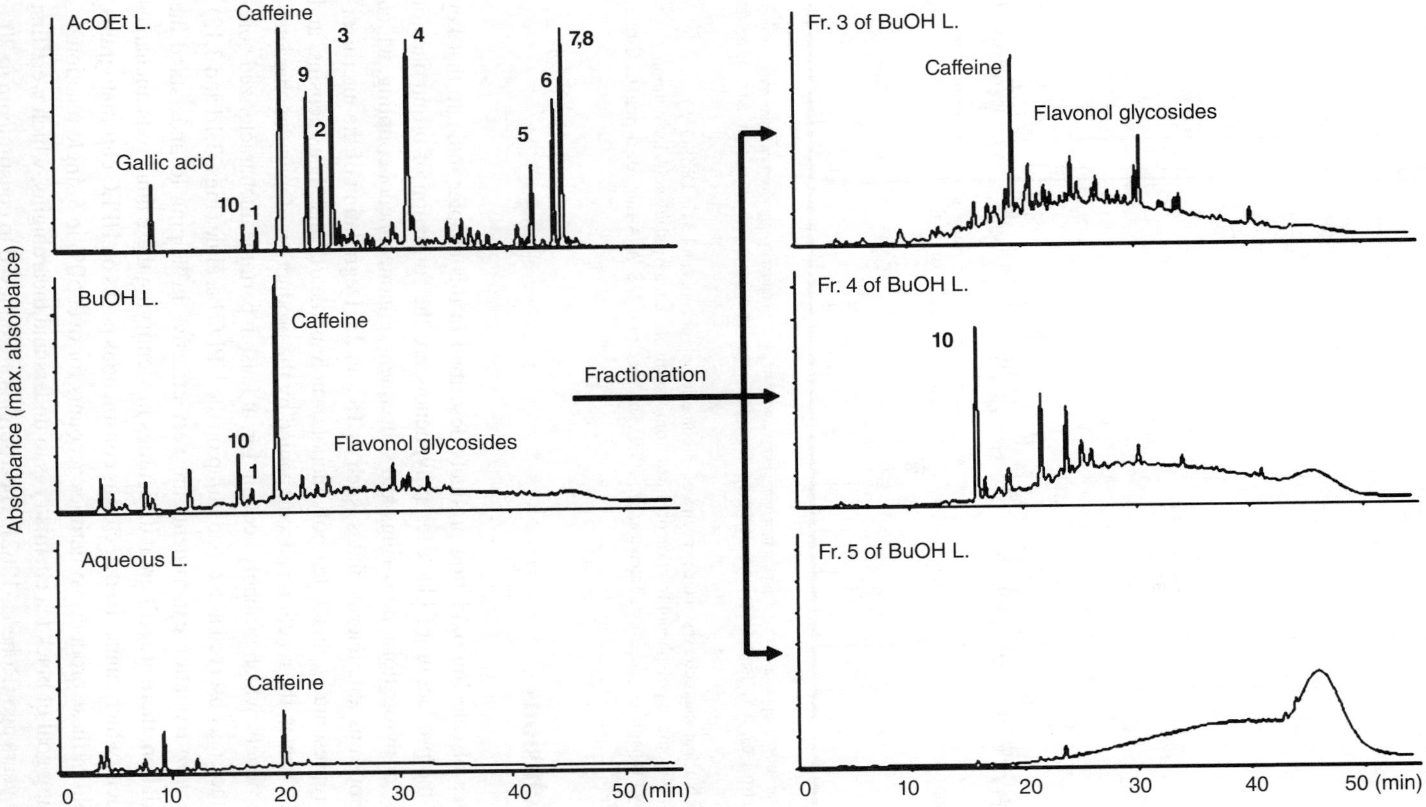

Fig. 3.13 Reversed-phase HPLC profiles of black tea fractions. Aqueous acetone extract of black tea was successively partitioned with ethyl acetate (AcOEt) and *n*-butanol (*n*-BuOH) (yields from black tea: AcOEt layer, 9%; *n*-BuOH layer, 17%; aqueous layer, 21%). The *n*-BuOH fraction was further fractionated by Sephadex LH-20 column chromatography into five fractions (yields from *n*-BuOH layer: Fr. 1 (caffeine), 43%; Fr. 2 (gallic acid and flavonol glycoside), 8.7%; Fr. 3, 8.7%; Fr. 4, 21%; Fr. 5, 16.4%). HPLC conditions were the same as described in Fig. 3.1.

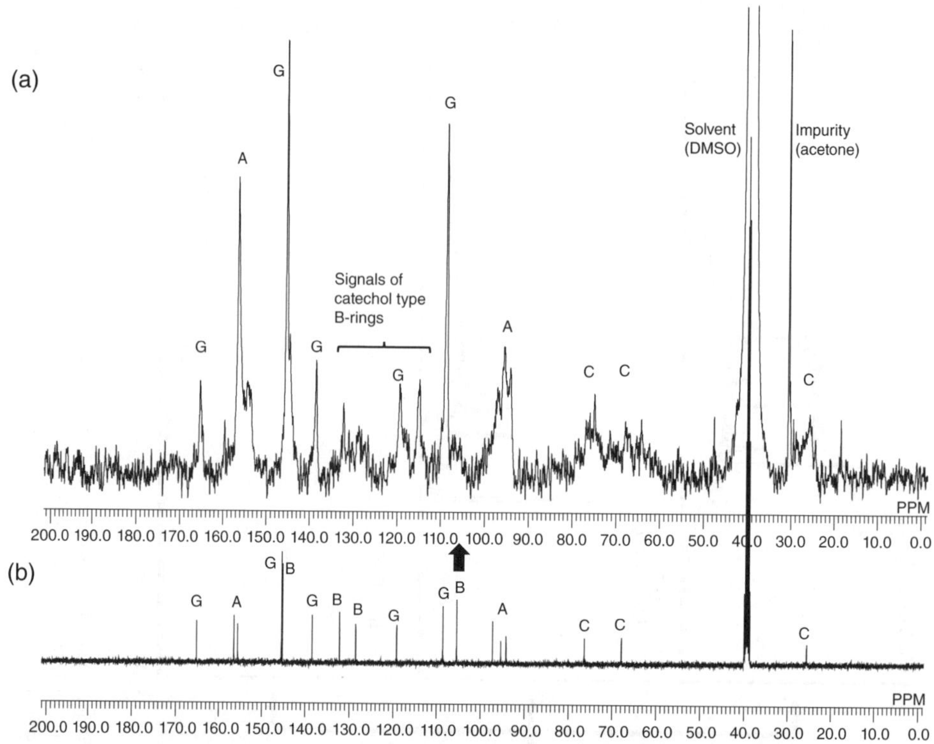

Fig. 3.14 ^{13}C Nuclear magnetic resonance spectra of oligomeric polyphenols and **3** (in DMSO-d_6). (a) Spectrum of oligomeric polyphenol fraction obtained from Assam black tea by size-exclusion column chromatography (Yanagida *et al.*, 2003; Tanaka *et al.*, 2009). (b) Spectrum of **3**. A, A-ring; B, B-ring; C, C-ring; G, galloyl group.

3.8 Conclusions

The structures of catechin oxidation products described in this chapter indicate that key reactions in the production of black tea polyphenols are the formation of *o*-quinones of catechol-type or pyrogallol-type B-rings and subsequent quinone–phenol coupling, which generate various unstable intermediates, such as **15** and **23**. Degradation of the intermediates affords representative black tea polyphenols, such as theaflavins, theasinensins, and theacitrins. However, the reaction is accompanied by the production of many uncharacterized minor products, which probably account for, at least in part, the uncharacterized polyphenols of black tea observed as a broad hump on the HPLC baseline (Figs. 3.1d and 3.13). Although *in vitro* oxidation experiments are very effective in helping to understand the catechin oxidation, there are still great difficulties in identifying these numerous uncharacterized products, which cannot be detected as conspicuous peaks on HPLC chromatograms. The total yield of these products sometimes accounts for over 50% in a single reaction.

A remaining goal of black tea chemistry is to understand thearubigins, which were first discussed 50 years ago (Roberts, 1962). Thearubigins are believed to constitute up to 20% of the black tea leaf and up to 60% of the solids in a black tea infusion (Haslam, 1998).

The oligomeric polyphenols obtained by solvent partitioning or SEC of black tea extract probably form the main body of thearubigins. Spectroscopic data indicated that the substance is mostly catechin oligomers produced by oxidative coupling of B-rings. However, the results of our *in vitro* experiments seemed to produce dimers and oligomers. There is no evidence that oligomeric polyphenols are produced by sequential elongation of the molecule (i.e., trimers, tetramers, and oligomers). Indeed, theasinensin A (**9**) was not oxidized under the conditions where epigallocatechin gallate (**3**) was oxidized. The production mechanism of oligomeric polyphenols may be independent of the production of theaflavins and theasinensins, and further studies by different strategies are probably needed to understand thearubigins. It should be noted here that recent studies of thearubigins using ESI–tandem and MALDI–TOF mass spectrometry have revealed new aspects about thearubigins, which indicated presence of polyhydroxylated catechin dimers in thearubigin fractions (Yassin *et al.*, 2014a). Furthermore, the results also suggested occurrence of intermolecular oxidative couplings between A-rings and pyrogallol-type B-rings (Yassin *et al.*, 2014b), which may explain production of oligomers from dimeric products.

Acknowledgments

This work was supported by Grants-in-Aid for Scientific Research Nos. 25870532, 22700734, and 18510189 from the Japan Society for the Promotion of Science.

References

Bryce, T., Collier, P.D., Fowlis, I., Thomas, P.E., Frost, D. and Wilkins, C.K. (1970) The structures of the theaflavins of black tea. *Tetrahedron Letters*, **11**, 2789–2792.

Collier, P.D., Bryce, T., Mallows, R. *et al.* (1973) The theaflavins of black tea. *Tetrahedron*, **29**, 125–142.

Davis, A.L., Lewis, J.R., Cai, Y. *et al.* (1997) A polyphenolic pigment from black tea. *Phytochemistry*, **46**, 1397–1402.

Drynan, J.W., Clifford, M.N., Obuchowicz, J. and Kuhnert, N. (2010) The chemistry of low molecular weight black tea polyphenols. *Natural Product Reports*, **27**, 417–462.

Ferretti, A., Flanagan, V.P., Bondarovich, H.A. and Gianturco, M.A. (1968) The chemistry of tea, structures of compounds A and B of Roberts and reactions of some model compounds. *Journal of Agricultural and Food Chemistry*, **16**, 759–761.

Guyot, S., Vercauteren, J. and Cheynier, V. (1996) Structural determination of colourless and yellow dimers resulting from (+)-catechin coupling catalysed by grape polyphenoloxidase. *Phytochemistry*, **42**, 1279–1288.

Guyot, S., Marnet, N., Sanoner, P. and Drilleau, J. (2003) Variability of the polyphenolic composition of cider apple (*Malus domestica*) fruits and juices. *Journal of Agricultural and Food Chemistry*, **51**, 6240–6247.

Hashimoto, F., Nonaka, G. and Nishioka, I. (1988) Tannins and related compounds. LXIX. Isolation and structure elucidation of B, B'-linked bisflavonoids, theasinensins D–G and oolongtheanin from oolong tea. (2). *Chemical and Pharmaceutical Bulletin*, **36**, 1676–1684.

Hashimoto, F., Nonaka, G. and Nishioka, I. (1989) Tannins and related compounds. LXXVII. Novel chalcan–flavan dimers, assamicains A, B and C, and a new flavan-3-ol and proanthocyanidins from the fresh leaves of *Camellia sinensis* L. var. *assamica* KITAMURA. *Chemical and Pharmaceutical Bulletin*, **37**, 77–85.

Hashimoto, F., Nonaka, G. and Nishioka, I. (1992) Tannins and related compounds. CXIV. Structures of novel fermentation products, theogallinin, theaflavonin and desgalloyl theaflavonin from black tea, and changes of tea leaf polyphenols during fermentation. *Chemical and Pharmaceutical Bulletin*, **40**, 1383–1389.

Haslam, E. (1998) Quinone tannin and oxidative polymerization, in *Practical Polyphenolics: from Structure to Molecular Recognition and Physiological Action* (ed E. Haslam), Cambridge University Press, Cambridge, pp. 335–373.

Haslam, E. (2003) Thoughts on thearubigins. *Phytochemistry*, **64**, 61–73.

Ho, C.T., Lin, J.K. and Shahidi, F. (2008) *Tea and tea products. Chemistry and health-promoting properties*, CRC Press, Boca Raton, FL.

Jiang, H.Y., Shii, T., Matsuo, Y., Tanaka, T., Jiang, Z.H. and Kouno, I. (2011) A new catechin oxidation product and polymeric polyphenols of post-fermented tea. *Food Chemistry*, **129**, 830–836.

Kusano, R., Tanaka, T., Matsuo, Y. and Kouno, I. (2007) Structures of epicatechin gallate trimer and tetramer produced by enzymatic oxidation. *Chemical and Pharmaceutical Bulletin*, **55**, 1768–1772.

Kusano, R., Andou, H., Fujieda, M., Tanaka, T., Matsuo, Y. and Kouno, I. (2008) Polymer-like polyphenols of black tea and their lipase and amylase inhibitory activities. *Chemical and Pharmaceutical Bulletin*, **56**, 266–272.

Lewis, J.R., Davis, A.L., Cai, Y., Davies, A.P., Wilkins, J.P.G. and Pennington, M. (1998) Theaflavate B, isotheaflavin-3'-*O*-gallate and neotheaflavin-3-*O*-gallate: three polyphenolic pigments from black tea. *Phytochemistry*, **49**, 2511–2519.

Li, Y., Tanaka, T. and Kouno, I. (2007) Oxidative coupling of the pyrogallol B-ring with a galloyl group during enzymatic oxidation of epigallocatechin 3-*O*-gallate. *Phytochemistry*, **68**, 1081–1088.

Li, Y., Shibahara, A., Matsuo, Y., Tanaka, T. and Kouno, I. (2010) Reaction of the black tea pigment theaflavin during enzymatic oxidation of tea catechins. *Journal of Natural Products*, **73**, 33–39.

Matsuo, Y., Tanaka, T. and Kouno, I. (2006) A new mechanism for oxidation of epigallocatechin and production of benzotropolone pigments. *Tetrahedron*, **62**, 4774–4783.

Matsuo, Y., Yamada, Y., Tanaka, T. and Kouno, I. (2008) Enzymatic oxidation of gallocatechin and epigallocatechin: effects of C-ring configuration on the reaction products. *Phytochemistry*, **69**, 3054–3061.

Matsuo, Y., Tanaka, T. and Kouno, I. (2009) Production mechanism of proepitheaflagallin, a precursor of benzotropolone-type black tea pigment, derived from epigallocatechin via a bicyclo[3.2.1] octane-type intermediate. *Tetrahedron Letters*, **50**, 1348–1351.

Matsuo, Y., Fujita, Y., Ohnishi, S. *et al.* (2010) Chemical constituents of the leaves of rabbiteye blueberry (*Vaccinium ashei*) and characterisation of polymeric proanthocyanidins containing phenylpropanoid units and A-type linkages. *Food Chemistry*, **121**, 1073–1079.

Matsuo, Y., Li, Y., Watarumi, S., Tanaka, T. and Kouno, I. (2011) Production and degradation mechanism of theacitrin C, a black tea pigment derived from epigallocatechin-3-*O*-gallate via a bicyclo[3.2.1]octane-type intermediate. *Tetrahedron*, **67**, 2051–2059.

Matsuo, Y., Hayashi, T., Saito, Y., Kouno, I. and Tanaka, T. (2013) Structures of enzymatic oxidation products of epigallocatechin. *Tetrahedron*, **69**, 8952–8958.

Matsuo, Y., Tadakuma, F., Shii, T., Saito, Y. and Tanaka, T. (2015) Selective oxidation of pyrogallol-type catechins with unripe fruit homogenate of *Citrus unshiu* and structural revision of oolongtheanins. *Tetrahedron*, **71**, 2540–2548.

Menet, M.C., Sang, S., Yang, C.S., Ho, C.T. and Rosen, R.T. (2004) Analysis of theaflavins and thearubigins from black tea extract by MALDI–TOF mass spectrometry. *Journal of Agricultural and Food Chemistry*, **52**, 2455–2461.

Morimoto, S., Nonaka, G. and Nishioka, I. (1987) Tannins and related compounds. LIX. Aesculitannins, novel proanthocyanidins with doubly-bonded structures from *Aesculus hippocastanum* L. *Chemical and Pharmaceutical Bulletin*, **35**, 4717–4729.

Nakai, M., Fukui, Y., Asami, A. *et al.* (2005) Inhibitory effects of oolong tea polyphenols on pancreatic lipase *in vitro*. *Journal of Agricultural and Food Chemistry*, **53**, 4593–4598.

Ricardo da Silva, J.M., Rigaud, J., Cheynier, V., Cheminat, A. and Moutounet, M. (1991) Procyanidin dimers and trimers from grape seeds. *Phytochemistry*, **30**, 1259–1264.

Roberts, E.A.H. (1957a) Oxidation–reduction potentials in tea fermentation. *Chemistry & Industry*, 1354–1355.

Roberts, E.A.H. (1957b) Oxidative condensation of flavanols in tea fermentation. *Chemistry & Industry*, 1355–1356.

Roberts, E.A.H. (1962) Economic importance of flavonoid substances: tea fermentation, in *The Chemistry of Flavonoid Compounds*, Pergamon Press, Oxford, pp. 468–512.

Roberts, E.A.H. and Myers, M. (1959) The phenolic substances of manufactured tea. IV. Enzymic oxidations of individual substrates. *Journal of the Science of Food and Agriculture*, **10**, 167–172.

Robertson, A. (1983) Effects of physical and chemical conditions on the *in vitro* oxidation of tea leaf catechins. *Phytochemistry*, **22**, 889–896.

Sang, S., Tian, S., Meng, X. *et al.* (2002) Theadibenzotropolone A, a new type pigment from enzymatic oxidation of (–)-epicatechin and (–)-epigallocatechin gallate and characterized from black tea using LC/MS/MS. *Tetrahedron Letters*, **43**, 7129–7133.

Sang, S., Tian, S., Stark, R.E., Yanga, C.S. and Ho, C.T. (2004) New dibenzotropolone derivatives characterized from black tea using LC/MS/MS. *Bioorganic & Medicinal Chemistry*, **12**, 3009–3017.

Shii, T., Miyamoto, M., Matsuo, Y., Tanaka, T. and Kouno, I. (2011a) Biomimetic one-pot preparation of a black tea polyphenol theasinensin A from epigallocatechin gallate by treatment with copper(II) chloride and ascorbic acid. *Chemical and Pharmaceutical Bulletin*, **59**, 1183–1185.

Shii, T., Tanaka, T., Watarumi, S. *et al.* (2011b) Polyphenol composition of a functional fermented tea obtained by tea-rolling processing of green tea and loquat leaves. *Journal of Agricultural and Food Chemistry*, **59**, 7253–7260.

Stodt, U. and Engelhardt, U.H. (2013) Progress in the analysis of selected tea constituents over the past 20 years. *Food Research International*, **53**, 636–648.

Takino, Y. and Imagawa, H. (1963) Studies on the oxidation of myricetin glycosides by tea oxidase. *Agricultural and Biological Chemistry*, **27**, 666–668.

Takino, Y., Imagawa, H., Horikawa, H. and Tanaka, A. (1964) Studies on the mechanism of the oxidation of tea leaf catechins. Part III. Formation of a reddish orange pigment and its spectral relationship to some benzotropolone derivatives. *Agricultural and Biological Chemistry*, **28**, 64–71.

Takino, Y., Ferretti, A., Flanagan, V., Gianturco, M. and Vogel, M. (1965) The structure of theaflavin, a polyphenol of black tea. *Tetrahedron Letters*, **6**, 4019–4025.

Tanaka, T., Takahashi, R., Kouno, I. and Nonaka, G. (1994) Chemical evidence for the de-astringency (insolubilization of tannins) of persimmon fruit. *Journal of the Chemical Society, Perkin Transactions* **1**, 3013–3022.

Tanaka, T., Kondou, K. and Kouno, I. (2000a) Oxidation and epimerization of epigallocatechin in banana fruits. *Phytochemistry*, **53**, 311–316.

Tanaka, T., Betsumiya, Y., Mine, C. and Kouno, I. (2000b) Theanaphthoquinone, a novel pigment oxidatively derived from theaflavin during tea-fermentation. *Chemical Communications*, 1365–1366.

Tanaka, T., Inoue, K., Betsumiya, Y., Mine, C. and Kouno, I. (2001) Two types of oxidative dimerization of the black tea polyphenol theaflavin. *Journal of Agricultural and Food Chemistry*, **49**, 5785–5789.

Tanaka, T., Mine, C., Inoue, K., Matsuda, M. and Kouno, I. (2002a) Synthesis of theaflavin from epicatechin and epigallocatechin by plant homogenates and role of epicatechin quinone in the synthesis and degradation of theaflavin. *Journal of Agricultural and Food Chemistry*, **50**, 2142–2148.

Tanaka, T., Mine, C. and Kouno, I. (2002b) Structures of two new oxidation products of green tea polyphenols generated by model tea fermentation. *Tetrahedron*, **58**, 8851–8856.

Tanaka, T., Mine, C., Watarumi, S. *et al.* (2002c) Accumulation of epigallocatechin quinone dimers during tea fermentation and formation of theasinensins. *Journal of Natural Products*, **65**, 1582–1587.

Tanaka, T., Watarumi, S., Matsuo, Y., Kamei, M. and Kouno, I. (2003) Production of theasinensins A and D, epigallocatechin gallate dimers of black tea, by oxidation–reduction dismutation of dehydrotheasinensin A. *Tetrahedron*, **59**, 7939–7947.

Tanaka, T., Inoue, K. and Kouno, I. (2005a) Regioselectivity in benzotropolone formation between catechins and proanthocyanidins. *Natural Product Research*, **19**, 731–737.

Tanaka, T., Matsuo, Y. and Kouno, I. (2005b) A novel black tea pigment and two new oxidation products of epigallocatechin-3-*O*-gallate. *Journal of Agricultural and Food Chemistry*, **53**, 7571–7578.

Tanaka, T., Miyata, Y., Tamaya, K. *et al.* (2009) Increase of theaflavin gallates and thearubigins by acceleration of catechin oxidation in a new fermented tea product obtained by the tea-rolling processing of loquat (*Eriobotrya japonica*) and green tea leaves. *Journal of Agricultural and Food Chemistry*, **57**, 5816–5822.

Tanaka, T., Nagai, S., Shii, T., Matsuo, Y. and Kouno, I. (2011) Isolation of 1,3-diphenylpropan-2-ols, identical to tea catechin metabolites produced by intestinal bacteria, and pyrogallol from Japanese post-fermented tea (Japanese). *Nippon Shokuhin Kagaku Gakkaishi*, **18**, 6–11.

Tanaka, T., Umeki, H., Nagai, S., Shii, T., Matsuo, Y. and Kouno, I. (2012) Transformation of tea catechins and flavonoid glycosides by treatment with Japanese post-fermented tea acetone powder. *Food Chemistry*, **134**, 276–281.

Valcic, A., Muders, A., Jacobsen, N.E., Liebler, D.C. and Timmermann, B.N. (1999) Antioxidant chemistry of green tea catechins. Identification of products of the reaction of (−)-epigallocatechin gallate with peroxyl radicals. *Chemical Research in Toxicology*, **12**, 382–386.

Valcic, S., Burr, J.A., Timmermann, B.N. and Liebler, D.C. (2000) Antioxidant chemistry of green tea catechins. New oxidation products of (−)-epigallocatechin gallate and (−)-epigallocatechin from their reactions with peroxyl radicals. *Chemical Research in Toxicology*, **13**, 801–810.

Wan, X., Nursten, H.E., Cai, Y., Davis, A.L., Wilkins, J.P.G. and Davies, A.P. (1997) A new type of tea pigment—from the chemical oxidation of epicatechin gallate and isolated from tea. *Journal of the Science of Food and Agriculture*, **74**, 401–408.

Wulandari, R.A., Amano, M., Yanagita, T. *et al.* (2011) New phenolic compounds from *Camellia sinensis* L. leaves fermented with *Aspergillus* sp. *Journal of Natural Medicines*, **65**, 594–597.

Yanagida, A., Shoji, T. and Shibusawa, Y. (2003) Separation of proanthocyanidins by degree of polymerization by means of size-exclusion chromatography and related techniques. *Journal of Biochemical and Biophysical Methods*, **56**, 311–322.

Yanase, E., Sawaki, K. and Nakatsuka, S. (2005) The isolation of a bicyclo[3.2.1] intermediate during formation of benzotropolones, a common nucleus found in black tea pigments: theaflavins. *Synlett*, **2005**, 2661–2663.

Yassin, G.H., Koek, J.H., Jayaraman, S. and Kuhnert, N. (2014a) Identification of novel homologous series of polyhydroxylated theasinensins and theanaphthoquinones in the SII fraction of black tea thearubigins using ESI/HPLC tandem mass spectrometry. *Journal of Agricultural and Food Chemistry*, **62**, 9848–9859.

Yassin, G.H., Koek, J.H. and Kuhnert, N. (2014b) Identification of trimeric and tetrameric flavan-3-ol derivatives in the SII black tea thearubigin fraction of black tea using ESI–tandem and MALDI–TOF mass spectrometry. *Food Research International*, **63**, 317–327.

Chapter 4

A Proteomic-Based Quantitative Analysis of the Relationship Between Monolignol Biosynthetic Protein Abundance and Lignin Content Using Transgenic *Populus trichocarpa*

Jack P. Wang, Sermsawat Tunlaya-Anukit, Rui Shi, Ting-Feng Yeh, Ling Chuang, Fikret Isik, Chenmin Yang, Jie Liu, Quanzi Li, Philip L. Loziuk, Punith P. Naik, David C. Muddiman, Joel J. Ducoste, Cranos M. Williams, Ronald R. Sederoff and Vincent L. Chiang

Abstract: As part of a long-term project to develop a predictive model of lignin biosynthesis in the stem differentiating xylem of *Populus trichocarpa*, we explored the quantitative relationships of gene-specific monolignol pathway proteins and the amount of the lignin polymer. We determined the absolute abundance of monolignol pathway proteins in wild-type (Nisqually-1) and 80 transgenic trees, downregulated for the expression of genes in the monolignol pathway. Total lignin content for wild-type and transgenics ranged from 9.4 to 24.2%. Comparison of protein variation with lignin content showed that the proteins are produced in several-fold excess, suggesting that the monolignol pathway is highly homeostatic. Strong downregulation of xylem-specific PtrPAL2, 4, and 5 showed the most direct relationship with lignin content. These results are consistent with the Predictive Kinetic Metabolic Flux (PKMF) model of Wang *et al.*, based on differential equations of mass action kinetics. Functional redundancy of multiple genes also moderates gene-specific effects. In addition, targeted production of specific proteins may affect the concentration of nontarget monolignol proteins, suggesting a yet-to-be-described level of coordinated control.

Keywords: segmented regression, Klason, LC–MS/MS, systems biology, predictive model, poplar, stem differentiating xylem

Recent Advances in Polyphenol Research, First Edition. Edited by Kumi Yoshida, Véronique Cheynier and Stéphane Quideau.

4.1 Introduction

Lignin is one of the most important biological polymers because of its roles in plant structure and function and in the carbon cycle. The lignin content of plants affects the yield and quality of the world's agriculture and wood product industries, in addition to the conversion efficiency and utility of plants for bioenergy (Sarkanen & Ludwig, 1971; Harada & Cote, 1985). Lignin, made by higher plants, provides structural support within and between their secondary cell walls (Harada & Cote, 1985). The hydrophobic surface of the lignified cell wall enables the vertical transport of water in herbaceous and woody plants, permitting trees to achieve heights of more than 100 m and wood volumes of 1743 m^3 (Carder, 1995; Koch *et al.*, 2004). Lignin is a high-molecular-mass phenolic polymer that is resistant to decay by bacteria or fungi (Vance *et al.*, 1980; Eriksson *et al.*, 1990). In part, as a result of the resistance to decay, trees may survive for hundreds to thousands of years (Curry, 1965). Lignin is an abundant component of wood and bark; therefore, it makes up one-fourth or more of the biomass of woody plants. Due to its resistance to decay in soils and the compound growth of wood in trees, lignin sequesters large amounts of carbon and is an important component of the terrestrial carbon cycle (Novaes *et al.*, 2010; Pan *et al.*, 2011).

The lignin polymer has a higher energy content than cellulose and hemicelluloses (Novaes *et al.*, 2010). A high lignin content is advantageous when biomass or wood is used directly for fuel (White, 1987). For liquid biofuel production by fermentation of cellulosics, lignin is a barrier to the release of sugars from wood (Hu *et al.*, 1999; Chiang, 2002; Ragauskas *et al.*, 2006; Chen & Dixon, 2007; Novaes *et al.*, 2010; Li *et al.*, 2014). Similarly, lignin is a barrier in the production of wood pulp and cellulose, requiring chemical treatments, high temperature, or mechanical disruption of the fibrous tissue of wood (Li *et al.*, 2003). Lignin also reduces the efficiency of utilization of plant biomass as forage (Jung & Deetz, 1993).

Because of its industrial significance, the structure and biosynthesis of lignin have been studied intensively for more than six decades (Freudenberg *et al.*, 1952; Wright *et al.*, 1958; Brown, 1961; Higuchi & Brown, 1963). Lignin is derived from phenylalanine through 7–10 enzymatic reactions that make the lignin precursors known as monolignols (Fig. 4.1; Wang *et al.*, 2014), which in turn undergo oxidative polymerization to form the high-molecular-mass lignin polymer (Ralph *et al.*, 2004).

The initial step in monolignol biosynthesis (Fig. 4.1) is also the beginning of general phenylpropanoid metabolism. Monolignol biosynthesis begins with the deamination of phenylalanine (**1**) to cinnamic acid (**2**), which in turn is hydroxylated at the ring-4 position to 4-coumaric acid (**3**). 4-Coumaric acid is activated as a CoA thioester (4-coumaroyl-CoA) (**8**). It can be subsequently reduced to 4-coumaraldehyde (**15**) and then to the monolignol, 4-coumaryl alcohol (**20**). Through an alternative route (Fig. 4.1), 4-coumaric acid is also converted by hydroxylation at the ring-3 position to caffeic acid (**4**). Caffeic acid undergoes CoA activation to become caffeoyl-CoA (**11**). 4-Coumaroyl-CoA (**8**) is alternatively converted to caffeoyl-CoA (**11**) through the formation of a shikimic acid ester (**9**) (4-coumaroyl shikimic acid), which is converted by 3-hydroxylation to caffeoyl shikimic acid (**10**) and then to caffeoyl-CoA (**11**) (Fig. 4.1). Caffeoyl-CoA is converted by methylation to feruloyl-CoA (**12**), which in turn is reduced to coniferaldehyde (**17**).

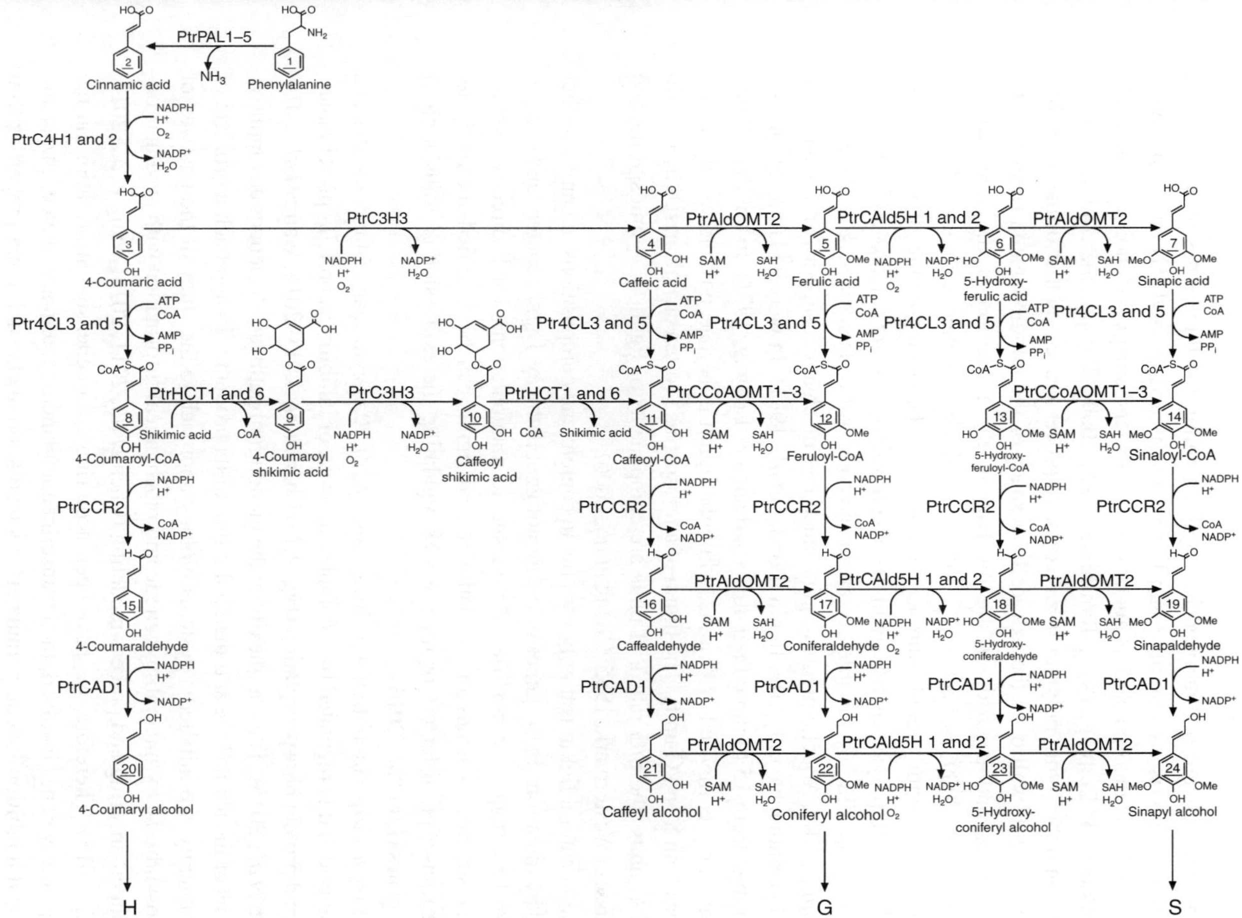

Fig. 4.1 The monolignol biosynthetic pathway in *Populus trichocarpa*. The pathway consists of 20 enzymes (phenylalanine ammonia-lyases, PtrPAL1-5; cinnamic acid 4-hydroxylases, PtrC4H1 and 2; cinnamic acid 3-hydroxylase, PtrC3H3; hydroxycinnamoyl-CoA shikimate hydroxycinnamoyl transferases, PtrHCT1 and 6; 4-coumaric acid:CoA ligases, Ptr4CL3 and 5; caffeoyl-CoA *O*-methyltransferases, PtrCCoAOMT1-3; cinnamoyl-CoA reductase, PtrCCR2; coniferaldehyde 5-hydroxylases, PtrCAld5H1 and 2; 5-hydroxyconiferaldehyde *O*-methyltransferase, PtrAldOMT2; cinnamyl alcohol dehydrogenase, PtrCAD1) and 24 metabolites (bold numbers) that convert phenylalanine to the monolignols. The cofactors are reduced nicotinamide adenine dinucleotide phosphate (NADPH), oxidized nicotinamide adenine dinucleotide phosphate (NADP+), coenzyme A (CoA), adenosine triphosphate (ATP), adenosine monophosphate (AMP), pyrophosphate (PPi), *S*-adenosyl methionine (SAM), and *S*-adenosyl homocysteine (SAH).

Coniferaldehyde is a branch point leading directly by reduction to coniferyl alcohol (**22**) or by further hydroxylation to 5-hydroxyconiferaldehyde (**18**) and then to sinapaldehyde (**19**) and to sinapyl alcohol (**24**). Ten enzyme families carry out these core reactions of monolignol biosynthesis in the Stem Differentiating Xylem (SDX) of *Populus trichocarpa* (Fig. 4.1 and Table 4.1).

The monolignols are polymerized into lignin forming the three types of lignin units. 4-Coumaryl alcohol is incorporated into lignin as 4-hydroxyphenyl (H) units. Similarly, coniferyl alcohol is incorporated as guaiacyl (G) units, and sinapyl alcohol as syringyl (S) units (Sarkanen & Ludwig, 1971; Terashima *et al.*, 2009). The polymerization of lignin monomers into the lignin polymer is an oxidative process based on the formation of free radicals in the cell wall or in the extracellular space of the middle lamella (Ralph *et al.*, 2004). Laccases and peroxidases are involved in the formation of the free radicals (Marjamaa *et al.*, 2009; Berthet *et al.*, 2012). The lignin polymer is composed of linear chains of subunits with variable numbers of cross-links between growing chains and with cross-links to wall polysaccharides within the cell walls (Ralph *et al.*, 2004, 2007).

Natural lignin is highly variable in content and composition (Sarkanen & Ludwig, 1971; Ralph *et al.*, 2007). Aquatic plants require much less mechanical support, and plants such as water hyacinth have only 10% lignin (Poddar *et al.*, 1991). In woody plants, lignin content can range from 15 to more than 40% (Sarkanen & Ludwig, 1971). In conifers, lignin is predominantly composed of G units, while the lignin in herbaceous and woody angiosperms has both S and G units. The lignin of angiosperms is often characterized by its ratio of S and G units (the S/G ratio). H units are components of lignin in gymnosperms and some grasses (Westermark, 1985; Zhang *et al.*, 2006).

Although natural lignin is highly variable in content and compositions, it can be further modified by mutations in the genes of the monolignol pathway. Lignin content and composition can be extensively modified by transgenic manipulation, which directly overexpresses or suppresses genes in the pathway, and lignin can be modified indirectly by affecting transcription factors or microRNAs regulating the expression or translation of pathway genes (Lu *et al.*, 2013).

The *P. trichocarpa* monolignol pathway consists of 20 enzymes and 24 metabolites in a metabolic grid and is regulated by a complex network of inhibitions, protein–protein interactions, and protein phosphorylation (Fig. 4.1) (Chen *et al.*, 2011, 2014; Wang *et al.*, 2014, 2015; Lin *et al.*, 2015). The complexity of the monolignol pathway challenges any intuitive predictions of the abundance and metabolic flux of its products. To approach a problem of such complexity, we initiated a systems biology approach to the study of the pathway of lignin biosynthesis, by integrating genetic and biochemical information with systems concepts from engineering and process modeling (Chen *et al.*, 2011, 2014; Wang *et al.*, 2014; Lin *et al.*, 2015). Therefore, we have been acquiring comprehensive information on the specific pathway genes involved, their transcript abundance, protein abundance, enzymatic reaction and inhibition kinetic parameters, metabolic products, and lignin properties (Shi *et al.*, 2010a, 2013; Chen *et al.*, 2011, 2013, 2014; Liu *et al.*, 2012; Wang *et al.*, 2012, 2014; Lin *et al.*, 2015).

The purpose of our approach was to integrate this information and to develop predictive mathematical models. Our first such model was a Predictive Kinetic Metabolic Flux

Table 4.1 Core proteins of monolignol biosynthesis in *Populus trichocarpa*.

Enzyme	Full name	Absolute abundance (nM)	Molecular mass (kDa)	GenBank accession number	JGI version 3 gene model ID
PtrPAL1	Phenylalanine ammonia-lyase 1	30.23	77.73	EU603319	Potri.006G126800
PtrPAL2	Phenylalanine ammonia-lyase 2	163.6	77.61	EU603321	Potri.008G038200
PtrPAL3	Phenylalanine ammonia-lyase 3	22.29	77.89	EU603318	Potri.016G091100
PtrPAL4/5	Phenylalanine ammonia-lyase 4/5	164.9	77.75/77.78	EU603322/EU603320	Potri.010G224100 Potri.010G224200
PtrC4H1	Cinnamic acid 4-hydroxylase 1	80.73	58.14	EU603304	Potri.013G157900
PtrC4H2	Cinnamic acid 4-hydroxylase 2	46.12	57.98	EU603302	Potri.019G130700
PtrC3H3	4-Coumaric acid 3-hydroxylase 3	91.21	57.43	EU603301	Potri.006G033300
Ptr4CL3	4-Coumaric acid:CoA ligase	880.6	59.02	EU603298	Potri.001G036900
Ptr4CL5	4-Coumaric acid:CoA ligase 5	105	59.31	EU603299	Potri.003G188500
PtrHCT1	Hydroxycinnamoyl-CoA:shikimate hydroxycinnamoyl transferase 1	125.6	47.82	EU603313	Potri.003G188500
PtrHCT6	Hydroxycinnamoyl-CoA:shikimate hydroxycinnamoyl transferase 6	34.68	47.98	EU603314	Potri.001G042900
PtrCCoAOMT1	Caffeoyl-CoA *O*-methyltransferase 1	431.3	27.81	EU603307	Potri.009G099800
PtrCCoAOMT2	Caffeoyl-CoA *O*-methyltransferase 2	301.7	27.96	EU603309	Potri.001G304800
PtrCCoAOMT3	Caffeoyl-CoA *O*-methyltransferase 3	61.27	26.38	EU603308	Potri.008G136600
PtrCCR2	Cinnamoyl-CoA reductase 2	118	37.07	EU603310	Potri.003G181400
PtrCAD1	Cinnamyl alcohol dehydrogenase 1	322.1	38.96	EU603306	Potri.009G095800
PtrCAld5H1	Coniferaldehyde 5-hydroxylase 1	86.75	58.53	EU603312	Potri.005G117500
PtrCAld5H2	Coniferaldehyde 5-hydroxylase 2	93.01	58.49	EU603311	Potri.007G016400
PtrAldOMT2	5-Hydroxyconiferaldehyde *O*-methyltransferase 2	6070	39.74	EU603317	Potri.012G006400

Absolute protein abundance estimates are those given by Shuford *et al.* (2012).

(PKMF) model of the monolignol biosynthetic pathway in SDX of *P. trichocarpa*. The PKMF model was assembled from a system of mechanistic nonlinear differential equations containing 104 Michaelis–Menten kinetic parameters, 85 inhibition kinetic parameters, and the absolute quantity of all monolignol proteins in *P. trichocarpa* SDX (Wang *et al.*, 2014). This PKMF model predicts how the individual pathway enzymes affect lignin content and composition and provides mechanistic explanations for the effects of genetic and transgenic perturbations of the pathway in *P. trichocarpa*.

The PKMF model also predicts that the supply and ratio of specific monolignols are the major determinants of lignin content and composition. This prediction supports a classical hypothesis of a "random" polymerization mechanism (Sederoff *et al.*, 1999; Ralph *et al.*, 2004) and does not support an alternative hypothesis requiring protein-template-based determination of lignin composition (Davin *et al.*, 1997). To test our metabolic model more rigorously, transgenic plants were established to perturb the expression of each gene as a test of the initial predictions based on a wild-type genotype, to verify or extend the original "base" model. In this way, we integrate genetic engineering and mathematical modeling, providing a system-based predictive model. Comprehensive understanding of the flux dynamics and regulation of the monolignol biosynthetic pathway will facilitate the strategic engineering of optimized lignin for plant growth and biomass utilization.

The relative quantities of gene transcripts involved in lignin biosynthesis have been studied intensively (Raes *et al.*, 2003; Hamberger *et al.*, 2007, Shi *et al.*, 2010a). It has been only recently that the proteins of the monolignol pathway have been measured by absolute quantification (Shuford *et al.*, 2012). With such quantitative measurements, it is now possible to investigate the quantitative relationships of the monolignol biosynthetic enzymes with other components of the pathway. Here, we present the results of a quantitative analysis of the relationship of the absolute protein quantities of specific monolignol pathway enzymes and the abundance of the lignin polymer made in SDX of *P. trichocarpa*.

4.2 Results

4.2.1 Production of transgenic trees downregulated for genes in monolignol biosynthesis

To study the functional significance of different genes associated with lignification, we generated transgenic trees with perturbed gene expression during lignin formation in SDX, the wood forming tissue of *P. trichocarpa*. Artificial microRNA (amiRNA) constructs were designed for gene-specific expression knockdown (Schwab *et al.*, 2006). RNA interference (RNAi) constructs were used for expression knockdown of multiple genes of a gene family or subgroup of a gene family. All of these constructs are designed for xylem-specific expression using a Ptr4CL3 promoter (Wang *et al.*, 2014) to drive expression of the transgenes and to ensure that the modified expression of the respective transgenes are limited to the SDX tissue.

Transcript abundance of target genes in SDX of 3-month-old wild-type and transgenic trees was first analyzed using qRT-PCR to detect modification of transcript abundance of the target genes. Then transgenic lines with different levels of target gene transcript abundance were selected for further analysis. Selected lines of trees were clonally propagated and grown in a greenhouse for 6 months to provide enough material for analysis of protein concentration and lignin content. We included 12 wild-type and 80 transgenic lines that target single genes, gene family subgroups, or whole gene families for *PtrPAL*, *PtrC4H*, *PtrC3H*, *PtrCCR*, and *PtrCAD*. For example, we have transgenic lines that specifically target the downregulation of individual *PtrPAL* genes (*PtrPAL1-5*), combinations of the *PtrPAL* genes, and transgenic lines that target the downregulation of all *PtrPAL* genes.

A single gene *PtrCCR2* is active in monolignol biosynthesis in SDX of *P. trichocarpa* (Shi *et al.*, 2010a). Three transgenic lines with different levels of downregulation of the *PtrCCR*2 transcripts were selected and analyzed. PtrCAD is the enzyme that produces the final monolignol precursors for polymerization. Two *PtrCAD* genes (*PtrCAD1* and *PtrCAD2*) were downregulated individually or together. The abundance of the PtrCAD2 protein is very low, precluding reliable detection, and therefore, PtrCAD2 was excluded from further analysis. The downregulated lines allow us to observe both direct and indirect effects of reduced gene expression. The direct effects result from presumed suppression of target gene expression, observed as reduced transcript abundance. We were able to test for indirect effects, where a reduction in transcript or protein of a target gene results in a modification in the expression of other genes in the monolignol pathway, either in the same gene family or in other gene families.

Although other genes of monolignol biosynthesis were not direct targets of our transgenic lines, nontargeted genes were also affected. Therefore, our plant material provides a basis for analysis of all monolignol gene products for their coordinated responses and the relation of such effects to their function in the pathway. All of the transgenic effects are compared to a series of wild-type trees to control for nongenetic variation.

In these transgenic plants, therefore, we capture genetic variation due to the specific constructs and the resulting variation in the protein abundance of the target genes. We also obtain protein abundance measurements for other genes in the monolignol pathway. In addition, due to replication and experimental design over several years of assays, we obtained many estimates that include environmental variation of gene-specific expression. The series of transgenic constructs we have obtained and analyzed does not include knockdowns or knockouts for all the genes in the pathway; nevertheless, they provide sufficient information to observe many broad features of the quantitative relationships of monolignol genes and lignin content, until a more comprehensive set of variants for all monolignol genes become available.

4.2.2 Absolute quantification of protein abundance

The quantitative analysis of the relationship of specific protein abundance and the amount of lignin polymer depends on the ability to determine absolute protein quantity. We have carried out such measurements using stable-isotope-labeled "surrogate" peptides added to tryptic digests of SDX extracts for quantification. The digests are subsequently analyzed by protein cleavage–isotope dilution mass spectrometry (PC-IDMS) (Shuford *et al.*, 2012). The core enzymes of the monolignol pathway found in SDX tissues of *P. trichocarpa* (Table 4.1) include 10 gene/enzyme families. The PtrPAL enzymes have five isoforms. PtrPAL4 and PtrPAL5 are so similar, no individual surrogate peptide can distinguish them, and therefore, their abundance is estimated together, as a sum, by a single surrogate peptide. Enzyme family PtrCCoAOMT has three isoforms, and four families have two isoforms expressed in SDX (PtrC4H, Ptr4CL, PtrHCT, and PtrCAld5H). Four enzymes are represented by a single isoform (PtrC3H, PtrCCR, PtrCAD, and PtrAldOMT). The most abundant protein is PtrAldOMT2 at 6.07 mM, which is the third most abundant protein in the SDX extract (Shuford *et al.*, 2012).

4.2.3 Variation in protein abundance in wild-type and transgenic plants

Before examining the relationship of protein abundance and lignin content, it is of interest to examine the extent of variation in protein abundance in wild-type and transgenic plants. It is important to estimate the extent of variation in each protein relative to wild type, and the extent of suppression of specific target gene effects in transgenic plants (Fig. 4.2). The bar graphs for individual constructs show the protein abundance, compared to its wild-type control, for each specific target gene, and in some cases, the abundance of the target gene in a construct where the expression of more than one gene is suppressed. For example, PtrPAL1 is suppressed specifically, but is also suppressed in a construct suppressing both PtrPAL1 and PtrPAL3. The range of variation within the same wild-type or specific construct defines a baseline and range of concentrations and estimates the extent of the relationships of protein concentration and lignin. In addition, it is important to point out whether the relative abundance of protein for each gene is independent of other genes and proteins in the pathway, or if there is regulation that might suggest an additional level of control. Our data on those genes that have transgenic constructs showed that the protein concentration of all PtrPAL isoforms, as single target genes, were substantially suppressed (Fig. 4.2). In some cases, constructs with multiple targets in the same gene family had stronger suppression. Variation in concentration of proteins may also include indirect non-target effects. These effects occur when the construct that is designed for suppression of a target gene has an indirect effect on the concentration of other proteins in the pathway. The effects were not significant for several genes, but 7 cases of 11 showed a similar trend.

4.2.4 Variation in lignin content

A difference in lignin content of transgenic and wild-type stems (Fig. 4.3a) is found in the plants analyzed in this experiment. The wild-type genotype is a *P. trichocarpa* clone (Nisqually-1). The best estimate of the lignin content of the wood of this clone at age 6 months was $22.03 \pm 0.53\%$. The lignin content of the Nisqually-1 wild-type ranged from 18.5 to 24.2%. This distribution reflects environmental (nongenetic) variation, which is attributable in part to differences between batches. Due to the size of these experiments and the numbers of plants to be tested, several batches of transgenics were established and tested over several years. Standardizing each batch of transgenics to their own group of

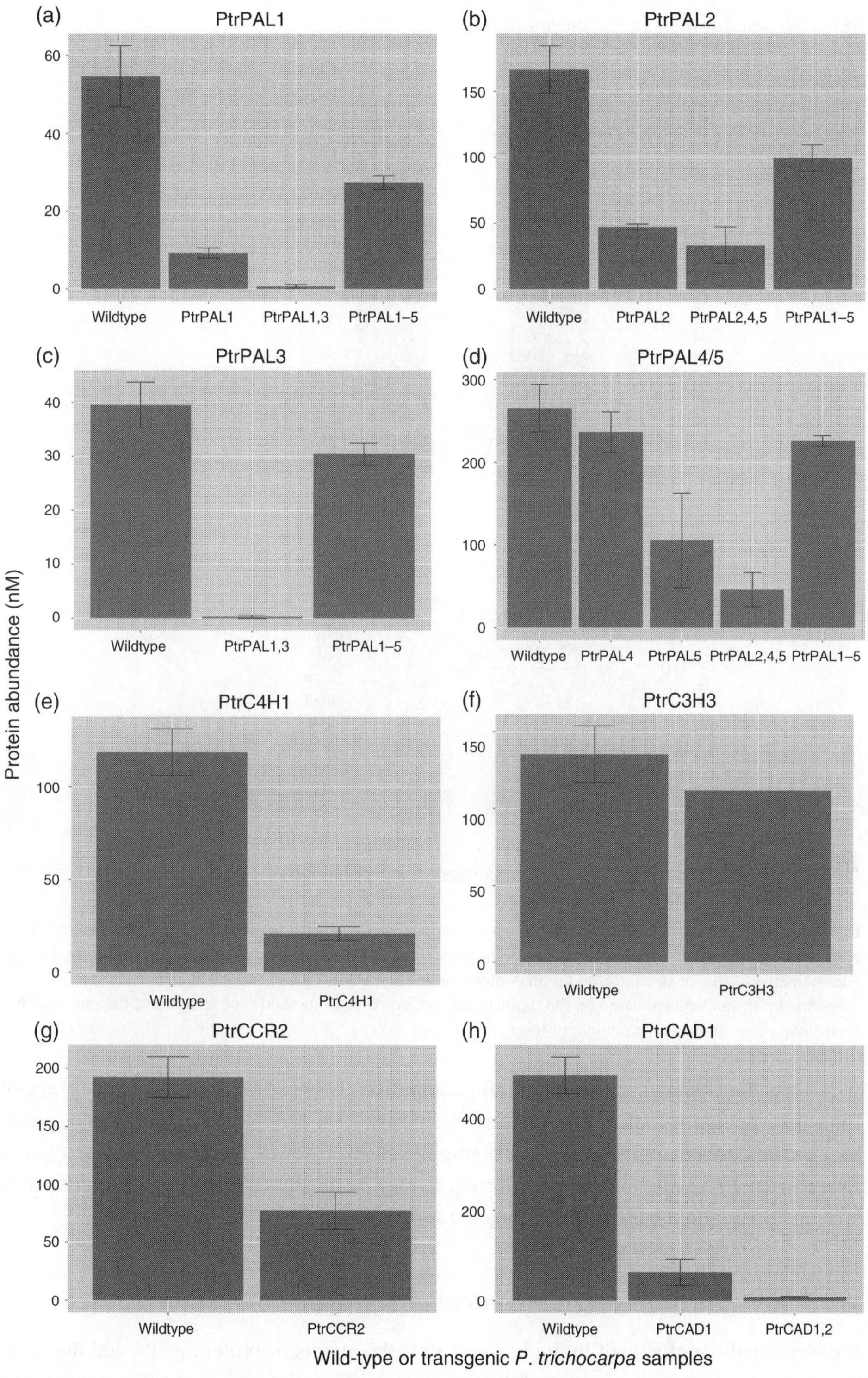

Fig. 4.2 Monolignol biosynthetic protein abundance in wild-type and transgenic *Populus trichocarpa*. The LC–MS/MS-based absolute protein abundance of (a) PtrPAL1, (b) PtrPAL2, (c) PtrPAL3, (d) PtrPAL4/5, (e) PtrC4H1, (f) PtrC3H3, (g) PtrCCR2, and (h) PtrCAD1 in stem differentiating xylem of the 92 wild-type and transgenic *P. trichocarpa* trees (6 months old). The *x*-axis specifies the target gene(s) for downregulation in each transgenic sample. Error bars represent one standard error of biological replicates.

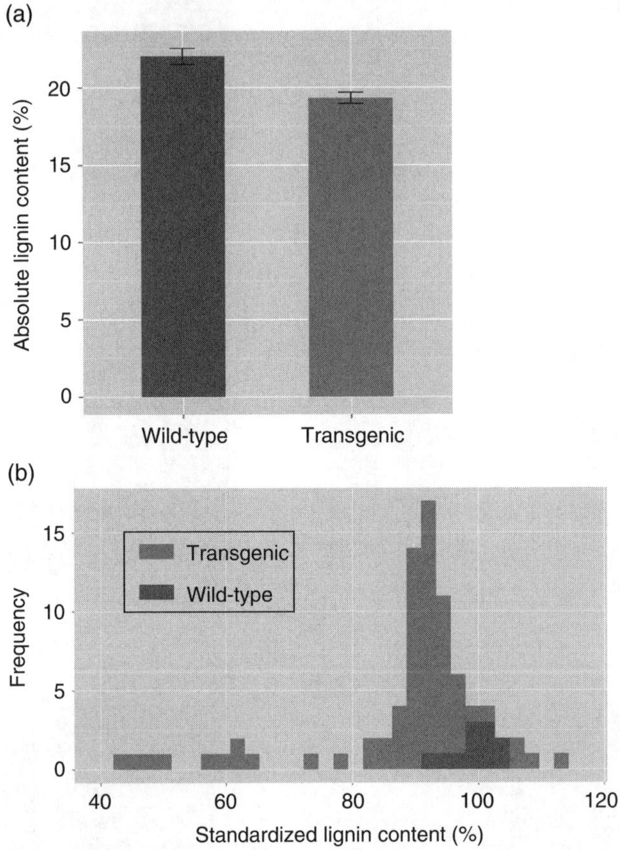

Fig. 4.3 Lignin content of wild-type and transgenic *Populus trichocarpa*. (a) The absolute lignin content of wild-type and transgenic *P. trichocarpa* as weight percentage of the extractive-free wood, determined by the Klason method. (b) The standardized lignin content for wild-type (dark gray) and transgenic (light gray) *P. trichocarpa*, standardized based on the lignin content of corresponding wild-type samples of the same batch. Frequency values indicate the number of trees.

wild-type plants allows more meaningful comparison between the batches. The average of the wild-type lignin content was set to 100% in Fig. 4.3b. The lignin content of transgenics had a much greater range and a distribution skewed to lower lignin contents and had a lower mean 19.32±0.36%. These differences may be attributed in part to a larger sample size, in addition to the effects of transgenic suppression.

4.2.5 Relationship of lignin content and protein abundance

We examined the effects of individual proteins, the effects of protein pairs, and the combined effects of multiple proteins of the same family. The relationship of lignin content and protein abundance was analyzed using segmented regression (Table 4.2 and Fig. 4.4). For at least some genes, due to redundancy or insufficient transgenic suppression, the initial slopes did not go through the origin. For many genes, protein abundance explained a large percentage of variation in lignin quantity. The relationship was linear and positive and

Table 4.2 Parameter estimates from segmented regression models predicting lignin content of *Populus trichocarpa* from protein abundance.

				Segmented regression				
Protein	Transgenic	Pr(>*F*)	R^2	Intercept	Breakpoint	Slope before breakpoint	Slope after breakpoint	Lignin at breakpoint
PtrPAL1	Yes	<0.001	0.31	16.4***	65.6***	0.092***	−0.059	22.4
PtrPAL2	Yes	<0.001	0.58	11.1***	14.8	0.562	0.007*	19.5
PtrPAL3	No	0.012	0.30	15.8***	28.9***	0.163***	0.018	20.5
PtrPAL4.5	Yes	<0.001	0.53	10.5***	22.4***	0.428***	0.001	20.1
PtrC4H1	Yes	0.004	0.14	15.4***	91.9***	0.064**	−0.023	21.3
PtrC4H2	No	<0.001	0.37	3.2	35.8***	0.457***	0.024	19.5
PtrC3H3	Yes	0.002	0.15	9.9**	64.4***	0.166*	−0.007	20.6
Ptr4CL3	No	0.093	0.19	5.1	78.9***	0.173	0.003**	18.8
Ptr4CL5	No	0.126	0.14	15.4***	20.4**	0.202	0.020*	19.5
PtrHCT1	No	0.145	0.05	13.5***	61.7***	0.109	−0.004	20.2
PtrHCT6	No	0.023	0.10	11.4**	24.0***	0.369	−0.010	20.3
PtrCCoAOMT1	No	0.005	0.16	13.3***	516.7***	0.016**	−0.005	21.4
PtrCCoAOMT2	No	0.029	0.11	11.0**	248.4***	0.036	0.000	19.9
PtrCCoAOMT3	No	<0.001	0.22	0.0	82.6***	0.253**	−0.015	20.9
PtrCCR2	Yes	0.001	0.20	12.3***	129.0***	0.063***	−0.004	20.4
PtrCAD1	Yes	0.377	0.02	20.6***	168.1	−0.011	0.002	18.8
PtrCAld5H1	No	<0.001	0.27	12.0***	61.3***	0.153***	−0.015	21.4
PtrCAld5H2	No	<0.001	0.22	13.3***	82.4***	0.093***	−0.011	21.0
PtrAldOMT2	No	<0.001	0.21	0.5	5130.3***	0.004*	0.000	19.8
PtrPAL family	Yes	<0.001	0.58	9.6***	100.4***	0.098***	0.003	19.5
PtrC4H family	Yes	<0.001	0.24	7.1*	92.2***	0.145**	−0.003	20.5
Ptr4CL family	No	0.097	0.19	5.1	88.4***	0.154	0.003**	18.8
PtrHCT family	No	0.081	0.06	12.6**	86.2***	0.089	−0.003	20.2
PtrCCoAOMT family	No	0.002	0.15	7.3	626.1***	0.021	0.000	20.4
PtrCAld5H family	No	<0.001	0.25	11.8***	130.2***	0.072**	−0.006	21.1

The slopes after the breakpoint are not different from 0, suggesting that increases in protein abundance in the system do not cause any change in lignin quantity after a saturation point, which may vary for each of the monolignol genes. The Pr>*F* value is the *p* value of an *F* test to compare the residual sum of squares of the segmented regression with a linear regression. R^2 is the sum squares of the segmented model divided by sum of squares of lignin content, and measures the proportion of variation of lignin explained by the model. The parameter estimates corresponding to the significant level are shown as * $p<0.05$, ** $p<0.01$, and *** $p<0.001$.

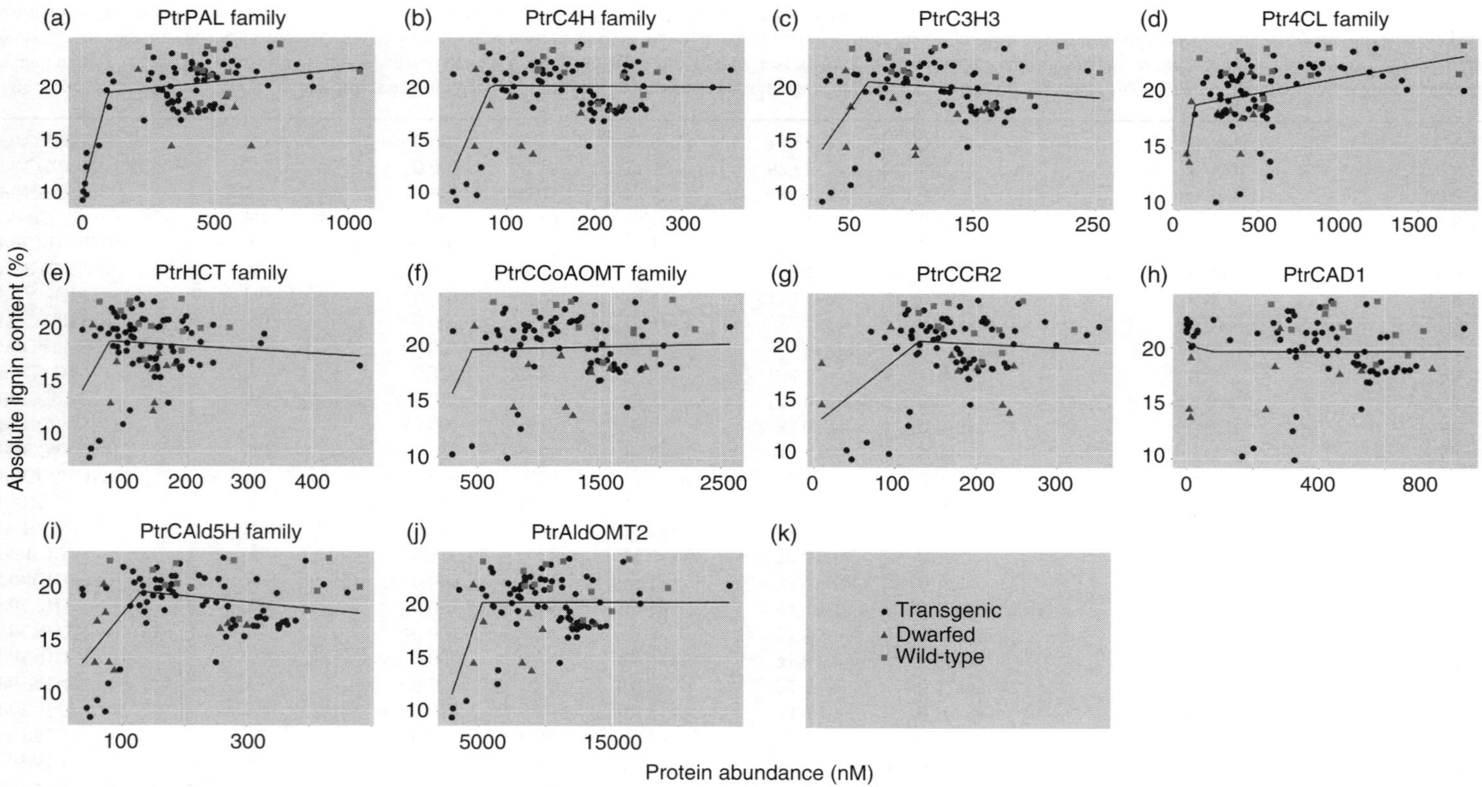

Fig. 4.4 The scatter plots show the relationship between the protein abundance and lignin content for monolignol biosynthetic protein families (a) PtrPAL, (b) PtrC4H, (c) PtrC3H, (d) Ptr4CL, (e) PtrHCT, (f) PtrCCoAOMT, (g) PtrCCR, (h) PtrCAD, (i) PtrCAld5H, and (j) PtrAldOMT. (k) The solid circles represent data from the transgenic trees. The triangles represent transgenic trees exhibiting dwarfed phenotype. The squares represent data from wild-type. The solid lines represent segmented regression models. Protein abundance (nM) was derived from concentration (μg/g of total protein) using the conversion factor of 0.74 ml/g (water content of SDX tissue) (Shuford et al., 2012).

flattened after a saturation point. A plateau indicates that specific proteins were in substantial excess and that the pathway is highly homeostatic. Some constructs with downregulated targets reached low levels of expression and this affected lignin content (Fig. 4.4).

The PtrPAL protein family has five members that are expressed in the *P. trichocarpa* monolignol pathway in SDX. Although *PtrPAL4* and *PtrPAL5* transcripts have the most specificity for differentiating xylem, the other PAL transcripts are also abundant, but are less xylem-specific (Shi *et al.*, 2010b). The plateau values were not different from wild-type. PtrPAL2 was the most abundant of the PtrPAL proteins, contributing 43% of the total PtrPAL (Table 4.1). The sum of the PtrPAL proteins relative to lignin content (Fig. 4.4a) showed a clear initial slope, breakpoint (saturation), and plateau. The total PtrPAL protein level had to be less than 100 nM before a reduction in lignin content was observed.

There are two functional isoforms for PtrC4H (PtrC4H1 and PtrC4H2). PtrC4H1 provides 64% of the PtrC4H protein. Downregulated transgenic plants were only available for PtrC4H1 for this analysis. While the individual data plots did not provide clear estimates of regression parameters, the combined data of both proteins indicated an initial slope, breakpoint, and plateau (Table 4.2). The Ptr4CL protein pair provided a less clear relationship, although no data from downregulated transgenic plants were available. These proteins have partially redundant functions, but show different effects of inhibitors and regulatory control as a Ptr4CL3/Ptr4CL5 protein complex (Chen *et al.*, 2013, 2014). Two other protein pairs, PtrCAld5H1/PtrCAld5H2 and PtrHCT1/PtrHCT6, show initial slopes and breakpoints that were not well defined, again due at least in part to the absence of downregulated transgenic constructs and functional redundancy (Fig. 4.4). Similarly, the PtrCCoAOMT1, 2, and 3 did not show a clear initial slope either as individual genes, or as a family, again most likely because of the absence of transgenic suppression.

Four proteins in the pathway analyzed here were found in SDX as single isoforms: PtrC3H3, PtrCAD1, PtrCCR2, and PtrAldOMT2. Of these four, only PtrAldOMT2 did not include transgenic suppression (Fig. 4.4). Even in the case of the transgenic suppressed lines, there was insufficient data to describe the effects of low protein levels on lignin content. In the case of PtrCAD1, there was a strong plateau, without a suggestion of a breakpoint or initial slope. Low abundance of PtrCAD2 protein could contribute sufficient activity to produce adequate levels of lignin.

4.3 Discussion

We produced a large number of transgenic *P. trichocarpa* plants generated as part of a project to develop a quantitative systems approach to better understand lignin biosynthesis in a model tree (Shi *et al.*, 2010a; Wang *et al.*, 2014). New methods for absolute quantification of specific enzymes involved in monolignol biosynthesis (Shuford *et al.*, 2012) enabled us to measure protein quantity produced in SDX and to examine the effects of protein abundance on lignin content. In these experiments, lignin content of wild-type *P. trichocarpa* ranged from 18.5 to 24.2%, while the standardized lignin content in the transgenics, downregulated for specific genes in the monolignol pathway ranged from 43.4 to 113% of wild-type (Fig. 4.3b). We also found a high level of variation in specific protein concentration among the samples (Fig. 4.2).

Results of analysis suggested that all monolignol proteins showed abundance levels that were far greater than a quantity needed to produce normal lignin content (Fig. 4.4), and a substantial reduction of their protein abundance to below that of the breakpoint level (Table 4.2) is required to reduce lignin. These results are consistent with the analysis and the predictions of the metabolic flux model of Wang *et al.* (2014). High protein levels and high variation were found for both single isoforms and multiple proteins in families. Some transgenics that substantially reduce specific protein levels did not have apparent effects on lignin content, as might be expected for genes in families that have functional redundancy. In some transgenic plants, when one target protein was suppressed, proteins from different (nontarget) families were reduced in abundance, suggesting some coordinated regulation of many proteins in the pathway.

The high level of variation for specific protein concentrations and for lignin content was unexpected. Learning the source of this variation requires further data. The high variation was observed in both wild-type (Nisqually-1) and in the transgenic lines, indicating high levels of environmental variation in biological replicates. Larger numbers of replicates would reduce variation and should be considered in the design of future experiments of this large scale.

Our objective is to develop predictive models for lignin biosynthesis based on the quantitative relationships of genes, transcripts, proteins, metabolites, and the lignin polymer. Based on the genome sequence (Tuskan *et al.*, 2006) and the relative abundance of transcripts for specific genes in differentiating xylem, the quantitative gene and transcript abundance relationships for the core monolignol pathway genes were defined (Shi *et al.*, 2010a). It should be possible to quantify the transcript to protein relationships for this set of genes and to integrate this information into a more comprehensive mathematical model. Similarly, more work is needed on the quantification of metabolites, the monolignol precursors, and the specific linkages/structures in the lignin polymer. Given the large number of transgenics generated in our project that downregulate the core genes in the monolignol pathway, much of the information for building a more extensive model should be available in the near future.

4.4 Materials and methods

4.4.1 Production of transgenic trees

P. trichocarpa genotype Nisqually-1 was used for all experiments. Wild-type and transgenic trees, grown in a greenhouse for 6 months, were harvested for analysis. To knockdown individual target genes, amiRNAs (21 nt) were designed using the online program WMD (http://wmd.weigelworld.org, Schwab *et al.*, 2006) based on the genome of *P. trichocarpa* version 1.0 (JGI). amiRNA* sequence and primer sets for integrating amiRNA and amiRNA* sequences into an MIR gene backbone were designed by the WMD program's oligo function under the vector option "RS3000 (MIR319a)." To integrate the designed amiRNA and amiRNA* sequences into the Ptr-MIR408 transgene backbone, primer sets were modified by replacing the vector sequence of Ath-MIR319a (RS300 vector) with that of Ptr-MIR408. pBI121-based amiRNA expression binary vectors were constructed following procedures described by Shi *et al.* (2010b), except that the primer for the promoter sequence was based on the xylem-specific promoter of the *Ptr4CL3* gene (Wang *et al.*, 2014).

To assemble the RNAi constructs that target multiple genes, such as genes in the same subgroup or gene family, one 150–300 bp sequence sharing over 70% sequence similarity between different target genes but less than 70% with untargeted genes was selected from one target gene and used as an RNAi sequence. Such an RNAi sequence fragment was amplified from a cloned gene transcript sequence (Shi *et al.*, 2010a). In one case, the RNAi sequence did not have sufficient similarity to all target genes (>70%); therefore, more than one similar sequence was prepared and then assembled by overlapping PCR.

RNAi fragments were amplified using appropriate primers, then digested by restriction enzymes, and assembled with a 600 bp GUS linker (GL) to form an antisense: GL: sense fragment as an RNAi transgene (Li *et al.*, 2011). The assembled RNAi transgene fragments were digested from an intermediate plasmid and then cloned into pBI121-4CLXP plasmid by replacing the GUS sequence (Wang *et al.*, 2014). All binary constructs were introduced into *Agrobacterium tumefaciens* C58 and used for plant transformation following Song *et al.* (2006).

To confirm positive transgenic insertions, genomic DNA of putative plantlets was extracted from young leaf tissue using the Qiagen DNeasy Plant Mini Kit (Invitrogen/Life Technologies, Grand Island, NY). PCR assays using promoter, forward primers along with transgene-specific reverse primers were conducted for the putative transformed plantlet along with a positive control, the transformation vector plasmid DNA, and a negative control, that is, genomic DNA from a wild-type tree.

From transgenic lines selected for different levels of gene knockdown expression, stem differentiating xylem (SDX) tissues of 3-month-old transgenic trees, grown in a greenhouse, were collected for extracting total RNA using Qiagen RNeasy Mini Kit (Invitrogen/Life Technologies). Target gene expression was analyzed using qRT-PCR as in our previous studies of transcript abundance (Shi *et al.*, 2010a, 2010b). Three lines with different levels of target gene knockdown were further propagated in a greenhouse using branch rooting. After 6 months, SDX tissue and wood was collected for analysis. Due to greenhouse space limitation, we could not grow all transgenic trees at the same time. Therefore, transgenic trees of the different constructs were grown and collected at different times/seasons each with wild-type controls, which were used to standardize data of trees from different batch conditions and collections.

4.4.2 Proteomic analysis

For absolute quantification of the 20 monolignol biosynthetic proteins in SDX of wild-type and transgenic *P. trichocarpa*, assays based on PC-IDMS were carried out as described by Shuford *et al.* (2012). This method is based on the premise that a protein quantity is equivalent to the concentration of its unique peptides produced via tryptic digestion, and the absolute peptide concentration can be determined by spiking in a known amount of its stable isotope-labeled (SIL) peptide analog as an internal standard (Brun *et al.*, 2009; Shuford *et al.*, 2012).

The calculated concentration of the native tryptic peptide is assumed to be approximately equal to the native protein concentration (Shuford *et al.*, 2012; Loziuk *et al.*, 2013).

Protein abundance (nM) was derived from concentration (μg/g of total protein) using the conversion factor of 0.74 ml/g (water content of SDX tissue) (Shuford *et al.*, 2012).

4.4.3 Lignin quantification

For lignin content determination, wood stem segments (10 cm) were collected from each transgenic line and wild-type (three biological replicates). The debarked segments were extracted with acetone for 2 days at room temperature. After air drying, the extractive-free wood segments were ground into wood meal (0.4–0.6 mm) in a Wiley mill and dried under vacuum over P_2O_5 overnight. The lignin content was determined using the modified Klason lignin method (Yeh *et al.*, 2005; Lu *et al.*, 2013). The dried wood meal was suspended in 72% H_2SO_4 at room temperature for 90 minutes and diluted with deionized water to 3% H_2SO_4 prior to heating to 121°C and held at 2 atm for 1 hour. The heated samples were then filtered through a crucible, dried, and the acid-insoluble lignin was determined by weight. The acid-soluble lignin content of the filtrate was determined by the absorbance at 205 nm using an extinction coefficient of 110 ($AUL{\cdot}g^{-1}{\cdot}cm^{-1}$). The acid-soluble and acid-insoluble lignin contents were combined as the total Klason lignin content (Yeh *et al.*, 2005).

4.4.4 Statistical analysis

Scatter plots of lignin content and abundance for the monolignol biosynthetic pathway proteins for most genes and gene families showed a positive linear relationship before, and a flat relationship after, a saturation point.

To describe the relationship between the lignin content and the absolute abundance of the monolignol biosynthetic pathway proteins, segmented regression analysis was performed based on 92 samples from transgenic and wild-type *P. trichocarpa*. We fit the segmented regression to explain the lignin quantity using the protein abundance. R package "segmented" was used for all the analysis (Muggeo, 2003, 2008). All the regression models included an intercept. The segmented regression model is

$$y_j = \beta_0 + \beta_{1i}x_{ij} + (\beta_{2i} - \beta_{1i})\max(x_{ij} - \beta_{pi}, 0) + \varepsilon_j$$

where y_j is the lignin content of sample j, x_{ij} is the concentration of protein i sample j, β_0 is the intercept of the protein i, β_{1i} is the slope before the breakpoint, β_{2i} is the slope after the breakpoint of protein i, β_{pi} is the breakpoint of changing slope from β_{1i} to β_{2i}, and ε_j is a random error of the model. Outliers were removed that had an absolute standardized residual greater than 3. The segmented regression was compared against the linear regression model using an F test to test the significance of the model.

References

Berthet, S., Thevenin, J., Baratiny, D. *et al.* (2012) Role of plant laccases in lignin polymerization. *Advances in Botanical Research*, **61**, 145–172.

Brown, S.A. (1961) Chemistry of lignification. *Science*, **134**, 305–313.

Brun, V., Masselon, C., Garin, J. *et al.* (2009) Isotope dilution strategies for absolute quantitative proteomics. *Journal of Proteomics*, **72**, 740–749.

Carder, A. (1995) *Forest Giants of the World: Past and Present*, Fitzhenry and Whiteside, Markham, ON.

Chen, F. and Dixon, R.A. (2007) Lignin modification improves fermentable sugar yields for biofuel production. *Nature Biotechnology*, **25**, 759–761.

Chen, H.C., Li, Q., Shuford, C.M. *et al.* (2011) Membrane protein complexes catalyze both 4- and 3-hydroxylation of cinnamic acid derivatives in monolignol biosynthesis. *Proceedings of the National Academy of Sciences of the United States of America*, **108**, 21253–21258.

Chen, H.C., Song, J., Williams, C.M. *et al.* (2013) Monolignol pathway 4-coumaric acid:coenzyme A ligases in *Populus trichocarpa*: novel specificity, metabolic regulation, and simulation of coenzyme A ligation fluxes. *Plant Physiology*, **161**, 1501–1516.

Chen, H.C., Song, J., Wang, J.P. *et al.* (2014) Systems biology of lignin biosynthesis in *Populus trichocarpa*: heteromeric 4-coumaric acid: coenzyme A ligase protein complex formation, regulation, and numerical modeling. *Plant Cell*, **26**, 876–893.

Chiang, V.L. (2002) From rags to riches. *Nature Biotechnology*, **20**, 557–558.

Curry, D.R. (1965) An ancient bristlecone pine stand in eastern Nevada. *Ecology*, **46**, 564–566.

Davin, L.B., Wang, H.B. and Crowell, A.L. (1997) Stereoselective biomolecular phenoxy radical coupling by an auxiliary (dirigent) protein without an active center. *Science*, **275**, 362–366.

Eriksson, K.-E.L., Blanchette, R.A. and Ander, P. (1990) *Microbial and Enzymatic Degradation of Wood and Wood Components*, Springer-Verlag, Berlin.

Freudenberg, K., Reznik, H. and Bocsenberg, H. (1952) The enzyme system participating in lignification. *Chemische Berichte*, **85**, 641–647.

Hamberger, B., Ellis, M., Friedmann, M. *et al.* (2007) Genome-wide analyses of phenylpropanoid-related genes in *Populus trichocarpa*, *Arabidopsis thaliana*, and *Oryza sativa*: the *Populus* lignin toolbox and conservation and diversification of angiosperm gene families. *Canadian Journal of Botany*, **85**, 1182–1201.

Harada, T. and Cote, W.A. (1985) Structure of wood, in *Biosynthesis and Biodegradation of Wood Components* (ed T. Higuchi), Academic Press, San Diego, CA.

Higuchi, T. and Brown, S.A. (1963) Studies on lignin biosynthesis using isotopic carbon. XI. Reactions relating to lignification in young wheat plants. *Canadian Journal of Biochemistry and Physiology*, **41**, 65–76.

Hu, W., Harding, S.A., Lung, J. *et al.* (1999) Repression of lignin biosynthesis promotes cellulose accumulation and growth in transgenic trees. *Nature Biotechnology*, **17**, 808–812.

Jung, H.G. and Deetz, D.A. (1993) Cell wall lignification and degradability, in *Forage Cell Wall Structure and Digestibility* (ed H.G. Jung), ACSESS, Madison, WI.

Koch, G.W., Sillett, S.C., Jennings, G.M. *et al.* (2004) The limits to tree height. *Nature Biotechnology*, **428**, 851–854.

Li, L., Zhou, Y., Cheng, X. *et al.* (2003) Combinatorial modification of multiple lignin traits in trees through multigene cotransformation. *Proceedings of the National Academy of Sciences of the United States of America*, **100**, 4939–4944.

Li, Q., Min, D., Wang, J.P. *et al.* (2011) Down-regulation of glycosyltransferase 8D genes in *Populus trichocarpa* caused reduced mechanical strength and xylan content in wood. *Tree Physiology*, **31**, 226–236.

Li, Q., Song, J., Peng, S. *et al.* (2014) Plant biotechnology for lignocellulosic biofuel production. *Plant Biotechnology Journal*, **12**, 1174–1192.

Lin, C.-Y., Wang, J.P., Li, Q. *et al.* (2015) 4-Coumaroyl and caffeoyl shikimic acids inhibit 4-coumaric acid:coenzyme A ligases and modulate metabolic flux for 3-hydroxylation in monolignol biosynthesis of *Populus trichocarpa*. *Molecular Plant*, **8**, 176–187.

Liu, J., Shi, R., Li, Q. *et al.* (2012) A standard reaction condition and a single HPLC separation system are sufficient for estimation of monolignol biosynthetic pathway enzyme activities. *Planta*, **236**, 879–885.

Loziuk, P.L., Wang, J., Li, Q. *et al.* (2013) Understanding the role of proteolytic digestion on discovery and targeted proteomic measurements using liquid chromatography tandem mass spectrometry and design of experiments. *Journal of Proteome Research*, **12**, 5820–5829.

Lu, S., Li, Q., Wei, H. *et al.* (2013) Ptr-miR397a is a negative regulator of laccase genes affecting lignin content in *Populus trichocarpa. Proceedings of the National Academy of Sciences of the United States of America*, **110**, 10848–10853.

Marjamaa, K., Kukkola, E.M. and Fagerstedt, K.V. (2009) The role of xylem class III peroxidases in lignification. *Journal of Experimental Botany*, **60**, 367–376.

Muggeo, V.M.R. (2003) Estimating regression models with unknown break-points. *Statistics in Medicine*, **22**, 3055–3071.

Muggeo, V.M.R. (2008) Segmented: an R package to fit regression models with broken-line relationships. *R News*, **8**, 20–25.

Novaes, E., Kirst, M., Chiang, V. *et al.* (2010) Lignin and biomass: a negative correlation for wood formation and lignin content in trees. *Plant Physiology*, **154**, 555–561.

Pan, Y., Birdsey, R. and Fang, J. (2011) A large and persistent carbon sink in the world's forests. *Science*, **333**, 988–993.

Poddar, K., Mandal, L. and Banerjee, G.C. (1991) Studies on water hyacinth (*Eichhornia crassipes*)—chemical composition of the plant and water from different habitats. *Indian Veterinary Journal*, **68**, 833–837.

Raes, J., Rohde, A., Christensen, J.H. *et al.* (2003) Genome-wide characterization of the lignification toolbox in *Arabidopsis. Plant Physiology*, **133**, 1051–1071.

Ragauskas, A.J., Williams, C.K., Davison, B.H. *et al.* (2006) The path forward for biofuels and biomaterials. *Science*, **311**, 484–489.

Ralph, J., Lundquist, K. and Brunow, G. (2004) Lignins: natural polymers from oxidative coupling of 4-hydroxyphenyl-propanoids. *Phytochemistry Reviews*, **3**, 29–60.

Ralph, J., Brunow, G. and Boerjan, W. (2007) Lignins. *Encyclopedia of Life Sciences*. doi: 10.1002/9780470015902.a0020104.

Sarkanen, K.V. and Ludwig, C.H. (1971) *Lignins: Occurrence, Formation, Structure and Reactions*, John Wiley & Sons, Inc., New York.

Schwab, R., Ossowski, S., Riester, M. *et al.* (2006) Highly specific gene silencing by artificial microRNAs in *Arabidopsis. Plant Cell*, **18**, 1121–1133.

Sederoff, R.R., Mackay, J.J., Ralph, J. *et al.* (1999) Unexpected variation in lignin. *Current Opinion in Plant Biology*, **2**, 145–152.

Shi, R., Sun, Y., Li, Q. *et al.* (2010a) Towards a systems approach for lignin biosynthesis in *Populus trichocarpa*: transcript abundance and specificity of the monolignol biosynthetic genes. *Plant Cell Physiology*, **51**, 144–163.

Shi, R., Yang, C., Lu, S. *et al.* (2010b) Specific down-regulation of PAL genes by artificial microRNAs in *Populus trichocarpa. Planta*, **6**, 1281–1288.

Shi, R., Shuford, C.M., Wang, J.P. *et al.* (2013) Regulation of phenylalanine ammonia-lyase (PAL) gene family in wood forming tissue of *Populus trichocarpa. Planta*, **238**, 487–497.

Shuford, C.M., Li, Q., Sun, Y. *et al.* (2012) Comprehensive quantification of monolignol-pathway enzymes in *Populus trichocarpa* by protein cleavage isotope dilution mass spectrometry. *Journal of Proteome Research*, **11**, 3390–3404.

Song, J., Lu, S., Chen, Z.Z. *et al.* (2006) Genetic transformation of *Populus trichocarpa* genotype Nisqually-1: a functional genomic tool for woody plants. *Plant Cell Physiology*, **47**, 1582–1589.

Terashima, N., Kitano, K., Kojima, M. *et al.* (2009) Nanostructural assembly of cellulose, hemicellulose, and lignin in the middle layer of secondary wall of ginkgo tracheid. *Journal of Wood Science*, **55**, 409–416.

Tuskan, G.A., Difazio, S., Jansson, S. *et al.* (2006) The genome of black cottonwood, *Populus trichocarpa* (Torr. & Gray). *Science*, **313**, 1596–1604.

Vance, C.P., Kirk, T.K. and Sherwood, R.T. (1980) Lignification as a mechanism of disease resistance. *Annual Review of Phytopathology*, **18**, 259–288.

Wang, J.P., Shuford, C.M., Li, Q. *et al.* (2012) Functional redundancy of the two 5-hydroxylases in monolignol biosynthesis of *Populus trichocarpa*: LC–MS/MS based protein quantification and metabolic flux analysis. *Planta*, **236**, 795–808.

Wang, J.P., Naik, P.P., Chen, H. *et al.* (2014) Complete proteomic-based enzyme reaction and inhibition kinetics reveal how monolignol biosynthetic enzyme families affect metabolic flux and lignin in *Populus trichocarpa*. *Plant Cell*, **26**, 894–914.

Wang, J.P., Chuang, L., Loziuk, P.L. *et al.* (2015) Phosphorylation is an on/off switch for 5-hydroxyconiferaldehyde *O*-methyltransferase activity in poplar monolignol biosynthesis. *Proceedings of the National Academy of Sciences of the United States of America*, **112**, 8481–8486.

Westermark, U. (1985) The occurrence of *p*-hydroxyphenylpropane units in the middle lamella of spruce (*Picea abies*). *Wood Science and Technology*, **19**, 223–232.

White, R.H. (1987) Effect of lignin content and extractives on the higher heating value of wood. *Wood and Fiber Science*, **19**, 446–452.

Wright, D., Brown, S.A. and Neish, A.C. (1958) Studies of lignin biosynthesis using isotopic carbon. VI. Formation of the side chain of the phenylpropane monomer. *Canadian Journal of Biochemistry and Physiology*, **36**, 1037–1045.

Yeh, T.F., Yamada, T., Capanema, E. *et al.* (2005) Rapid screening of wood chemical component variations using transmittance near-infrared spectroscopy. *Journal of Agricultural and Food Chemistry*, **53**, 3328–3332.

Zhang, K., Qian, Q., Huang, Z. *et al.* (2006) GOLD HULL and INTERNODE2 encodes a primarily multifunctional cinnamyl-alcohol dehydrogenase in rice. *Plant Physiology*, **140**, 972–983.

Chapter 5
Monolignol Biosynthesis and Regulation in Grasses

Peng Xu and Laigeng Li

Abstract: Lignin is a main phenolic component of cell walls that provides plant tissues with strength and rigidness. The nature of inherent hydrophobicity and extensive cross-links with wall-bound polysaccharides makes lignin extremely recalcitrant to mechanical disruption and enzymatic degradation. Grasses encompass most of the world's economical cereal crops and fodder species that supply human society with dietary grains, fiber, and forage. Manipulation of lignin composition and structure in grasses has a major impact on the nutritional quality of animal feed and the yield of fermentable sugars generating from bulk plant biomass. As lignin in grass species displays some characteristic difference from that in dicot plants, evidences from biochemical and genetic studies have been accumulated for the understanding of the monolignol biosynthesis in grasses. Here, we summarize the advancements of the grass lignin synthesis studies, which would aid in a comprehensive understanding of the monolignol biosynthesis and regulation as well as confer implications for bioengineering strategies of cell walls in various plants.

Keywords: monocot, dicot, cell walls, monolignol biosynthesis, vascular system.

5.1 Introduction

Lignin, a major cell wall component that bestows plant cells with structural integrity and stiffness, represents up to 30% of the total biomass produced by the land plants (Boerjan *et al.*, 2003). Lignin is mainly polymerized by hydroxycinnamyl alcohols or monolignols that are derived from the phenylpropanoid pathway. Monolignols include *p*-coumaryl alcohol, coniferyl alcohol, and sinapyl alcohol that give rise to hydroxyphenyl (H), guaiacyl (G), and syringyl (S) lignin units during polymerization at the intercellular space (Boerjan *et al.*, 2003). The composition of lignin exhibits notable differences in a variety of plant species. All tracheophytes are believed to contain G lignin, while deposition of S lignin is

Recent Advances in Polyphenol Research, First Edition. Edited by Kumi Yoshida, Véronique Cheynier and Stéphane Quideau.

only restricted to certain lineages. S lignin is largely absent in fern and gymnosperm species but universally present in angiosperms. The detection of S lignin in some lycophyte species such as *Selaginella moellendorffii* suggests that the ability to produce S lignin may have been lost or evolved many times during tracheophyte evolution (Weng *et al.*, 2008; Weng & Chapple, 2010). Compared with nonvascular plants, biosynthesis of lignin is considered to be unique for tracheophyte species. Bryophytes contain orthologs of the entire set of monolignol biosynthesis genes, but the output of phenylpropanoid pathway in bryophytes is the defense metabolites against damaging UV-B radiation, rather than lignin compounds (Xu *et al.*, 2009). Similarly with the lignin compounds, plants produce dimeric lignans through coupling of hydroxycinnamyl alcohols, allylphenols, or hydroxycinnamic acids. Lignans in bryophytes merely consist of hydroxycinnamyl acids, while a variety of hydroxycinnamyl alcohols were contained in the lignans of tracheophytes (Umezawa, 2003). This observation suggests that metabolic flux of phenylpropanoid pathway is redirected from hydroxycinnamyl acids to the level of alcohols in the evolution of tracheophytes. In addition to the three typical monolignols, the lignin polymers also have several noncanonical units, including ferulate, coniferaldehyde, and acylated monolignols (Sederoff *et al.*, 1999; Ralph *et al.*, 2004, 2007).

Current understanding of the monolignol biosynthesis and regulation is mainly built on biochemical and genetic studies in *Arabidopsis* and *Populus*, the two research models of dicot plants. The knowledge gained from the dicot species has been proven instructive in manipulating lignin content and structure. Monocots, on the other hand, encompass a variety of valuable cereal crops and fodder species that make a major contribution to the supply of dietary grains, fiber, and forage for human society. In particular, switchgrass and *Miscanthus* have been developed as potential bioenergy crops dedicated for biofuel production. Lignin composition and structure are different in these grasses compared with dicot species. For example, a higher level of H lignin unit is present in grasses and extensive covalent linkages between lignin and ferulate enhance strength of their secondary cell walls. Recent research studies have exerted great efforts to reveal molecular mechanisms underlying these unique aspects. Here, we summarize the up-to-date knowledge of the monolignol biosynthesis and regulation in grasses, which would help facilitate modification of grass cell walls.

5.2 Unique cell walls in grasses

Two types of primary cell walls have been categorized in flowering plants according to their composition (Carpita & Gibeaut, 1993; Carpita, 1996). Type I cell walls are universally present in gymnosperms, dicots, and noncommelinoid monocots, featured with xyloglucans, pectins, and structural proteins embedded in the cellulose microfibril matrix. Type II cell walls are restricted to the commelinoid monocots, such as grasses, sedges, rushes, and gingers. They contain much lower levels of xyloglucans, pectins, and structural proteins, but higher amounts of heteroxylans and hydroxycinnamyl acids. Arabinose and glucuronic acid are widely attached to the β-1,4-linked xylose backbone in grass cell walls, making arabinoxylans and glucuronoarabinoxylans as main heteroxylan components.

These xylans provide a major structural support in grasses through extensively binding to cellulose microfibrils and other cell wall components. For hydroxycinnamyl acids, up to 4% ferulate (in maize) and 3% *p*-coumarate (in sorghum) are contained in grass cell walls (Hatfield *et al.*, 1999; Saulnier *et al.*, 1999). Other cell wall components, for example, (1,3;1,4)-β-glucans, are also unique components to grasses.

Secondary cell walls are much thicker and more rigid than the thin and flexible primary cell walls. Lignin constitutes a 7–10% of secondary cell wall content in grasses (Ishii, 1997). In contrast with dicot secondary cell walls, grasses contain a significantly larger amount of H lignin with a proportion varying from 4% of mature maize stalks to 15% of rice straw (Baucher *et al.*, 1998). Lignin in grasses can be covalently linked to ferulate via an ether linkage. Meanwhile, ferulate could be attached to arabinoxylans via an ester linkage between its carboxylic acid group and the C5-hydroxyl of the arabinose side chain (de O Buanafina, 2009). Oxidative coupling of ferulate leads to ferulate dimers or even oligoferulates, which cross-link adjacent arabinoxylans and create ester–ether bridges between arabinoxylans and lignin (Grabber *et al.*, 1995; Jacquet *et al.*, 1995; Ralph *et al.*, 1995; Wende & Fry, 1997; Saulnier *et al.*, 1999; Fry *et al.*, 2000). The covalent connections between ferulate, arabinoxylans, and lignin greatly strengthen secondary cell walls in grasses. In addition, ferulate in grass cell walls is suggested to serve as initiation sites for lignin formation (Jacquet *et al.*, 1995; Ralph *et al.*, 1995; Grabber *et al.*, 2002). Apart from ferulate, grass stem tissues also contained a high amount of *p*-coumarate esters. Most of the *p*-coumarate in grasses is attached to lignin fraction, although certain amount of *p*-coumarate is esterified to arabinoxylans (Morrison *et al.*, 1998). It was demonstrated that *p*-coumarate is acylated to the γ-position of syringyl unit prior to their incorporation into lignin (Ralph *et al.*, 1994; Grabber *et al.*, 1996).

5.3 Lignin deposition in grasses

Deposition of lignin is mainly confined to sclerenchyma cells and strictly regulated throughout plant growth and development. Vascular tissues of angiosperms contain vessel cells and fiber cells, which develop thickened secondary cell walls. As lignin provides vessel cells with strength to bear the negative pressure generated from transpiration, disruption of lignin biosynthesis leads to collapsed vessels with a thinner cell wall (Jones *et al.*, 2001). Compared with G-lignin-enriched vessel cells, fiber cells are more abundant in S lignin, suggesting that biosynthesis of lignin monomers is strictly controlled during development of these cell types (Chapple *et al.*, 1992; Li *et al.*, 2001). When there is a lack lignified cell walls in fiber cells, plant stem is fragile and difficult to stand erect (Zhong *et al.*, 1997, 2007b; Mitsuda *et al.*, 2007). Apart from vascular tissues, deposition of lignin in endothecium is required for proper anther dehiscence (Yang *et al.*, 2007; Gui *et al.*, 2011). Lignin is also present in valve margin tissue of seed pods allowing plants to shatter and disperse the seeds (Dinneny, 2005).

Compared with cylindrical arrangement of vascular bundles in dicots, grasses develop scattered vascular bundles throughout the internode stem. Lignin amount in vascular tissues increases during maturation of the grass internode (Morrison *et al.*, 1998). In the young tissues of switchgrass, deposition of lignin is mainly restricted in the vascular bundles. In contrast, both parenchyma and interfascicular fiber cells are lignified when plant grows older (Shen *et al.*, 2009). The ratio of ester-linked *p*-coumarate to ferulate is often employed to reflect the overall extent of lignification and cell wall digestibility, because *p*-coumarate is mainly esterified to the syringyl units, reflecting overall lignification in grass tissues, and saponification of ferulate declines as cell walls undergo lignification (Morrison *et al.*, 1998; Grabber *et al.*, 2004; Shen *et al.*, 2009).

5.4 Monolignol biosynthesis in grasses

5.4.1 Proposed pathway for monolignol biosynthesis

Lignin biosynthesis involves highly regulated processes of monolignol biosynthesis, transportation, and polymerization (Boerjan *et al.*, 2003; Bonawitz & Chapple, 2010; Vanholme *et al.*, 2010). The biosynthesis of monolignols generally begins with phenylalanine, although tyrosine was also suggested as a precursor in grass species (Fig. 5.1) (Rosler *et al.*, 1997). Deamination of phenylalanine by phenylalanine ammonia-lyase (PAL) to form cinnamate is the first committed step in monolignol biosynthesis (Sewalt *et al.*, 1997; Rasmussen & Dixon, 1999). Conversion of cinnamate into *p*-coumarate is mediated by cinnamate 4-hydroxylase (C4H). *p*-Coumaroyl-CoA could be produced through activation of *p*-coumarate to a CoA thioester by 4-coumarate:CoA ligase (4CL) (Lee *et al.*, 1997). *p*-Coumaroyl-CoA is a general precursor for biosynthesis of lignin and other phenylpropanoids. It can be reduced to *p*-coumaryl alcohol by cinnamoyl-CoA reductase (CCR) and cinnamyl alcohol dehydrogenase (CAD) (Chabannes *et al.*, 2001; Jones *et al.*, 2001) or channeled to produce caffeoyl-CoA via the shikimate shunt pathway. Hydroxycinnamoyl-CoA:shikimate hydroxycinnamoyl transferase (HCT) catalyzes the synthesis of *p*-coumaroyl shikimate using *p*-coumaroyl-CoA as acyl group donors (Hoffmann *et al.*, 2003). *p*-Coumarate 3-hydroxylase (C3H) can hydroxylate *p*-coumaroyl

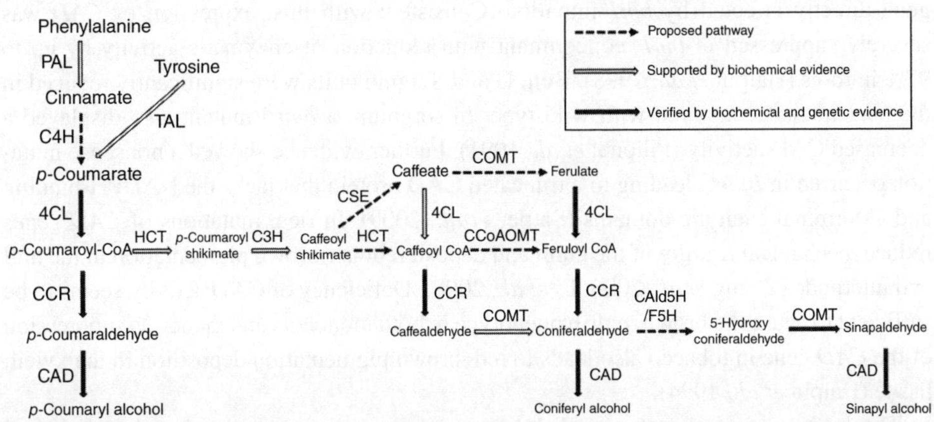

Fig. 5.1 The monolignol biosynthesis pathway in grasses.

shikimate into the corresponding caffeate conjugate, which is then converted to caffeoyl-CoA by HCT (Schoch *et al.*, 2001). Recent study identified a caffeoyl shikimate esterase (CSE), which hydrolyzes caffeoyl shikimate into caffeate, bypassing the last catalytic step by HCT (Vanholme *et al.*, 2013). The resulted caffeate could be converted to caffeoyl-CoA by 4CLs (Lee *et al.*, 1997).

Further methylation and hydroxylation of caffeoyl-CoA is required to produce coniferyl and sinapyl alcohol. Methylation of caffeoyl-CoA at 3-hydroxyl group by Caffeoyl CoA *O*-methyltransferase (CCoAOMT) leads to feruloyl-CoA (Zhong *et al.*, 1998; Li *et al.*, 1999). Feruloyl-CoA could be reduced to coniferaldehyde by CCR, and further reduction of coniferaldehyde by CAD generates coniferyl alcohol. Meanwhile, coniferaldehyde can be 5-hydroxylated into 5-hydroxyconiferaldehyde by coniferaldehyde 5-hydroxylase (CAld5H) or ferulate 5-hydroxylase (F5H) (Humphreys *et al.*, 1999; Osakabe *et al.*, 1999). Methylation of the 5-hydroxyconiferaldehyde by caffeic acid *O*-methyltransferase (COMT) at 5-hydroxyl group leads to sinapaldehyde, which can be reduced by CAD to generate sinapyl alcohol (Li, 2000; Guo *et al.*, 2001).

5.4.2 Monolignol biosynthetic genes in grasses

Several mutants associated with lignin biosynthesis have been discovered in grasses since 1924. The *brown midrib* (*bm*) mutants in maize were first identified for the reddish brown pigmentation in the leaf midrib and stalk pith (Barrière *et al.*, 2004). Four *bm* mutants, namely *bm1–4*, were identified in maize in 1924–1947, and two new *bm* mutants *bm5* and *bm6* were discovered recently from the Maize Genetic Stock Center and confirmed by the tests of allelism (Ali *et al.*, 2010). Although genes responsible for *bm* phenotypes segregate as a simple Mendelian recessive character, genetic information of the *bm* mutants is currently limited to *bm1*, *bm2*, and *bm3*. In addition to maize, *bm* mutants were also identified in sorghum and millet through chemically induced mutagenesis (Porter *et al.*, 1978; Cherney *et al.*, 1988). Some mutants exhibit reduced lignin content, increased cell wall digestibility, and enhanced glucose yield from cell wall digestion (Dien *et al.*, 2009).

The *bm1* locus co-segregates with the *CAD* gene in maize, suggesting that *CAD* is the gene directly affected by *bm1* mutation. Consistent with this, expression of *CAD* was severely suppressed in *bm1*, concomitant with reduction of enzymatic activity by up to 97% in roots (Halpin *et al.*, 1998). Both G and S lignin units were significantly reduced in the *bm1* mutant compared with wild-type. In sorghum, a *bmr6* mutant also displayed a decreased CAD activity (Pillonel *et al.*, 1991). Further evidence showed a nonsense mutation occurred in *Bmr6*, leading to a truncated CAD protein that lacks the NADPH-binding and C-terminal catalytic domains (Sattler *et al.*, 2009). In rice, mutations of *CAD* genes reduce mechanical rigidity of the culm and deposit reddish-brown pigmentation in the hull and internode (Zhang *et al.*, 2006; Li *et al.*, 2008). Deficiency of CAD activity seems to be sufficient to cause the brown midrib phenotype in both monocots and dicots, as suppression of the *CAD* gene in tobacco also leads to red–brown pigmentation deposition in the xylem tissue (Halpin *et al.*, 1994).

The *bm2* mutant exhibits a reddish-brown pigmentation on both adaxial and abaxial surfaces of the leaf midrib. Compared with wild-type maize, approximately 93% of the

Klason lignin is accumulated in the postanthesis stalks of *bm2* mutant, with decreased amounts of both S and G lignin (Tang *et al.*, 2014). A map-based cloning revealed that *bm2* encodes a methylenetetrahydrofolate reductase (MTHFR), which catalyzes formation of 5-methyltetrahydrofolate during methionine biosynthesis. MTHFR possibly involves in the biosynthesis of *S*-adenosyl-L-methionine, a methyl donor for CCoAOMT and COMT enzymes during monolignol biosynthesis. In *Arabidopsis*, suppression of the gene encoding *S*-adenosylmethionine synthetase also reduced lignin level (Shen *et al.*, 2002), suggesting that the adenosylmethionine-related pathway is important for lignin biosynthesis.

The maize *bm3* mutant has attracted extensive investigations because it contains more easily digested cell walls with a special composition of lignin. In *bm3* mutants, total lignin content was reduced by 25–40%, with heavily depleted S lignin units but nearly unaffected G units (Barrière *et al.*, 2004). A novel lignin unit, 5-hydroxyconiferyl alcohol, was shown to form benzodioxane structures during polymerization (Ralph *et al.*, 2000). The *bm3* plants contain mutations of the COMT gene, which catalyzes methylation of the 5-position of monolignol precursors (Vignols *et al.*, 1995). Consistently, transgenic maize with antisense *COMT* cDNA exhibited a lignin composition profile resembling the *bm3* phenotype with reduced Klason lignin content and decreased syringyl units (Piquemal, 2002). Functions of *COMT* are largely conserved in other grass species, such as sorghum, tall fescue, and perennial ryegrass (Bout & Vermerris, 2003; Chen *et al.*, 2004; Tu *et al.*, 2010).

Apart from the *bm* mutants, monolignol biosynthetic genes *4CL* and *CCR* were also investigated in grasses through biochemical and genetic approaches. Compared with *Arabidopsis* and *Populus*, *4CL* genes form a separate clade in rice, suggesting an evolutionarily divergent pattern (Gui *et al.*, 2011). Although 4CLs in dicots use *p*-coumarate as the most effective substrate *in vitro* (Lee *et al.*, 1997; Chen *et al.*, 2013), the rice 4CL enzymes exhibit a different enzymatic property with the highest activities toward ferulate. Consistent with this, downregulation of the dominant 4CL gene *Os4CL3* led to significant reduced lignin content, with concomitant accumulation of wall-bound and soluble *p*-coumarate and ferulate (Gui *et al.*, 2011). These results suggest that both *p*-coumarate and ferulate are *in vivo* substrates of Os4CL3. Suppression of *Os4CL3* also caused abnormal plant growth including reduced panicle fertility, suggesting that lignin biosynthesis is required for rice anther development. In switchgrass, 4CL is efficient to catalyze both *p*-coumarate and ferulate, and silencing expression of *4CL* gene significantly increased the efficiency of fermentable sugar release (Xu *et al.*, 2011).

Functions of CCR have been investigated in a broad range of grasses such as rice, wheat, maize, switchgrass, and perennial ryegrass. Similarly with dicots, lignin-related CCRs use feruloyl-CoA as the best substrate in grasses (Lauvergeat *et al.*, 2001; Ma, 2007; Escamilla-Treviño *et al.*, 2010). Suppression of *CCR*s in maize and perennial ryegrass caused a reduction of lignin content (Tu *et al.*, 2010; Tamasloukht *et al.*, 2011). As up to 50% of H lignin was decreased in the CCR-suppressed transgenic plants, the results also suggest that CCRs make a significant contribution to the biosynthesis of *p*-coumaryl alcohol in grasses. Deficiency of CCR enzymes also leads to increased cell wall digestibility without compromising plant growth and development. Interestingly, downregulation of *CCR* in ryegrass results in a reduced level of caffealdehyde (Tu *et al.*, 2010), indicating that caffeoyl CoA is a substrate of CCR in the species.

5.4.3 Functional genomics of monolignol biosynthesis in grass species

As numerous genome sequences are available in grass species, comparative genomic studies have been conducted to identify putative monolignol biosynthetic genes in these species (Barrière *et al.*, 2007; Andersen *et al.*, 2008; Xu *et al.*, 2009; Shen *et al.*, 2013). Phylogenetic analysis revealed that lignin-related gene families are expanded after the divergence of monocots and dicots (Xu *et al.*, 2009). Different expression patterns could be detected in the paralogs of monolignol biosynthetic genes, suggesting that functional divergence likely occurred after gene duplication. A recent study performed whole-genome microarray analysis in switchgrass and cataloged the candidate genes for monolignol biosynthesis (Shen *et al.*, 2013). Through summarizing these studies, we proposed a general map of monolignol biosynthesis pathway, which contains all of the key biosynthetic enzymes (Fig. 5.1). Currently, only *4CL*, *COMT*, *CCR*, and *CAD* have been characterized with biochemical and genetic evidence in grass species, and enzymatic activities of *PAL*, *TAL*, *HCT*, and *C3H* have been detected through *in vitro* approaches (Rosler *et al.*, 1997; Escamilla-Treviño *et al.*, 2013). Although phylogenetic and transcriptome results support the presence of *C4H* and *CCoAOMT*, suppression of their gene expression in switchgrass caused no significant change in lignin content or composition, which may be due to gene redundancy or complex regulations of lignin biosynthesis in the species (Shen *et al.*, 2013). So far, information about biological functions of *CSE* and *CAld5H/F5H* is scarce in grasses.

Grasses possess a unique cell wall structure, including a significant proportion of ferulate that cross-links with arabinoxylan and lignin network. However, molecular mechanisms underlying the biosynthesis of ferulated arabinoxylan remain to be investigated in grasses. In *Arabidopsis*, ferulate could be synthesized from coniferaldehyde via a hydroxycinnamaldehyde dehydrogenase (HCALDH) (Nair *et al.*, 2004). However, ferulate may not be a direct donor substrate for arabinoxylan feruloylation in grasses as transfer of the feruloyl group onto polysaccharides possibly occurs through feruloyl-CoA (de O Buanafina, 2009). In rice, activities of a feruloyltransferase were detected to catalyze the transfer of ferulate from feruloyl-CoA to arabinoxylan trisaccharide (Yoshida-Shimokawa *et al.*, 2001). Another feature of grass species is that their cell walls contain a high amount of *p*-coumarate attached to lignin fraction and arabinoxylans. Recent progresses suggest that *p*-coumaroylation in grass cell walls is mediated by BAHD family genes. Overexpression of a BAHD acyl-CoA transferase in rice resulted in threefold increase of cell wall bound *p*-coumarate that is ester-linked to arabinoxylan (Bartley *et al.*, 2013). Another BAHD acyl-CoA transferase protein was able to catalyze ester linkage formation between *p*-coumarate-CoA and sinapyl alcohol *in vitro* (Withers *et al.*, 2012).

5.5 Regulation of monolignol biosynthesis in grasses

5.5.1 Lignin regulation in secondary cell wall biosynthesis

Lignin is a major component of the secondary cell walls that predominately develop in vascular tissues. In *Arabidopsis* vascular fiber cells, secondary cell wall formation is redundantly regulated by NAC transcription factor SECONDARY WALL ASSOCIATED

NAC DOMAIN PROTEIN 1 (SND1) and its homolog NAC SECONDARY WALL THICKENING PROMOTING FACTOR 1 (NST1). Overexpression of these transcription factors induced ectopic deposition of secondary cell wall in normally unlignified cells such as stem parenchyma cells, flower cells, and leaf epidermal and mesophyll cells (Zhong *et al.*, 2006, 2007b; Mitsuda *et al.*, 2007). Double mutant of *SND1* and *NST1* completely suppressed secondary cell wall thickening in fibers cells. In *Arabidopsis*, secondary cell walls in silique valve margins are also regulated by these related NAC transcription factors. Expression of *NST1* was restricted to valve margins. Knockout mutant of *NST1* suppressed secondary wall formation in the region, and *snd1/nst1* double mutant displayed a loss of secondary walls in the whole silique (Mitsuda & Ohme-Takagi, 2008). *NST1* and its homolog *NST2* are responsible for the secondary wall thickening in anther endothecium (Mitsuda, 2005). A MYB transcription factor *AtMYB26* also regulates secondary wall biosynthesis in the anther endothecium, which is possibly through *NST1*- and *NST2*-mediated program (Yang *et al.*, 2007). Notably, secondary cell walls in vessel cells are not affected in the *snd1/nst1* double mutant, suggesting that different regulatory mechanisms exist in vessel cell formation. Consistently, another group of NAC transcription factors *VND6* and *VND7* were demonstrated to regulate development of vessel cells in *Arabidopsis* root (Kubo, 2005). *VND6* and *VND7* could directly active gene expressions associated with programmed cell death during vessel cell formation (Ohashi-Ito *et al.*, 2010; Zhong *et al.*, 2010; Yamaguchi *et al.*, 2011). *VND6* and *VND7* could also rescue secondary cell wall defects of the *snd1/nst1* double mutants through activating the secondary cell wall biosynthesis program (Zhong *et al.*, 2010).

The secondary-wall-associated NACs directly regulate targets of cell wall biosynthetic genes and a group of downstream transcription factors (Zhong *et al.*, 2008, 2010; Ohashi-Ito *et al.*, 2010; Yamaguchi *et al.*, 2011). Those targets include *MYB46* and *MYB83* that are specifically expressed during secondary wall formation (Zhong *et al.*, 2007a; McCarthy *et al.*, 2009). *MYB46* could activate expression of biosynthetic genes of cellulose, hemicellulose, and lignin during secondary cell wall formation. It can also regulate other transcription factors including *MYB58* and *MYB63*, which are particularly in charge of lignin biosynthesis (Ko *et al.*, 2009; Zhong & Ye, 2011; Kim *et al.*, 2012). MYB58 could directly activate expression of monolignol biosynthetic genes (*PAL, 4CL, C3H, CCoAOMT, CCR,* and *CAD*) through binding to the AC element, the conserved promoter sequence in most monolignol biosynthetic genes (Raes *et al.*, 2003; Fornalé *et al.*, 2006). Overexpression of *MYB58* and *MYB63* caused ectopic deposition of lignin in the normally unlignified cells (Zhou *et al.*, 2009). *MYB103* is another transcription factor downstream SND1 and NST1 and is required for S lignin production. *Arabidopsis myb103* mutant exhibited more than 70% decrease in S lignin, concomitant with a strong decrease of *CAld5H/F5H* gene (Öhman *et al.*, 2013). However, *CAld5H/F5H* is not a direct target of MYB103 or MYB58. In contrast, it can be directly regulated by SND1 and MYB46, suggesting that S lignin biosynthetic genes involves specific regulations, which are different from G lignin production (Zhao *et al.*, 2010; Ko *et al.*, 2014). Apart from the secondary-wall-associated NAC transcription network, *NtLIM1* is another transcriptional activator that can recognize promoters of lignin biosynthetic genes (Kawaoka *et al.*, 2000). Transgenic tobacco with antisense *NtLIM1* displayed approximately 27% reduction of lignin content accompanied by decreased expression levels of *PAL, 4CL,* and *CAD*.

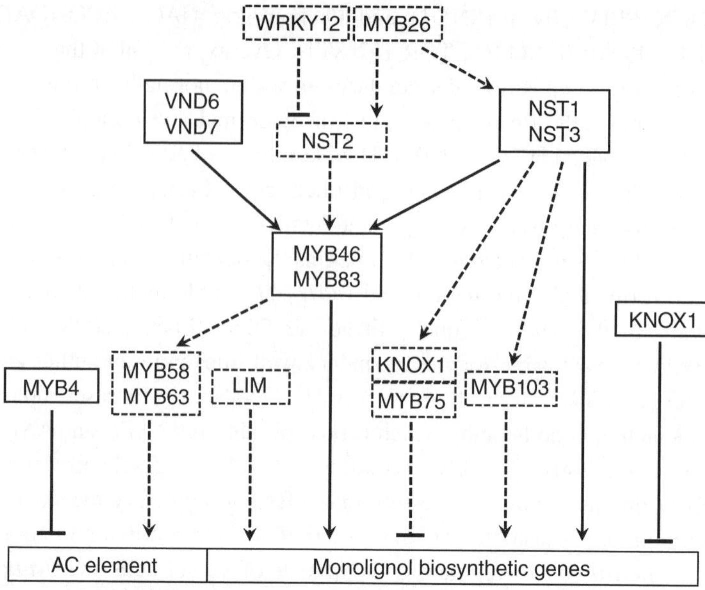

Fig. 5.2 Regulation of monolignol biosynthesis in grasses. Genes in boxes are regulators of monolignol biosynthesis. Dashed lines indicate that gene functions have not been verified in grasses.

Current analyses revealed a similar mechanism in regulation of lignin biosynthesis in grasses (Fig. 5.2). Coexpression analysis demonstrated that expression patterns of lignin biosynthetic genes are conserved in both dicots and monocots (Ruprecht & Persson, 2012). Using large-scale microarray analysis, recent study identified 107 conserved vascular gene groups, which are predominantly expressed in the stem tissues of *Arabidopsis*, *Populus*, and rice (Xu *et al.*, 2013). The conserved vascular gene groups include genes responsible for lignin biosynthesis, such as *PAL*, *4CL*, *C3H*, *CCR*, *CAD*, and *COMT*, and secondary-wall-associated transcription factors such as NAC, MYB, and KNOX genes (Xu *et al.*, 2013). Consistent with this, secondary-wall-associated NACs in rice and maize (*OsSWN*s and *ZmSWN*s) were sufficient to complement secondary cell wall defects of the *Arabidopsis snd1/nst1* double mutant (Zhong *et al.*, 2011). Overexpression of *OsSWN*s and *ZmSWN*s resulted in ectopic deposition of cellulose, xylan, and lignin in *Arabidopsis*. Moreover, OsSWNs and ZmSWNs could recognize consensus *cis*-acting elements in the promoters of *OsMYB46* and *ZmMYB46*, which are orthologs of *MYB46*. Overexpression of *OsMYB46* and *ZmMYB46* was also able to rescue secondary wall defects of the vessels in the *myb4/ myb83* double mutant of *Arabidopsis*. Collectively, these evidences suggest that regulations of secondary-wall-associated NAC-mediated network are conserved in dicots and grasses. Recently, transgenic experiments of NAC transcription factors in *Brachypodium* further support the conclusion. Overexpression of *BdSWN5* in *Brachypodium* resulted in ectopic secondary cell wall formation in roots and shoots, and secondary cell wall biosynthetic genes and cell death related genes were upregulated in the overexpression lines (Valdivia *et al.*, 2013).

Although conserved mechanism in secondary cell biosynthesis has been uncovered, the possibility that some regulations are unique to grasses could not be excluded. In *Arabidopsis*, *SHN* clade genes encode plant AP2/EREBP transcription factors and their gain-of-function mutants caused sixfold increase of cuticular wax levels (Aharoni, 2004). Overexpression of *SHN* genes in *Arabidopsis* led to various changes in plant morphology, including varied leaf and petal epidermal cell structure, altered trichome number, and stomatal index. However, when *AtSHN2* was overexpressed in rice, lignin biosynthetic genes were downregulated while cellulose and other cell-wall-associated genes were upregulated (Ambavaram *et al.*, 2010). Gel mobility shift assays demonstrated that AtSHN2 was capable of binding to the promoters of rice cell-wall-associated transcription factors, such as NAC and MYB regulators. The unexpected biological functions of *AtSHN2* in rice suggested different regulatory mechanisms between *Arabidopsis* and rice.

5.5.2 Repressor genes of monolignol biosynthesis in grasses

Lignin represents the major output of phenylpropanoid pathway that accumulated in specific types of plant cells. Suppression of lignin deposition in unlignified tissues is crucial for plant normal growth and development. Most of the characterized lignin repressor genes belong to subgroup 4 of R2R3-MYB transcription factor family, such as *AtMYB4* in *Arabidopsis*, *EgMYB1* in *Eucalyptus*, *AmMYB308* and *AmMYB330* in *Antirrhinum*. In *Arabidopsis*, mutant of *AtMYB4* is more tolerant for UV-B irradiation (Jin *et al.*, 2000). More sinapate esters are accumulated in leaves of the mutant, concomitant with the enhanced expression level of *C4H*. *AmMYB308* is an ortholog gene of *AtMYB4* in *Antirrhinum*. When *AmMYB308* was overexpressed in tobacco, expressions of *C4H*, *4CL*, and *CAD* were downregulated and 17% of lignin was reduced (Tamagnone *et al.*, 1998). *EgMYB1* is another ortholog gene of *AtMYB4* that could directly bind to the promoters of *CCR* and *CAD* genes in *Eucalyptus* (Legay *et al.*, 2007, 2010). Apart from subgroup 4 of R2R3-MYB gene family, lignin repressors also include other MYB subgroup members such as *PttMYB21a* from *Populus* and *VvMYB5a* from grapevine (Karpinska *et al.*, 2004; Deluc *et al.*, 2006).

Repression of lignification is critical for growth and differentiation of parenchyma cells. *BREVIPEDICELLUS* (*BP*) is a class I *KNOTTED1*-like homeobox (KNOX) gene that plays an important role in meristem maintenance of *Arabidopsis*. BP could repress lignin biosynthetic genes *COMT* and *CCoAOMT* through direct binding to their promoters (Mele, 2003). Consistently, aberrant lignin deposition was detected in discrete stem regions of the *bp* mutant. Repression of lignification in parenchyma cells of the cortex and pith regions requires a WRKY family transcription factor *WRKY12* (Wang *et al.*, 2010). Plant *wrky12* mutants exhibit increased secondary wall components, including cellulose, xylan as well as lignin. EMSA results showed that WRKY12 directly binds to the *NST2* promoter fragment. In *Arabidopsis* leaf protoplasts, promoter activity of *NST2* is repressed when overexpressing *WRKY12*, consistent with its functions repressing secondary cell wall formation. Other transcription factors such as *MYB75* and *KNAT7* were also demonstrated as repressors for secondary cell wall thickening (Bhargava *et al.*, 2010; Li *et al.*, 2012).

Compared with dicots, only a few lignin repressors have been characterized in grasses (Fig. 5.2). *ZmMYB31* and *ZmMYB42* belong to subgroup 4 MYB family of maize (Fornalé *et al.*, 2006). Overexpression of *ZmMYB31* and *ZmMYB42* in *Arabidopsis* reduced lignin content (Sonbol *et al.*, 2009; Fornalé *et al.*, 2010). ZmMYB31 could repress expression of lignin biosynthetic genes *ZmCOMT* and *ZmF5H* through directly binding to AC element in the promoters. Chromatin immunoprecipitation experiment confirmed this interaction *in vivo*. A switchgrass MYB gene *PvMYB4* was also demonstrated to bind to the AC elements of monolignol pathway genes and repress the expression of these genes (Shen *et al.*, 2012). Ectopic overexpression of *PvMYB4* in switchgrass resulted in reduced lignin content and increased sugar release efficiency. In maize, *KNOX1* genes were also characterized for lignin repression. Lignin content was reduced when overexpressing maize *KNOX1* gene *KN1* (Townsley *et al.*, 2013). Overexpression of *KN1* in tobacco also suppressed lignin formation due to reduced expressions of several lignin biosynthetic genes, suggesting that *KNOX1* genes mediate negative regulation of lignin biosynthesis in both dicots and grasses.

5.5.3 Regulation of monolignol biosynthesis under stress

Lignin provides mechanical support for plant growth and functions as a physical barrier against infection of pathogens. When treated with an autoclaved cell wall preparation of *Rhizosphaera kalkhoffii* as elicitor, lignin was rapidly induced in spruce cultured cells with an increase of lignin biosynthetic genes (Messner & Boll, 1993; Lange *et al.*, 1995). Biosynthesis of lignin is also responsive to abiotic stimuli. For example, copper excess can induce lignin accumulation in *Arabidopsis* root (Lequeux *et al.*, 2010), and leaning stems in woody plants deposit a high amount of lignin (Moura *et al.*, 2010). Another intriguing phenomenon is that primary cell wall disruption is often accompanied with ectopic lignin deposition, a process likely associated with a cellular signaling pathway, which guards cell wall integrity. *AtCTL1* is a ubiquitous expressed chitinase-like protein that is involved in cellulose biosynthesis in primary cell walls (Sanchez-Rodriguez *et al.*, 2012). Mutant of *AtCTL1* in *Arabidopsis* displayed aberrant cell wall shapes and produced ectopic lignin in the pith of inflorescence stems (Zhong *et al.*, 2002). *CESA3* encodes a cellulose synthase subunit that is responsible for cellulose microfibrils biosynthesis in primary cell walls. Mutations in *CESA3* altered cell expansion, activated aberrant lignification, and invoked defense responses through jasmonate and ethylene signaling pathways (Cano-Delgado *et al.*, 2000, 2003). *CCR2* is merely expressed during *Arabidopsis* normal development, but its expression was drastically increased in the *cesa3* mutant, suggesting that ectopic lignin formation may involve a different set of biosynthetic genes.

Pathogen infections in grasses often invoke expression of lignin biosynthetic genes such as CCR and CCR-like genes. *OsRac1* encodes a small GTPase of Rac/Rop family that plays important roles in rice pathogen responses (Kawasaki *et al.*, 2005). OsRac1 interacts with OsCCR1 in a GTP-dependent manner, which induced enzymatic activity of OsCCR1 and promoted lignin accumulation in transgenic cell cultures. Thus, OsCCR1 functions as an effector of OsRac1 and regulates lignin production in pathogen response. Apart from *OsCCR1*, a rice CCR-like gene *Snl6* was identified through screening mutants of positive regulators of immunity (Bart *et al.*, 2010). *Snl6* was demonstrated to be required for pathogenesis-related gene expression during plant resistance to bacterial pathogen. Disruption of *Snl6* reduced lignin content and increased sugar extractability, suggesting a connection between lignin formation and pathogen defense.

5.6 Remarks

As a main component for integrity and rigidity of cell walls, lignin has attracted extensive investigations in its biosynthesis and regulation for a few decades. In dicots, significant progresses have been achieved in uncovering the genes responsible for core monolignol biosynthesis pathway. Several genes that exhibit functional conservation in grasses have been characterized through biochemical and genetic approaches. However, a complete picture of monolignol biosynthesis pathways in grasses has not been mapped yet. This may be largely due to the fact that several rounds of genome duplications in grasses have aggravated gene redundancy for lignin biosynthesis, reducing the efficiency of gene knockdown experiments. In this case, it is intriguing to investigate whether monolignol biosynthetic genes have evolved distinct functions accompanied with grass species diversification. Since covalent linkages between ferulate and arabinoxylan–lignin network greatly strengthen grass cell walls, elucidating how grasses synthesize feruloylated arabinoxylans is also significant. Although recent study revealed that a BAHD acyl-CoA transferase mediates *p*-coumaroylation of arabinoxylan, the feruloyltransferase that transfers ferulate to arabinoxylan has not been identified in grasses. Considering lignin as an effective defense barrier against pathogen infection, the nature of defensive lignins and the corresponding stress-related signals is another important aspect of lignin synthesis. It would be interesting to investigate whether a different set of biosynthetic genes are employed in grasses during defensive lignin formation.

Lignin deposition is spatiotemporally regulated in vascular tissues throughout plant growth and development. Studies have identified conserved gene regulators for monolignol biosynthesis, which may facilitate transfer of the acquired knowledge from dicots to grasses. However, grasses develop a special vascular system with the lignin composition different from dicots. Thus, a unique scenario of regulatory network likely underlies the different cell wall structures. Unraveling these distinctive aspects of monolignol biosynthesis needs more comprehensive genomics and genetic studies. Genome-wide association analysis is capable to dissect allelic variations in cultivated varieties with different cell wall compositions. Employing this approach in lignin researches would accelerate to discover more molecules regulating monolignol biosynthesis, which may assist innovation of bioengineering strategies to improve crops as well as biomass feedstock production.

Acknowledgments

This work was supported by the National Natural Science Foundation of China (31130012) and the National Key Basic Research Program of China (grant no. 2013CB127002).

References

Aharoni, A. (2004) The SHINE clade of AP2 domain transcription factors activates wax biosynthesis, alters cuticle properties, and confers drought tolerance when overexpressed in *Arabidopsis*. *Plant Cell*, **16**, 2463–2480.

Ali, F., Scott, P., Bakht, J., Chen, Y. and Lubberstedt, T. (2010) Identification of novel brown midrib genes in maize by tests of allelism. *Plant Breeding*, **129**, 724–726.

Ambavaram, M.M.R., Krishnan, A., Trijatmiko, K.R. and Pereira, A. (2010) Coordinated activation of cellulose and repression of lignin biosynthesis pathways in rice. *Plant Physiology*, **155**, 916–931.

Andersen, J., Zein, I., Wenzel, G. *et al.* (2008) Characterization of phenylpropanoid pathway genes within European maize (*Zea mays* L.) inbreds. *BMC Plant Biology*, **8**, 2.

Barrière, Y., Ralph, J., Méchin, V. *et al.* (2004) Genetic and molecular basis of grass cell wall biosynthesis and degradability. II. Lessons from *brown-midrib* mutants. *Comptes Rendus Biologies*, **327**, 847–860.

Barrière, Y., Riboulet, C., Méchin, V. *et al.* (2007) Genetics and genomics of lignification in grass cell walls based on maize as model species. *Genes, Genomes, and Genomics*, **1**, 133–156.

Bart, R.S., Chern, M., Vega-Sanchez, M.E., Canlas, P. and Ronald, P.C. (2010) Rice Snl6, a cinnamoyl-CoA reductase-like gene family member, is required for NH1-mediated immunity to *Xanthomonas oryzae pv. oryzae*. *PLoS Genetics*, **6**, e1001123.

Bartley, L.E., Peck, M.L., Kim, S.R. *et al.* (2013) Overexpression of a BAHD acyltransferase, OsAt10, alters rice cell wall hydroxycinnamic acid content and saccharification. *Plant Physiology*, **161**, 1615–1633.

Baucher, M., Monties, B., Montagu, M.V. and Boerjan, W. (1998) Biosynthesis and genetic engineering of lignin. *Critical Reviews in Plant Sciences*, **17**, 125–197.

Bhargava, A., Mansfield, S.D., Hall, H.C., Douglas, C.J. and Ellis, B.E. (2010) MYB75 functions in regulation of secondary cell wall formation in the *Arabidopsis* inflorescence stem. *Plant Physiology*, **154**, 1428–1438.

Boerjan, W., Ralph, J. and Baucher, M. (2003) Lignin biosynthesis. *Annual Review of Plant Biology*, **54**, 519–546.

Bonawitz, N.D. and Chapple, C. (2010) The genetics of lignin biosynthesis: connecting genotype to phenotype. *Annual Review of Genetics*, **44**, 337–363.

Bout, S. and Vermerris, W. (2003) A candidate-gene approach to clone the sorghum Brown midrib gene encoding caffeic acid *O*-methyltransferase. *Molecular Genetics and Genomics*, **269**, 205–214.

Cano-Delgado, A.I., Metzlaff, K. and Bevan, M.W. (2000) The *eli1* mutation reveals a link between cell expansion and secondary cell wall formation in *Arabidopsis thaliana*. *Development*, **127**, 3395–3405.

Cano-Delgado, A., Penfield, S., Smith, C., Catley, M. and Bevan, M. (2003) Reduced cellulose synthesis invokes lignification and defense responses in *Arabidopsis thaliana*. *Plant Journal*, **34**, 351–362.

Carpita, N.C. (1996) Structure and biogenesis of the cellwall of grasses. *Annual Review of Plant Physiology and Plant Molecular Biology*, **47**, 445–476.

Carpita, N.C. and Gibeaut, D.M. (1993) Structural models of primary cell walls in flowering plants: consistency of molecular structure with the physical properties of the walls during growth. *Plant Journal*, **3**, 1–30.

Chabannes, M., Barakate, A., Lapierre, C. *et al.* (2001) Strong decrease in lignin content without significant alteration of plant development is induced by simultaneous down-regulation of cinnamoyl CoA reductase (CCR) and cinnamyl alcohol dehydrogenase (CAD) in tobacco plants. *Plant Journal*, **28**, 257–270.

Chapple, C.C., Vogt, T., Ellis, B.E. and Somerville, C.R. (1992) An *Arabidopsis* mutant defective in the general phenylpropanoid pathway. *Plant Cell*, **4**, 1413–1424.

Chen, L., Auh, C.-K., Dowling, P., Bell, J., Lehmann, D. and Wang, Z.-Y. (2004) Transgenic downregulation of caffeic acid *O*-methyltransferase (COMT) led to improved digestibility in tall fescue (*Festuca arundinacea*). *Functional Plant Biology*, **31**, 235–245.

Chen, H.C., Song, J., Williams, C.M. *et al.* (2013) Monolignol pathway 4-coumaric acid:coenzyme A ligases in *Populus. trichocarpa*: novel specificity, metabolic regulation, and simulation of coenzyme A ligation fluxes. *Plant Physiology*, **161**, 1501–1516.

Cherney, J.H., Axtell, J.D., Hassen, M.M. and Anliker, K.S. (1988) Forage quality characterization of a chemically induced *brown-midrib* mutant in pearl millet. *Crop Science*, **28**, 783–787.

Deluc, L., Barrieu, F., Marchive, C. *et al.* (2006) Characterization of a grapevine R2R3-MYB transcription factor that regulates the phenylpropanoid pathway. *Plant Physiology*, **140**, 499–511.

Dien, B.S., Sarath, G., Pedersen, J.F. *et al.* (2009) Improved sugar conversion and ethanol yield for forage sorghum (*Sorghum bicolor* L. Moench) lines with reduced lignin contents. *BioEnergy Research*, **2**, 153–164.

Dinneny, J.R. (2005) A genetic framework for fruit patterning in *Arabidopsis thaliana*. *Development*, **132**, 4687–4696.

Escamilla-Treviño, L.L., Shen, H., Uppalapati, S.R. *et al.* (2010) Switchgrass (*Panicum virgatum*) possesses a divergent family of cinnamoyl CoA reductases with distinct biochemical properties. *New Phytologist*, **185**, 143–155.

Escamilla-Treviño, L.L., Shen, H., Hernandez, T., Yin, Y., Xu, Y. and Dixon, R.A. (2013) Early lignin pathway enzymes and routes to chlorogenic acid in switchgrass (*Panicum virgatum* L.) *Plant Molecular Biology*, **84**, 565–576.

Fornalé, S., Sonbol, F.-M., Maes, T. *et al.* (2006) Down-regulation of the maize and *Arabidopsis thaliana* caffeic acid *O*-methyl-transferase genes by two new maize R2R3-MYB transcription factors. *Plant Molecular Biology*, **62**, 809–823.

Fornalé, S., Shi, X., Chai, C. *et al.* (2010) ZmMYB31 directly represses maize lignin genes and redirects the phenylpropanoid metabolic flux. *Plant Journal*, **64**, 633–644.

Fry, S.C., Willis, S.C. and Paterson, A.E.J. (2000) Intraprotoplasmic and wall-localised formation of arabinoxylan-bound diferulates and larger ferulate coupling-products in maize cell-suspension cultures. *Planta*, **211**, 679–692.

Grabber, J.H., Hatfield, R.D., Ralph, J., Zon, J. and Amrhein, N. (1995) Ferulate cross-linking in cell walls isolated from maize cell suspensions. *Phytochemistry*, **40**, 1077–1082.

Grabber, J.H., Quideau, S. and Ralph, J. (1996) *p*-Coumaroylated syringyl units in maize lignin: implications for β-ether cleavage by thioacidolysis. *Phytochemistry*, **43**, 1189–1194.

Grabber, J.H., Ralph, J. and Hatfield, R.D. (2002) Model studies of ferulate–coniferyl alcohol cross-product formation in primary maize walls: implications for lignification in grasses. *Journal of Agricultural and Food Chemistry*, **50**, 6008–6016.

Grabber, J.H., Ralph, J., Lapierre, C. and Barrière, Y. (2004) Genetic and molecular basis of grass cell-wall degradability. I. Lignin–cell wall matrix interactions. *Comptes Rendus Biologies*, **327**, 455–465.

Gui, J., Shen, J. and Li, L. (2011) Functional characterization of evolutionarily divergent 4-coumarate:coenzyme A ligases in rice. *Plant Physiology*, **157**, 574–586.

Guo, D., Chen, F., Inoue, K., Blount, J.W. and Dixon, R.A. (2001) Downregulation of caffeic acid 3-*O*-methyltransferase and caffeoyl CoA 3-*O*-methyltransferase in transgenic alfalfa: impacts on lignin structure and implications for the biosynthesis of G and S lignin. *Plant Cell*, **13**, 73–88.

Halpin, C., Knight, M.E., Foxon, G.A. *et al.* (1994) Manipulation of lignin quality by downregulation of cinnamyl alcohol dehydrogenase. *Plant Journal*, **6**, 339–350.

Halpin, C., Holt, K., Chojecki, J. *et al.* (1998) *Brown-midrib* maize (*bm1*)—a mutation affecting the cinnamyl alcohol dehydrogenase gene. *Plant Journal*, **14**, 545–553.

Hatfield, R.D., Wilson, J.R. and Mertens, D.R. (1999) Composition of cell walls isolated from cell types of grain sorghum stems. *Journal of the Science of Food and Agriculture*, **79**, 891–899.

Hoffmann, L., Maury, S., Martz, F., Geoffroy, P. and Legrand, M. (2003) Purification, cloning, and properties of an acyltransferase controlling shikimate and quinate ester intermediates in phenylpropanoid metabolism. *Journal of Biological Chemistry*, **278**, 95–103.

Humphreys, J.M., Hemm, M.R. and Chapple, C. (1999) New routes for lignin biosynthesis defined by biochemical characterization of recombinant ferulate 5-hydroxylase, a multifunctional cytochrome P450-dependent monooxygenase. *Proceedings of the National Academy of Sciences of the United States of America*, **96**, 10045–10050.

Ishii, T. (1997) Structure and functions of feruloylated polysaccharides. *Plant Science*, **127**, 111–127.

Jacquet, G., Pollet, B., Lapierre, C., Mhamdi, F. and Rolando, C. (1995) New ether-linked ferulic acid–coniferyl alcohol dimers identified in grass straws. *Journal of Agricultural and Food Chemistry*, **43**, 2746–2751.

Jin, H., Cominelli, E., Bailey, P. *et al.* (2000) Transcriptional repression by AtMYB4 controls production of UV-protecting sunscreens in *Arabidopsis*. *EMBO Journal*, **19**, 6150–6161.

Jones, L., Ennos, A.R. and Turner, S.R. (2001) Cloning and characterization of irregular xylem4 (irx4): a severely lignin-deficient mutant of *Arabidopsis*. *Plant Journal*, **26**, 205–216.

Karpinska, B., Karlsson, M., Srivastava, M. *et al.* (2004) MYB transcription factors are differentially expressed and regulated during secondary vascular tissue development in hybrid aspen. *Plant Molecular Biology*, **56**, 255–270.

Kawaoka, A., Kaothien, P., Yoshida, K., Endo, S., Yamada, K. and Ebinuma, H. (2000) Functional analysis of tobacco LIM protein Ntlim1 involved in lignin biosynthesis. *Plant Journal*, **22**, 289–301.

Kawasaki, T., Koita, H., Nakatsubo, T. *et al.* (2005) Cinnamoyl-CoA reductase, a key enzyme in lignin biosynthesis, is an effector of small GTPase Rac in defense signaling in rice. *Proceedings of the National Academy of Sciences of the United States of America*, **103**, 230–235.

Kim, W.-C., Ko, J.-H. and Han, K.-H. (2012) Identification of a *cis*-acting regulatory motif recognized by MYB46, a master transcriptional regulator of secondary wall biosynthesis. *Plant Molecular Biology*, **78**, 489–501.

Ko, J.-H., Kim, W.-C. and Han, K.-H. (2009) Ectopic expression of MYB46 identifies transcriptional regulatory genes involved in secondary wall biosynthesis in *Arabidopsis*. *Plant Journal*, **60**, 649–665.

Ko, J.-H., Jeon, H.W., Kim, W.C., Kim, J.Y. and Han, K.H. (2014) The MYB46/MYB83-mediated transcriptional regulatory programme is a gatekeeper of secondary wall biosynthesis. *Annals of Botany*, **114**, 1099–1107.

Kubo, M. (2005) Transcription switches for protoxylem and metaxylem vessel formation. *Genes & Development*, **19**, 1855–1860.

Lange, B.M., Lapierre, C. and Sandermann, H., Jr (1995) Elicitor-induced spruce stress lignin (structural similarity to early developmental lignins). *Plant Physiology*, **108**, 1277–1287.

Lauvergeat, V., Lacomme, C., Lacombe, E., Lasserre, E., Roby, D. and Grima-Pettenati, J. (2001) Two cinnamoyl-CoA reductase (CCR) genes from *Arabidopsis thaliana* are differentially expressed during development and in response to infection with pathogenic bacteria. *Phytochemistry*, **57**, 1187–1195.

Lee, D., Meyer, K., Chapple, C. and Douglas, C.J. (1997) Antisense suppression of 4-coumarate:coenzyme A ligase activity in *Arabidopsis* leads to altered lignin subunit composition. *Plant Cell*, **9**, 1985–1998.

Legay, S., Lacombe, E., Goicoechea, M. *et al.* (2007) Molecular characterization of EgMYB1, a putative transcriptional repressor of the lignin biosynthetic pathway. *Plant Science*, **173**, 542–549.

Legay, S., Sivadon, P., Blervacq, A.-S. *et al.* (2010) EgMYB1, an R2R3 MYB transcription factor from eucalyptus negatively regulates secondary cell wall formation in *Arabidopsis* and poplar. *New Phytologist*, **188**, 774–786.

Lequeux, H., Hermans, C., Lutts, S. and Verbruggen, N. (2010) Response to copper excess in *Arabidopsis thaliana*: impact on the root system architecture, hormone distribution, lignin accumulation and mineral profile. *Plant Physiology and Biochemistry*, **48**, 673–682.

Li, L. (2000) 5-Hydroxyconiferyl aldehyde modulates enzymatic methylation for syringyl monolignol formation, a new view of monolignol biosynthesis in angiosperms. *Journal of Biological Chemistry*, **275**, 6537–6545.

Li, L., Osakabe, Y., Joshi, C.P. and Chiang, V.L. (1999) Secondary xylem-specific expression of caffeoyl-coenzyme A 3-*O*-methyltransferase plays an important role in the methylation pathway associated with lignin biosynthesis in loblolly pine. *Plant Molecular Biology*, **40**, 555–565.

Li, L., Cheng, X.F., LeshkevichE, J., Umezawa, T., Harding, S.A. and Chiang, V.L. (2001) The last step of syringyl monolignol biosynthesis in angiosperms is regulated by a novel gene encoding sinapyl alcohol dehydrogenase. *Plant Cell*, **13**, 1567–1586.

Li, X., Yang, Y., Yao, J. *et al.* (2008) FLEXIBLE CULM 1 encoding a cinnamyl-alcohol dehydrogenase controls culm mechanical strength in rice. *Plant Molecular Biology*, **69**, 685–697.

Li, E., Bhargava, A., Qiang, W. *et al.* (2012) The class II KNOX gene KNAT7 negatively regulates secondary wall formation in *Arabidopsis* and is functionally conserved in *Populus*. *New Phytologist*, **194**, 102–115.

Ma, Q.H. (2007) Characterization of a cinnamoyl-CoA reductase that is associated with stem development in wheat. *Journal of Experimental Botany*, **58**, 2011–2021.

McCarthy, R.L., Zhong, R. and Ye, Z.H. (2009) MYB83 is a direct target of SND1 and acts redundantly with MYB46 in the regulation of secondary cell wall biosynthesis in *Arabidopsis*. *Plant and Cell Physiology*, **50**, 1950–1964.

Mele, G. (2003) The knotted1-like homeobox gene BREVIPEDICELLUS regulates cell differentiation by modulating metabolic pathways. *Genes & Development*, **17**, 2088–2093.

Messner, B. and Boll, M. (1993) Elicitor-mediated induction of enzymes of lignin biosynthesis and formation of lignin-like material in a cell suspension culture of spruce (*Picea abies*). *Plant Cell, Tissue and Organ Culture*, **34**, 261–269.

Mitsuda, N. (2005) The NAC Transcription factors NST1 and NST2 of *Arabidopsis* regulate secondary wall thickenings and are required for anther dehiscence. *Plant Cell*, **17**, 2993–3006.

Mitsuda, N. and Ohme-Takagi, M. (2008) NAC transcription factors NST1 and NST3 regulate pod shattering in a partially redundant manner by promoting secondary wall formation after the establishment of tissue identity. *Plant Journal*, **56**, 768–778.

Mitsuda, N., Iwase, A., Yamamoto, H. *et al.* (2007) NAC transcription factors, NST1 and NST3, are key regulators of the formation of secondary walls in woody tissues of *Arabidopsis*. *Plant Cell*, **19**, 270–280.

Morrison, T.A., Jung, H.G., Buxton, D.R. and Hatfield, R.D. (1998) Cell-wall composition of maize internodes of varying maturity. *Crop Science*, **38**, 455–460.

Moura, J.C.M.S., Bonine, C.A.V., De Oliveira Fernandes Viana, J., Dornelas, M.C. and Mazzafera, P. (2010) Abiotic and biotic stresses and changes in the lignin content and composition in plants. *Journal of Integrative Plant Biology*, **52**, 360–376.

Nair, R.B., Bastress, K.L., Ruegger, M.O., Denault, J.W. and Chapple, C. (2004) The *Arabidopsis thaliana* REDUCED EPIDERMAL FLUORESCENCE1 gene encodes an aldehyde dehydrogenase involved in ferulic acid and sinapic acid biosynthesis. *Plant Cell*, **16**, 544–554.

de O Buanafina, M.M. (2009) Feruloylation in grasses: current and future perspectives. *Molecular Plant*, **2**, 861–872.

Ohashi-Ito, K., Oda, Y. and Fukuda, H. (2010) *Arabidopsis* VASCULAR-RELATED NAC-DOMAIN6 directly regulates the genes that govern programmed cell death and secondary wall formation during xylem differentiation. *Plant Cell*, **22**, 3461–3473.

Öhman, D., Demedts, B., Kumar, M. *et al.* (2013) MYB103 is required for FERULATE-5-HYDROXYLASE expression and syringyl lignin biosynthesis in *Arabidopsis* stems. *Plant Journal*, **73**, 63–76.

Osakabe, K., Tsao, C.C., Li, L. *et al.* (1999) Coniferyl aldehyde 5-hydroxylation and methylation direct syringyl lignin biosynthesis in angiosperms. *Proceedings of the National Academy of Sciences of the United States of America*, **96**, 8955–8960.

Pillonel, C., Mulder, M.M., Boon, J.J., Forster, B. and Binder, A. (1991) Involvement of cinnamyl-alcohol dehydrogenase in the control of lignin formation in *Sorghum bicolor* L. Moench. *Planta*, **185**, 538–544.

Piquemal, J. (2002) Down-regulation of caffeic acid *O*-methyltransferase in maize revisited using a transgenic approach. *Plant Physiology*, **130**, 1675–1685.

Porter, K.S., Axtell, J.D., Lechtenberg, V.L. and Colenbrander, V.F. (1978) Phenotype, fiber composition, and *in vitro* dry matter disappearance of chemically induced *brown midrib* (*bmr*) mutants of sorghum. *Crop Science*, **18**, 205–208.

Raes, J., Rohde, A., Christensen, J.H., Van de Peer, Y. and Boerjan, W. (2003) Genome-wide characterization of the lignification toolbox in *Arabidopsis*. *Plant Physiology*, **133**, 1051–1071.

Ralph, J., Hatfield, R.D., Quideau, S., Helm, R.F., Grabber, J.H. and Jung, H.-J.G. (1994) Pathway of *p*-coumaric acid incorporation into maize lignin as revealed by NMR. *Journal of the American Chemical Society*, **116**, 9448–9456.

Ralph, J., Grabber, J.H. and Hatfield, R.D. (1995) Lignin–ferulate cross-links in grasses: active incorporation of ferulate polysaccharide esters into ryegrass lignins. *Carbohydrate Research*, **275**, 167–178.

Ralph, J., Lapierre, C., Lu, F. *et al.* (2000) NMR evidence for benzodioxane structures resulting from incorporation of 5-hydroxyconiferyl alcohol into lignins of *O*-methyltransferase-deficient poplars. *Journal of Agricultural and Food Chemistry*, **49**, 86–91.

Ralph, J., Lundquist, K., Brunow, G. *et al.* (2004) Lignins: natural polymers from oxidative coupling of 4-hydroxyphenyl-propanoids. *Phytochemistry Reviews*, **3**, 29–60.

Ralph, J., Kim, H., Lu, F. *et al.* (2007) Identification of the structure and origin of a thioacidolysis marker compound for ferulic acid incorporation into angiosperm lignins (and an indicator for cinnamoyl CoA reductase deficiency). *Plant Journal*, **53**, 368–379.

Rasmussen, S. and Dixon, R.A. (1999) Transgene-mediated and elicitor-induced perturbation of metabolic channeling at the entry point into the phenylpropanoid pathway. *Plant Cell*, **11**, 1537–1552.

Rosler, J., Krekel, F., Amrhein, N. and Schmid, J. (1997) Maize phenylalanine ammonia-lyase has tyrosine ammonia-lyase activity. *Plant Physiology*, **113**, 175–179.

Ruprecht, C. and Persson, S. (2012) Co-expression of cell-wall related genes: new tools and insights. *Frontiers in Plant Science*, **3**.

Sanchez-Rodriguez, C., Bauer, S., Hematy, K. *et al.* (2012) CHITINASE-LIKE1/POM–POM1 and its homolog CTL2 are glucan-interacting proteins important for cellulose biosynthesis in *Arabidopsis*. *Plant Cell*, **24**, 589–607.

Sattler, S.E., Saathoff, A.J., Haas, E.J. *et al.* (2009) A nonsense mutation in a cinnamyl alcohol dehydrogenase gene is responsible for the sorghum brown midrib6 phenotype. *Plant Physiology*, **150**, 584–595.

Saulnier, L., Crépeau, M.-J., Lahaye, M. *et al.* (1999) Isolation and structural determination of two 5,5′-diferuloyl oligosaccharides indicate that maize heteroxylans are covalently cross-linked by oxidatively coupled ferulates. *Carbohydrate Research*, **320**, 82–92.

Schoch, G., Goepfert, S., Morant, M. *et al.* (2001) CYP98A3 from *Arabidopsis thaliana* is a 3′-hydroxylase of phenolic esters, a missing link in the phenylpropanoid pathway. *Journal of Biological Chemistry*, **276**, 36566–36574.

Sederoff, R.R., Mackay, J.J., Ralph, J. and Hatfield, R.D. (1999) Unexpected variation in lignin. *Current Opinion in Plant Biology*, **2**, 145–152.

Sewalt, V., Ni, W., Blount, J.W. *et al.* (1997) Reduced lignin content and altered lignin composition in transgenic tobacco down-regulated in expression of l-phenylalanine ammonia-lyase or cinnamate 4-hydroxylase. *Plant Physiology*, **115**, 41–50.

Shen, B., Li, C. and Tarczynski, M.C. (2002) High free-methionine and decreased lignin content result from a mutation in the *Arabidopsis S*-adenosyl-l-methionine synthetase 3 gene. *Plant Journal*, **29**, 371–380.

Shen, H., Fu, C., Xiao, X. *et al.* (2009) Developmental control of lignification in stems of lowland switchgrass variety *Alamo* and the effects on saccharification efficiency. *BioEnergy Research*, **2**, 233–245.

Shen, H., He, X., Poovaiah, C.R. *et al.* (2012) Functional characterization of the switchgrass (*Panicum virgatum*) R2R3-MYB transcription factor PvMYB4 for improvement of lignocellulosic feedstocks. *New Phytologist*, **193**, 121–136.

Shen, H., Mazarei, M., Hisano, H. *et al.* (2013) A genomics approach to deciphering lignin biosynthesis in switchgrass. *Plant Cell*, **25**, 4342–4361.

Sonbol, F.-M., Fornalé, S., Cappellades, M. *et al.* (2009) The maize ZmMYB42 represses the phenylpropanoid pathway and affects the cell wall structure, composition and degradability in *Arabidopsis thaliana*. *Plant Molecular Biology*, **70**, 283–296.

Tamagnone, L., Merida, A., Parr, A. *et al.* (1998) The AmMYB308 and AmMYB330 transcription factors from *Antirrhinum* regulate phenylpropanoid and lignin biosynthesis in transgenic tobacco. *Plant Cell*, **10**, 135–154.

Tamasloukht, B., Wong Quai Lam, M.S.J., Martinez, Y. *et al.* (2011) Characterization of a cinnamoyl-CoA reductase 1 (CCR1) mutant in maize: effects on lignification, fibre development, and global gene expression. *Journal of Experimental Botany*, **62**, 3837–3848.

Tang, H.M., Liu, S., Hill-Skinner, S. *et al.* (2014) The maizebrown midrib2 (bm2) gene encodes a methylenetetrahydrofolate reductase that contributes to lignin accumulation. *Plant Journal*, **77**, 380–392.

Townsley, B.T., Sinha, N.R. and Kang, J. (2013) KNOX1 genes regulate lignin deposition and composition in monocots and dicots. *Frontiers in Plant Science*, **4**.

Tu, Y., Rochfort, S., Liu, Z. *et al.* (2010) Functional analyses of caffeic acid *O*-methyltransferase and cinnamoyl-CoA-reductase genes from perennial ryegrass (*Lolium perenne*). *Plant Cell*, **22**, 3357–3373.

Umezawa, T. (2003) Diversity in lignan biosynthesis. *Phytochemistry Reviews*, **2**, 371–390.

Valdivia, E.R., Herrera, M.T., Gianzo, C. *et al.* (2013) Regulation of secondary wall synthesis and cell death by NAC transcription factors in the monocot *Brachypodium distachyon*. *Journal of Experimental Botany*, **64**, 1333–1343.

Vanholme, R., Demedts, B., Morreel, K., Ralph, J. and Boerjan, W. (2010) Lignin biosynthesis and structure. *Plant Physiology*, **153**, 895–905.

Vanholme, R., Cesarino, I., Rataj, K. *et al.* (2013) Caffeoyl shikimate esterase (CSE) is an enzyme in the lignin biosynthetic pathway. *Science*, **341**, 1103–1106.

Vignols, F., Rigau, J., Torres, M.A., Capellades, M. and Puigdom Nech, P. (1995) The brown midrib3 (bm3) mutation in maize occurs in the gene encoding caffeic acid *O*-methyltransferase. *Plant Cell*, **7**, 407–416.

Wang, H., Avci, U., Nakashima, J., Hahn, M.G., Chen, F. and Dixon, R.A. (2010) Mutation of WRKY transcription factors initiates pith secondary wall formation and increases stem biomass in dicotyledonous plants. *Proceedings of the National Academy of Sciences of the United States of America*, **107**, 22338–22343.

Wende, G. and Fry, S.C. (1997) 2-*O*-β-d-Xylopyranosyl-(5-*O*-feruloyl)-l-arabinose, a widespread component of grass cell walls. *Phytochemistry*, **44**, 1019–1030.

Weng, J.-K. and Chapple, C. (2010) The origin and evolution of lignin biosynthesis. *New Phytologist*, **187**, 273–285.

Weng, J.-K., Li, X., Stout, J. and Chapple, C. (2008) Independent origins of syringyl lignin in vascular plants. *Proceedings of the National Academy of Sciences of the United States of America*, **105**, 7887–7892.

Withers, S., Lu, F., Kim, H., Zhu, Y., Ralph, J. and Wilkerson, C.G. (2012) Identification of grass-specific enzyme that acylates monolignols with *p*-coumarate. *Journal of Biological Chemistry*, **287**, 8347–8355.

Xu, Z., Zhang, D., Hu, J. *et al.* (2009) Comparative genome analysis of lignin biosynthesis gene families across the plant kingdom. *BMC Bioinformatics*, **10**, S3.

Xu, B., Escamilla-Treviño, L.L., Sathitsuksanoh, N. *et al.* (2011) Silencing of 4-coumarate:coenzyme A ligase in switchgrass leads to reduced lignin content and improved fermentable sugar yields for biofuel production. *New Phytologist*, **192**, 611–625.

Xu, P., Kong, Y., Li, X. and Li, L. (2013) Identification of molecular processes needed for vascular formation through transcriptome analysis of different vascular systems. *BMC Genomics*, **14**, 217.

Yamaguchi, M., Mitsuda, N., Ohtani, M., Ohme-Takagi, M., Kato, K. and Demura, T. (2011) VASCULAR-RELATED NAC-DOMAIN 7 directly regulates the expression of a broad range of genes for xylem vessel formation. *Plant Journal*, **66**, 579–590.

Yang, C., Xu, Z., Song, J., Conner, K., Vizcay Barrena, G. and Wilson, Z.A. (2007) *Arabidopsis* MYB26/MALE STERILE35 regulates secondary thickening in the endothecium and is essential for anther dehiscence. *Plant Cell*, **19**, 534–548.

Yoshida-Shimokawa, T., Yoshida, S., Kakegawa, K. and Ishii, T. (2001) Enzymic feruloylation of arabinoxylan-trisaccharide by feruloyl-CoA:arabinoxylan-trisaccharide *O*-hydroxycinnamoyl transferase from *Oryza sativa*. *Planta*, **212**, 470–474.

Zhang, K., Qian, Q., Huang, Z. *et al.* (2006) GOLD HULL AND INTERNODE2 encodes a primarily multifunctional cinnamyl-alcohol dehydrogenase in rice. *Plant Physiology*, **140**, 972–983.

Zhao, Q., Wang, H., Yin, Y., Xu, Y., Chen, F. and Dixon, R.A. (2010) Syringyl lignin biosynthesis is directly regulated by a secondary cell wall master switch. *Proceedings of the National Academy of Sciences of the United States of America*, **107**, 14496–14501.

Zhong, R. and Ye, Z.H. (2011) MYB46 and MYB83 bind to the SMRE sites and directly activate a suite of transcription factors and secondary wall biosynthetic genes. *Plant and Cell Physiology*, **53**, 368–380.

Zhong, R., Taylor, J.J. and Ye, Z.H. (1997) Disruption of interfascicular fiber differentiation in an *Arabidopsis* mutant. *Plant Cell*, **9**, 2159–2170.

Zhong, R., III, W.H., Negrel, J. and Ye, Z.H. (1998) Dual methylation pathways in lignin biosynthesis. *Plant Cell*, **10**, 2033–2046.

Zhong, R., Kays, S.J., Schroeder, B.P. and Ye, Z.-H. (2002) Mutation of a chitinase-like gene causes ectopic deposition of lignin, aberrant cell shapes, and overproduction of ethylene. *Plant Cell*, **14**, 165–179.

Zhong, R., Demura, T. and Ye, Z.H. (2006) SND1, a NAC domain transcription factor, is a key regulator of secondary wall synthesis in fibers of *Arabidopsis*. *Plant Cell*, **18**, 3158–3170.

Zhong, R., Richardson, E.A. and Ye, Z.H. (2007a) The MYB46 transcription factor is a direct target of SND1 and regulates secondary wall biosynthesis in *Arabidopsis*. *Plant Cell*, **19**, 2776–2792.

Zhong, R., Richardson, E.A. and Ye, Z.H. (2007b) Two NAC domain transcription factors, SND1 and NST1, function redundantly in regulation of secondary wall synthesis in fibers of *Arabidopsis*. *Planta*, **225**, 1603–1611.

Zhong, R., Lee, C., Zhou, J., McCarthy, R.L. and Ye, Z.H. (2008) A battery of transcription factors involved in the regulation of secondary cell wall biosynthesis in *Arabidopsis*. *Plant Cell*, **20**, 2763–2782.

Zhong, R., Lee, C. and Ye, Z.H. (2010) Global analysis of direct targets of secondary wall NAC master switches in *Arabidopsis*. *Molecular Plant*, **3**, 1087–1103.

Zhong, R., Lee, C., McCarthy, R.L., Reeves, C.K., Jones, E.G. and Ye, Z.H. (2011) Transcriptional activation of secondary wall biosynthesis by rice and maize NAC and MYB transcription factors. *Plant and Cell Physiology*, **52**, 1856–1871.

Zhou, J., Lee, C., Zhong, R. and Ye, Z.H. (2009) MYB58 and MYB63 are transcriptional activators of the lignin biosynthetic pathway during secondary cell wall formation in *Arabidopsis*. *Plant Cell*, **21**, 248–266.

Chapter 6

Creation of Flower Color Mutants Using Ion Beams and a Comprehensive Analysis of Anthocyanin Composition and Genetic Background

Yoshihiro Hase

Abstract: Mutagenesis is an effective approach to expand flower color variation in ornamental plants. Ion beams generated by accelerators are useful mutagens in plant mutation breeding, although the mutability often differs, probably depending on the genetic background of the parental plants. This chapter summarizes the analyses of flower color mutants of cyclamen and carnation obtained by ion beams. Accumulated knowledge regarding the relationship among flower colors, pigments, and associated genes will help our understanding of their genetic background and the choice of suitable starting materials. Our research demonstrated that mutants with altered anthocyanin compositions corresponded to mutation in genes in the biosynthetic pathway and that further understanding of biosynthetic genes associated with the expression of flower colors is important to create novel flower varieties.

Keywords: ion beam, mutation, flower color, anthocyanin, fragrant cyclamen, carnation.

6.1 Introduction

After artificially induced mutations by X-rays were first demonstrated in plants (Stadler, 1928), thousands of mutant plants have been isolated using ionizing radiation and chemical mutagens. The induced mutants have contributed to many fields of plant biology, as well as plant breeding. To date, more than 3200 mutant varieties have been registered in the mutant variety database (Joint FAO/IAEA Programme). Most of those mutant varieties were obtained using gamma rays. The Takasaki Ion Accelerators for

Recent Advances in Polyphenol Research, First Edition. Edited by Kumi Yoshida, Véronique Cheynier and Stéphane Quideau.

Advanced Radiation Application (TIARA) was established in 1991 to promote applied research using ion beams. The ion beam is a charged particle beam that produces dense ionization along the track of the ion particles (Kraft *et al.*, 1992; Goodhead, 1995). The amount of energy released along the unit length of the track is represented by the linear energy transfer (LET). Ion beams are termed high-LET radiation, because they have relatively high LET values. By contrast, gamma rays or X-rays are low-LET radiation, because they produce ionization only sparsely along their track. In terms of the spatial distribution of ionization within cells, gamma rays produce an almost homogeneous energy deposition, while ion beams produce much more localized energy deposition. These differences in energy deposition should result in different biological effects of ion beams and gamma rays. The biological effects of ion beams have been studied extensively in mammalian cells. Ion beams have higher biological effects than low-LET radiations per unit dose, such as growth inhibition, cytological damage, and mutation induction (Yatagai, 2004). However, only a few studies examined the biological effects of ion beams in plants before the 1990s (Hirono *et al.*, 1970; Kranz *et al.*, 1990). We have been studying the mutagenic effects of ion beams and their use as a new mutagen in plant research (Tanaka *et al.*, 2010). Mutation breeding using ion beams has been termed "ion beam breeding" and is becoming a practical tool in plant breeding.

More than half of the mutant varieties obtained using our facility are ornamental plants. This is probably because flower color variation is affected by many factors, including pigment composition or vacuolar pH, and flower color is not necessary for plant survival. It is not hard to anticipate that many genes are involved in the expression of flower color and a mutation in any one of those genes could result in altered color. Mutation frequencies at the percent level are often observed in flower colors, particularly in vegetatively propagated ornamental plants, which are thought to be highly heterogeneous compared with seed-propagated crops. In addition, ornamental plants with various flower colors are subject to changing fashions, and there is a continuous demand for new varieties. Thus, flower color has been one of the major targets in plant breeding using induced mutations.

Increasing evidence indicates that the ion beam is a useful mutagen in plant breeding. However, it is not always effective, because mutability often differs, even between varieties of the same plant species. Taking flower color variation as an example, the chrysanthemum cultivar "Taihei" provided various mutants, as will be described later (Nagatomi *et al.*, 1997), while the chrysanthemum cultivar "Natsuyasumi" was hardly changed (Sato *et al.*, 2006). Thus, the mutant traits that will be obtained are significantly affected by the genetic background of the material plants. It is thought that a greater understanding of a plant's genetic background is important to obtain desirable mutants efficiently. Flower color is an easily screenable target and flavonoid biosynthesis is well studied in higher plants at the molecular basis. In this context, our recent studies were carried out in cyclamen and carnation to accumulate the knowledge regarding the relationship among flower color, pigments, and the associated genes, as well as creating new varieties.

6.2 Induction of flower color mutants by ion beams

The pioneering work with ion beams on flower color mutations was carried out using the pink-colored chrysanthemum (*Chrysanthemum morifolium*) cultivar "Taihei" (Nagatomi *et al.*, 1997). Cultured leaf and petal explants were irradiated with carbon ion beams. Regenerated plants from the irradiated tissue were grown in the field, and various flower color mutants were obtained. The mutation spectrum was compared between carbon ion irradiation and gamma irradiation. Interestingly, gamma rays mostly induced light-pink mutants, while carbon ions induced a wider range of colors, including white, orange, and yellow mutants (Table 6.1). In addition, complex types and striped flowers, which were not obtained by gamma rays, were frequently obtained by carbon ions. Finally, five mutants, "Ion-no-komyo," "Ion-no-seiko," "Ion-no-hatsune," "Ion-no-koki," and "Ion-no-maho," were first applied for registration as mutant varieties obtained by ion beams (Tanaka *et al.*, 2010). A similar experiment was carried out in the spray-type carnation (*Dianthus caryophyllus*) cultivar "Vital" (Okamura *et al.*, 2003). Again, carbon ions showed a wider mutation spectrum compared with ethyl methane sulfonate (EMS), X-rays, and gamma rays (Table 6.2). Moreover, flower shape mutants were frequently obtained in the case of carbon ions. Although the underlying molecular mechanism remains to be explained, these studies suggested that the ion beams could induce novel mutant phenotypes that are difficult to obtain by low-LET radiations.

Generally, mutation breeding is a useful tool to improve a single or a few traits, because only a limited number of genes are affected among the entire genome. From this point of view, ion beams are more effective than low-LET radiations, because the ion beams deposit high energy locally along their track. In the case of carnation described earlier, the cherry pink color of the parental variety "Vital" was changed to red by ion beam irradiation and a new variety, "Red Vital," was developed (Okamura *et al.*, 2006). Additionally, the serrated edge of the petals was changed to a round edge and a new variety, "Beam Cherry," was developed. The new varieties retained the good agricultural traits of the parental variety. Thus, the flower color and petal shape mutations did not affect other aspects of growth and development. Ion beams affect a limited number of genes; therefore, mutants could be improved by a second round of irradiation, without appreciable effects on important agricultural traits. A novel deep orange-colored mutant variety "Vient Labios," which was not

Table 6.1 Frequency of flower mutants induced by gamma rays and carbon ions in the chrysanthemum cultivar "Taihei" (Pink flowers).

	Mutant frequency (%)					
Mutagen	White	Light pink	Dark pink	Orange	Yellow	Complex/stripe
Control	0	0.3	0	0	0	0
Gamma rays	0	27.7	2.1	0	0	0
Carbon ions	0.3	4.6	0.3	0.3	0.2	10.2

Data in this table are from Nagatomi *et al.* (1997). Light pink and dark pink indicate a lighter and a darker color than the pink color of the original cultivar, respectively.

Table 6.2 Frequency of flower mutants induced by carbon ions, EMS, gamma rays, and X-rays in the carnation cultivar "Vital" (cherry pink flowers).

	Mutation frequency ($\times 10^{-1}$ %)										
	Flower color									Flower shape	
Mutagen	Light pink	Pink	Dark pink	Red	Salmon	Yellow	Cream	Stripe	Complex	Round petal	Dianthus-type petal
Carbon ions	3.5	4.7	1.2	3.5	2.4	1.2	1.2	3.5	2.4	4.7	2.4
EMS	0	5.2	0	1.0	0	0	0	3.1	0	0	0
X-rays	1.7	8.4	0	3.4	0	0	0	0	0	0	0
Gamma rays	9.6	9.6	0	1.7	0	0	0	0	0	0.9	0

Data in this table are from Okamura *et al.* (2003). Light pink and dark pink indicate a lighter and a darker color than the cherry pink color of the original cultivar, respectively.

found among the existing varieties, was obtained after two rounds of carbon-ion irradiation from the original yellow-colored variety "Lemon Symphony" (Okada *et al.*, 2009). Thus, reirradiation of mutants with ion beams is thought to be effective to widen the mutation spectrum.

6.3 Mutagenic effects and the molecular nature of the mutations

The mutagenic effects of ion beams were examined in detail in *Arabidopsis thaliana* using marker genes whose mutations confer a visible phenotype (Shikazono *et al.*, 2003, 2005). Mutation in either one of the *TRANSPARENT TESTA* (*TT*) genes, which are involved in pigment biosynthesis, resulted in pale-yellow or light-brown seeds (Lepiniec *et al.*, 2006). *GLABRA* (*GL*) genes, which are required to develop trichomes, were also used (Oppenheimer *et al.*, 1991). Wild-type seeds were irradiated with carbon ions (LET 113 keV/μm) and an electron beam (0.2 keV/μm) at the same effective dose for seed survival, and *tt* and *gl* mutants were screened (Tanaka *et al.*, 1997). The mutation frequency of carbon ions was 2.8×10^{-4}/locus, which was four times higher than the mutation frequency of the electron beam (Shikazono *et al.*, 2003). This suggested that the carbon ions effectively induce mutations with a relatively low-dose range. The *Arabidopsis* genome has about 26,000 genes; therefore, this result also suggested that carbon ions induced about seven gene mutations on average in the *Arabidopsis* genome.

The molecular nature of the ion beam induced mutations that lead to *tt* and *gl* phenotypes has been identified (Shikazono *et al.*, 2001, 2005). Of the 29 mutant alleles obtained by carbon ions, 14 were intragenic mutations, such as base changes and small deletions, and 15 were large structural alterations, including large deletions, translocations, and inversions. In the case of the electron beam, nine were intragenic mutations and three were large structural alterations. These results suggested that the carbon ions induce large structural alterations more frequently than the electron beam. In addition, the features of the intragenic mutations and rejoined sites of the structural alterations suggested that carbon ions preferentially induce deletions.

6.4 Comprehensive analyses of flower color, pigments, and associated genes in fragrant cyclamen

Ordinary cyclamen cultivars have been developed by breeding with the wild ancestor of *Cyclamen percicum*. Although various flower colors, such as purple, pink, red, pale yellow, and white, are available in cyclamen cultivars, those flowers emit woody- or powdery-like scents that are targets for improvement. Three cultivars of fragrant cyclamen, "Kaori-no-mai (KM)" (flower color: purple, Fig. 6.1m), "Kokou-no-kaori (KO)" (light purple, Fig. 6.1n), and "Uruwashi-no-kaori (UR)" (pink, Fig. 6.1d), were created by crossing cyclamen cultivars with the scented cyclamen species *Cyclamen purpurascens* (Fig. 6.1o)

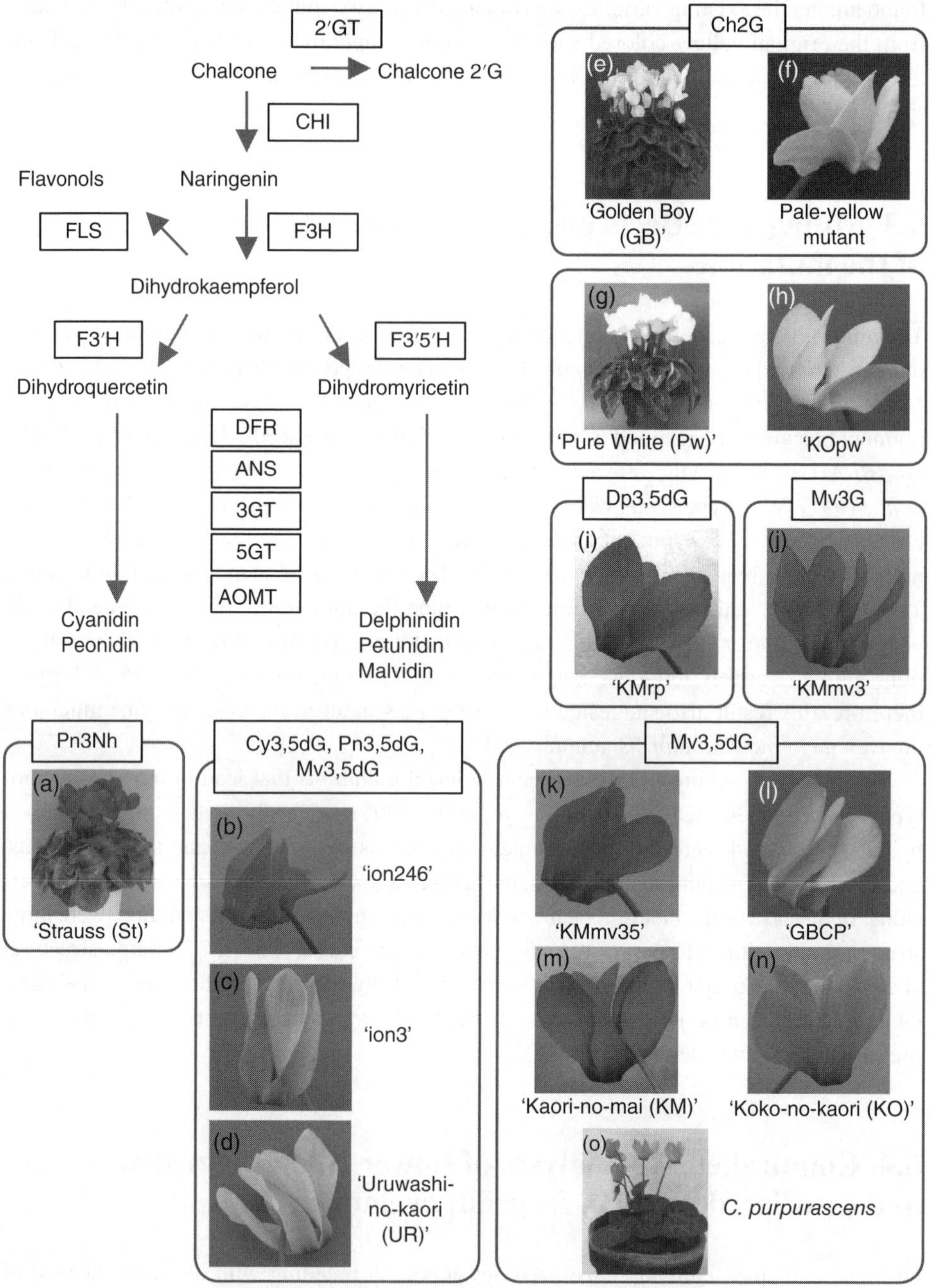

Fig. 6.1 Cyclamen cultivars, wild species, and flower color mutants shown beside the anthocyanin biosynthesis pathway. Flowers are grouped according to their major anthocyanins. (a) *Cyclamen persicum* cultivar "Strauss (St)." (b–d) Deeper color mutants "ion246" (b) and "ion3" (c) obtained from fragrant cyclamen cultivar "Uruwashi-no-kaori (UR)" (d). (e) *C. persicum* cultivar "Golden Boy (GB)." (f) A pale yellow-flowered mutant obtained from a hybrid (GB × *Cyclamen purpurascens*) "GBCP." (g) *C. persicum* cultivar "Pure White (Pw)." (h) A white-flowered mutant obtained from fragrant cyclamen cultivar "Koko-no-kaori (KO)." (i–k) Red–purple-flowered mutants "KMrp" (i), "KMmv3" (j), and "KMmv35" (k) obtained from fragrant cyclamen cultivar

(Ishizaka, 2008). Flowers of fragrant cyclamen emit rose- or hyacinth-like scents, similar to the flowers of *C. purpurascens*. This sweet fragrance adds significant ornamental value; however, the flower colors of fragrant cyclamen were limited to a range of pink to purple.

Cyclamen cultivars with red flowers ("Strauss (St)," Fig. 6.1a) and white flowers ("Pure white (Pw)," Fig. 6.1g) were used to cross with *C. purpurascens*, and crosses between the siblings were also performed in a breeding process. Therefore, the three fragrant cyclamens have a genome from St, Pw, and *C. purpurascens*. St has peonidin 3-neohesperidoside (Pn3Nh) as the major anthocyanin, along with quercetin and kaempferol as major flavonols (Takamura *et al.*, 2005). Pw has no detectable anthocyanin and accumulates quercetin and kaempferol (Takamura *et al.*, 1995). *C. purpurascens* accumulates malvidin 3,5-diglucoside (Mv3,5dG) as a major anthocyanin, along with quercetin and kaempferol as major flavonols (Takamura *et al.*, 2005). For the fragrant cyclamen, the major anthocyanin of KM and KO is Mv3,5dG. The major anthocyanins of UR are cyanidin 3,5-diglucoside (Cy3,5dG), peonidin 3,5-diglucoside (Pn3,5dG), and Mv3,5dG (Ishizaka *et al.*, 2012). Thus, among the three fragrant cyclamen, only UR has the ability to synthesize cyanidin and peonidin, which was probably inherited from the St genome. All of the anthocyanins of the fragrant cyclamen are the 3,5dG type. This suggested that the 3,5dG-type anthocyanin is genetically dominant to the 3Nh type. It is likely that the anthocyanin 5-glucosyltransferase (5GT) activity came from the *C. purpurascens* genome. As mentioned earlier, the flower color of the three cultivars of fragrant cyclamen is limited to a range of pink to purple. This likely reflects the introgression of the *C. purpurascens* genome. It is expected that new flower colors, such as yellow, red, and white, could be produced by inactivating anthocyanin biosynthetic genes from *C. purpurascens*. We carried out screening of flower color mutants to create new varieties and to promote greater understanding of the genetic background of fragrant cyclamen.

6.5 Mutagenesis and screening

To date, we have irradiated seeds, cultured petals, etiolated petioles, and adventitious embryos with ion beams. The etiolated petiole is thought to be the best among them because of its high regeneration frequency. In addition, petioles are relatively uniform in thickness, and this is advantageous for homogeneous irradiation with ion beams. The irradiated petioles were transferred to fresh medium with appropriate phytohormones to induce adventitious shoot and roots, and then the regenerated plantlets were acclimatized and grown to

"kaori-no-mai (KM)." (l) A hybrid "GBCP" derived from GB × *C. purpurascens*. (m and n) Fragrant cyclamen cultivars KM (m) and KO (n). (o) Scented wild species *C. purpurascens*. Fragrant cyclamen cultivars were created by a cross between *C. persicum* cultivars and *C. purpurascens*. Flower color mutants were obtained from mutagenized populations of fragrant cyclamen cultivars irradiated with ion beams. ANS, anthocyanidin synthase; AOMT, anthocyanin *O*-methyltransferase; Ch2G, chalcone 2′-glucoside; CHI, chalcone isomerase; Cy3,5dG, cyanidin 3,5-diglucoside; DFR, dihydroflavonol 4-reductase; Dp3,5dG, delphinidin 3,5-diglucoside; F3H, flavanone 3-hydroxylase; F3′H, flavonoid 3′-hydroxylase; F3′5′H, flavonoid 3′5′-hydroxylase; FLS, flavonol synthase; 2′GT, chalcone 2′-glucosyltransferase; 3GT, anthocyanidin 3-glucosyltransferase; 5GT, anthocyanin 5-glucosyltransferase; Mv3G, malvidin 3-glucoside; Mv3,5dG, malvidin 3,5-diglucoside; Pn3Nh, peonidin 3-neohesperidoside; Pn3,5dG, peonidin 3,5-diglucoside. (*See insert for color representation of the figure*)

maturity in a greenhouse to screen for flower color mutants (Kondo *et al.*, 2009). Interestingly, cultured petals often generated petal-like tissue in culture bottles; therefore, regenerated shoots were not efficiently obtained.

The commercialized fragrant cyclamen are amphidiploid with cyclamen cultivars (*Cyclamen persicum*) and *C. purpurascens* genomes. In mutation breeding, we expect recessive mutations that are caused by the loss of a certain gene function. From this viewpoint, it is more difficult to induce mutant traits in plants with higher ploidy levels because of genetic redundancy. Therefore, dihaploid plants, which were generated by anther culture, were mostly used in our experiments, rather than amphidiploid plants. As will be described later, red–purple mutants were obtained from a mutagenized population of purple-flowered KM. In the case of the irradiation of dihaploid tissue, red–purple mutants were obtained directly from the regenerated plants. By contrast, in the case of the irradiation to amphidiploid tissue, mutants were obtained only in the next generation obtained by self-pollination of the regenerated plants. These observations are consistent with the notion that the mutant trait is caused by a recessive mutation. Thus, etiolated petioles of dihaploid plants are thought to be the most efficient material to obtain flower color mutants in fragrant cyclamen.

All of the obtained flower color mutants were chimeric plants that had both mutant flowers and parental flowers (Ishizaka *et al.*, 2012). To obtain solid mutants, mutant tissues were propagated using tissue culture. The obtained solid mutants were usually sterile because they were dihaploid. Induction of chromosome doubling by colchicine treatment *in vitro* was used to produce fertile amphidiploid plants. The fertile plants could be distinguished by their relatively larger flowers and pollen production. Interestingly, spontaneous chromosome doubling occasionally produced fertile plants. For example, in the case of the irradiation to dihaploid tissue of the salmon-pink mutant "ion3," 93 plants out of 560 plants could produce pollen (Ishizaka *et al.*, unpublished data). Thus, ion beam irradiation might influence the frequency of somaclonal chromosome doubling.

6.5.1 Yellow mutants

To develop a yellow-flowered fragrant cyclamen, the cyclamen cultivar "Golden Boy (GB)" (pale yellow, Fig. 6.1e) was crossed with *C. purpurascens*. The *CHI* gene of GB was inactivated by an insertion sequence and the flowers of GB accumulated chalcone 2′-glucoside (Ch2G) (Miyajima *et al.*, 1991). The hybrid plants "GBCP" (Fig. 6.1l) had light-pink flowers that accumulated Mv3,5dG, because they have a functional *CHI* gene derived from *C. purpurascens*. Yellow-flowered mutants would be obtained by inactivating the functional *CHI* gene. The etiolated petioles of the dihaploid hybrid were irradiated with carbon ions and 1306 regenerated plants were grown. In eight of them, a part of the plant had yellow-green leaves, as opposed to the parental leaves that accumulate pigments. As anticipated, pale-yellow flowers were developed near the yellow-green leaves (Fig. 6.1f). The pale-yellow flowers were confirmed to accumulate Ch2G. The nonfunctional *CHI* from GB and the functional *CHI* from *C. purpurascens* are distinguishable by PCR amplification (Miyajima *et al.*, 1991). The functional *CHI* gene sequence was amplified from normal-leaf DNA but not from the yellow-green leaf DNA in some of the chimeric mutant plants (Hase *et al.*, 2012). These results suggested that the functional *CHI* from *C. purpurascens* was inactivated by ion beam irradiation and that disruption of CHI activity resulted in the accumulation of Ch2G via the authentic glucosyltransferase activity, which converts chalcone to Ch2G, giving a yellow flower (Togami *et al.*, 2011).

6.5.2 Red–purple mutants

Three types of red–purple mutants were obtained in a mutagenized population from KM (Fig. 6.1i–k). Although they were indistinguishable from each other by appearance, they had different types of major anthocyanins. The first one, "KMrp" (Fig. 6.1i), accumulated delphinidin 3,5-diglucoside (Dp3,5dG) as a major anthocyanin (Kondo *et al.*, 2009). As far as we know, this is the first identified plant to accumulate delphinidin-type anthocyanin in the genus *Cyclamen*. The major anthocyanin of parental KM is Mv3,5dG; therefore, this mutant was thought to have a mutation in the methylation step from delphinidin to malvidin. To identify the gene responsible, four homologs of anthocyanin *O*-methyltransferase (AOMT) were identified from KM, based on the sequence homology with previously identified AOMTs in other plant species (Akita *et al.*, 2011). Two of them, *CkmOMT1* and *CKmOMT2*, showed petal-specific expression, while the other two were expressed in both petals and leaves. This suggested that *CkmOMT1* and *CKmOMT2* are associated with anthocyanin biosynthesis in petals. *CkmOMT2* was not expressed in KMrp petals and genomic PCR failed to amplify the *CkmOMT2* coding sequence from KMrp. This suggested that the *CkmOMT2* coding region was deleted in KMrp. Furthermore, the recombinant CkmOMT2 protein was confirmed to have methylation activity *in vitro*. These observations suggested strongly that the mutation in *CkmOMT2* was associated with disruption of methylation activity from delphinidin to malvidin in KMrp.

The second red–purple mutant, "KMmv3" (Fig. 6.1j), accumulated malvidin 3-glucoside (Mv3G), suggesting a defective 5GT activity. The major anthocyanin of the last red–purple mutant, "KMmv35," (Fig. 6.1k) was Mv3,5dG, which is the same as the major anthocyanin of parental KM. While the anthocyanin concentration in the petals was not significantly different among KM, KMmv3, and KMmv35, the flavonol content was greatly reduced in KMmv35 compared with KM and KMmv3 (Nakayama *et al.*, 2012). Thin-layer chromatography confirmed that some of the flavonols contribute to the purple coloration as copigments for Mv3,5dG, but not for Mv-3G. These results suggested that the original color of Dp3,5dG, Mv3G, and Mv3,5dG is red purple in fragrant cyclamen, and only Mv3,5dG expresses purple coloration via copigmentation with some flavonols.

6.5.3 White mutants

One could argue that we have a better chance of getting a white-flowered mutant than the other flower color mutants, because white mutants could be generated by any mutation that blocks pigment biosynthesis. A white-flowered mutant, "KOpw" (Fig. 6.1h), was obtained from a mutagenized population from KO. The major anthocyanin of KO is Mv3,5dG, while KOpw petals had no detectable anthocyanins. The flavonol content was not significantly different between KO and KOpw.

Fragrant cyclamen is characterized by a sweet fragrance missing from ordinary cyclamen cultivars. KOpw appeared to have a slightly different fragrance compared with the commercialized fragrant cyclamen. The major fragrance components in fragrant cyclamen are citronellol and cinnamyl alcohol (Ishizaka *et al.*, 2002; Ishizaka, 2008); however, cinnamyl alcohol were not detected in the fragrance component of KOpw. KOpw still has a significant amount of citronellol and the fact remains that KOpw has a good fragrance. Although the gene responsible for KOpw has yet to be identified, KOpw may have a mutation that affects both anthocyanin biosynthesis and the biosynthesis of fragrance compounds.

6.5.4 Deeper color mutants

In the mutagenized population from pink-colored UR, we expected to obtain a red-colored mutant by inactivating the malvidin pathway. Although a red-colored mutant was not obtained, a salmon-red mutant "ion3" was (Fig. 6.1c). Although the major anthocyanins of ion3 were Cy3,5dG, Pn3,5dG, and Mv3,5dG, which are in common with parental UR, the anthocyanin concentration was greatly increased. In addition, the concentration of flavonols was greatly decreased. This suggested that certain types of mutation altered the biosynthetic flow in ion3. Generally, petal tissues contain much greater amounts of flavonols compared with the amount of anthocyanins. It is conceivable that a decrease in biosynthetic activity into flavonols occurred in ion3 and that this resulted in activation of metabolism into anthocyanins. To obtain more reddish mutants, reirradiation of the ion3 mutant was carried out, and the deep salmon-red mutant "ion246" (Fig. 6.1b) was obtained. These experiments demonstrated that deeper color mutants could be generated by ion beam irradiation by affecting the biosynthetic flow of anthocyanins. The deeper and lighter flower colors are one of the most frequent mutant phenotypes obtained in mutation breeding of ornamental plants. As illustrated in the studies in chrysanthemum and carnation (Tables 6.1 and 6.2), lighter color mutants are generally more frequent than deeper color mutants. These observations are consistent with the hypothesis that induced mutations often inhibit a certain biosynthetic pathway and sometimes result in the accumulation of another secondary metabolite.

6.6 Genetic background and the obtained mutants

Through mutant screening, red–purple mutants (Fig. 6.1i–k) were obtained from KM (Fig. 6.1m), a white mutant (Fig. 6.1h) was obtained from KO (Fig. 6.1n), salmon-red mutants (Fig. 6.1b and c) were obtained from UR (Fig. 6.1d), and a pale-yellow mutant (Fig. 6.1f) was obtained from GBCP (Fig. 6.1l). These results suggested that the genetic background of the parental plant limits the mutant traits obtained by ion beam irradiation.

As mentioned earlier, the commercialized fragrant cyclamens KM, KO, and UR have a genome from cyclamen cultivar (*C. persicum*, $2n=2x=48$) and *C. purpurascens* ($2n=2x=34$). Therefore, the amphidiploid fragrant cyclamen is thought to have 82 chromosomes. In pollen mother cells of the amphidiploid fragrant cyclamen, 41 bivalents were observed frequently (Ishizaka, 2008). This suggested that both genomes are distantly related and the chromosomes from *C. persicum* and *C. purpurascens* behave independently of each other. The red-flowered cultivar St and the white-flowered Pw were used for crossing with *C. purpurascens*, and crosses between siblings were also performed during the breeding process of fragrant cyclamen (Fig. 6.1). Recombination between St and Pw chromosomes could happen; therefore KO, KM, and UR are thought to have different genetic backgrounds with regard to the genome from *C. persicum*. This difference might be closely related to the types of mutant traits obtained from each parental variety.

The KMrp mutant (Fig. 6.1i), which accumulates delphinidin, was suggested to be caused by a mutation in the *CkmOMT2*, one of the AOMT homologs identified in KM (Akita *et al.*, 2011). Two of the four identified AOMT homologs showed petal-specific expression. Comparisons of their coding sequences and intron sizes suggested that *CkmOMT1* is from *C. purpurascens* and *CkmOMT2* is from *C. persicum* (Akita *et al.*, personal communication). The *C. purpurascens* petals accumulate malvidin and, therefore, definitely exhibit an AOMT activity against delphinidin. *CkmOMT1* was expressed in KMrp petals at a comparable level to KM petals, and no mutation was found in the *CkmOMT1* coding sequence. Thus, a mutation in *CkmOMT2* alone could not fully explain why KMrp lost methylation activity against delphinidin. In the deduced amino acid sequence, two amino acid residues differ between CkmOMT1 and CkmOMT2. For CkmOMT2, methylation activity was confirmed by an *in vitro* enzyme assay. However, the methylation activity of CkmOMT1 could not be examined, because we failed to isolate a soluble recombinant protein. The two amino acid differences might be important for the proteins' characteristics. The CkmOMT2 recombinant protein showed relatively higher methylation activity against delphinidin than cyanidin. CkmOMT1 and CkmOMT2 might have different substrate specificities. In addition, we could not exclude the possibility that *C. purpurascens* has another, as yet unidentified, functional AOMT. KMrp and KM are interesting materials that require further analysis to determine the regulation of anthocyanin biosynthesis.

6.7 Carnations with peculiar glittering colors

Carnations are one of the most important ornamental plants. We focused on creating carnations with peculiar glittering colors, ranging from red to purple (herein referred to as peculiar color), which are very rare among existing varieties. A few carnation cultivars, such as "Nazareno," display a peculiar purple color with some metallic hues (Fig. 6.2a). In contrast to ordinary carnations, in which the pigment is uniformly distributed over the vacuoles of the epidermal cells, the petal epidermal cells of "Nazareno" have an aggregation of anthocyanin in vacuoles, termed anthocyanin vacuolar inclusions (AVIs) (Markham *et al.*, 2000). Although AVIs have been studied in carnations, grapes, and sweet potato, the mechanism of their formation and their physiological role remain poorly understood (Mizuno *et al.*, 2006; Zhang *et al.*, 2006). To increase our knowledge concerning the expression of peculiar colors and AVI formation, crossing experiments were performed using carnation cultivars with peculiar colors as well as experimental strains that produced descendants with a peculiar color by

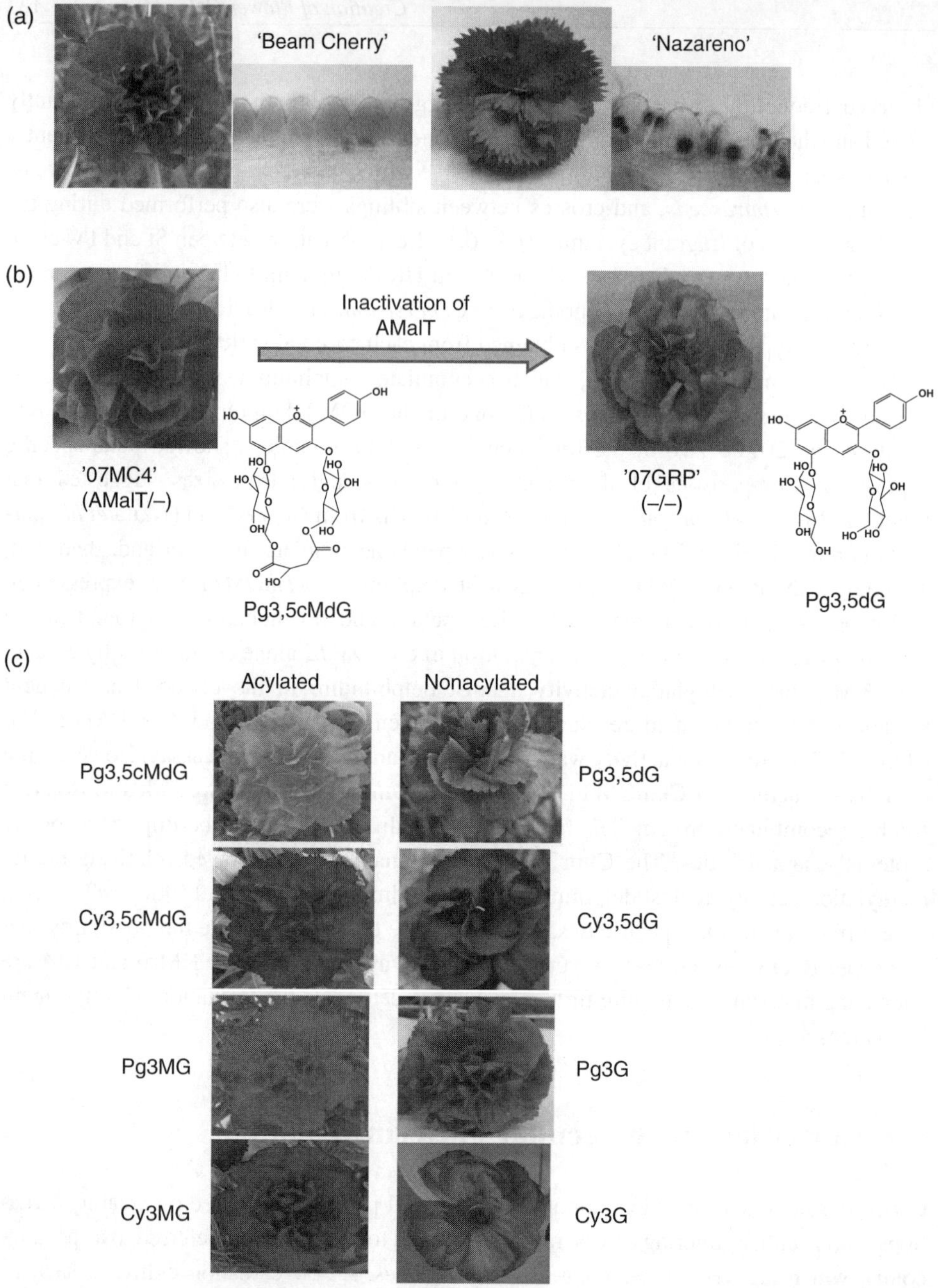

Fig. 6.2 Carnation with peculiar colors that contain nonacylated anthocyanins. (a) Flowers and transverse sections of petal epidermis of carnation cultivar "Beam Cherry (left)" and "Nazareno (right)." (b) The experimental strain "07MC4" and the mutant "07GRP" with peculiar colors obtained by inactivation of AMalT activity with carbon ion irradiation. The genotype for AMalT locus is shown in parenthesis. AMalT and a bar represent the active and inactive AMalT locus, respectively. The major anthocyanins are shown beside each flower. (c) Flowers of all four types of acylated and nonacylated-type anthocyanin in carnation. The major anthocyanins are shown beside each flower. AMalT, anthocyanin malyltransferase; Cy3,5cMdG, cyanidin 3,5-cyclic malyl diglucoside; Cy3,5dG, cyanidin 3,5-diglucoside; Cy3G, cyanidin 3-glucoside; Cy3MG, cyanidin 3-malyl glucoside; Pg3,5cMdG, pelargonidin 3,5-cyclic malyl diglucoside; Pg3,5dG, pelargonidin 3,5-diglucoside; Pg3G, pelargonidin 3-glucoside; Pg3MG, pelargonidin 3-malyl glucoside. (*See insert for color representation of the figure*)

self-pollination. The results suggested that the expression of a peculiar color was tightly linked with AVI formation and that this is controlled by a single recessive factor (Okamura *et al.*, 2012, 2013). Generally, carnations accumulate either one of four types of anthocyanins having glucose moieties acylated by malic acid: pelargonidin 3-malylglucoside (Pg3MG), cyanidin 3-malylglucoside (Cy3MG), pelargonidin 3,5-cyclicmalyldiglucoside (Pg3,5cMdG), and cyanidin 3,5-cyclicmalyldiglucoside (Cy3,5cMdG) (Fig. 6.2c left, Nakayama *et al.*, 2000). In contrast, "Nazareno" and the carnation strain "01MA1," with a peculiar color, accumulated a nonacylated-type anthocyanin, Pg3,5dG (Okamura *et al.*, 2013). Accumulation of nonacylated-type anthocyanin was also linked to AVI formation and peculiar color in the segregating population. These observations indicated that a nonacylated-type anthocyanin is closely linked with the manifestation of peculiar colors. Carnations have a single copy of the anthocyanin malyltransferase (AMalT) gene (Abe *et al.*, 2008). Cultured tissue of experimental strain "07MC4," which is thought to be heterozygous for AMalT, was mutagenized with carbon ions, and a mutant "07GRP" with a peculiar color was obtained successfully from about 1800 regenerated plants (Fig. 6.2b). The parental "07MC4" accumulated Pg3,5cMdG, while the mutant "07GRP" accumulated Pg3,5dG. This result also supported the hypothesis that inactivation of AMalT confers peculiar coloration. To diversify the peculiar colors, another experimental strain with a glittering bluish purple color, which contains Pg3,5dG, was mutagenized with ion beams. Three new strains with peculiar colors (reddish purple, crimson, and red) were obtained (Okamura *et al.*, 2013). The major anthocyanin of the reddish purple mutant was Pg3,5dG, and the color alteration was thought to be caused by reduced amounts of anthocyanin. The crimson and red mutants contained Pg3G, which was consistent with the lack of 5GT expression in these mutants. The parental bluish-purple line was confirmed to be heterozygous for functional 5GT; therefore, 5GT disruptants were obtained at a reasonable frequency. The red mutant had a lower amount of anthocyanin than the crimson mutant. This is consistent with the concept that a lower amount of anthocyanin results in a brighter color tone. Finally, carnation strains that contained nonacylated-type anthocyanins for each of all four types of anthocyanin were obtained, and all of them had AVIs and showed peculiar colors (Fig. 6.2c right, Nakayama *et al.*, 2012). Thus, carnations with peculiar colors could be obtained by focusing on AVIs and nonacylated-type anthocyanins, at least in our materials. However, other plant species that contain nonacylated-type anthocyanin do not always have AVIs. Moreover, a grape cultivar with a high density of AVIs had a high percentage of acylated-type anthocyanin (Mizuno *et al.*, 2006). In our experiment, we observed that the extent of inclusion differed between the adaxial and abaxial epidermis and between strains. These observations suggested that there are many unidentified factors that affect AVI formation. Novel flower colors are expected to be produced by identifying those factors and by controlling AVI formation.

6.8 Conclusions

Our research demonstrated that mutants with altered anthocyanin composition could be obtained in accordance with the proposed biosynthetic pathway. However, it remains hard to predict what types of mutants would be obtained, because the final modification of anthocyanins greatly affects the flower color. Although the current mutagenesis strategies still largely depend on chance, our research suggested that further understanding of genetic information associated with flower color is important to create novel plant varieties and to conduct more efficient and logical mutant screening.

Acknowledgments

The work on fragrant cyclamen and carnation was conducted with Drs Akita Y., Kitamura S., and Tanaka A. (JAEA). We also collaborated with Drs Okamura M. (KIRIN), Ishizaka H. (Saitama Pref.), Ozeki Y. (TUAT), and Nakayama M. (NIFS, NARO), with support by a grant from the Research and Development Program for New Bio-industry Initiatives of the Bio-oriented Technology Research Advancement Institution (BRAIN).

References

Abe, Y., Tera, M., Sasaki, N. *et al.* (2008) Detection of 1-*O*-malylglucose: pelargonidin 3-*O*-glucose-6″-*O*-malyltransferase activity in carnation (*Dianthus caryophyllus*). *Biochemical and Biophysical Research Communications*, **373**, 473–477.

Akita, Y., Kitamura, S., Hase, Y. *et al.* (2011) Isolation and characterization of the fragrant cyclamen *O*-methyltransferase involved in flower coloration. *Planta*, **234**, 1127–1136.

Goodhead, D.T. (1995) Molecular and cell models of biological effects of heavy ion radiation. *Radiation and Environmental Biophysics*, **34**, 67–72.

Hase, Y., Akita, Y., Kitamura, S., Narumi, I. and Tanaka, A. (2012) Development of an efficient mutagenesis technique using ion beams: toward more controlled mutation breeding. *Plant Biotechnology*, **29**, 193–200.

Hirono, Y., Smith, H.H., Lyman, J.T., Thompson, K.H. and Baum, J.W. (1970) Relative biological effectiveness of heavy ions in producing mutations, tumors, and growth inhibition in the crucifer plant, *Arabidopsis*. *Radiation Research*, **44**, 204–223.

Ishizaka, H. (2008) Interspecific hybridization by embryo rescue in the genus *Cyclamen*. *Plant Biotechnology*, **25**, 511–519.

Ishizaka, H., Yamada, H. and Sasaki, K. (2002) Volatile compounds in the flowers of *Cyclamen persicum*, *C. purpurascens* and their hybrids. *Scientia Horticulturae*, **94**, 125–135.

Ishizaka, H., Kondo, M. and Kameari, N. (2012) Production of novel flower color mutants from the fragrant cyclamen (*Cyclamen persicum*×*C. purpurascens*) by ion-beam irradiation. *Plant Biotechnology*, **29**, 201–208.

Joint FAO/IAEA Programme. *Mutant Variety Database (MVD)*, https://mvd.iaea.org/ (accessed June 24, 2016).

Kondo, E., Nakayama, M., Kameari, N. *et al.* (2009) Red–purple flower due to delphinidin 3,5-diglucoside, a novel pigment for *Cyclamen* spp., generated by ion-beam irradiation. *Plant Biotechnology*, **26**, 565–569.

Kraft, G., Kramer, M. and Scholz, M. (1992) LET, track structure and models. *Radiation and Environmental Biophysics*, **31**, 161–180.

Kranz, A.R., Bork, U., Bücker, H. and Reitz, G. (1990) Biological damage induced by ionizing cosmic rays in dry *Arabidopsis* seeds. *Nuclear Tracks and Radiation Measurements*, **17**, 155–165.

Lepiniec, L., Debeaujon, I., Routaboul, J.M. *et al.* (2006) Genetics and biochemistry of seed flavonoids. *Annual Review of Plant Biology*, **57**, 405–430.

Markham, K.R., Gould, K.S., Winefield, C.S., Mitchell, K.A., Bloor, S.J. and Boase, M.R. (2000) Anthocyanin vacuolar inclusions—their nature and significance in flower colouration. *Phytochemistry*, **55**, 327–336.

Miyajima, I., Maehara, T., Kage, T. and Fujieda, K. (1991) Identification of the main agent causing yellow color of yellow-flowered cyclamen mutant. *Journal of the Japanese Society for Horticultural Science*, **60**, 409–414.

Mizuno, H., Hirano, K. and Okamoto, G. (2006) Effect of anthocyanin composition in grape skin on anthocyanic vacuolar inclusion development and skin coloration. *Vitis*, **45**, 173–177.

Nagatomi, S., Tanaka, A., Watanabe, H., and Tano, S. (1997) Chrysanthemum mutants regenerated from *in vitro* explants irradiated with $^{12}C^{5+}$ ion beam. Institute of Radiation Breeding, *Technical News No. 60*, http://www.nias.affrc.go.jp/newsletter/tech_news/pdf/TechnicalNews60.pdf (accessed June 24, 2016).

Nakayama, M., Koshioka, M., Yoshida, H. *et al.* (2000) Cyclic malyl anthocyanins in *Dianthus caryophyllus*. *Phytochemistry*, **55**, 937–939.

Nakayama, M., Tanikawa, N., Morita, Y. and Ban, Y. (2012) Comprehensive analyses of anthocyanin and related compounds to understand flower color change in ion-beam mutants of cyclamen (*Cyclamen* spp.) and carnation (*Dianthus caryophyllus*). *Plant Biotechnology*, **29**, 215–221.

Okada, T., Iizuka, M., Yoshihara, R., *et al.* (2009) Development of commercial variety of *Osteospermum* by a stepwise mutagenesis by ion beam irradiation. *JAEA-Review 2009-041/JAEA Takasaki Annual Report 2008*, p. 73, http://dx.doi.org/10.11484/jaea-review-2009-041 (accessed June 24, 2016).

Okamura, M., Yasuno, N., Ohtsuka, M., Tanaka, A., Shikazono, N. and Hase, Y. (2003) Wide variety of flower-color and -shape mutants regenerated from leaf cultures irradiated with ion beams. *Nuclear Instruments and Methods in Physics Research B*, **206**, 574–578.

Okamura, M., Tanaka, A., Momose, M., Umemoto, N., da Silva, J.A.T. and Toguri, T. (2006) Advances of mutagenesis in flowers and their industrialization, in Floriculture, Ornamental and Plant Biotechnology Volume I (ed J.A.T. da Silva), Global Science Books, Ltd, Middlesex, pp. 619–628.

Okamura, M., Umemoto, N. and Onishi, N. (2012) Breeding glittering carnations by an efficient mutagenesis system. *Plant Biotechnology*, **29**, 209–214.

Okamura, M., Nakayama, M., Umemoto, N. *et al.* (2013) Crossbreeding of a metallic color carnation and diversification of the peculiar coloration by ion-beam irradiation. *Euphytica*, **191**, 45–56.

Oppenheimer, D.J., Herman, P.L., Sivakumaran, S., Esch, J. and Marks, M.D. (1991) A *myb* gene required for leaf trichome differentiation in *Arabidopsis* is expressed in stipules. *Cell*, **67**, 483–493.

Sato, T., Ohya, Y., Hase, Y., and Tanaka, A. (2006) Studies on flower color and morphological mutations from chrysanthemum *in vitro* explants irradiated with ion beams. *JAEA-Review 2005-001/ TIARA Annual Report 2004*, pp. 74–75, 10.11484/jaea-review-2005-001 (accessed June 24, 2016).

Shikazono, N., Tanaka, A., Watanabe, H. and Tano, S. (2001) Rearrangements of the DNA in carbon ion-induced mutants of *Arabidopsis thaliana*. *Genetics*, **157**, 379–387.

Shikazono, N., Yokota, Y., Kitamura, S. *et al.* (2003) Mutation rate and novel *tt* mutants of *Arabidopsis thaliana* induced by carbon ions. *Genetics*, **163**, 1449–1455.

Shikazono, N., Suzuki, C., Kitamura, S., Watanabe, H., Tano, S. and Tanaka, A. (2005) Analysis of mutations induced by carbon ions in *Arabidopsis thaliana*. *Journal of Experimental Botany*, **56**, 587–596.

Stadler, L.J. (1928) Mutations in barley induced by X-rays and radium. *Science*, **68**, 186–187.

Takamura, T., Tomihama, T. and Miyajima, I. (1995) Inheritance of yellow-flowered characteristic and yellow pigments in diploid cyclamen (*Cyclamen persicum* Mill.) cultivars. *Scientia Horticulturae*, **64**, 55–63.

Takamura, T., Nakayama, M. and Ishizaka, H. (2005) Inheritance of flower color pigment in crosses between cyclamen cultivars and *Cyclamen purpurascens*. *Acta Horticulturae*, **673**, 437–441.

Tanaka, A., Shikazono, N., Yokota, Y., Watanabe, H. and Tano, S. (1997) Effects of heavy ions on the germination and survival of *Arabidopsis thaliana*. *International Journal of Radiation Biology*, **72**, 121–127.

Tanaka, A., Shikazono, N. and Hase, Y. (2010) Studies on biological effects of ion beams on lethality, molecular nature of mutation, mutation rate, and spectrum of mutation phenotype for mutation breeding in higher plants. *Journal of Radiation Research*, **51**, 223–233.

Togami, J., Okuhara, H., Nakamura, N. *et al.* (2011) Isolation of cDNAs encoding tetrahydroxychalcone 2′-glucosyltransferase activity from carnation, cyclamen, and catharanthus. *Plant Biotechnology*, **28**, 231–238.

Yatagai, F. (2004) Mutations induced by heavy charged particles. *Biological Sciences in Space*, **18**, 224–234.

Zhang, H., Wang, L., Deroles, S., Bennett, R. and Davis, K. (2006) New insight into the structures and formation of anthocyanic vacuolar inclusions in flower petals. *BMC Plant Biology*, **6**. doi: 10.1186/1471-2229-6-29.

Chapter 7
Flavonols Regulate Plant Growth and Development through Regulation of Auxin Transport and Cellular Redox Status

Sheena R. Gayomba, Justin M. Watkins and Gloria K. Muday

Abstract: Interest in the roles of flavonoids in fundamental biological processes has spurred an extensive characterization of their biosynthesis and demonstrated that specific metabolites in this pathway modulate a diversity of plant developmental and stress responses. *Arabidopsis thaliana* mutants with defects in flavonol biosynthesis have been used to demonstrate that flavonols modulate gravitropism, root elongation, root and inflorescence branching, and guard cell signaling to modulate stomatal aperture. In several cases, specific functions have been linked to individual flavonol derivatives. Mutants in flavonoid biosynthesis in other species share common phenotypes, but have also led to the identification of other functions of flavonols not found in *Arabidopsis*, such as fertilization and pollen viability. The molecular action of flavonols includes inhibition of auxin transport and antioxidant activity. The extensive environmental and hormonal regulation of flavonol biosynthesis, through transcriptional controls of genes encoding pathway enzymes, provides a mechanism to regulate the physiological processes modulated by flavonols.

Keywords: flavonols, flavonoids, root development, gravitropism, guard cells, auxin, ethylene, transcriptional controls, auxin transport, antioxidant.

7.1 Introduction

This chapter explores the role of flavonoids in controlling plant growth and development, with a focus on flavonols. The ability of plants to rapidly induce the synthesis of flavonoids in response to a variety of stimuli allows local synthesis of these molecules at sites of their action. Genetic approaches have provided insight into the function of these metabolites by

Recent Advances in Polyphenol Research, First Edition. Edited by Kumi Yoshida, Véronique Cheynier and Stéphane Quideau.

using plants that are deficient in all flavonoids or a subset of pathway metabolites to identify physiological and developmental processes that differ between these mutants and wild-type parents. The chapter also introduces the flavonols and the machinery that synthesizes these molecules, how this synthesis is regulated and linked to their physiological effects, and summarizes the current insight into the direct functional targets of flavonols that may mediate plant responses to these important molecules.

7.2 The flavonoids and their biosynthetic pathway

The flavonoid biosynthetic pathway is an attractive metabolic sequence because of the elegant biochemistry and the beauty of the anthocyanin products, which are the major sources of red, blue, and purple pigments in plants (Shirley *et al.*, 1995; Winkel-Shirley, 2001a). The presence and absence in mutants of the colored products of this pathway have facilitated identification of the genes encoding the key metabolic and regulatory proteins controlling flavonoid biosynthesis using genetic approaches (Winkel-Shirley, 2001a). The pioneering work of Maarten Koornneef led to the identification of mutants in *Arabidopsis thaliana* named *transparent testa* (*tt*). These mutants have reduced pigmentation in the seed coat due to defects in genes encoding the central enzymes and regulatory molecules of flavonoid metabolism (Winkel-Shirley, 2001a). The colored flavonoid products have also been useful markers to understand basic genetic concepts used by Gregor Mendel and Barbara McClintock to uncover genetic inheritance and transposable elements and more recently in cosuppression of transgenes (Que *et al.*, 1997; Winkel-Shirley, 2001b). Although the core biosynthetic pathway is conserved across higher plants, there are unique derivatives of the pathway found in some plant lineages, such as isoflavonoids in legumes, which function in nodulation (Subramanian *et al.*, 2007; Zhang *et al.*, 2009).

The genes encoding enzymes and regulatory factors controlling flavonoid metabolism have been identified in *Arabidopsis* and mutants in each step of the pathway have been isolated (Winkel-Shirley, 2001a; Lepiniec *et al.*, 2006). Figure 7.1 illustrates the metabolites and enzymes of this pathway and the names of *Arabidopsis* mutants at each step, as well as defines the abbreviations that are used for both genes and protein products. The first committed step of flavonoid synthesis is catalyzed by chalcone synthase (CHS), which acts on 4-coumaryl Co-A, produced from phenylalanine via the common phenylpropanoid pathway, and malonyl Co-A, produced during fatty acid metabolism, to produce naringenin chalcone. The central biosynthetic pathway for flavonoids produces chalcones, flavones, flavonols, flavandiols, anthocyanins, flavan-3-ols, and condensed tannins (Winkel-Shirley, 2001a; Koes *et al.*, 2005). The enzymes in the early pathway that control carbon flow into the pathway and flavonol synthesis are encoded by *CHS, CHALCONE ISOMERASE* (*CHI*), *FLAVONONE 3-HYDROXYLASE* (*F3H*), *FLAVONOL SYNTHASE* (*FLS*), and *FLAVONOID 3′,5′-HYDROXYLASE (F3′H)*, with the branch-point enzyme encoded by *DIHYDROFLAVONOL REDUCTASE* (*DFR*) controlling conversion of dihydroflavonol precursors to anthocyanins. The relative levels of early and late pathway enzymes control the metabolite profiles, as illustrated by the absence of anthocyanins in roots, linked to the absence of detectable *DFR* transcript in this tissue (Lewis *et al.*, 2011b). The transcriptional

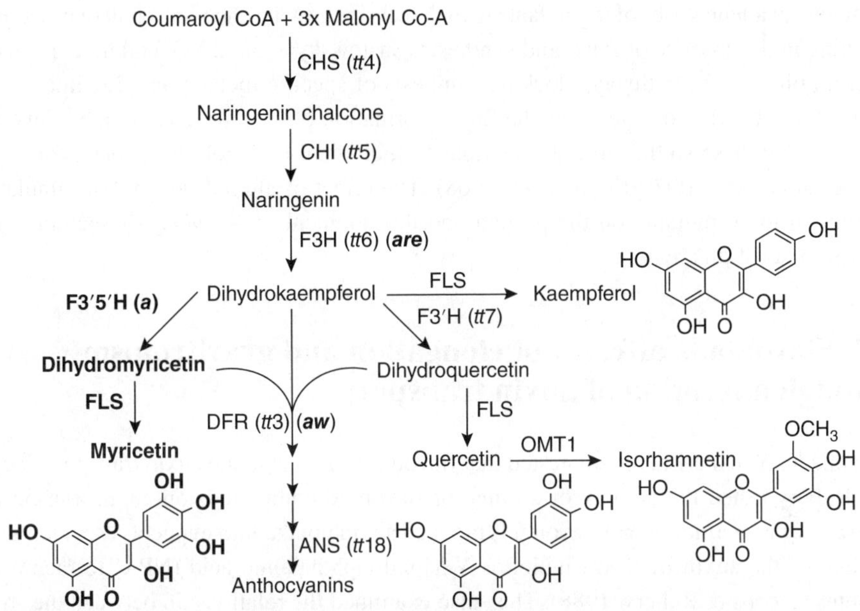

Fig. 7.1 The flavonol biosynthetic pathway is illustrated, showing enzymatic steps and indicating *Arabidopsis transport testa* (*tt*) and tomato anthocyanin mutant names. Bolded sections of the pathway are specific to tomato and petunia. CHI, chalcone isomerase; CHS, chalcone synthase; F3H, flavanone 3-hydroxylase; F3′H, flavonoid 3′-hydroxylase; F3′5′H, flavonoid 3′,5′-hydroxylase; FLS, flavonol synthase; DFR, dihydroflavonol reductase; OMT1, flavone 3′-*O*-methyltransferase 1. Tomato mutants are *anthocyaninless* (*a*), *anthocyanin reduced* (*are*), and *anthocyanin without* (*aw*).

machinery that controls the expression of genes encoding enzymes of flavonoid synthesis has been extensively studied and requires complexes between R2R3-MYB, basic helix–loop–helix (bHLH), and WD40-type transcription factors (Hichri *et al.*, 2011; Xu *et al.*, 2015). Unique combinations of transcription factors control expression of the pathway in distinct tissues, modulating synthesis in response to a broad set of stimuli, with examples that link flavonol synthesis to growth and development summarized as follows.

The chapter focuses on the function of flavonols, which include kaempferol, quercetin, and isorhamnetin in all plant species, but also myricetin in some species, such as in *Solanum lycopersicum* (tomato) (Ballester *et al.*, 2010). Figure 7.1 illustrates how different species produce anthocyanins from different dihydroflavonol derivatives. The DFR enzyme preferentially acts on dihydroquercetin in *Arabidopsis* (Dong *et al.*, 2001) and dihydromyricetin in tomato (Bovy *et al.*, 2002; Butelli *et al.*, 2008) as the first step in the production of cyanidin-type or delphinidin-type anthocyanins, which have either reddish or purplish pigmentation, respectively. In addition to the synthesis of the core aglycone flavonols outlined in this figure, flavonols are glycosylated to modulate structure and function, with dozens of different glycosylated flavonols in every plant (35 have been identified in *Arabidopsis*) (Saito *et al.*, 2013). As aglycones are rarely detected *in planta*, glycosylated flavonols are the molecules that function *in vivo* (Ringli *et al.*, 2008; Saito *et al.*, 2013; Yin *et al.*, 2013). Examination of the physiological functions of flavonoids has been aided by the availability of *tt* mutants, including an allelic series of *tt4* mutants with defects in the gene encoding chalcone synthase, resulting in the absence of flavonoid synthesis (Saslowsky *et al.*, 2000) and mutants in key branch points in the pathway, blocking synthesis of specific metabolites (Lepiniec *et al.*, 2006). Identification of mutations that block formation of specific glycosylated flavonols has extended these studies to link functions to individual flavonol glycosides (Yonekura-Sakakibara *et al.*, 2007; Ringli *et al.*, 2008). The effect of altered flavonol accumulation profiles in these mutants on the growth and development of *Arabidopsis* seedlings is a central focus of this review.

7.3 Flavonoids affect root elongation and gravitropism through alteration of auxin transport

Several classic experiments suggested that flavonoids may regulate auxin transport. Jacobs and Rubery tested the activity of a range of flavonoid compounds on auxin transport *in vitro* and *in vivo*, finding that flavonoids blocked auxin uptake into hypocotyl segments and binding of the auxin transport inhibitor *N*-naphthylphthalamic acid (NPA) to membrane proteins (Jacobs & Rubery, 1988). They also examined the relationship between the structure of these molecules and their function, finding that the most active compounds were flavonols. The flavonol quercetin was the most effective at blocking auxin transport and at reducing tritiated NPA binding (Jacobs & Rubery, 1988). These experiments demonstrated that exogenously applied flavonoids inhibit auxin transport and suggested the testable hypothesis that mutants with impaired flavonoid synthesis might have elevated auxin transport and changes in dependent physiological processes.

Experiments using *tt4* mutant alleles, which produce no flavonoids due to defects in the gene encoding CHS, revealed the first evidence for a role of flavonoids in *Arabidopsis* growth and development and linkage of these phenotypes to altered auxin transport. Enhanced root elongation and increased numbers of lateral and adventitious roots were identified in the *tt4* (2YY6) line (which has the *tt4-2* mutation) (Brown *et al.*, 2001). The root gravitropic response is delayed in five different *tt4* alleles (Buer & Muday, 2004; Buer *et al.*, 2006; Lewis *et al.*, 2011b; Buer *et al.*, 2013), and this defect is reversed by treatment with the flavonol precursor, naringenin, which restored flavonol accumulation to wild-type levels (Buer & Muday, 2004). Consistent with a role of flavonoids in root gravity response, treatment of an agravitropic mutant harboring a defect in the gene encoding the auxin transport protein, PIN-FORMED2 (PIN2), with exogenous flavonoids restored gravitropic response (Santelia *et al.*, 2008).

Two *tt4* mutant alleles were reported to have elevated transport of naturally occurring auxin, indole-3-acetic acid (IAA) in inflorescences, hypocotyls, and roots (Murphy *et al.*, 2000; Brown *et al.*, 2001) including shootward (or basipetal) auxin transport in roots that is required for gravitropic responses (Buer & Muday, 2004; Lewis *et al.*, 2011b). Enhanced IAA transport is reversed by treatment with naringenin (Murphy *et al.*, 2000; Brown *et al.*, 2001). The increased IAA transport in inflorescences, hypocotyls, and roots of *tt4* is consistent with the absence of a negative regulator causing defects in auxin transport-dependent processes, such as root gravitropism.

The distribution of flavonols in young wild-type seedlings has been visualized using diphenylboric acid 2-aminoethyl ester (DPBA), a dye that becomes fluorescent upon binding flavonols (Sheahan & Rechnitz, 1992). The specificity of DPBA is evident by the absence of fluorescence in flavonol-deficient *tt4* seedlings and by different fluorescent properties when bound to kaempferol and quercetin (Murphy *et al.*, 2000; Peer *et al.*, 2001; Lewis *et al.*, 2011b). In wild-type plants, flavonols were localized to tissues in which auxin is transported, including the epidermis at the root tip and the elongation zone, the root–shoot junction, and shoot apex (Peer *et al.*, 2001; Buer & Muday, 2004). Consistent with a requirement for flavonoids for gravitropic response, the abundance of flavonols transiently increases at the root tip and in the elongation zone after reorientation of the root relative to gravity with kinetics that precede gravitropic curvature (Buer & Muday, 2004; Santelia *et al.*, 2008). An important question is why the absence of a negative regulator of auxin transport in *tt4* reduces the gravitropic response. A model drawn in Fig. 7.2 illustrates how localized synthesis of flavonols in the elongation zone of vertically grown roots may modulate auxin redistribution to facilitate formation of auxin gradients across the root tip. In *tt4*, the zone of auxin-induced gene expression is elongated as auxin moves a greater distance in the absence of flavonols (Buer & Muday, 2004), which may be linked to higher levels of *PIN2* transcript (Peer *et al.*, 2004). Gravitropic reorientation increases accumulation of flavonols in the elongation zone of wild-type plants (Santelia *et al.*, 2008) and may block auxin movement away from the tip, thereby concentrating auxin in the distal elongation zone where differential growth occurs with inhibited elongation on the lower side of the root. In *tt4* mutants, higher levels of *PIN2* transcripts (and presumably protein) and the absence of flavonols concentrating auxin in the root tip may result in a weaker gradient of auxin forming across the root to reduce the rate of gravitropic curvature.

Experiments with *tt4* demonstrated that flavonoids regulate root growth and gravitropism and shootward auxin transport, but not which flavonoids are important in the process. Examination of these processes in plants with mutations in genes encoding other pathway enzymes defined the active molecules. The specific role of quercetin derivatives in regulating shootward transport of IAA and root gravitropism was demonstrated using the *tt7* mutant, which has a defect in the gene encoding F3′H and synthesizes kaempferol, but not quercetin (Lewis *et al.*, 2011b). The gravitropic defects and increases in auxin transport are identical in the Columbia (Col-0) alleles, *tt4-2* and *tt7-2*, consistent with quercetin being the active transport inhibitor in this tissue (Lewis *et al.*, 2011b). In contrast, a glycosylated kaempferol is the active flavonoid in regulating rootward auxin transport in the *Arabidopsis* inflorescence (Yin *et al.*, 2013), and kaempferol derivatives have been reported to regulate auxin transport from hypocotyls into roots (Peer *et al.*, 2004). Transport of auxin in hypocotyls is increased in *tt4*, which produces no flavonols, and decreased in *tt3*, which has elevated levels of flavonols due to a defect in *DFR* (Peer *et al.*, 2004). These investigators found that auxin transport in *tt3-1* and *tt7-1* are similar, consistent with elevated levels of kaempferol derivatives being sufficient to reduce IAA transport through hypocotyls (Peer *et al.*, 2004). These results are consistent with specific flavonoids controlling distinct aspects of growth and development and auxin transport in a tissue-specific fashion.

The synthesis and function of flavonoids in regulation of gravitropic curvature have also been linked to ethylene levels in roots. Ethylene reduces the rate of root gravitropic

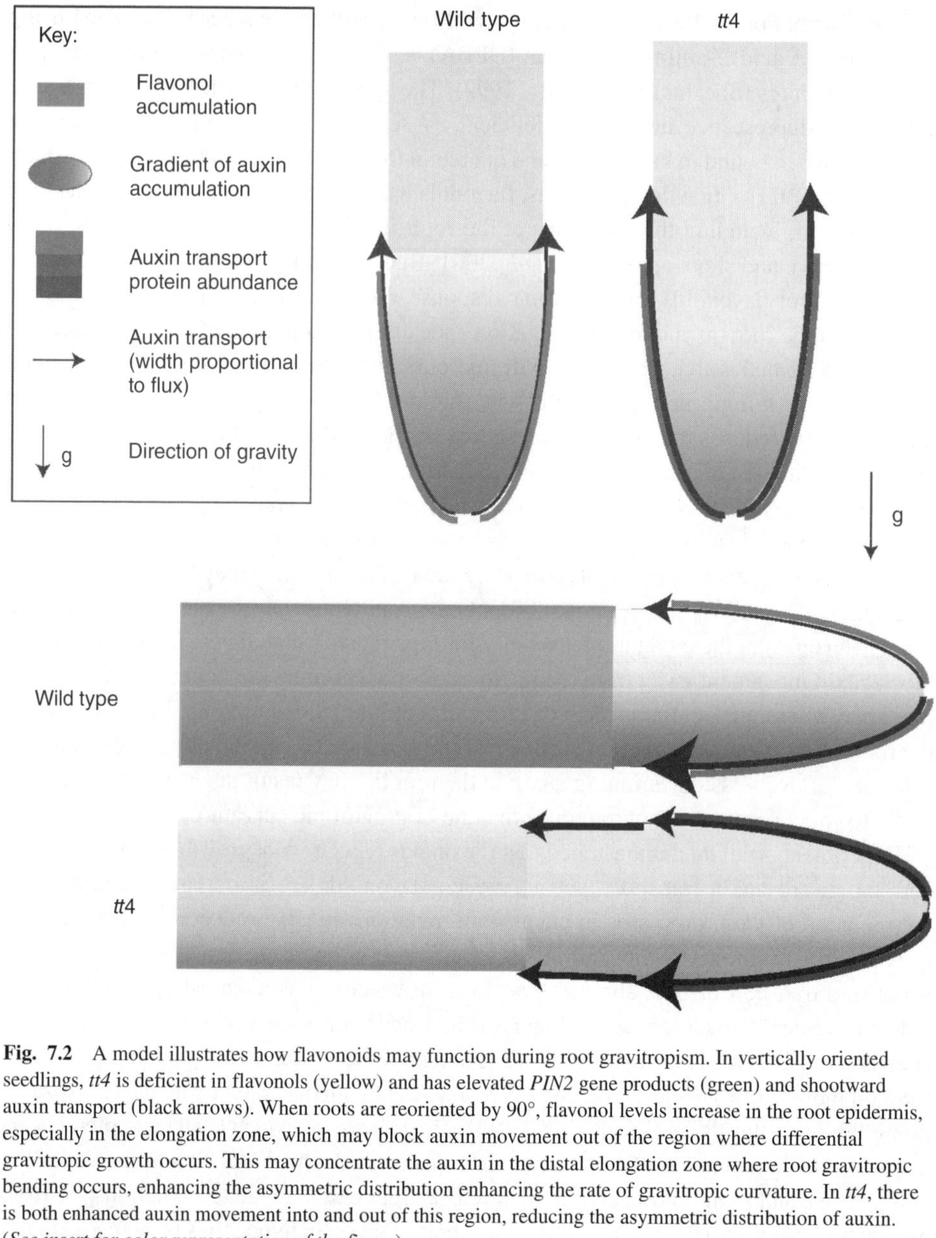

Fig. 7.2 A model illustrates how flavonoids may function during root gravitropism. In vertically oriented seedlings, *tt4* is deficient in flavonols (yellow) and has elevated *PIN2* gene products (green) and shootward auxin transport (black arrows). When roots are reoriented by 90°, flavonol levels increase in the root epidermis, especially in the elongation zone, which may block auxin movement out of the region where differential gravitropic growth occurs. This may concentrate the auxin in the distal elongation zone where root gravitropic bending occurs, enhancing the asymmetric distribution enhancing the rate of gravitropic curvature. In *tt4*, there is both enhanced auxin movement into and out of this region, reducing the asymmetric distribution of auxin. (*See insert for color representation of the figure*)

curvature and root elongation through the canonical ethylene signaling pathway mediated by ETR1, an ethylene receptor, and EIN2, an ethylene signaling protein (Buer *et al.*, 2006). This ethylene response requires flavonoid accumulation, as the lag in initiation of root curvature in response to elevated ethylene is lost in multiple *tt4* alleles (Buer *et al.*, 2006; Lewis *et al.*, 2011b). Ethylene delays the transient gravity-induced flavonoid accumulation in root tips (Buer *et al.*, 2006), while enhancing flavonol synthesis in the elongation zone

(Lewis *et al.*, 2011b; Muday *et al.*, 2012). Treatment with the ethylene precursor, 1-aminocyclopropane-1-carboxylic acid (ACC), also affects rootward or acropetal IAA transport, which is mediated by increases in synthesis of two auxin transport proteins, PIN3 and PIN7 (Lewis *et al.*, 2011a). This effect of ethylene is independent of flavonols.

Flavonoids may also play a role in modulation of primary root development by regulating the stem cell niche at the root apical meristem. In studies to understand the role of WRKY23 transcription factor (TF) in controlling primary root development, constructs that either increased or decreased *WRKY23* expression led to altered organization of the root stem cell niche (Grunewald *et al.*, 2012). WRKY23 is a positive regulator of flavonol synthesis, leading to the significant increases in *F3′H* transcript abundance and quercetin synthesis. The developmental defects resulting in shorter primary roots in lines transformed with *WRKY23* overexpression constructs were lost when those constructs were crossed into a *tt4-8* mutant (Grunewald *et al.*, 2012). Consistent with WRKY23-dependent flavonoid synthesis reducing auxin transport, auxin accumulation at the root tip was increased and rootward auxin transport was decreased in *WRKY23* overexpression lines. These results suggest that WRKY23-dependent localized synthesis of flavonoids at the root tip modulates auxin transport and localized auxin accumulation that controls stem cell activity and primary root development (Grunewald *et al.*, 2012).

7.4 Mechanisms by which flavonols regulate IAA transport

The mechanism by which flavonoids reduce auxin transport has been explored in several studies, providing evidence for multiple mechanisms. Some of the early insights into auxin transport came from identification of chemicals that inhibited auxin efflux, such as NPA, suggesting the simple model that flavonoids may bind to these same sites (Jacobs & Rubery, 1988). The high concentration of quercetin needed to block NPA binding (Jacobs & Rubery, 1988; Brunn *et al.*, 1992) combined with substantial structural differences between these two molecules suggests that quercetin may not bind to the same binding site as synthetic auxin transport inhibitors (Bures *et al.*, 1991). The identification of auxin transport proteins (Zazimalova *et al.*, 2010) allowed direct tests of whether flavonoids regulate the activity of these proteins. The auxin transport capacity of ATP-binding cassette type B (ABCB) families (Zazimalova *et al.*, 2010) expressed in a heterologous system is reduced by quercetin (Geisler *et al.*, 2005; Bouchard *et al.*, 2006; Blakeslee *et al.*, 2007). Additionally, the effect of *tt4* on root gravitropism is lost in an *abcb4 tt4* double mutant, consistent with flavonoids and ABCB4 acting in the same pathway *in vivo* (Lewis *et al.*, 2007). Quercetin, and to a lesser extent kaempferol, disrupts the protein complex between ABCB1 and TWISTED DWARF1 (TWD1), an immunophilin that regulates ABCB1 transport activity (Bailly *et al.*, 2008), suggesting that flavonols may inhibit auxin transport by altering protein complexes needed for maximal auxin transport.

Additional experiments suggest that interactions between auxin and flavonoids at the root tip are more complex than the simple model of flavonoids acting as inhibitors of auxin transport proteins. Treatments to raise auxin concentrations increase synthesis and accumulation of flavonols (Buer *et al.*, 2006; Santelia *et al.*, 2008; Lewis *et al.*, 2011b) by increasing accumulation of transcripts encoding pathway enzymes (Lewis *et al.*, 2011b; Grunewald *et al.*, 2012). Consistent with a positive role of auxin on flavonoid synthesis, the accumulation of flavonols in vertical and gravity stimulated roots is reduced in the *pin2* mutant (Santelia *et al.*, 2008), which is agravitropic and has reduced shootward transport of auxin. Further supporting the role of flavonoids in gravity response is the restoration of gravitropic response and a gradient of DR5-GFP fluorescence across the roots of *pin2* when flavonoids are exogenously applied to these roots (Santelia *et al.*, 2008). The restoration of gravitropic curvature after flavonol treatment of *pin2* mutants was associated with increased synthesis and altered localization of PIN1 protein at the *pin2* root tip, consistent with flavonoids regulating *PIN1* synthesis and localization (Santelia *et al.*, 2008). This finding is consistent with another report that flavonoids affect transcription of genes encoding auxin transport proteins of the PIN family (Peer *et al.*, 2004). Experiments using both mutants in flavonoid synthesis and exogenous treatment with flavonol precursors indicated that flavonoids affect *PIN1* and *PIN2* oppositely at the transcriptional level and promote the asymmetric localization of PIN1 (Peer *et al.*, 2004). Together these results highlight a complex transcriptional circuit that links flavonols and auxin transport to controls of gravitropic curvature.

Flavonols have been shown to be potent kinase inhibitors (Agullo *et al.*, 1997), and it has been hypothesized that flavonols might regulate auxin transport through modulation of phosphorylation status of auxin transport proteins (Peer & Murphy, 2007). Phosphorylation of PIN and ABCB proteins by AGCVIII kinases (including PINOID, WAG, and D6 protein kinase) or phosphatidylinositol monophosphate 5-kinases 2 regulates auxin transport and dependent growth responses (Michniewicz *et al.*, 2007; Mei *et al.*, 2012; Barbosa & Schwechheimer, 2014; Barbosa *et al.*, 2014). An equally important role of protein dephosphorylation in controlling auxin transport has also been reported (Garbers *et al.*, 1996; Deruère *et al.*, 1999; Rashotte *et al.*, 2001; Michniewicz *et al.*, 2007; Sukumar *et al.*, 2009). Demonstration of a direct relationship between phosphorylation status of auxin transport proteins and flavonoids awaits further experimentation.

A final mechanism by which flavonoids might affect auxin transport is through altering the redox state of the cell and thereby regulating auxin transport or auxin homeostasis (Fernández-Marcos *et al.*, 2013). Flavonols may function as antioxidants in plants and in the animals that consume them (Hernandez *et al.*, 2009), thereby modulating reactive oxygen species (ROS) signaling and preventing oxidative damage (Hernandez *et al.*, 2009). Although the antioxidant activity of flavonols *in planta* has been debated (Hernandez *et al.*, 2009; Pollastri & Tattini, 2011; Agati *et al.*, 2012), previous studies have shown that the antioxidant capacity of these compounds *in vitro* is greater than that of vitamin C or E, which are two well-documented antioxidants *in planta* (Rice-Evans *et al.*, 1995). ROS have been reported to increase levels of oxidized IAA (ox-IAA) (Peer *et al.*, 2013) and modulate polar auxin transport (Fernández-Marcos *et al.*, 2013). Specifically, polar auxin transport is impaired in plants with increased ROS accumulation. Plants with defects in genes encoding thioredoxin reductase (*ntra* and *ntrb*) and an enzyme of glutathione biosynthesis, *cad2*, exhibit altered ROS homeostasis due to the absence of these antioxidant proteins (Bashandy *et al.*, 2010). The *ntra ntrb cad2* triple mutant has impaired auxin transport and developmental phenotypes consistent with altered auxin transport, including PIN-shaped inflorescences and reduced lateral root formation (Bashandy *et al.*, 2010). Decreased transcript levels for auxin efflux and influx carriers in response to elevated ROS suggest that ROS may modulate polar auxin transport at the level of synthesis of transport proteins (Blomster *et al.*, 2011; Tognetti *et al.*, 2012). *PIN1*, *PIN2*, *PIN3*, and *PIN4* transcripts were downregulated after addition of buthionine sulfoximine (BSO), a specific inhibitor of glutathione biosynthesis (Bashandy *et al.*, 2010) or alloxan, a compound able to generate ROS in the presence of intracellular thiols (Pasternak *et al.*, 2005). These results raise the possibility that flavonols reduce ROS levels, thereby indirectly regulating auxin levels and/or transport (Fernández-Marcos *et al.*, 2013). The potential for flavonols to act as antioxidants regulating auxin-independent signaling mechanisms is examined later, in regulation of guard cell function and root hair formation.

7.5 Lateral root formation

Understanding the relationship between flavonoids and lateral root formation is complicated by the plasticity of this developmental process and the regulation of both root development and flavonoid synthesis by light and sucrose levels. The root developmental phenotypes of *tt* mutant alleles vary depending on the mutant allele and growth conditions. Plants grown in the presence of sucrose have increased flux through the flavonoid pathway (Teng *et al.*, 2005; Solfanelli *et al.*, 2006) and greater lateral root formation, with a 2.5-fold increase in lateral root density in wild-type Landsberg *erecta* (L*er*) and a 10-fold increase in *tt4-1* (Buer & Djordjevic, 2009). The *tt4-1* and *tt4-2* mutants in L*er* and Col-0 background, respectively, have elevated numbers of lateral roots and greater lateral root density (number of lateral roots/primary root length) compared to their parental lines when grown on media containing sucrose (Brown *et al.*, 2001; Buer & Djordjevic, 2009). When lateral root formation is compared between Col-0 and *tt4-2* in the presence and absence of sucrose and phosphate, increased number of lateral roots in *tt4-2* occurs when sucrose is present and phosphate is limited (Jain *et al.*, 2007). Elevation of flavonol levels in plants with altered expression of the TF TCP3 results in reduced numbers of lateral roots in plants grown in the presence of sucrose (Li & Zachgo, 2013). Thus, in conditions where flavonoids are abundant and carbohydrates are not limiting, flavonoids negatively regulate root branching in Arabidopsis.

In contrast, when grown in media without sucrose and with a plate at a 45° angle, which induces root waving and lateral root formation on the convex side of root waves, *tt4-1* shows reduced number of lateral roots relative to L*er* (Buer & Djordjevic, 2009). Surprisingly, mutations affecting subsequent steps in this biosynthetic pathway, *tt5-1* (defective in CHI) and *tt6-1* (defective in F3H), which should also reduce levels of flavonols, result in greater numbers of lateral roots under these conditions (Buer & Djordjevic, 2009). Long-term growth of *tt4-1*, *tt5-1*, and *tt7-1* in sucrose-free media also resulted in formation of lateral root primordia with defects in structure and with multiple primordia forming from the same position on the primary root (Buer *et al.*, 2013). The observation of these phenotypes in single *tt* mutant alleles and without either genetic (Maloney *et al.*, 2014) or chemical complementation by treatment with flavonols or their precursors (Brown *et al.*, 2001; Buer & Muday, 2004; Ringli *et al.*, 2008) makes these sucrose-independent and long-term growth analyses more difficult to interpret. Taken together with the studies performed on sucrose-containing media, these data raise the intriguing possibility that flavonoids may be involved in tailoring the development of the root system to suit the available carbohydrates produced *in planta* or supplied through the growth medium.

The effect of flavonol distribution on lateral root development has also been examined in tomato. The *anthocyanin reduced* (*are*) mutant has a point mutation in *F3H* and no detectable F3H enzyme activity (Yoder, 1994; Maloney *et al.*, 2014). Roots of the *are* mutant accumulate more than sevenfold higher levels of naringenin and less than 10% of the flavonols kaempferol and quercetin than wild-type. The reduced abundance of anthocyanins in the stem and reduction in the number of lateral roots in *are* mutant plants are evident in Fig. 7.3, with roots grown on nutrient agar plates or in Turface, a soil substitute. The *are* mutant transformed with the wild-type copy of the *F3H* gene restored flavonol accumulation profiles and lateral root development to wild-type levels (Maloney *et al.*, 2014). To ask if flavonol levels are correlated with lateral rooting in other tomato mutants, the levels of flavonols and lateral root formation in the *anthocyanin without* (*aw*) mutant, where the mutation is linked to the *DFR* gene, were determined. A 100-fold increase in kaempferol in roots of *aw* over the parental line was detected by LCMS, and a substantial increase in lateral root formation was observed (Maloney *et al.*, 2014). Since both *are* and *aw* have decreased anthocyanin accumulation, but opposite effects on flavonol accumulation, the difference in root formation is strongly correlated with flavonol levels.

7.6 Cotyledon, trichome, and root hair development

Insight into the role of flavonoids in cotyledon morphology came from the isolation of the *rol1* mutant, which has a defect in a gene encoding rhamnose synthase, *RHAMNOSE BIOSYNTHESIS 1* (*RHM1*), resulting in an altered flavonoid glycoside composition (Ringli *et al.*, 2008; Kuhn *et al.*, 2011). This mutant has hyponastic cotyledons, altered shape of pavement and guard cells, and deformed trichomes (Ringli *et al.*, 2008; Kuhn *et al.*, 2011). In a double mutant with *rol1* and either *tt4-2*, *fls*, or *myb111* (a TF required for flavonol synthesis) mutants, these shoot phenotypes are lost, consistent with developmental roles of flavonoid glycosides (Ringli *et al.*, 2008; Kuhn *et al.*, 2011). Treatment of the *rol1 tt4-2* double mutant with naringenin restored the *rol1* phenotype (Ringli *et al.*, 2008). In contrast, a *tt7-1 rol1* double mutant, which produces no quercetin derivatives, has similar phenotypes to *rol1*, implicating kaempferol derivatives in this phenotype (Ringli *et al.*, 2008). These shoot phenotypes are mediated by asymmetric expression of both *RHM* and *FLS* on the upper and lower surfaces of *Arabidopsis* leaves (Kuhn *et al.*, 2011). Although the cotyledon growth phenotype of *rol1* can be induced by treatment with the auxin transport inhibitor, NPA, the pavement cell and trichome phenotypes are not induced by NPA (Ringli *et al.*, 2008). The *rol1* mutant has reduced export of the synthetic auxin 1-naphthaleneacetic acid (NAA) from protoplasts, which is reversed in the *rol1 fls* double mutant (Kuhn *et al.*, 2011). Although the *rol1* mutant has shorter primary roots and increased root hair density, this phenotype does not require a functional CHS gene, indicating that this phenotype is not mediated through flavonoid levels (Kuhn *et al.*, 2011).

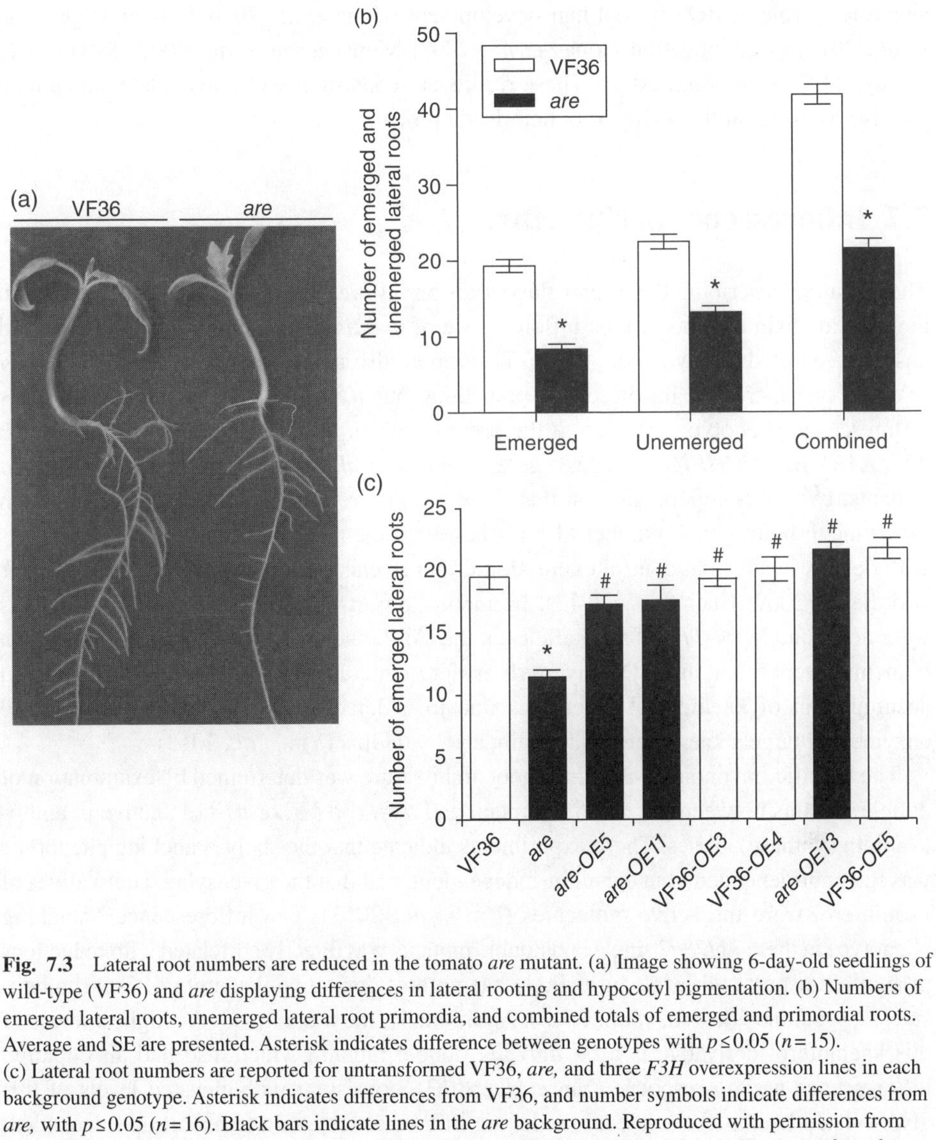

Fig. 7.3 Lateral root numbers are reduced in the tomato *are* mutant. (a) Image showing 6-day-old seedlings of wild-type (VF36) and *are* displaying differences in lateral rooting and hypocotyl pigmentation. (b) Numbers of emerged lateral roots, unemerged lateral root primordia, and combined totals of emerged and primordial roots. Average and SE are presented. Asterisk indicates difference between genotypes with $p \leq 0.05$ ($n = 15$).
(c) Lateral root numbers are reported for untransformed VF36, *are,* and three *F3H* overexpression lines in each background genotype. Asterisk indicates differences from VF36, and number symbols indicate differences from *are,* with $p \leq 0.05$ ($n = 16$). Black bars indicate lines in the *are* background. Reproduced with permission from Maloney *et al.* (2014). © American Society of Plant Biologists (www.plantphysiol.org). (*See insert for color representation of the figure*)

In tomato, flavonoids may regulate root hair formation through their antioxidant activity. The *are* mutant, with reduced flavonol accumulation, has enhanced root hair formation (Maloney *et al.*, 2014). The root hairs show higher levels of reactive oxygen species, as judged by the generic ROS sensor 2′,7′dihydrodichloro-fluorescein diacetate (DCF) (Maloney *et al.*, 2014). When roots are treated with an exogenous antioxidant, ascorbate, root hair number and DCF signals are decreased in *are* relative to wild-type, while treatment of wild-type roots with hydrogen peroxide increases root hair numbers and DCF fluorescence to levels equivalent to those in *are* (Maloney *et al.*, 2014). Previous reports have suggested a role of ROS in root hair development (Duan *et al.*, 2010; Sundaravelpandian *et al.*, 2013) and elongation (Jones *et al.*, 2007; Monshausen *et al.*, 2007; Swanson & Gilroy, 2010; Shin *et al.*, 2011). These results are consistent with flavonols dampening a positive ROS signal that drives root hair development.

7.7 Inflorescence architecture

The literature describing the role of flavonoids on inflorescence branching is complex. An increase in auxin transport in the inflorescence of two *tt4* alleles in the Col-0 background has been reported (Brown *et al.*, 2001). This report also noted that one of these alleles, *tt4* (2YY6), had increased inflorescence branching, but it was later demonstrated that this inflorescence phenotype was due to the presence of an unlinked mutation in the *MORE AXILLARY BRANCHING 4* (*MAX4*) gene (Bennett *et al.*, 2006). Using a series of L*er tt* mutants, two other reports suggest that there are differences in inflorescence elongation rate, time to bolting, and number of branches, with the most profound effects in mutants with defects in TFs that control expression of genes encoding pathway enzymes (Buer & Djordjevic, 2009; Buer *et al.*, 2013). In another report, inflorescence branching defects were not found in *tt4-11* and *tt7-3* alleles in the Col-0 background, but striking changes in branching were found in a UDP-glycosyltransferase mutant, *ugt78d2*, which had an altered accumulation of kaempferol glycosides due to a defect in a UDP-glycosyltransferase enzyme that targets kaempferol, quercetin, and cyanidins (Yin *et al.*, 2013).

The specific flavonoid controlling shoot architecture was determined by examination of double mutants between the *ugt78d2* mutant and *tt4-11*, *tt7-3*, or *tt18* (defective in anthocyanidin synthase) alleles. These experiments indicate that the shoot branching phenotype was flavonol-dependent, but cyanidin-independent, and that the glycosylated derivatives of kaempferol were the active molecules (Yin *et al.*, 2013). The inflorescence branching phenotype in these *ugt78d2* single and double mutants was directly correlated with reductions in auxin transport and levels of a single kaempferol derivative, kaempferol 3-*O*-rhamnoside-7-*O*-rhamnoside (Yin *et al.*, 2013). Additionally, the mutant with the highest levels of this kaempferol derivative is a *tt7 ugt78d2* double mutant, which also had an impaired inflorescence gravity response (Yin *et al.*, 2013), consistent with elevated levels of this glycosylated kaempferol molecule reducing inflorescence auxin transport and dependent physiological processes.

7.8 Fertility and pollen development

Flavonoids have been implicated in the maintenance of proper anther and pollen development in many, but not all, species (Taylor & Grotewold, 2005). Flavonoids typically make up a very large part of the total metabolite pool in anthers and pollen (Yonekura-Sakakibara *et al.*, 2014). The presence of flavonols is required for the normal progression of pollen development, dehiscence, and germination in species including maize (*Zea mays*) (Mo *et al.*, 1992), tobacco (*Nicotiana tabacum*) (Ylstra *et al.*, 1992; Qin *et al.*, 2011), petunia

(*Petunia hybrida*) (Pollak *et al.*, 1993), and tomato (Schijlen *et al.*, 2007). In maize, a mutant with a nonfunctional *CHS* is male sterile and cannot produce functional pollen tubes (Mo *et al.*, 1992). Transgenic tomato and tobacco lines with lowered flavonoid content showed similar defects, with fruit producing less seeds and decreased pollen tube germination frequency (Ylstra *et al.*, 1992; Schijlen *et al.*, 2007). Similarly, a tomato line expressing a *CHS* RNAi construct is parthenocarpic and has defects in pollen tube growth (Schijlen *et al.*, 2007). In both maize and tobacco, these pollen tube germination phenotypes were reversed by addition of the flavonols kaempferol and quercetin (Mo *et al.*, 1992; Ylstra *et al.*, 1992). In contrast, several studies report that fertility and seed production are not dependent on the presence of flavonoids in *Arabidopsis*, as *tt4* mutants are fully fertile (Burbulis *et al.*, 1996; Ylstra *et al.*, 1996; Qin *et al.*, 2011). One study uncovered a pollen germination defect in *tt4-2* pollen under elevated UV-B irradiation (Hsieh & Huang, 2007), suggesting that flavonols regulate pollen development and germination with roles that are species-specific and sensitive to environmental context.

7.9 Flavonols modulate ROS signaling in guard cells to regulate stomatal aperture

The observation of accumulation of flavonols in guard cells suggested the testable hypothesis that these molecules may modulate ROS-dependent guard cell closure. Studies in fava bean (*Vicia faba*), pea (*Pisum sativa*), and onion (*Allium cepa*) reported that flavonols accumulated in guard cells and surrounding pavement cells on the epidermal layer of leaves (Schnabl *et al.*, 1986; Weissenboeck & Scharf, 1986; Weissenboeck *et al.*, 1987; Takahama, 1988). Using confocal microscopy and DPBA, a more recent study in *Arabidopsis* showed flavonol accumulation localized to the guard cells, but not the surrounding pavements cells on the abaxial or bottom surface of leaves (Watkins *et al.*, 2014). The fluorescence of DPBA was visible in wild-type guard cells; however, no fluorescence was present in guard cells of plants with the *tt4-2* mutation (Fig. 7.4). Additionally, the product of a *CHS* promoter-GUS fusion reporter construct also accumulated in guard cells (Watkins *et al.*, 2014), as did fluorescence of a GFP-FLS fusion protein (Kuhn *et al.*, 2011), consistent with flavonol synthesis in guard cells.

ROS can act as signaling molecules that regulate plant physiology and development (De Tullio *et al.*, 2010; Swanson & Gilroy, 2010; Mittler *et al.*, 2011; Schmidt & Schippers, 2015) raising the question of whether flavonoids modulate guard cell signaling through acting as antioxidants. Elevated ABA triggers H_2O_2 production through activation of respiratory burst oxidase (RBOH) enzymes (Kwak *et al.*, 2003; Murata *et al.*, 2015). During signaling of stomatal closure, ROS must reach a certain threshold to oxidize their targets, which likely include ion channel proteins that control ion concentrations that mediate stomatal closure (Pei *et al.*, 2000; Song *et al.*, 2014). Yet, to prevent ROS from reaching concentrations that cause oxidative damage, ROS levels must be tightly regulated by small molecule and/or protein antioxidants (Conklin *et al.*, 1996; Apel & Hirt, 2004; Munne-Bosch *et al.*, 2013), with the importance of glutathione demonstrated in guard cells (Okuma *et al.*, 2011; Munemasa *et al.*, 2013).

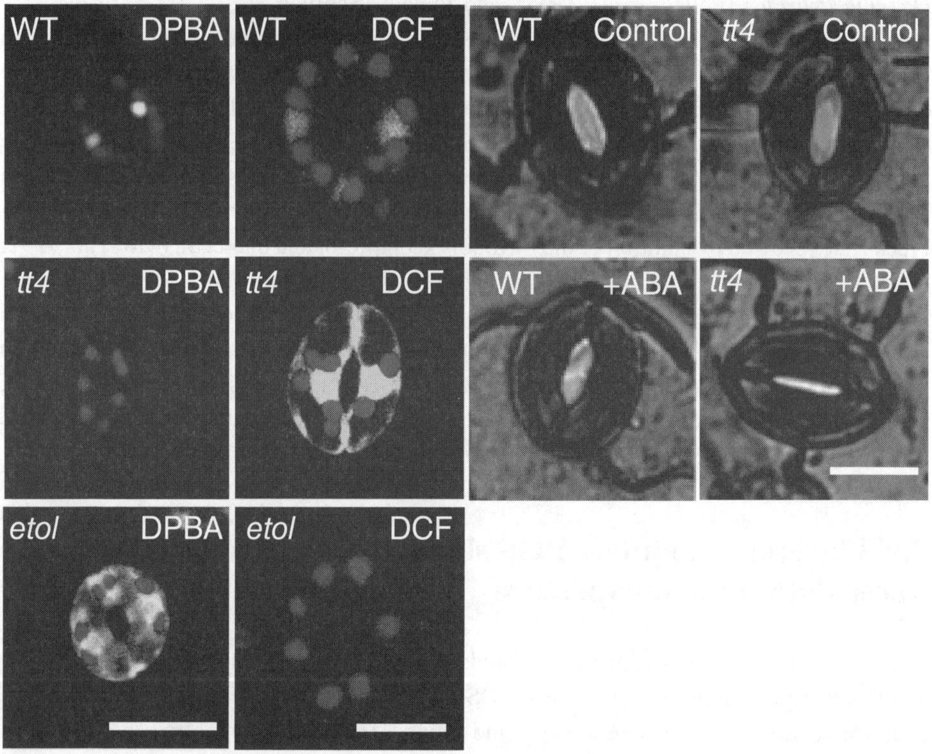

Fig. 7.4 Flavonol accumulation is absent in *tt4* leading to increased ROS and more rapid stomatal closure. Flavonols are visualized by diphenylboric acid 2-aminoethyl ester (DPBA) (yellow), and ROS is visualized by 2′,7′dihydrodichloro-fluorescein diacetate (DCF) (green), with chlorophyll in blue. Ethylene overproduction in *etol* leads to enhanced flavonol accumulation and less ROS. Bar = 15 μm. Guard cells of wild-type and *tt4-2* were examined by light microscopy immediately before treatment with ABA and 45 minutes after treatment with 20 μM ABA. Bar = 15 μm. Watkins *et al.* (2014). © The fluorescent image is reprinted with permission from American Society of Plant Biologists (www.plantphysiol.org). (*See insert for color representation of the figure*)

If flavonols act as antioxidants *in planta,* then guard cells of *tt4-2* mutants should have increased levels of ROS, making them more sensitive to ABA-induced stomatal closure. ROS levels were visualized by DCF fluorescence in wild-type and *tt4-2* guard cells, as shown in Fig. 7.4. Consistent with flavonols acting as antioxidants, the levels of DCF fluorescence (green) were elevated in *tt4-2* guard cells relative to wild-type. In accordance to enhanced ROS signaling in *tt4-2*, ABA-induced stomatal closure was greater in this mutant (Fig. 7.4) (Watkins *et al.*, 2014).

If flavonols act as antioxidants in guard cells and prevent ROS-induced stomatal closure, then there is likely a mechanism by which the plant can control flavonol accumulation in this cell type. Elevation of ethylene levels leads to increased flavonol accumulation in roots (Lewis *et al.*, 2011b). The ability of exogenous ethylene to increase flavonol levels in *Arabidopsis* guard cells was also demonstrated (Watkins *et al.*, 2014). A similar increase in DPBA fluorescence was also observed in guard cells of plants of the ethylene overproducing *etol* mutant (Chae *et al.*, 2003), consistent with ethylene-induced flavonol accumulation in this tissue (Fig. 7.4). When flavonol synthesis is elevated in guard cells, there is a substantial reduction in ROS accumulation as shown in the DCF-stained guard cells of *eto1* (Watkins *et al.*, 2014). Consequently, elevated flavonols and decreased ROS in *eto1* guard cells resulted in decreased rates of ABA-induced stomatal closure when compared to wild-type and *tt4-2* (Watkins *et al.*, 2014). Two other studies indicate that ethylene promotes stomatal closure in peanut (*Arachis hypogaea*) and *Arabidopsis* (Pallas & Kays, 1982; Desikan *et al.*, 2006). Ethylene has also been reported to promote stomatal opening in carnation (*Dianthus caryophyllus*), tomato, and *Arabidopsis* (Madhavan *et al.*, 1983; Tanaka *et al.*, 2005). These studies suggest that ethylene modulation of stomatal closure varies among species and with the methods used for ethylene treatment and plant growth.

In addition to their function in guard cell signaling, flavonols may also modulate development of stomata and surrounding pavement cells. The altered flavonol glycosylation profiles in the *rol1* mutant are responsible for multiple phenotypes, including pavement and guard cell development, with larger stomata and decreased stomatal density (Ringli *et al.*, 2008). These phenotypes were rescued when the *rol1* mutation was crossed into the flavonoid-deficient *tt4-2* mutant, verifying that these phenotypes were linked to elevated glycosylated flavonols (Ringli *et al.*, 2008). Additionally, soybean isolines containing elevated concentrations of kaempferol tri-*O*-glucoside showed decreased stomatal density in response to blue light (Buttery & Buzzell, 1987; Liu-Gitz *et al.*, 2000). As the stomatal phenotype cannot be mimicked by auxin transport inhibitors, the role of flavonoids has been suggested to regulate these developmental patterns via an auxin-independent mechanism (Ringli *et al.*, 2008). In petunia and tomato, flavonol-deficient mutants or RNAi lines have altered pavement cell development and/or stomatal density (Mo *et al.*, 1992; Liu-Gitz *et al.*, 2000; Taylor & Grotewold, 2005; Schijlen *et al.*, 2007).

7.10 Transcriptional machinery that controls synthesis of flavonoids

Individual flavonoid metabolites accumulate in specific tissues in response to a variety of external and internal stimuli, consistent with the function of these molecules described earlier. The precise regulation of flavonoid synthesis is mediated by tissue-specific, hormone-regulated, and light-dependent networks of transcription factors that control not just when flavonoids are synthesized, but which metabolites accumulate (Lepiniec *et al.*, 2006; Hichri *et al.*, 2011). In many species, the flavonoid biosynthetic pathway is divided into the "early pathway," which produces flavonols, and the "late pathway" where dihydroflavonols are converted into anthocyanins and condensed tannins (Winkel-Shirley, 2001a; Lepiniec *et al.*, 2006). This review focuses on flavonol function and synthesis; however, controls of both early and late pathway enzymes are needed to alter flavonol accumulation. Increased flavonol synthesis can occur via increased transcription of early pathway genes, *CHS, CHI, F3H, FLS*, and *F3′H*, to increase carbon flow into the flavonol biosynthetic pathway. If the activity of late pathway enzymes is elevated in proportion, the flux through the pathway may increase without altering flavonol accumulation. In order for flavonol accumulation to increase, expression or activity of late pathway enzymes beginning with the branch-point enzyme, DFR must be blocked or independently regulated from early pathway genes.

TFs controlling the expression of genes encoding the enzymes involved in the early pathway have been identified, along with the environmental factors that control synthesis of these TFs (Lepiniec *et al.*, 2006; Hichri *et al.*, 2011), as summarized in Table 7.1. This table and associated text focus on TFs in *Arabidopsis* and *Solanaceae* species, where developmental phenotypes linked to flavonol levels are most thoroughly studied. The transcriptional machinery in maize has been reviewed elsewhere (Falcone Ferreyra *et al.*, 2012). In *Arabidopsis*, the role of the R2R3 MYB family TFs in controlling flavonoid synthesis has been extensively studied (Matsui *et al.*, 2004; Mehrtens *et al.*, 2005; Lepiniec *et al.*, 2006; Stracke *et al.*, 2007, 2010a, 2010b; Dubos *et al.*, 2008; Albert *et al.*, 2014; Fornalé *et al.*, 2014; Xu *et al.*, 2015). Genes encoding early pathway genes are regulated by R2R3-MYB TFs acting independently, such as MYB12 and MYB111 that bind to the promoters of *CHS*, *CHI*, *F3H*, and *FLS* to activate flavonol synthesis in roots and

Table 7.1 Summary of transcription factors that have been demonstrated to control levels of transcripts of genes encoding pathway enzymes of flavonoid biosynthesis and the organism and tissue in which this regulation has been observed.

	CHS	CHI	F3H	F3′H	FLS	DFR
Activators						
Whole seedlings (*Arabidopsis*)	TCP3[a]	ANAC078[b]	TCP3[a] ANAC078[b]		TCP3[a] ANAC078[b]	ANAC078[b]
Roots (*Arabidopsis*)	MYB12[c]	MYB12[c]	MYB12[c]	WRKY23[d]	MYB12[c]	NE
Cotyledons (*Arabidopsis*)	MYB111[c]	MYB111[c]	MYB111[c]		MYB111[c]	PAP1[e]
Shoots/stem (*Arabidopsis*)	SPL9[f]	SPL9[f]				PAP1[g]
Seeds/siliques (*Arabidopsis*)	PAP1[h] TT1[i]					TT1[i]
Fruit flesh (tomato)	MYB12[j]	MYB12[j]	MYB12[j] ROS1[k] DEL1[k]	MYB12[j]	MYB12[j]	
Fruit peel (tomato)			ROS1[k] DEL[k]			
Repressors						
Whole seedlings (*Arabidopsis*)	MYBL2[l,m]		MYBL2[l]	MYBL2[l] MYB7[n]		MYB7[n] MYBL2[m]
Roots (*Arabidopsis*)						MYBL2[m]
Shoots/stem (*Arabidopsis*)	MYBL2[m]			SPL9[f]		SPL9[f,o] MYBL2[m]
Seeds/siliques (*Arabidopsis*)	MYBL2[l,m]		MYBL2[l]	MYBL2[l]		MYBL2[l,m]
(Leaves/flowers) Petunia		MYB27[p]	MYB27[p]			MYB27[p]

NE, not expressed.
[a] Li and Zachgo (2013). [b] Morishita *et al.* (2009). [c] Stracke *et al.* (2007). [d] Grunewald *et al.* (2012). [e] Teng *et al.* (2005). [f] Gou *et al.* (2011). [g] Borevitz *et al.* (2000). [h] Matsui *et al.* (2004). [i] Appelhagen *et al.* (2011). [j] Ballester *et al.* (2010). [k] Butelli *et al.* (2008). [l] Dubos *et al.* (2008). [m] Matsui *et al.* (2008). [n] Fornalé *et al.* (2014). [o] SPL9 can bind to the promoter of *DFR,* but its action of repressing *DFR* expression *in planta* is through disruption of MBW complex. [p] Albert *et al.* (2014).

cotyledons, respectively (Mehrtens *et al.*, 2005; Stracke *et al.*, 2007, 2010b). Genes encoding late pathway enzymes are regulated by R2R3-MYB TFs that form ternary complexes with bHLH and WD40-type TFs called "MBW complexes" (Gonzalez *et al.*, 2008). PAP1/MYB75 is part of an MBW complex that increases abundance of transcripts encoding both early (CHS) and late enzymes (DFR), leading to elevated levels of anthocyanins in immature siliques and shoots (Borevitz *et al.*, 2000; Matsui *et al.*, 2004) and also interacts with other TFs that affect expression of genes in the early pathway: SPL9, TCP3, and TT1 (Matsui *et al.*, 2004; Gonzalez *et al.*, 2008; Gou *et al.*, 2011; Li & Zachgo, 2013). Other MYB TFs act as repressors of genes encoding flavonoid synthesis: an R2R3-MYB TF MYB7 suppresses *F3′H* and *DFR* (Fornalé *et al.*, 2014), while an R3-MYB-like protein, MYBL2, which contains a repressor motif of the R2 domain, suppresses *PAP1*, *CHS*, *F3H*, *F3′H*, and most strongly *DFR*, in seedlings and seeds by interacting with PAP1 and a bHLH TT8 to either sequester TFs or create an inactive complex (Dubos *et al.*, 2008; Matsui *et al.*, 2008). This machinery has the potential to uncouple flavonol synthesis from anthocyanin synthesis by repressing late pathway genes until anthocyanin synthesis is needed (Albert *et al.*, 2014).

Flavonol regulation by MYB TFs has been observed in other species: MYB27 represses *CHI*, *F3H*, *F3′5′H*, and *DFR* in leaves and flowers of petunia, reducing flux into the early pathway and preventing anthocyanin synthesis in the absence of environmental stress (Albert *et al.*, 2014). The authors suggest that when petunia plants are grown under stressful conditions and anthocyanins are needed, this repressive MYB is replaced by an MYB activator (Albert *et al.*, 2014). The ortholog of *MYB12* in tomato, *SlMYB12*, is responsible for the red color phenotype of ripe fruit by inducing expression of all genes in the flavonoid early pathway (Ballester *et al.*, 2010). Simultaneous overexpression of snapdragon MYB-related and bHLH TFs, *ROS1* and *DEL*, increased expression of *F3H* and *F3′5′H* and accumulation of flavonol derivatives and anthocyanins in the flesh and peel of tomato fruit (Butelli *et al.*, 2008).

In addition to the MYB TFs, flavonol regulation in *Arabidopsis* is under the control of other TF families involved in a suite of processes regulating plant development, signaling, and stress responses (Lepiniec *et al.*, 2006; Xu *et al.*, 2015). The TF TEOSINTE BRANCHED-CYCLOIDEA-PROLIFERATING CELL FACTOR 3 (TCP3) interacts with MYB12 and MYB111, as well as PAP1/MYB75. This interaction increases flux throughout the flavonoid pathway by increasing abundance of *CHS*, *F3H*, *DFR*, and *PAP1* transcripts, resulting in increased flavonol accumulation at the root tip, root–hypocotyl junction, and the cotyledonary node (Gou *et al.*, 2011; Li & Zachgo, 2013). The SQUAMOSA PROMOTER BINDING PROTEIN-LIKE 9 (SPL9) increases flavonol levels in stems by repressing anthocyanin levels through repression of *F3′H* and *DFR* and allowing increased expression of *MYB12*, *MYB111*, *CHS*, and *CHI* (Gou *et al.*, 2011). Interestingly, although SPL9 binds to the promoter of *DFR*, its mode of action on anthocyanin accumulation is in disrupting the MBW complex through interaction with PAP1 (Gou *et al.*, 2011). Additionally, an NAC TF family member, ANAC078, regulates synthesis of *PAP1*, *CHI*, *F3H*, *F3′H*, and *DFR* expression under high light intensity, which should lead to increases in flux through this pathway, leading to elevated anthocyanins (Morishita *et al.*, 2009).

7.11 Hormonal controls of flavonoid synthesis

Flavonol synthesis is positively regulated by auxin levels (Lewis *et al.*, 2011b; Grunewald *et al.*, 2012). The auxins, IAA and NAA, increase abundance of *CHS*, *CHI*, *F3′H*, and *FLS* transcripts and flavonol accumulation in roots, since the genes encoding late pathway enzymes are not expressed in this tissue. In particular, quercetin accumulates at higher levels than kaempferol in response to IAA after 8 hours of treatment (Lewis *et al.*, 2011b; Grunewald *et al.*, 2012). NAA treatment increases the expression of a member of the WRKY TF family, *WRKY23*, in the root tip in a region spanning from quiescent center to the elongation zone, and this response contributes to quercetin accumulation in roots by increasing *F3′H* transcript accumulation (Grunewald *et al.*, 2012). Expression of *MYB12* and *PAP1* increases in response to IAA treatment, with *MYB12* induction occurring with faster and more sustained kinetics than *PAP1* (Lewis *et al.*, 2011b). In the auxin signaling mutant, *tir1*, auxin-dependent increases in *CHS, CHI, F3′H, FLS*, and *MYB12* transcripts are lost, and this mutant accumulates fewer flavonols than wild-type after IAA treatment, demonstrating that flavonol accumulation occurs through the auxin signaling pathway (Lewis *et al.*, 2011b).

Misexpression of TFs that control flavonol synthesis alters flavonol profiles and auxin-related growth, consistent with a feedback loop between flavonols and auxin transport. Plants engineered to have higher levels of *TCP3* transcripts have increased flavonol levels in the root tip and hypocotyl–root junction, which results in lower *pPIN1:PIN1-GFP* fluorescence and fewer lateral roots (Li & Zachgo, 2013). Disruption of the gene encoding the light-responsive *HY5* TF, which controls induction of *PAP1* and *MYB12*, reduces flavonol content and enhances shootward auxin transport, resulting in enhanced root growth (Oyama *et al.*, 1997; Sibout *et al.*, 2006; Stracke *et al.*, 2010a; Shin *et al.*, 2013).

Ethylene, similar to auxin, modulates flavonol accumulation in *Arabidopsis* roots (Buer *et al.*, 2006; Lewis *et al.*, 2011b). Enhanced ethylene levels in plants treated with the ethylene precursor, ACC, promoted flavonol synthesis through transcript accumulation of *CHS*, *CHI*, and *FLS* (Lewis *et al.*, 2011b). In contrast to plants treated with IAA, ACC did not increase expression of *F3′H* and production of kaempferol remained higher than that of quercetin (Lewis *et al.*, 2011b). This result illustrates how hormonal control of synthesis of branch-point enzymes can partition flux of the early pathway to favor production of specific flavonols. These transcript accumulation changes in response to ACC are lost in ethylene signaling mutants *ein2* and *etr1*, but are not affected in the *tir1* mutant, while IAA responses are lost in the *tir1* mutant, but not in *ein2* and *etr1*, consistent with independent hormonal regulation of these targets (Lewis *et al.*, 2011b). Yet, both the IAA and ACC effects are lost in *myb12*, consistent with conserved transcriptional machinery mediating these two different hormonal responses (Lewis *et al.*, 2011b).

Elevated ethylene levels positively regulate flavonol accumulation in roots grown on nutrient media containing sucrose and in guard cells of soil grown plants (Lewis *et al.*, 2011b; Watkins *et al.*, 2014). However, ethylene suppresses anthocyanin accumulation in shoots of plants grown in the presence of sucrose (Jeong *et al.*, 2010). Sucrose concentrations of 60 mM (2%) or above increases expression of *CHS* and genes in the late pathway in a PAP1-dependent manner (Matsui *et al.*, 2004; Teng *et al.*, 2005; Solfanelli *et al.*, 2006;

Jeong *et al.*, 2010). Increased expression of *CHS* increases carbon flow into the flavonoid pathway, while transcript accumulation of genes encoding late pathway enzymes directs the flux of flavonoid biosynthesis to favor anthocyanin production in shoots, thereby reducing flavonol levels in hypocotyl and cotyledon tissue (Solfanelli *et al.*, 2006). However, in roots that do not make anthocyanins, this leads to increased flavonol concentrations. Anthocyanin accumulation is higher in shoots of ethylene signaling mutants (*etr1*, *etr2*, *ers1*, *ers2*, *ein2* single mutants, and an *ein3 eil1* double mutant) compared to wild-type plants (although flavonols levels may be lower in ethylene signaling mutants compared to wild-type plants); while anthocyanin content in the constitutive ethylene signaling mutant, *ctr1*, is similar to wild-type plants when plants are grown in the absence of sucrose (Gibson *et al.*, 2001; Jeong *et al.*, 2010). The expression of *CHS*, *PAP1*, the branch-point gene *DFR*, and genes in the late pathway was increased in the shoots of *etr1-1*, *ein2-1*, and *ein3 eil1* mutants, suggesting that ethylene suppresses accumulation of transcripts encoding TFs and genes in the early pathway in the presence of high sucrose in this tissue (Jeong *et al.*, 2010). Addition of inhibitors to disrupt ethylene binding or synthesis in wild-type plants increases anthocyanin accumulation, suggesting that ethylene reduces high sucrose signaling (Jeong *et al.*, 2010). These results suggest the tantalizing possibility that ethylene has opposite effects on the regulation of the flavonoid biosynthetic pathway in roots and shoots.

7.12 Flavonoid synthesis is regulated by light

A growing body of evidence suggests that flavonols may function as antioxidants, as described earlier, a role that is particularly important as prolonged exposure to light and other abiotic stresses may increase their synthesis while decreasing activity of antioxidant proteins such as superoxide dismutases (SODs) and ascorbate peroxidases (APXs) (Fini *et al.*, 2011). These environmental stresses include nitrogen or phosphate deficiency and drought, but the specific signaling pathways linking abiotic stress to flavonoid accumulation are beyond the scope of this chapter (Tattini *et al.*, 2004; Yang *et al.*, 2009; Xu *et al.*, 2015). Flavonoids are not produced in dark-grown seedlings, but transfer of seedlings to light rapidly increases accumulation of transcripts encoding pathway enzymes, their protein products, and flavonoid metabolites regulating light responses (Chory & Peto, 1990; Kubasek *et al.*, 1992; Winkel-Shirley, 2002). The difference in auxin transport between wild-type and *tt4* mutants is only evident in light-grown seedlings, consistent with the dependency of flavonol synthesis on light (Buer & Muday, 2004). Light increases the accumulation of *PAP1*, *MYB12*, and *CHS* transcripts in a HY5-dependent manner, increases the expression of the positive regulator of flavonol synthesis, *MYB111*, and reduces transcript levels of the negative regulator, *MYBL2* (Lee *et al.*, 2007; Dubos *et al.*, 2008; Stracke *et al.*, 2010a; Shin *et al.*, 2013).

Flavonoids and sinapate esters have been identified as UV screening molecules that prevent photoxidative damage (Fornalé *et al.*, 2014). Transcripts of genes encoding flavonoid enzymes and flavonol levels accumulate in proportion to light exposure, particularly when plants are transitioned from shade to full sunlight (Besseau *et al.*, 2007; Stracke *et al.*, 2010a; Agati *et al.*, 2013). Studies of mutants with either elevated or reduced levels of flavonols, but similar or higher levels of anthocyanins compared to wild-type, support the protective function of flavonols against UV-B damage. The *myb11myb12myb111* triple mutant, which has low flavonol but high anthocyanin content, is sensitive to UV-B light compared to plants overexpressing *MYB12* that have elevated levels of flavonols (Stracke *et al.*, 2010a). Increased flavonol levels in *Arabidopsis* plants transformed with *FLS* from *Z. mays* (maize) resulted in higher tolerance to UV-B-induced DNA damage, less photosynthetic damage, and reduced lipid peroxidation despite similar levels of anthocyanins between transgenic and wild-type plants (Emiliani *et al.*, 2013).

Increases in *CHS* transcripts occur in response to multiple light qualities (Chory & Peto, 1990; Wade *et al.*, 2001; Cominelli *et al.*, 2008). UV-A and blue light induce expression of *CHS* through the cryptochrome blue light receptor, CRY1 (Fuglevand *et al.*, 1996). UV-A-mediated induction of *CHS* by CRY1 is dependent on phytochrome B (PHYB), a receptor for red/far-red light (Wade *et al.*, 2001). Phytochrome A (PHYA) and PHYB both play a role in the enhancement of *CHS* induction when plants undergo preillumination with red light before exposure to UV-A light (Wade *et al.*, 2001). Interestingly, PHYB represses *CHS* induction by UV-B light, as *phyb* mutants accumulate *CHS* transcripts at higher levels compared to wild-type when exposed to blue/UV-B or UV-A/UV-B light (Wade *et al.*, 2001). The potential importance of flavonoids as UV sunscreens may explain why the activities of both phytochromes and cryptochromes are required to control the synthesis of these molecules (Jenkins *et al.*, 2001).

7.13 Conclusions

The experiments summarized in this chapter indicate the important roles of flavonols in a range of plant growth and developmental processes, as summarized in Fig. 7.5. Genetic approaches have been used to demonstrate these roles and define the functional molecules in *Arabidopsis* and other species. The transcriptional machinery that controls flavonol and anthocyanin synthesis has been extensively studied and allows environmental and hormonal regulation of carbon flow into the pathway leading to tissue-specific accumulation profiles of pathway metabolites that are linked to their developmental functions. As illustrated in Fig. 7.5, ethylene, auxin, sucrose, and light all positively affect flavonol biosynthesis. Flavonols modulate developmental processes through controls of auxin transport and oxidant status in cells. The mechanisms by which these molecules function are now receiving extensive study and future research will focus on identification of the molecular mechanisms by which flavonols control signaling pathways, such as ROS signaling, to modulate growth and developmental responses.

Acknowledgments

We appreciate the editorial suggestions from Drs Joëlle Mühlemann, Daniel Lewis, and Brenda Winkel. This work was supported by grants from the United States Department of Agriculture National Institute of Food and Agriculture program to GKM 2009-65116-20436 and 2015-09300, to SRG 2014-67012-22405, and to JWM 2014-67011-22277 and to GKM

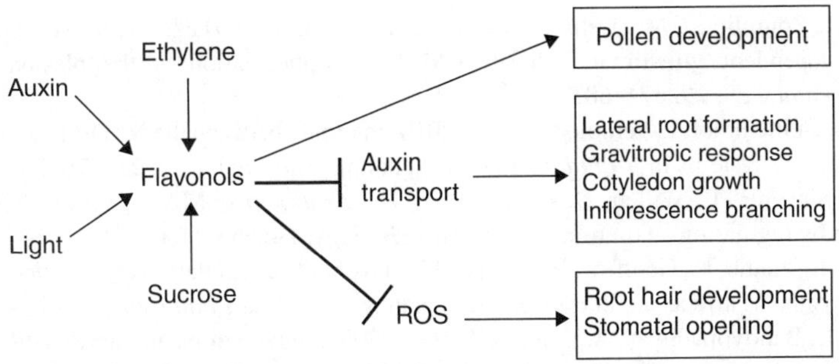

Fig. 7.5 A model of controls of flavonol synthesis and their function. This figure illustrates the positive effects of ethylene, auxin, sucrose, and light on flavonol synthesis. The action of flavonols in regulating root growth and other developmental processes are summarized with those shown to be mediated through effects on auxin transport and on levels of reactive oxygen species (ROS) indicated.

from the National Science Foundation (IOS-0820717 and IOS-1558046) and a graduate fellowship to JMW from the Wake Forest University Center for Molecular Communication and Signaling.

References

Agati, G., Azzarello, E., Pollastri, S. *et al.* (2012) Flavonoids as antioxidants in plants: location and functional significance. *Plant Science*, **196**, 67–76.

Agati, G., Brunetti, C., Di Ferdinando, M. *et al.* (2013) Functional roles of flavonoids in photoprotection: new evidence, lessons from the past. *Plant Physiology and Biochemistry*, **72**, 35–45.

Agullo, G., Gamet-Payrastre, L., Manenti, S. *et al.* (1997) Relationship between flavonoid structure and inhibition of phosphatidylinositol 3-kinase: a comparison with tyrosine kinase and protein kinase C inhibition. *Biochemical Pharmacology*, **53**, 1649–1657.

Albert, N.W., Davies, K.M., Lewis, D.H. *et al.* (2014) A conserved network of transcriptional activators and repressors regulates anthocyanin pigmentation in eudicots. *The Plant Cell*, **26**, 962–980.

Apel, K. and Hirt, H. (2004) Reactive oxygen species: metabolism, oxidative stress, and signal transduction. *Annual Review of Plant Biology*, **55**, 373–399.

Appelhagen, I., Huep, G., Lu, G.H. *et al.* (2010) Weird fingers: functional analysis of WIP domain proteins. *FEBS Letters*, **584**, 3116–3122.

Appelhagen, I., Lu, G.H., Huep, G. *et al.* (2011) TRANSPARENT TESTA1 interacts with R2R3-MYB factors and affects early and late steps of flavonoid biosynthesis in the endothelium of *Arabidopsis thaliana* seeds. *The Plant Journal*, **67**, 406–419.

Bailly, A., Sovero, V., Vincenzetti, V. *et al.* (2008) Modulation of P-glycoproteins by auxin transport inhibitors is mediated by interaction with immunophilins. *Journal of Biological Chemistry*, **283**, 21817–21826.

Ballester, A.R., Molthoff, J., de Vos, R. *et al.* (2010) Biochemical and molecular analysis of pink tomatoes: deregulated expression of the gene encoding transcription factor SlMYB12 leads to pink tomato fruit color. *Plant Physiology*, **152**, 71–84.

Barbosa, I.C. and Schwechheimer, C. (2014) Dynamic control of auxin transport-dependent growth by AGCVIII protein kinases. *Current Opinion in Plant Biology*, **22**, 108–115.

Barbosa, I.C., Zourelidou, M., Willige, B.C. *et al.* (2014) D6 PROTEIN KINASE activates auxin transport-dependent growth and PIN-FORMED phosphorylation at the plasma membrane. *Developmental Cell*, **29**, 674–685.

Bashandy, T., Guilleminot, J., Vernoux, T. *et al.* (2010) Interplay between the NADP-linked thioredoxin and glutathione systems in *Arabidopsis* auxin signaling. *The Plant Cell*, **22**, 376–391.

Bennett, T., Sieberer, T., Willett, B. *et al.* (2006) The *Arabidopsis* MAX pathway controls shoot branching by regulating auxin transport. *Current Biology*, **16**, 553–563.

Besseau, S., Hoffmann, L., Geoffroy, P. *et al.* (2007) Flavonoid accumulation in *Arabidopsis* repressed in lignin synthesis affects auxin transport and plant growth. *The Plant Cell*, **19**, 148–162.

Blakeslee, J.J., Bandyopadhyay, A., Lee, O.R. *et al.* (2007) Interactions among PIN-FORMED and P-glycoprotein auxin transporters in *Arabidopsis*. *The Plant Cell*, **19**, 131–147.

Blomster, T., Salojarvi, J., Sipari, N. *et al.* (2011) Apoplastic reactive oxygen species transiently decrease auxin signaling and cause stress-induced morphogenic response in *Arabidopsis*. *Plant Physiology*, **157**, 1866–1883.

Borevitz, J.O., Xia, Y., Blount, J. *et al.* (2000) Activation tagging identifies a conserved MYB regulator of phenylpropanoid biosynthesis. *The Plant Cell*, **12**, 2383–2394.

Bouchard, R., Bailly, A., Blakeslee, J.J. *et al.* (2006) Immunophilin-like TWISTED DWARF1 modulates auxin efflux activities of *Arabidopsis* P-glycoproteins. *Journal of Biological Chemistry*, **281**, 30603–30612.

Bovy, A., de Vos, R., Kemper, M. *et al.* (2002) High-flavonol tomatoes resulting from the heterologous expression of the maize transcription factor genes Lc and C1. *The Plant Cell*, **14**, 2509–2526.

Brown, D.E., Rashotte, A.M., Murphy, A.S. *et al.* (2001) Flavonoids act as negative regulators of auxin transport *in vivo* in *Arabidopsis*. *Plant Physiology*, **126**, 524–535.

Brunn, S.A., Muday, G.K. and Haworth, P. (1992) Auxin transport and the interaction of phytotropins: probing the properties of a phytotropin binding protein. *Plant Physiology*, **98**, 101–107.

Buer, C.S. and Djordjevic, M.A. (2009) Architectural phenotypes in the transparent testa mutants of *Arabidopsis thaliana*. *Journal of Experimental Botany*, **60**, 751–763.

Buer, C.S. and Muday, G.K. (2004) The *transparent testa4* mutation prevents flavonoid synthesis and alters auxin transport and the response of *Arabidopsis* roots to gravity and light. *The Plant Cell*, **16**, 1191–1205.

Buer, C.S., Sukumar, P. and Muday, G.K. (2006) Ethylene modulates flavonoid accumulation and gravitropic responses in roots of *Arabidopsis*. *Plant Physiology*, **140**, 1384–1396.

Buer, C.S., Kordbacheh, F., Truong, T.T. *et al.* (2013) Alteration of flavonoid accumulation patterns in transparent testa mutants disturbs auxin transport, gravity responses, and imparts long-term effects on root and shoot architecture. *Planta*, **238**, 171–189.

Burbulis, I.E., Iacobucci, M. and Shirley, B.W. (1996) A null mutation in the first enzyme of flavonoid biosynthesis does not affect male fertility in *Arabidopsis*. *The Plant Cell*, **8**, 1013–1025.

Bures, M.G., Black-Schaefer, C. and Gardner, G. (1991) The discovery of novel auxin transport inhibitors by molecular modeling and three-dimensional pattern analysis. *Journal of Computer-Aided Molecular Design*, **5**, 323–334.

Butelli, E., Titta, L., Giorgio, M. *et al.* (2008) Enrichment of tomato fruit with health-promoting anthocyanins by expression of select transcription factors. *Nature Biotechnology*, **26**, 1301–1308.

Buttery, B.R. and Buzzell, R.I. (1987) Leaf traits associated with flavonol glycoside genes in soybean. *Plant Physiology*, **85**, 20–21.

Chae, H.S., Faure, F. and Kieber, J.J. (2003) The *eto1*, *eto2*, and *eto3* mutations and cytokinin treatment increase ethylene biosynthesis in *Arabidopsis* by increasing the stability of ACS protein. *The Plant Cell*, **15**, 545–559.

Chory, J. and Peto, C.A. (1990) Mutations in the *DET1* gene affect cell-type-specific expression of light-regulated genes and chloroplast development in *Arabidopsis*. *Proceedings of the National Academy of Sciences of the United States of America*, **87**, 8776–8780.

Cominelli, E., Gusmaroli, G., Allegra, D. *et al.* (2008) Expression analysis of anthocyanin regulatory genes in response to different light qualities in *Arabidopsis thaliana*. *Journal of Plant Physiology*, **165**, 886–894.

Conklin, P.L., Williams, E.H. and Last, R.L. (1996) Environmental stress sensitivity of an ascorbic acid-deficient *Arabidopsis* mutant. *Proceedings of the National Academy of Sciences of the United States of America*, **93**, 9970–9974.

De Tullio, M.C., Jiang, K. and Feldman, L.J. (2010) Redox regulation of root apical meristem organization: connecting root development to its environment. *Plant Physiology and Biochemistry*, **48**, 328–336.

Deruère, J., Jackson, K., Garbers, C. *et al.* (1999) The *RCN1*-encoded A subunit of protein phosphatase 2A increases phosphatase activity *in vivo*. *The Plant Journal*, **20**, 389–399.

Desikan, R., Last, K., Harrett-Williams, R. *et al.* (2006) Ethylene-induced stomatal closure in *Arabidopsis* occurs via AtrbohF-mediated hydrogen peroxide synthesis. *The Plant Journal*, **47**, 907–916.

Dong, X., Braun, E.L. and Grotewold, E. (2001) Functional conservation of plant secondary metabolic enzymes revealed by complementation of *Arabidopsis* flavonoid mutants with maize genes. *Plant Physiology*, **127**, 46–57.

Duan, Q., Kita, D., Li, C. *et al.* (2010) FERONIA receptor-like kinase regulates RHO GTPase signaling of root hair development. *Proceedings of the National Academy of Sciences of the United States of America*, **107**, 17821–17826.

Dubos, C., Le Gourrierec, J., Baudry, A. *et al.* (2008) MYBL2 is a new regulator of flavonoid biosynthesis in *Arabidopsis thaliana*. *The Plant Journal*, **55**, 940–953.

Emiliani, J., Grotewold, E., Falcone Ferreyra, M.L. *et al.* (2013) Flavonols protect *Arabidopsis* plants against UV-B deleterious effects. *Molecular Plant*, **6**, 1376–1379.

Falcone Ferreyra, M.L., Rius, S.P. and Casati, P. (2012) Flavonoids: biosynthesis, biological functions, and biotechnological applications. *Frontiers in Plant Science*, **3**, 222.

Fernández-Marcos, M., Sanz, L., Lewis, D.R. *et al.* (2013) Control of auxin transport by reactive oxygen and nitrogen species, in *Polar Auxin Transport, Signaling and Communication in Plants*, vol. **17** (eds R. Chen and F. Baluska), Springer-Verlag, Berlin, pp. 103–117.

Fini, A., Brunetti, C., Di Ferdinando, M. *et al.* (2011) Stress-induced flavonoid biosynthesis and the antioxidant machinery of plants. *Plant Signaling & Behavior*, **6**, 709–711.

Fornalé, S., Lopez, E., Salazar-Henao, J.E. *et al.* (2014) AtMYB7, a new player in the regulation of UV-sunscreens in *Arabidopsis thaliana*. *Plant and Cell Physiology*, **55**, 507–516.

Fuglevand, G., Jackson, J.A. and Jenkins, G.I. (1996) UV-B, UV-A, and blue light signal transduction pathways interact synergistically to regulate chalcone synthase gene expression in *Arabidopsis*. *The Plant Cell*, **8**, 2347–2357.

Garbers, C., DeLong, A., Deruère, J. *et al.* (1996) A mutation in protein phosphatase 2A regulatory subunit A affects auxin transport in *Arabidopsis*. *The EMBO Journal*, **15**, 2115–2124.

Geisler, M., Blakeslee, J.J., Bouchard, R. *et al.* (2005) Cellular efflux of auxin catalyzed by the *Arabidopsis* MDR/PGP transporter AtPGP1. *The Plant Journal*, **44**, 179–194.

Gibson, S.I., Laby, R.J. and Kim, D. (2001) The sugar-insensitive1 (sis1) mutant of *Arabidopsis* is allelic to ctr1. *Biochemical and Biophysical Research Communications*, **280**, 196–203.

Gonzalez, A., Zhao, M., Leavitt, J.M. *et al.* (2008) Regulation of the anthocyanin biosynthetic pathway by the TTG1/bHLH/Myb transcriptional complex in *Arabidopsis* seedlings. *The Plant Journal*, **53**, 814–827.

Gou, J.Y., Felippes, F.F., Liu, C.J. *et al.* (2011) Negative regulation of anthocyanin biosynthesis in *Arabidopsis* by a miR156-targeted SPL transcription factor. *The Plant Cell*, **23**, 1512–1522.

Grunewald, W., De Smet, I., Lewis, D.R. *et al.* (2012) Transcription factor WRKY23 assists auxin distribution patterns during *Arabidopsis* root development through local control on flavonol biosynthesis. *Proceedings of the National Academy of Sciences of the United States of America*, **109**, 1554–1559.

Hernandez, I., Alegre, L., Van Breusegem, F. *et al.* (2009) How relevant are flavonoids as antioxidants in plants? *Trends in Plant Science*, **14**, 125–132.

Hichri, I., Barrieu, F., Bogs, J. *et al.* (2011) Recent advances in the transcriptional regulation of the flavonoid biosynthetic pathway. *Journal of Experimental Botany*, **62**, 2465–2483.

Hsieh, K. and Huang, A.H. (2007) Tapetosomes in *Brassica tapetum* accumulate endoplasmic reticulum-derived flavonoids and alkanes for delivery to the pollen surface. *The Plant Cell*, **19**, 582–596.

Jacobs, M. and Rubery, P.H. (1988) Naturally-occurring auxin transport regulators. *Science*, **241**, 346–349.

Jain, A., Poling, M.D., Karthikeyan, A.S. *et al.* (2007) Differential effects of sucrose and auxin on localized phosphate deficiency-induced modulation of different traits of root system architecture in *Arabidopsis*. *Plant Physiology*, **144**, 232–247.

Jenkins, G.I., Long, J.C., Wade, H.K. *et al.* (2001) UV and blue light signalling: pathways regulating chalcone synthase gene expression in Arabidopsis. *New Phytologist*, **151**, 121–131.

Jeong, S.W., Das, P.K., Jeoung, S.C. *et al.* (2010) Ethylene suppression of sugar-induced anthocyanin pigmentation in *Arabidopsis*. *Plant Physiology*, **154**, 1514–1531.

Jones, M.A., Raymond, M.J., Yang, Z. *et al.* (2007) NADPH oxidase-dependent reactive oxygen species formation required for root hair growth depends on ROP GTPase. *Journal of Experimental Botany*, **58**, 1261–1270.

Koes, R., Verweij, W. and Quattrocchio, F. (2005) Flavonoids: a colorful model for the regulation and evolution of biochemical pathways. *Trends in Plant Science*, **10**, 236–242.

Kubasek, W.L., Shirley, B.W., McKillop, A. *et al.* (1992) Regulation of flavonoid biosynthetic genes in germinating *Arabidopsis* seedlings. *The Plant Cell*, **4**, 1229–1236.

Kuhn, B.M., Geisler, M., Bigler, L. *et al.* (2011) Flavonols accumulate asymmetrically and affect auxin transport in Arabidopsis. *Plant Physiology*, **156**, 585–595.

Kwak, J.M., Mori, I.C., Pei, Z.M. *et al.* (2003) NADPH oxidase AtrbohD and AtrbohF genes function in ROS-dependent ABA signaling in *Arabidopsis*. *The EMBO Journal*, **22**, 2623–2633.

Lee, J., He, K., Stolc, V. *et al.* (2007) Analysis of transcription factor HY5 genomic binding sites revealed its hierarchical role in light regulation of development. *The Plant Cell*, **19**, 731–749.

Lepiniec, L., Debeaujon, I., Routaboul, J.M. *et al.* (2006) Genetics and biochemistry of seed flavonoids. *Annual Review of Plant Biology*, **57**, 405–430.

Lewis, D.R., Miller, N.D., Splitt, B.L. *et al.* (2007) Separating the roles of acropetal and basipetal auxin transport on gravitropism with mutations in two *Arabidopsis* multidrug resistance-like ABC transporter genes. *The Plant Cell*, **19**, 1838–1850.

Lewis, D.R., Negi, S., Sukumar, P. *et al.* (2011a) Ethylene inhibits lateral root development, increases IAA transport and expression of PIN3 and PIN7 auxin efflux carriers. *Development*, **138**, 3485–3495.

Lewis, D.R., Ramirez, M.V., Miller, N.D. *et al.* (2011b) Auxin and ethylene induce flavonol accumulation through distinct transcriptional networks. *Plant Physiology*, **156**, 144–164.

Li, S. and Zachgo, S. (2013) TCP3 interacts with R2R3-MYB proteins, promotes flavonoid biosynthesis and negatively regulates the auxin response in *Arabidopsis thaliana*. *The Plant Journal*, **76**, 901–913.

Liu-Gitz, L., Britz, S. and Wergin, W. (2000) Blue light inhibits stomatal development in soybean isolines containing kaempferol-3-*O*-2G-glycosyl-gentiobioside (K9), a unique flavonoid glycoside. *Plant and Cell Physiology*, **23**, 883–891.

Madhavan, S., Chrominiski, A. and Smith, B. (1983) Effect of ethylene on stomatal opening in tomato and carnation leaves. *Plant and Cell Physiology*, **24**, 569–572.

Maloney, G.S., DiNapoli, K.T. and Muday, G.K. (2014) The anthocyanin reduced tomato mutant demonstrates the role of flavonols in tomato lateral root and root hair development. *Plant Physiology*, **166**, 614–631.

Matsui, K., Tanaka, H. and Ohme-Takagi, M. (2004) Suppression of the biosynthesis of proanthocyanidin in *Arabidopsis* by a chimeric PAP1 repressor. *Plant Biotechnology Journal*, **2**, 487–493.

Matsui, K., Umemura, Y. and Ohme-Takagi, M. (2008) AtMYBL2, a protein with a single MYB domain, acts as a negative regulator of anthocyanin biosynthesis in *Arabidopsis*. *The Plant Journal*, **55**, 954–967.

Mehrtens, F., Kranz, H., Bednarek, P. *et al.* (2005) The *Arabidopsis* transcription factor MYB12 is a flavonol-specific regulator of phenylpropanoid biosynthesis. *Plant Physiology*, **138**, 1083–1096.

Mei, Y., Jia, W.J., Chu, Y.J. *et al.* (2012) *Arabidopsis* phosphatidylinositol monophosphate 5-kinase 2 is involved in root gravitropism through regulation of polar auxin transport by affecting the cycling of PIN proteins. *Cell Research*, **22**, 581–597.

Michniewicz, M., Zago, M.K., Abas, L. *et al.* (2007) Antagonistic regulation of PIN phosphorylation by PP2A and PINOID directs auxin flux. *Cell*, **130**, 1044–1056.

Mittler, R., Vanderauwera, S., Suzuki, N. *et al.* (2011) ROS signaling: the new wave? *Trends in Plant Science*, **16**, 300–309.

Mo, Y., Nagel, C. and Taylor, L.P. (1992) Biochemical complementation of chalcone synthase mutants defines a role for flavonols in functional pollen. *Proceedings of the National Academy of Sciences of the United States of America*, **89**, 7213–7217.

Monshausen, G.B., Bibikova, T.N., Messerli, M.A. *et al.* (2007) Oscillations in extracellular pH and reactive oxygen species modulate tip growth of *Arabidopsis* root hairs. *Proceedings of the National Academy of Sciences of the United States of America*, **104**, 20996–21001.

Morishita, T., Kojima, Y., Maruta, T. *et al.* (2009) *Arabidopsis* NAC transcription factor, ANAC078, regulates flavonoid biosynthesis under high-light. *Plant and Cell Physiology*, **50**, 2210–2222.

Muday, G.K., Rahman, A. and Binder, B.M. (2012) Auxin and ethylene: collaborators or competitors? *Trends in Plant Science*, **17**, 181–195.

Munemasa, S., Muroyama, D., Nagahashi, H. *et al.* (2013) Regulation of reactive oxygen species-mediated abscisic acid signaling in guard cells and drought tolerance by glutathione. *Frontiers in Plant Science*, **4**, 472.

Munne-Bosch, S., Queval, G. and Foyer, C.H. (2013) The impact of global change factors on redox signaling underpinning stress tolerance. *Plant Physiology*, **161**, 5–19.

Murata, Y., Mori, I.C. and Munemasa, S. (2015) Diverse stomatal signaling and the signal integration mechanism. *Annual Review of Plant Biology*, **66**, 369–392.

Murphy, A., Peer, W.A. and Taiz, L. (2000) Regulation of auxin transport by aminopeptidases and endogenous flavonoids. *Planta*, **211**, 315–324.

Okuma, E., Jahan, M.S., Munemasa, S. *et al.* (2011) Negative regulation of abscisic acid-induced stomatal closure by glutathione in *Arabidopsis*. *Journal of Plant Physiology*, **168**, 2048–2055.

Oyama, T., Shimura, Y. and Okada, K. (1997) The *Arabidopsis* HY5 gene encodes a bZIP protein that regulates stimulus-induced development of root and hypocotyl. *Genes & Development*, **11**, 2983–2995.

Pallas, J.E. and Kays, S.J. (1982) Inhibition of photosynthesis by ethylene—a stomatal effect. *Plant Physiology*, **70**, 598–601.

Pasternak, T., Potters, G., Caubergs, R. *et al.* (2005) Complementary interactions between oxidative stress and auxins control plant growth responses at plant, organ, and cellular level. *Journal of Experimental Botany*, **56**, 1991–2001.

Peer, W.A. and Murphy, A.S. (2007) Flavonoids and auxin transport: modulators or regulators? *Trends in Plant Science*, **12**, 556–563.

Peer, W.A., Brown, D.E., Tague, B.W. *et al.* (2001) Flavonoid accumulation patterns of transparent testa mutants of *Arabidopsis*. *Plant Physiology*, **126**, 536–548.

Peer, W.A., Bandyopadhyay, A., Blakeslee, J.J. *et al.* (2004) Variation in expression and protein localization of the PIN family of auxin efflux facilitator proteins in flavonoid mutants with altered auxin transport in *Arabidopsis thaliana*. *The Plant Cell*, **16**, 1898–1911.

Peer, W.A., Cheng, Y. and Murphy, A.S. (2013) Evidence of oxidative attenuation of auxin signalling. *Journal of Experimental Botany*, **64**, 2629–2639.

Pei, Z.M., Murata, Y., Benning, G. *et al.* (2000) Calcium channels activated by hydrogen peroxide mediate abscisic acid signalling in guard cells. *Nature*, **406**, 731–734.

Pollak, P.E., Vogt, T., Mo, Y. *et al.* (1993) Chalcone synthase and flavonol accumulation in stigmas and anthers of *Petunia hybrida*. *Plant Physiology*, **102**, 925–932.

Pollastri, S. and Tattini, M. (2011) Flavonols: old compounds for old roles. *Annals of Botany*, **108**, 1225–1233.

Qin, Y., Wysocki, R.J., Somogyi, A. *et al.* (2011) Sulfinylated azadecalins act as functional mimics of a pollen germination stimulant in *Arabidopsis* pistils. *The Plant Journal*, **68**, 800–815.

Que, Q., Wang, H.Y., English, J.J. *et al.* (1997) The frequency and degree of cosuppression by sense chalcone synthase transgenes are dependent on transgene promoter strength and are reduced by premature nonsense codons in the transgene coding sequence. *The Plant Cell*, **9**, 1357–1368.

Rashotte, A.M., DeLong, A. and Muday, G.K. (2001) Genetic and chemical reductions in protein phosphatase activity alter auxin transport, gravity response, and lateral root growth. *The Plant Cell*, **13**, 1683–1697.

Rice-Evans, C.A., Miller, N.J., Bolwell, P.G. *et al.* (1995) The relative antioxidant activities of plant-derived polyphenolic flavonoids. *Free Radical Research*, **22**, 375–383.

Ringli, C., Bigler, L., Kuhn, B.M. *et al.* (2008) The modified flavonol glycosylation profile in the *Arabidopsis rol1* mutants results in alterations in plant growth and cell shape formation. *The Plant Cell*, **20**, 1470–1481.

Saito, K., Yonekura-Sakakibara, K., Nakabayashi, R. *et al.* (2013) The flavonoid biosynthetic pathway in *Arabidopsis*: structural and genetic diversity. *Plant Physiology and Biochemistry*, **72**, 21–34.

Santelia, D., Henrichs, S., Vincenzetti, V. *et al.* (2008) Flavonoids redirect PIN-mediated polar auxin fluxes during root gravitropic responses. *Journal of Biological Chemistry*, **283**, 31218–31226.

Saslowsky, D.E., Dana, C.D. and Winkel-Shirley, B. (2000) An allelic series for the chalcone synthase locus in *Arabidopsis*. *Gene*, **255**, 127–138.

Schijlen, E.G., de Vos, C.H., Martens, S. *et al.* (2007) RNA interference silencing of chalcone synthase, the first step in the flavonoid biosynthesis pathway, leads to parthenocarpic tomato fruits. *Plant Physiology*, **144**, 1520–1530.

Schmidt, R. and Schippers, J.H. (2015) ROS-mediated redox signaling during cell differentiation in plants. *Biochimica et Biophysica Acta*, **1850**, 1497–1508.

Schnabl, H., Weissenboeck, G. and Scharf, H. (1986) *In vivo*-microspectrophotometric characterization of flavonol glycosides in *Vicia faba* guard and epidermal cells. *Journal of Experimental Botany*, **37**, 61–72.

Sheahan, J.J. and Rechnitz, G.A. (1992) Flavonoid-specific staining of *Arabidopsis thaliana*. *Biotechniques*, **13**, 880–883.

Shin, L.J., Huang, H.E., Chang, H. *et al.* (2011) Ectopic ferredoxin I protein promotes root hair growth through induction of reactive oxygen species in *Arabidopsis thaliana*. *Journal of Plant Physiology*, **168**, 434–440.

Shin, D.H., Choi, M., Kim, K. *et al.* (2013) HY5 regulates anthocyanin biosynthesis by inducing the transcriptional activation of the MYB75/PAP1 transcription factor in *Arabidopsis*. *FEBS Letters*, **587**, 1543–1547.

Shirley, B.W., Kubasek, W.L., Storz, G. *et al.* (1995) Analysis of *Arabidopsis* mutants deficient in flavonoid biosynthesis. *The Plant Journal*, **8**, 659–671.

Sibout, R., Sukumar, P., Hettiarachchi, C. *et al.* (2006) Opposite root growth phenotypes of *hy5* versus *hy5 hyh* mutants correlate with increased constitutive auxin signaling. *PLoS Genetics*, **2**, e202.

Solfanelli, C., Poggi, A., Loreti, E. *et al.* (2006) Sucrose-specific induction of the anthocyanin biosynthetic pathway in *Arabidopsis*. *Plant Physiology*, **140**, 637–646.

Song, Y., Miao, Y. and Song, C.P. (2014) Behind the scenes: the roles of reactive oxygen species in guard cells. *New Phytologist*, **201**, 1121–1140.

Stracke, R., Ishihara, H., Huep, G. *et al.* (2007) Differential regulation of closely related R2R3-MYB transcription factors controls flavonol accumulation in different parts of the *Arabidopsis thaliana* seedling. *The Plant Journal*, **50**, 660–677.

Stracke, R., Favory, J.J., Gruber, H. *et al.* (2010a) The *Arabidopsis* bZIP transcription factor HY5 regulates expression of the PFG1/MYB12 gene in response to light and ultraviolet-B radiation. *Plant, Cell & Environment*, **33**, 88–103.

Stracke, R., Jahns, O., Keck, M. *et al.* (2010b) Analysis of PRODUCTION OF FLAVONOL GLYCOSIDES-dependent flavonol glycoside accumulation in *Arabidopsis thaliana* plants reveals MYB11-, MYB12- and MYB111-independent flavonol glycoside accumulation. *New Phytologist*, **188**, 985–1000.

Subramanian, S., Stacey, G. and Yu, O. (2007) Distinct, crucial roles of flavonoids during legume nodulation. *Trends in Plant Science*, **12**, 282–285.

Sukumar, P., Edwards, K.S., Rahman, A. *et al.* (2009) PINOID kinase regulates root gravitropism through modulation of PIN2-dependent basipetal auxin transport in *Arabidopsis*. *Plant Physiology*, **150**, 722–735.

Sundaravelpandian, K., Chandrika, N.N. and Schmidt, W. (2013) PFT1, a transcriptional mediator complex subunit, controls root hair differentiation through reactive oxygen species (ROS) distribution in *Arabidopsis*. *New Phytologist*, **197**, 151–161.

Swanson, S. and Gilroy, S. (2010) ROS in plant development. *Physiologia Plantarum*, **138**, 384–392.

Takahama, U. (1988) Oxidation of flavonols by hydrogen peroxide in epidermal and guard cells of *Vicia faba L. Plant and Cell Physiology*, **29**, 433–438.

Tanaka, Y., Sano, T., Tamaoki, M. *et al.* (2005) Ethylene inhibits abscisic acid-induced stomatal closure in *Arabidopsis*. *Plant Physiology*, **138**, 2337–2343.

Tattini, M., Garlardi, C., Pinelli, P. *et al.* (2004) Differential accumulation of flavonoids and hydroxycinnamates in leaves of *Ligustrum vulgare* under excess light and drought stress. *New Phytologist*, **163**, 547–561.

Taylor, L.P. and Grotewold, E. (2005) Flavonoids as developmental regulators. *Current Opinion in Plant Biology*, **8**, 317–323.

Teng, S., Keurentjes, J., Bentsink, L. *et al.* (2005) Sucrose-specific induction of anthocyanin biosynthesis in *Arabidopsis* requires the MYB75/PAP1 gene. *Plant Physiology*, **139**, 1840–1852.

Tognetti, V.B., Muhlenbock, P. and Van Breusegem, F. (2012) Stress homeostasis—the redox and auxin perspective. *Plant, Cell & Environment*, **35**, 321–333.

Wade, H.K., Bibikova, T.N., Valentine, W.J. *et al.* (2001) Interactions within a network of phytochrome, cryptochrome and UV-B phototransduction pathways regulate chalcone synthase gene expression in *Arabidopsis* leaf tissue. *The Plant Journal*, **25**, 675–685.

Watkins, J.M., Hechler, P.J. and Muday, G.K. (2014) Ethylene-induced flavonol accumulation in guard cells suppresses reactive oxygen species and moderates stomatal aperture. *Plant Physiology*, **164**, 1707–1717.

Weissenboeck, G., Hedrich, R., Sachs, G. *et al.* (1986) Secondary phenolic products in isolated guard cell, epidermal cell and mesophyll cell protoplasts from pea (*Pisum sativum L.*) leaves: distribution and determination. *Protoplasma*, **134**, 141–148.

Weissenboeck, G., Schnable, H., Scharf, H. *et al.* (1987) On the propoerties of fluorescing compounds in gaurd and epidermal cells of *Allium cepa* L. *Planta*, **171**, 88–95.

Winkel-Shirley, B. (2001a) Flavonoid biosynthesis. A colorful model for genetics, biochemistry, cell biology, and biotechnology. *Plant Physiology*, **126**, 485–493.

Winkel-Shirley, B. (2001b) It takes a garden. How work on diverse plant species has contributed to an understanding of flavonoid metabolism. *Plant Physiology*, **127**, 1399–1404.

Winkel-Shirley, B. (2002) Biosynthesis of flavonoids and effects of stress. *Current Opinion in Plant Biology*, **5**, 218–223.

Xu, W., Dubos, C. and Lepiniec, L. (2015) Transcriptional control of flavonoid biosynthesis by MYB–bHLH–WDR complexes. *Trends in Plant Science*, **20**, 176–185.

Yang, H., Liu, H., Li, G. *et al.* (2009) Reduction of root flavonoid level and its potential involvement in lateral root emergence in *Arabidopsis thaliana* grown under low phosphate supply. *Functional Plant Biology*, **36**, 564–573.

Yin, R., Han, K., Heller, W. *et al.* (2013) Kaempferol 3-*O*-rhamnoside-7-*O*-rhamnoside is an endogenous flavonol inhibitor of polar auxin transport in *Arabidopsis* shoots. *New Phytologist*, **201**, 466–475.

Ylstra, B., Touraev, A., Moreno, R.M. *et al.* (1992) Flavonols stimulate development, germination, and tube growth of tobacco pollen. *Plant Physiology*, **100**, 902–907.

Ylstra, B., Muskens, M. and Van Tunen, A.J. (1996) Flavonols are not essential for fertilization in *Arabidopsis thaliana*. *Plant Molecular Biology*, **32**, 1155–1158.

Yoder, J.I., Belzile, F., Tong, Y. and Goldsbrough, A. (1994) Visual markers for tomato derived from the anthocyanin biosynthetic pathway. *Euphytica*, **79**, 163–167.

Yonekura-Sakakibara, K., Tohge, T., Niida, R. *et al.* (2007) Identification of a flavonol 7-*O*-rhamnosyltransferase gene determining flavonoid pattern in *Arabidopsis* by transcriptome coexpression analysis and reverse genetics. *Journal of Biological Chemistry*, **282**, 14932–14941.

Yonekura-Sakakibara, K., Nakabayashi, R., Sugawara, S. *et al.* (2014) A flavonoid 3-*O*-glucoside: 2″-*O*-glucosyltransferase responsible for terminal modification of pollen-specific flavonols in *Arabidopsis thaliana*. *The Plant Journal*, **79**, 769–782.

Zazimalova, E., Murphy, A.S., Yang, H.B. *et al.* (2010) Auxin transporters—why so many? *Cold Spring Harbor Perspectives in Biology*, **2**, a00152.

Zhang, J., Subramanian, S., Stacey, G. *et al.* (2009) Flavones and flavonols play distinct critical roles during nodulation of *Medicago truncatula* by *Sinorhizobium meliloti*. *The Plant Journal*, **57**, 171–183.

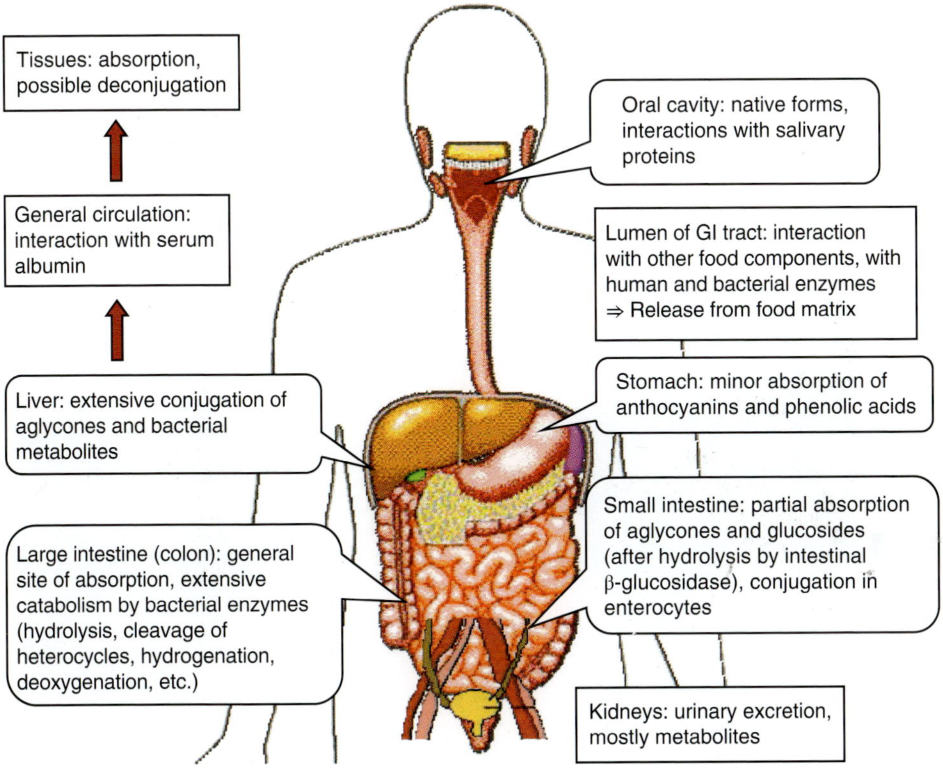

Fig. 1.1 A simplified view of polyphenol bioavailability.

Recent Advances in Polyphenol Research, First Edition. Edited by Kumi Yoshida, Véronique Cheynier and Stéphane Quideau.

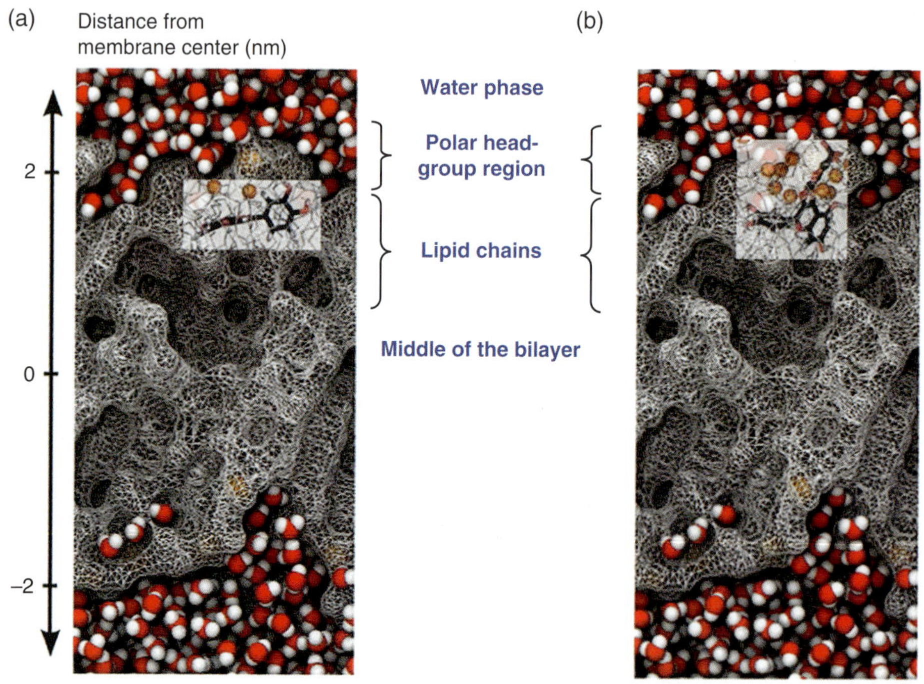

Fig. 1.4 Snapshots of the location of quercetin (a) and quercetin 3-*O*-β-D-glucuronide (b) in a 1,2-dioleoyl-*sn*-glycero-3-phosphatidylcholine lipid bilayer. Spheres represent the phosphatidyl groups (Source: Adapted from Kosinova *et al*. 2012).

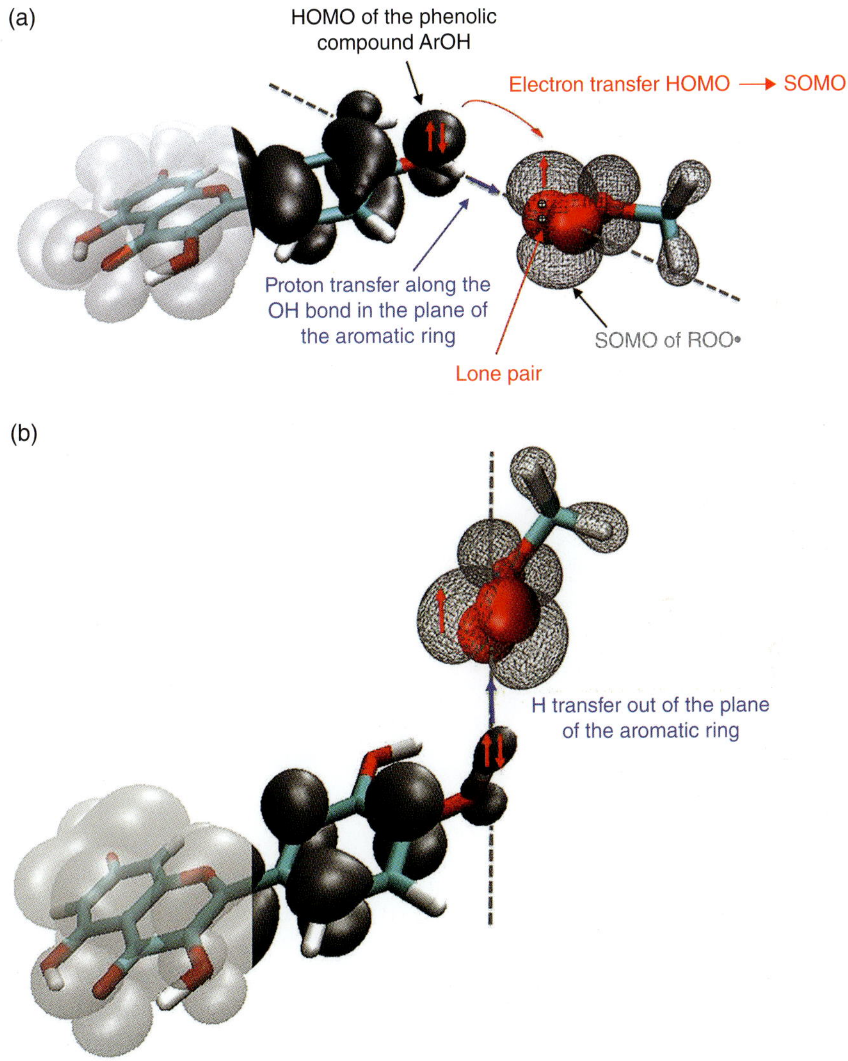

Fig. 1.5 Schematic description of (a) PCET involving five electrons and (b) HAT.

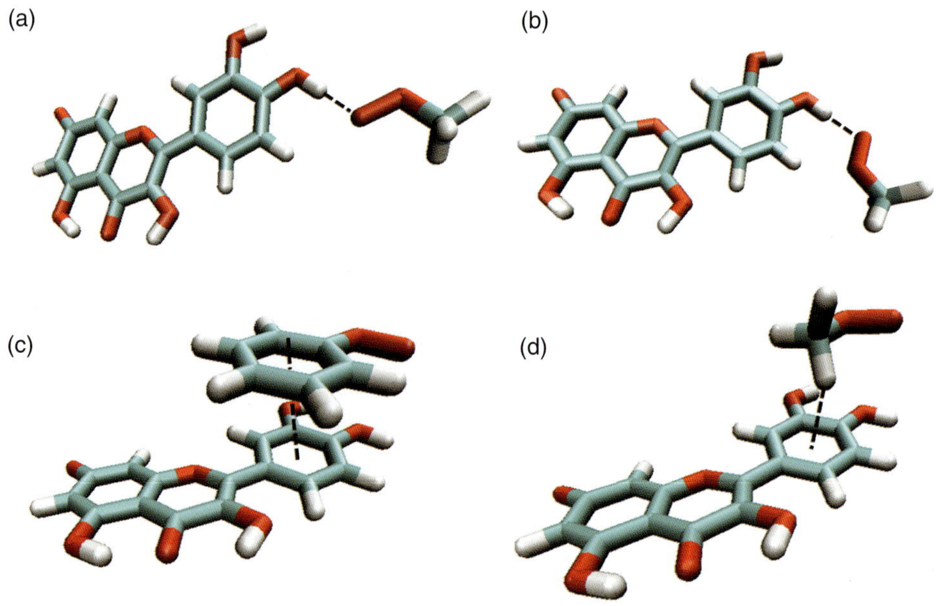

Fig. 1.6 Various geometries of noncovalent ArOH---ROO$^{\bullet}$ arrangements. (a) *transoid* H-bonding, (b) *cisoid* H-bonding, (c) π–π stacking, and (d) XH–π stacking complexes.

Fig. 2.1 Marchantia polymorpha.

Fig. 2.2 Marchantia paleacea var. diptera.

Fig. 2.3 Pellia epiphylla.

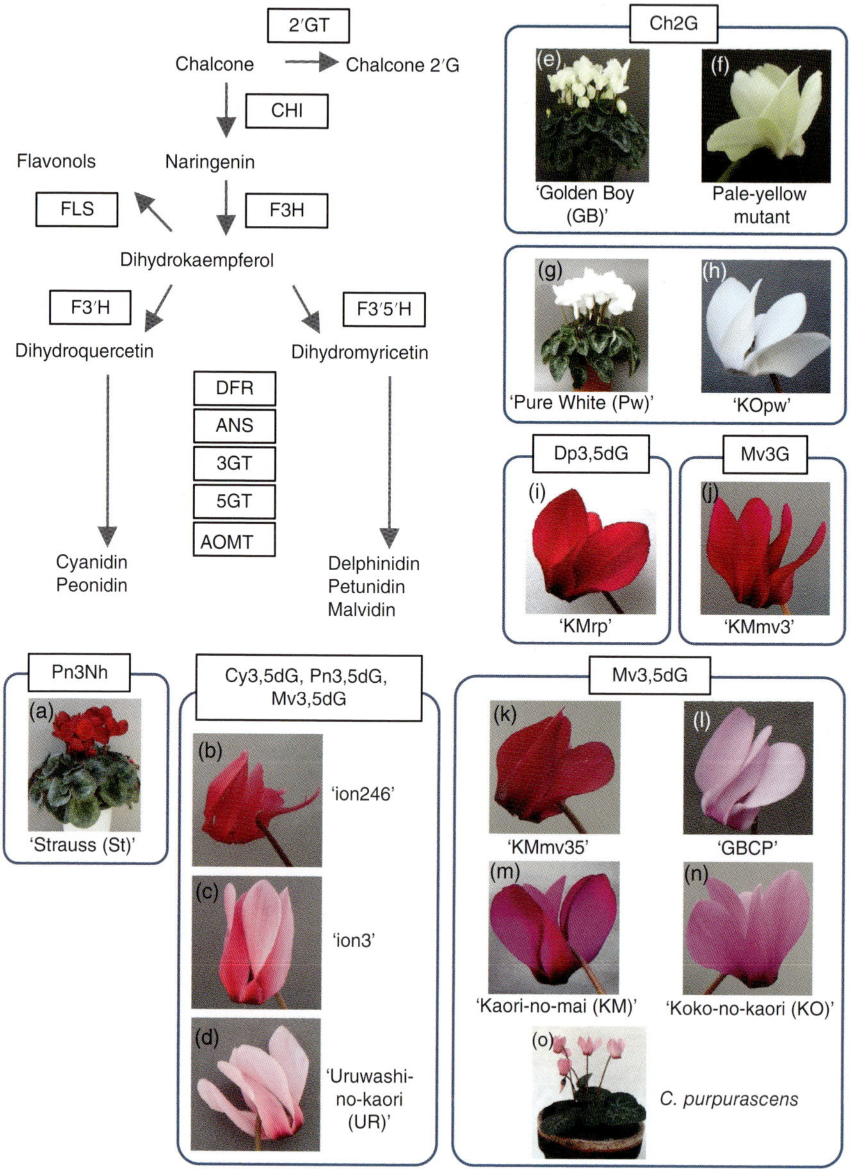

Fig. 6.1 Cyclamen cultivars, wild species, and flower color mutants shown beside the anthocyanin biosynthesis pathway. Flowers are grouped according to their major anthocyanins. (a) *Cyclamen persicum* cultivar "Strauss (St)." (b–d) Deeper color mutants "ion246" (b) and "ion3" (c) obtained from fragrant cyclamen cultivar "Uruwashi-no-kaori (UR)" (d). (e) *C. persicum* cultivar "Golden Boy (GB)." (f) A pale yellow-flowered mutant obtained from a hybrid (GB × *Cyclamen purpurascens*) "GBCP." (g) *C. persicum* cultivar "Pure White (Pw)." (h) A white-flowered mutant obtained from fragrant cyclamen cultivar "Koko-no-kaori (KO)." (i–k) Red–purple-flowered mutants "KMrp" (i), "KMmv3" (j), and "KMmv35" (k) obtained from fragrant cyclamen cultivar "kaori-no-mai (KM)." (l) A hybrid "GBCP" derived from GB × *C. purpurascens*. (m and n) Fragrant cyclamen cultivars KM (m) and KO (n). (o) Scented wild species *C. purpurascens*. Fragrant cyclamen cultivars were created by a cross between *C. persicum* cultivars and *C. purpurascens*. Flower color mutants were obtained from mutagenized populations of fragrant cyclamen cultivars irradiated with ion beams. ANS, anthocyanidin synthase; AOMT, anthocyanin *O*-methyltransferase; Ch2G, chalcone 2′-glucoside; CHI, chalcone isomerase; Cy3,5dG, cyanidin 3,5-diglucoside; DFR, dihydroflavonol 4-reductase; Dp3,5dG, delphinidin 3,5-diglucoside; F3H, flavanone 3-hydroxylase; F3′H, flavonoid 3′-hydroxylase; F3′5′H, flavonoid 3′5′-hydroxylase; FLS, flavonol synthase; 2′GT, chalcone 2′-glucosyltransferase; 3GT, anthocyanidin 3-glucosyltransferase; 5GT, anthocyanin 5-glucosyltransferase; Mv3G, malvidin 3-glucoside; Mv3,5dG, malvidin 3,5-diglucoside; Pn3Nh, peonidin 3-neohesperidoside; Pn3,5dG, peonidin 3,5-diglucoside.

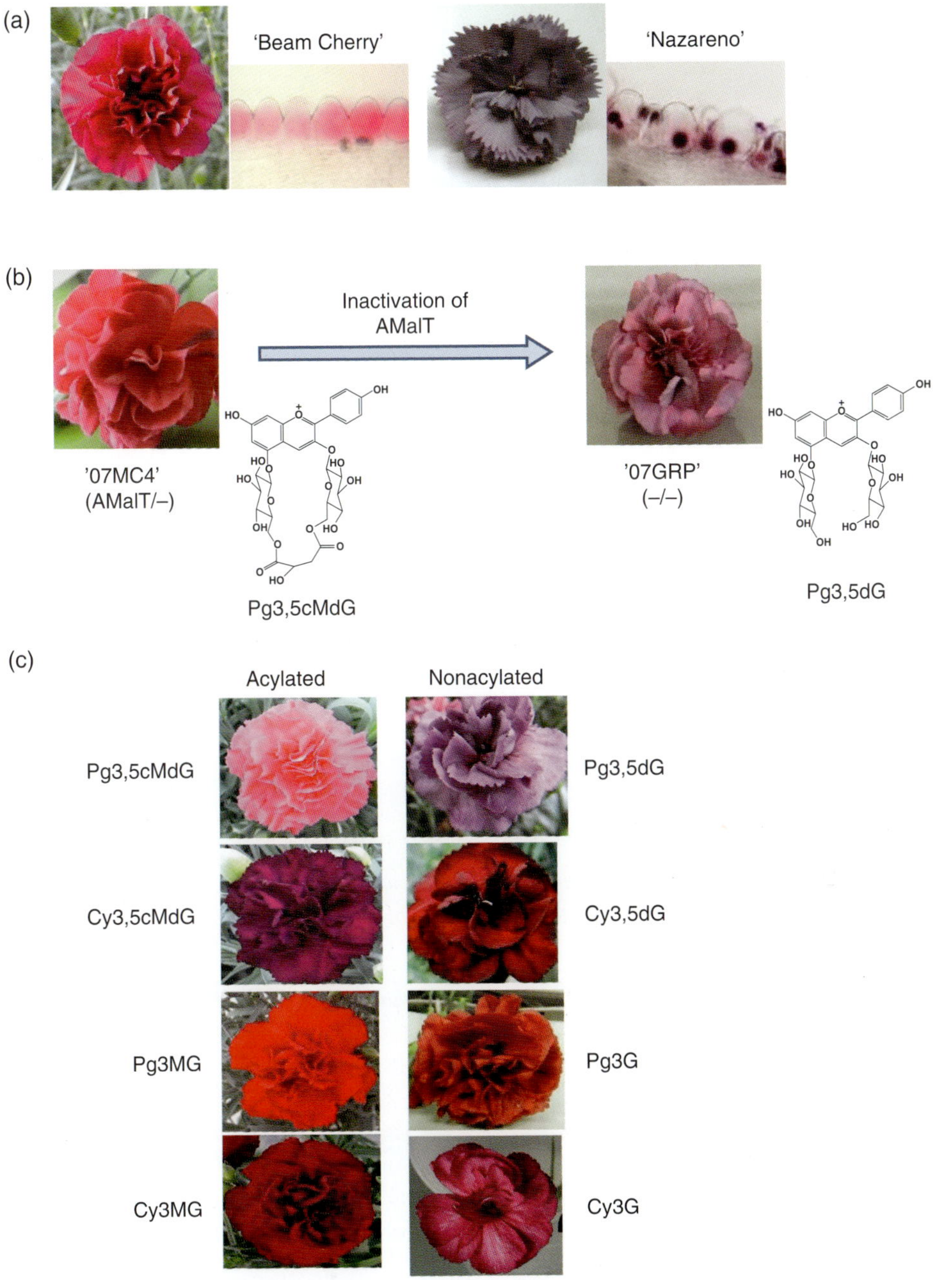

Fig. 6.2 Carnation with peculiar colors that contain nonacylated anthocyanins. (a) Flowers and transverse sections of petal epidermis of carnation cultivar "Beam Cherry (left)" and "Nazareno (right)." (b) The experimental strain "07MC4" and the mutant "07GRP" with peculiar colors obtained by inactivation of AMalT activity with carbon ion irradiation. The genotype for AMalT locus is shown in parenthesis. AMalT and a bar represent the active and inactive AMalT locus, respectively. The major anthocyanins are shown beside each flower. (c) Flowers of all four types of acylated and nonacylated-type anthocyanin in carnation. The major anthocyanins are shown beside each flower. AMalT, anthocyanin malyltransferase; Cy3,5cMdG, cyanidin 3,5-cyclic malyl diglucoside; Cy3,5dG, cyanidin 3,5-diglucoside; Cy3G, cyanidin 3-glucoside; Cy3MG, cyanidin 3-malyl glucoside; Pg3,5cMdG, pelargonidin 3,5-cyclic malyl diglucoside; Pg3,5dG, pelargonidin 3,5-diglucoside; Pg3G, pelargonidin 3-glucoside; Pg3MG, pelargonidin 3-malyl glucoside.

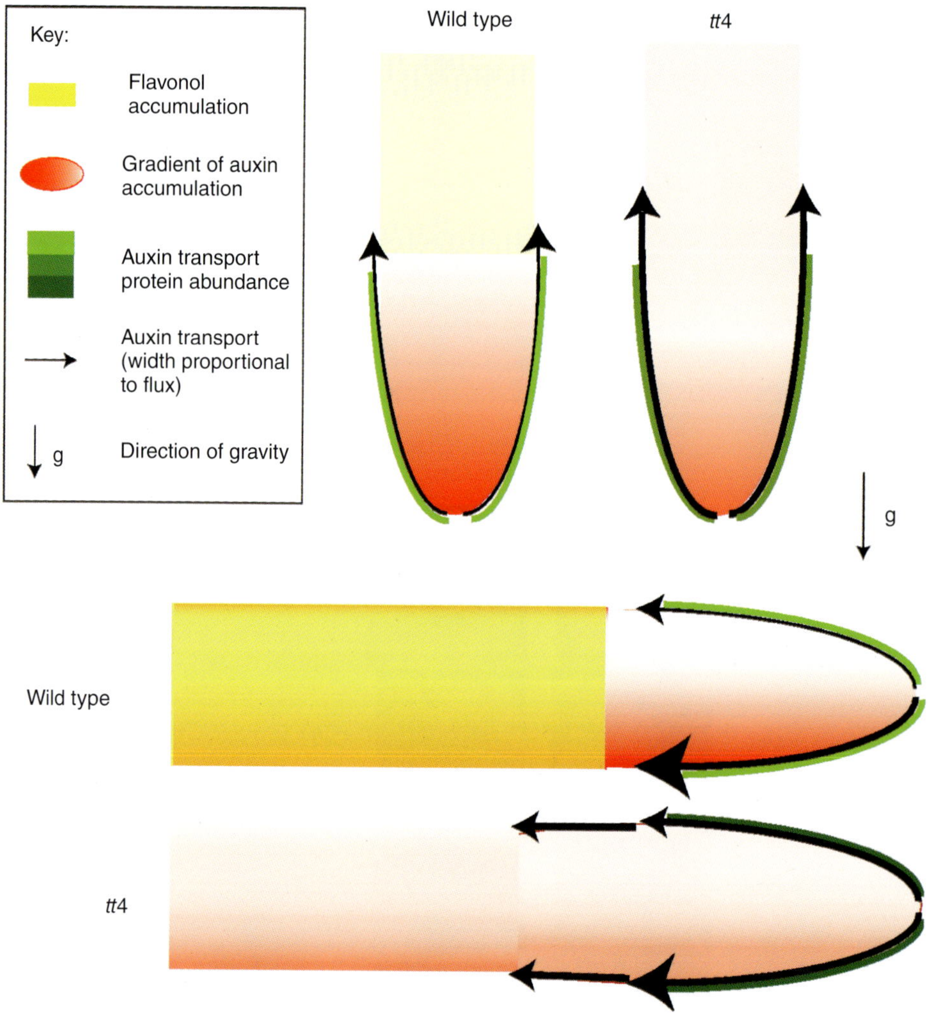

Fig. 7.2 A model illustrates how flavonoids may function during root gravitropism. In vertically oriented seedlings, *tt4* is deficient in flavonols (yellow) and has elevated *PIN2* gene products (green) and shootward auxin transport (black arrows). When roots are reoriented by 90°, flavonol levels increase in the root epidermis, especially in the elongation zone, which may block auxin movement out of the region where differential gravitropic growth occurs. This may concentrate the auxin in the distal elongation zone where root gravitropic bending occurs, enhancing the asymmetric distribution enhancing the rate of gravitropic curvature. In *tt4*, there is both enhanced auxin movement into and out of this region, reducing the asymmetric distribution of auxin.

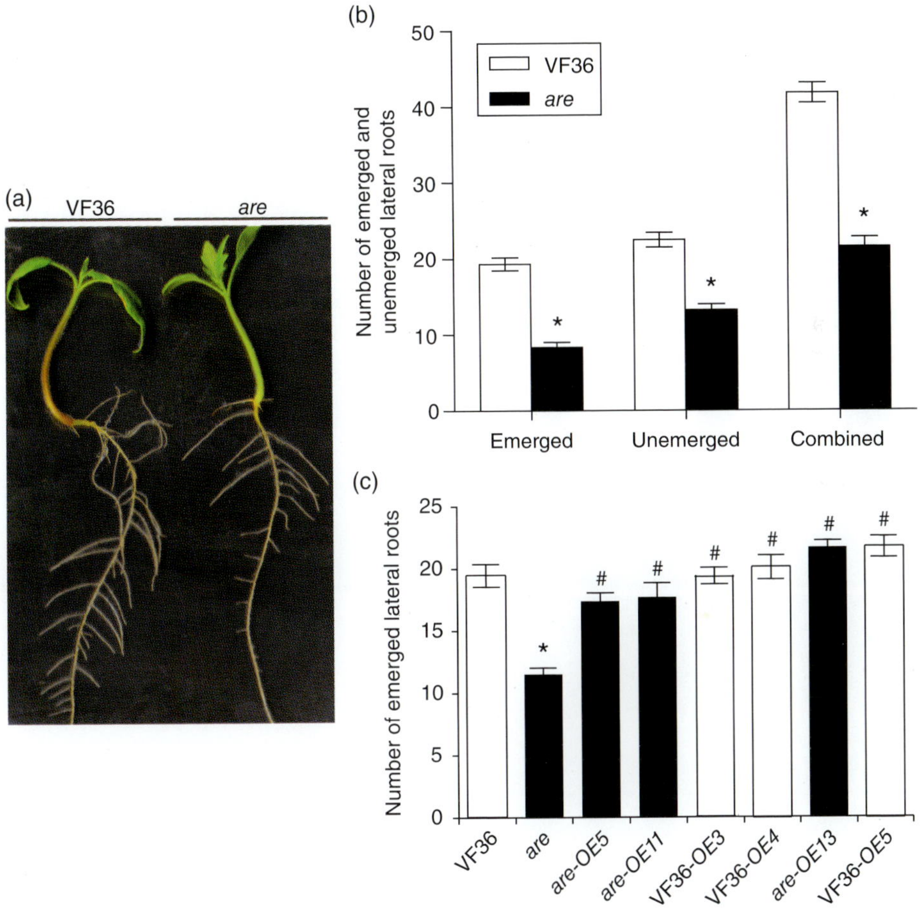

Fig. 7.3 Lateral root numbers are reduced in the tomato *are* mutant. (a) Image showing 6-day-old seedlings of wild-type (VF36) and *are* displaying differences in lateral rooting and hypocotyl pigmentation. (b) Numbers of emerged lateral roots, unemerged lateral root primordia, and combined totals of emerged and primordial roots. Average and SE are presented. Asterisk indicates difference between genotypes with $p \leq 0.05$ ($n = 15$). (c) Lateral root numbers are reported for untransformed VF36, *are,* and three *F3H* overexpression lines in each background genotype. Asterisk indicates differences from VF36, and number symbols indicate differences from *are,* with $p \leq 0.05$ ($n = 16$). Black bars indicate lines in the *are* background. Reproduced with permission from Maloney *et al.* (2014). © American Society of Plant Biologists (www.plantphysiol.org).

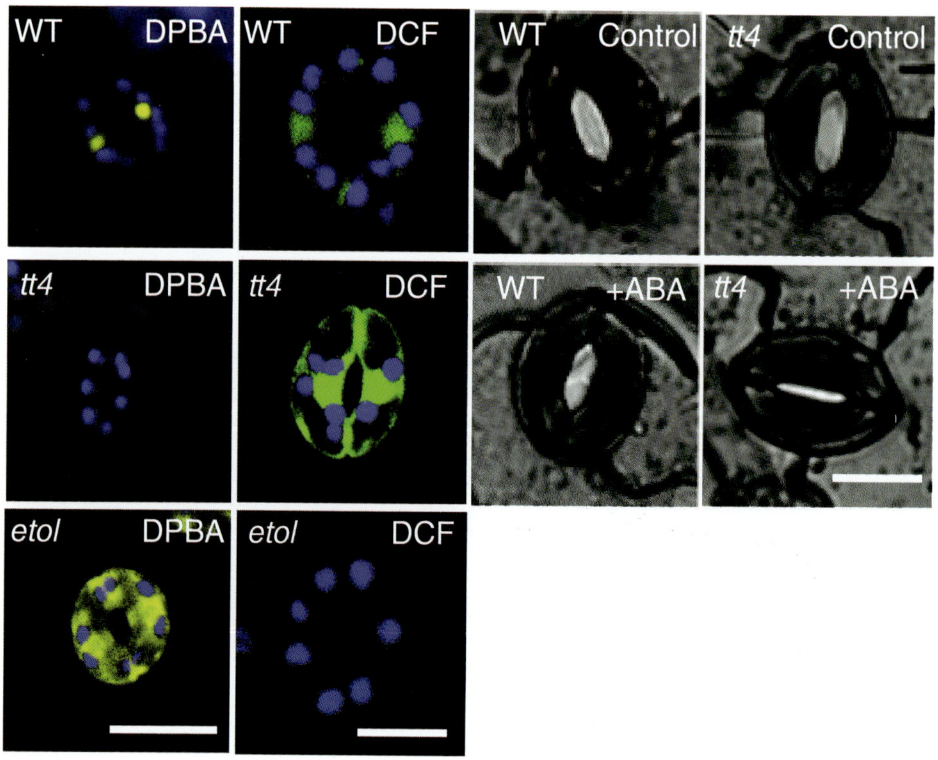

Fig. 7.4 Flavonol accumulation is absent in *tt4* leading to increased ROS and more rapid stomatal closure. Flavonols are visualized by diphenylboric acid 2-aminoethyl ester (DPBA) (yellow), and ROS is visualized by 2′,7′dihydrodichloro-fluorescein diacetate (DCF) (green), with chlorophyll in blue. Ethylene overproduction in *eto1* leads to enhanced flavonol accumulation and less ROS. Bar = 15 μm. Guard cells of wild-type and *tt4-2* were examined by light microscopy immediately before treatment with ABA and 45 minutes after treatment with 20 μM ABA. Bar = 15 μm. Watkins *et al.* (2014). © The fluorescent image is reprinted with permission from American Society of Plant Biologists (www.plantphysiol.org).

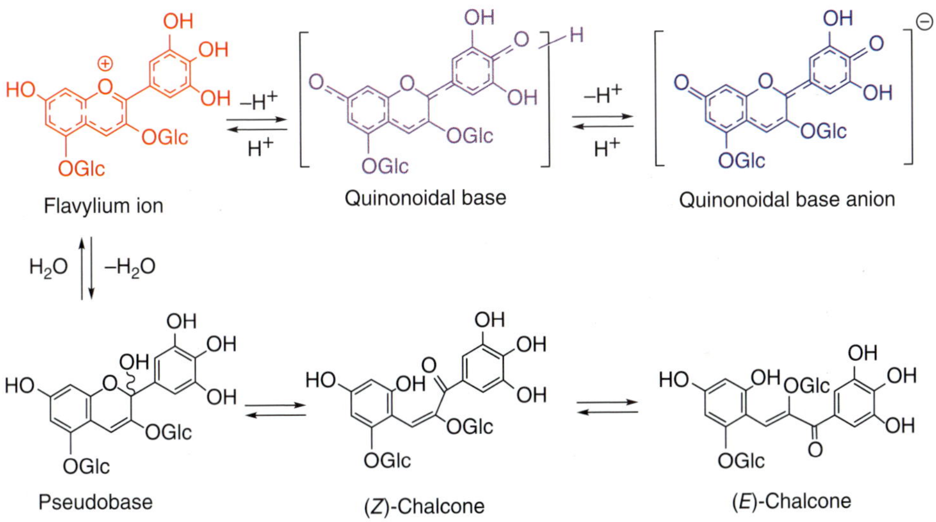

Fig. 8.1 Structure of anthocyanin forms, Flavylium ion-quinonoidal base-pseudobase-(*Z*)-chalcone-(*E*)-chalcone formed at various pH values.

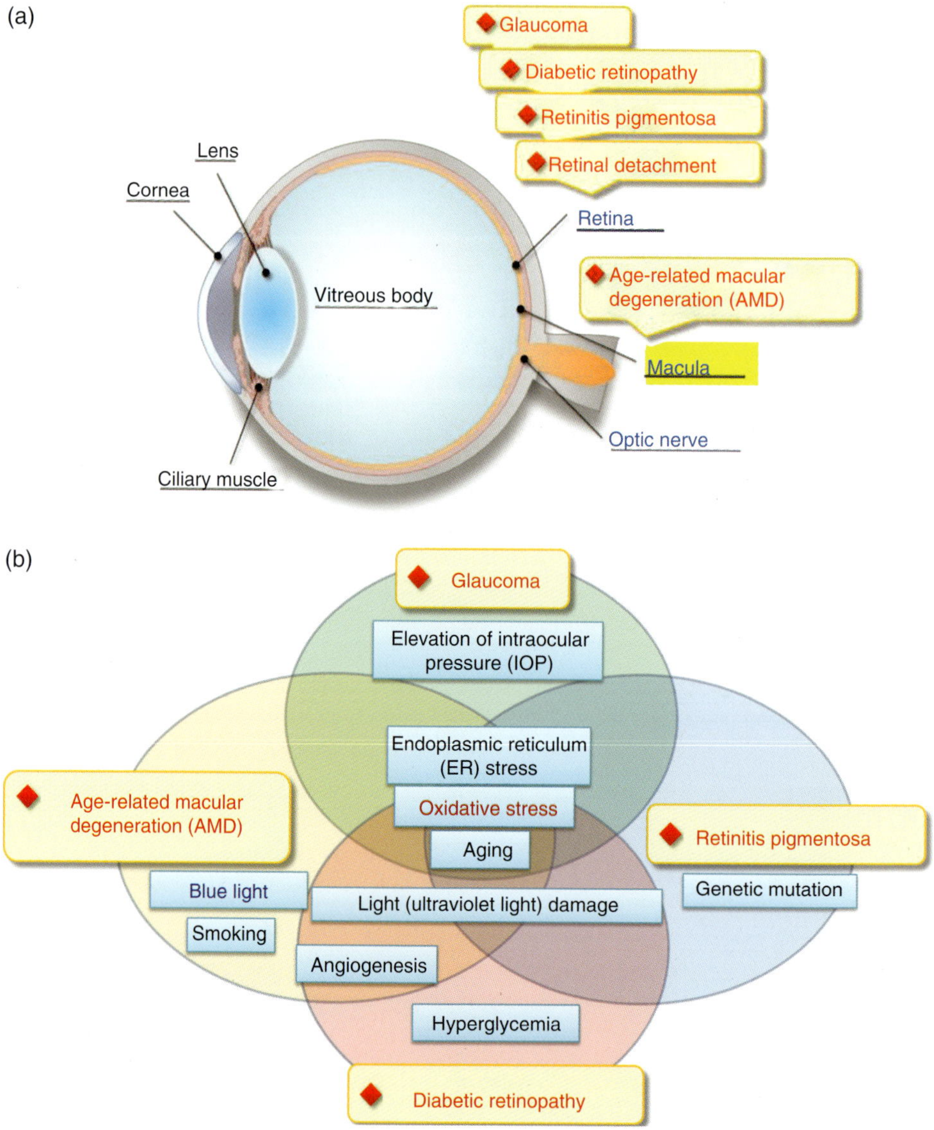

Fig. 9.1 Image showing posterior eye diseases and their common causes. (a) Image of eye structure and each posterior eye disease. (b) Mapping common causes of posterior eye diseases.

Fig. 9.2 Anthocyanin-rich foods used in our experiments.

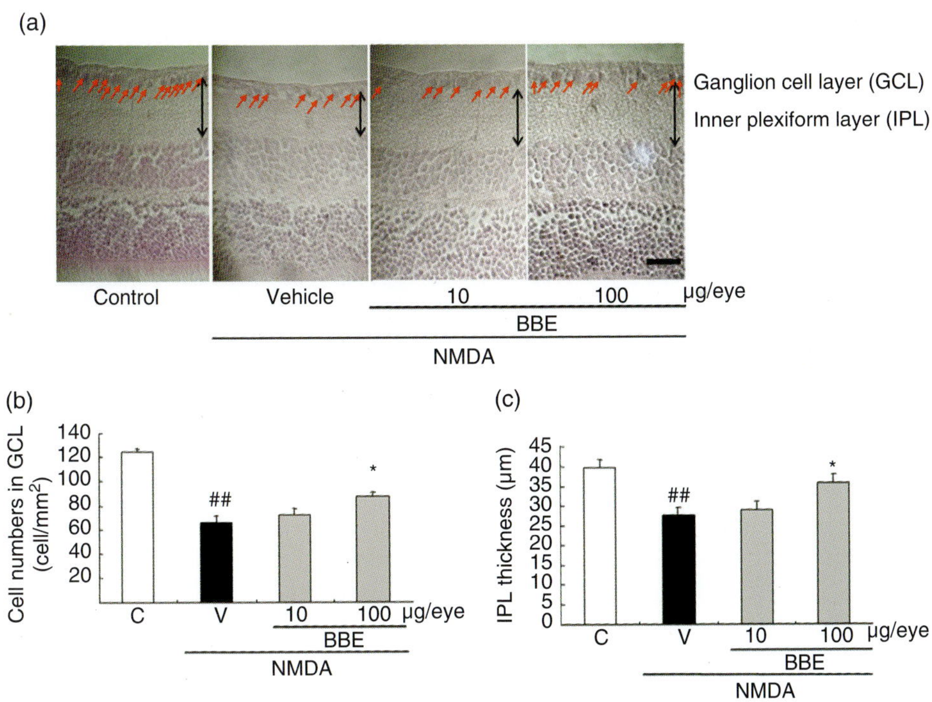

Fig. 9.3 Effect of bilberry extract (BBE) on *N*-methyl-D-aspartate receptor (NMDA)-induced retinal damage in mice. Hematoxylin and eosin staining of sections (thickness 5 µm) obtained from mice 7 days after intravitreal injection of NMDA. (a) Representative photographs show nontreated control retina, NMDA-injected vehicle-treated retina, retina treated with NMDA injection plus BBE at 10 µg/eye, and retina treated with NMDA injection plus BBE at 100 µg/eye. Horizontal bar represents 25 µm. Vertical bar and arrows indicate inner plexiform layer (IPL) and retinal ganglion cells (RGC), respectively. (b and c) Cell numbers in RGC and thickness of IPL. Data are shown as mean ± SEM (n = 6–11). C, control; V, vehicle; $^{\#\#}p < 0.01$ versus control; $^{*}p < 0.05$ versus vehicle. These results were cited in Matsunaga *et al.* (2009).

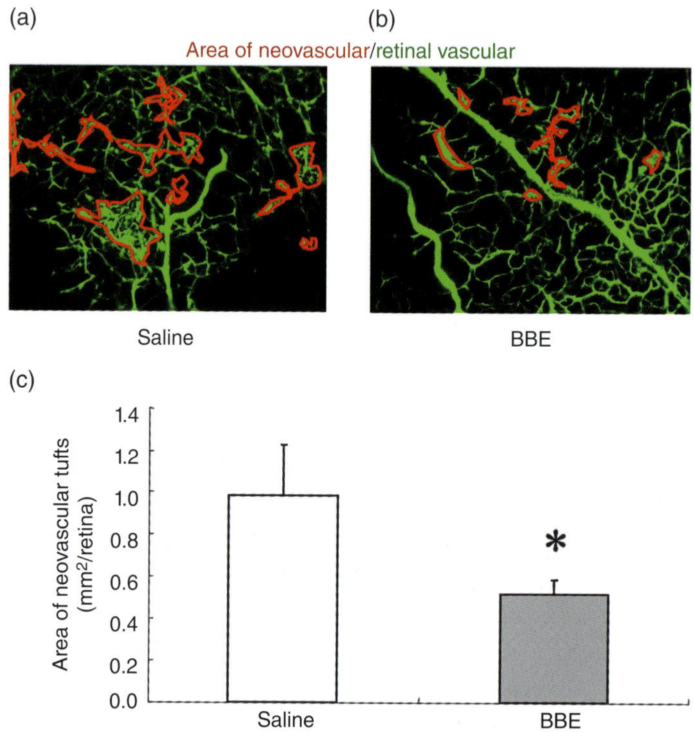

Fig. 9.5 Bilberry extract (BBE) inhibited neovascular tufts on oxygen-induced retinopathy in mice. Retinal flat mounts were examined by FITC–dextran angiography. Representative photographs of retina from saline-treated eye (a) and BBE-treated eye (b). Scale bar = 100 μm. (c) Areas of neovascular tufts in saline- and BBE-treated eyes. Each column and bar represent mean ± SEM ($n = 9$). $^{*}p < 0.05$ versus saline (paired t-test). These results were cited in Matsunaga *et al.* (2010a).

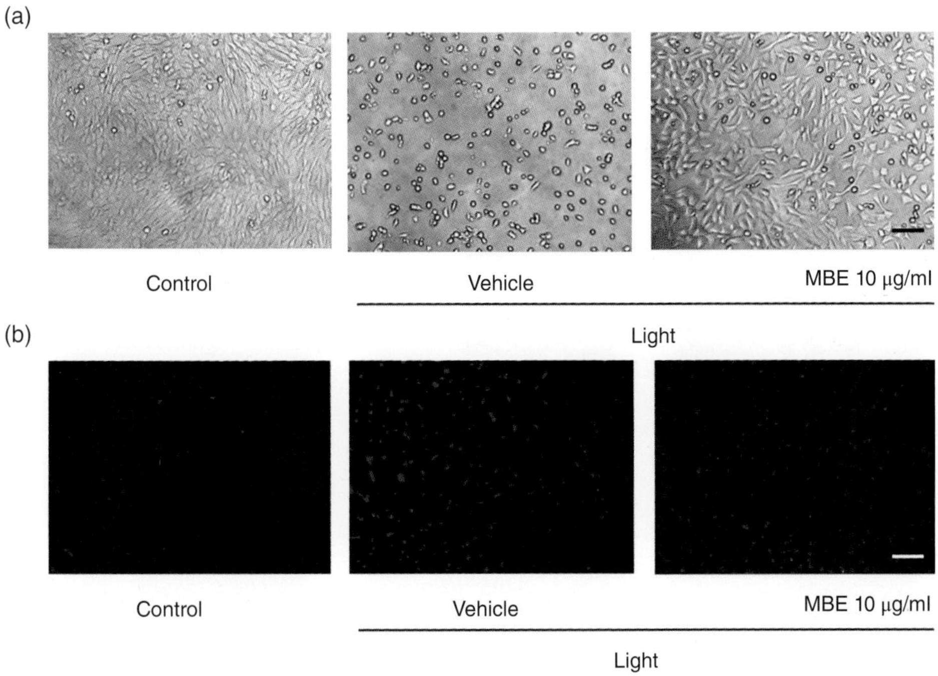

Fig. 9.6 Effects of maqui berry extract (MBE) on visible-light-induced 661W cell damage. (a) Representative microscopy at 24 hours after light exposure. (b) Representative fluorescence microscopy of Hoechst 33342 (blue color, cell nuclear stained) and propidium iodide (PI; red color, dead cell stained) fluorescence staining at 24 hours after light exposure. Each scale bar represents 250 μm. These results were cited in Tanaka *et al.* (2013).

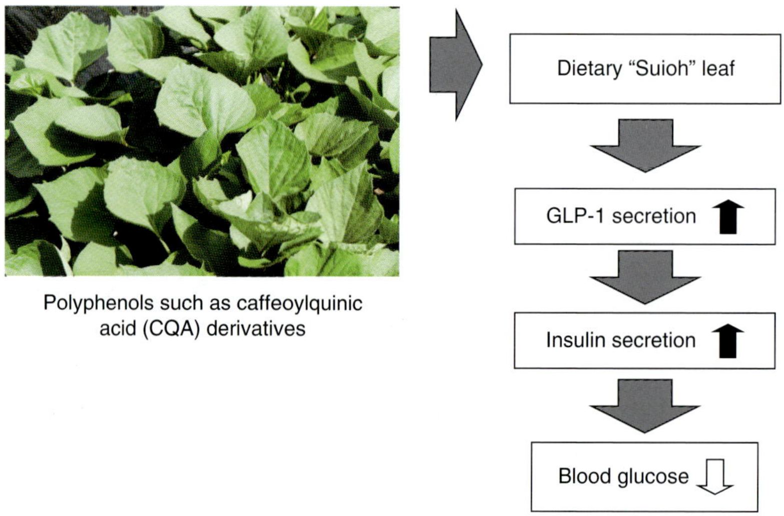

Fig. 10.10 Proposed mechanism for amelioration of hyperglycemia by dietary sweet potato leaves, Suioh, *in vivo*. Nagamine *et al.* (2014). Reproduced with permission from the Royal Society of Chemistry.

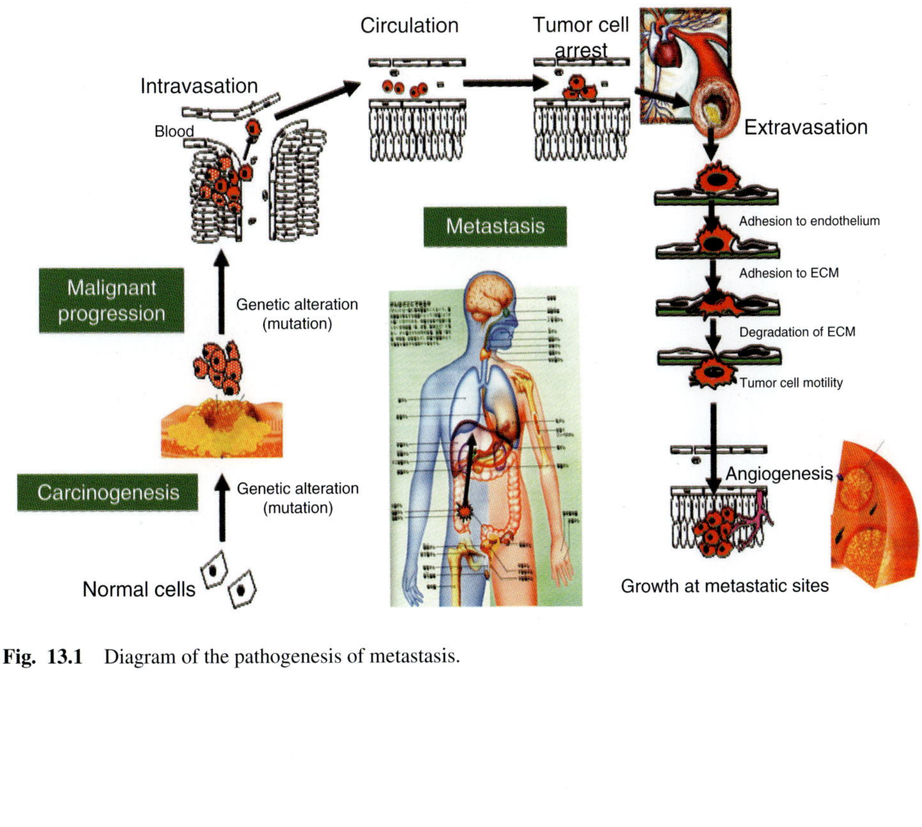

Fig. 13.1 Diagram of the pathogenesis of metastasis.

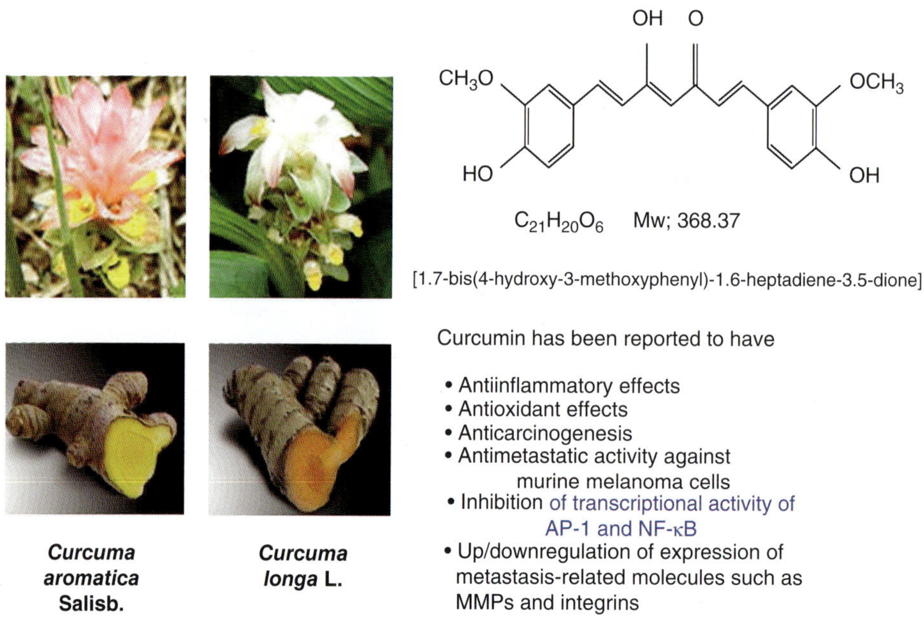

Fig. 13.2 Chemical structure of curcumin and its biological properties.

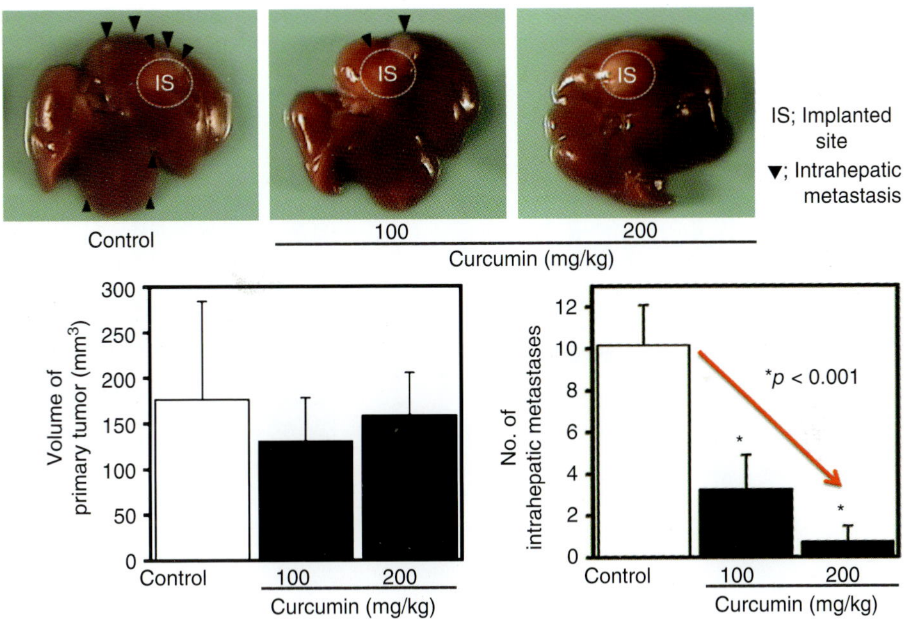

Fig. 13.3 Effect of curcumin on intrahepatic metastasis *in vivo*.

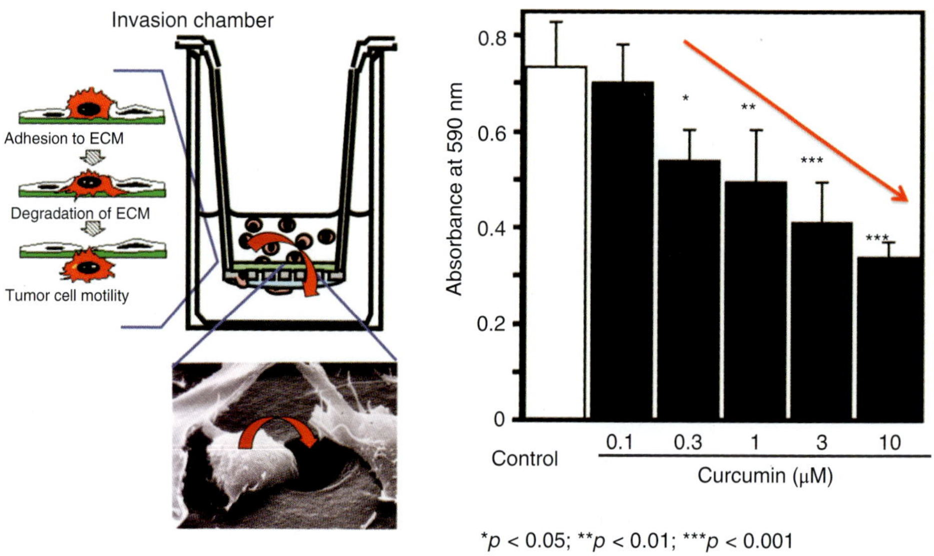

Fig. 13.4 Effect of curcumin on the invasion of CBO140C12 cells.

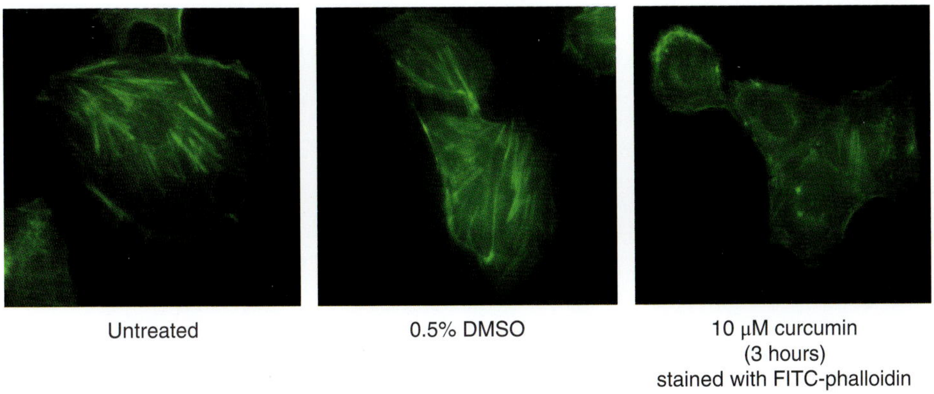

Fig. 13.6 Effect of curcumin on the formation of actin stress fiber on fibronectin-coated substrate.

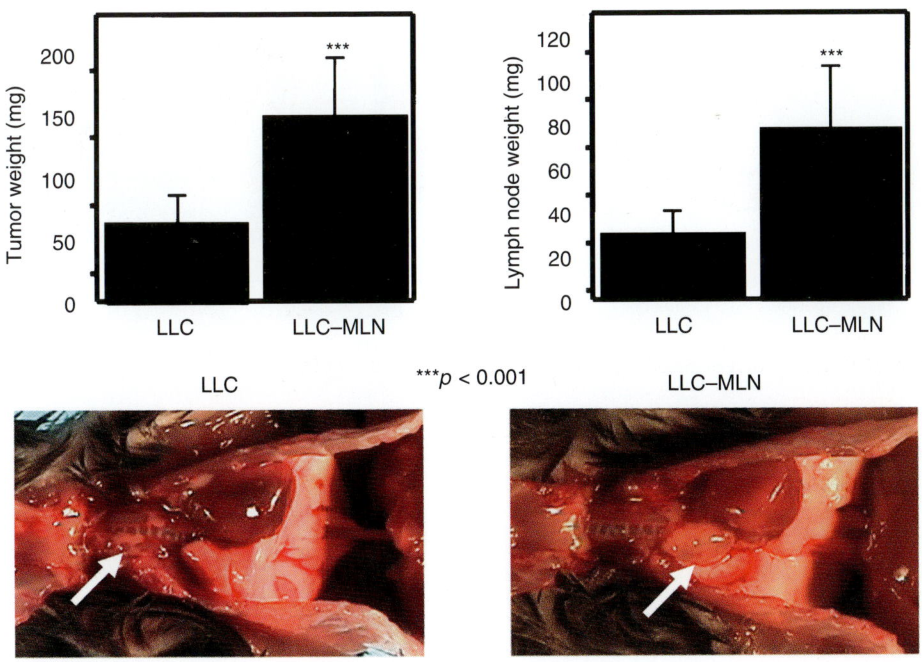

Fig. 13.7 Tumor growth and lymph node metastasis on day 15 after orthotopic implantation of LLC or LLC-MLN cells.

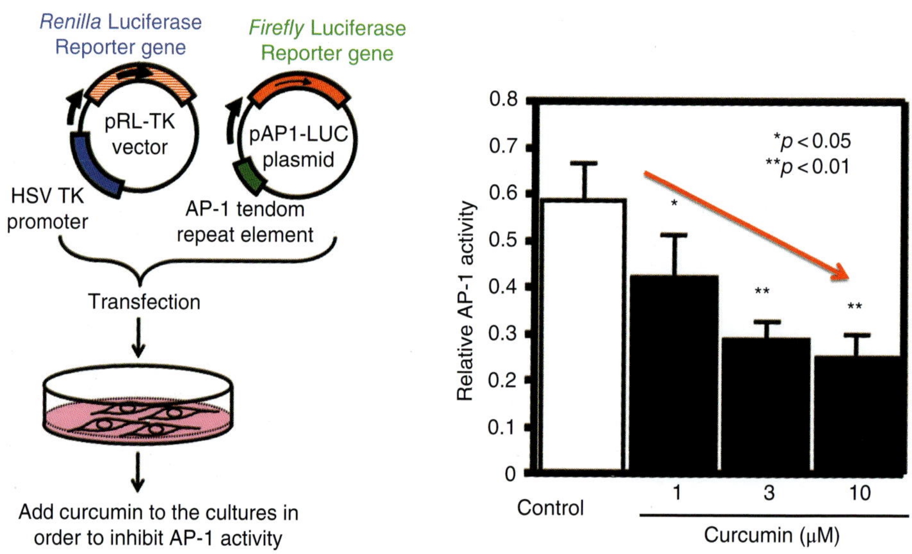

Fig. 13.10 Dual reporter assay for evaluation of anti-AP-1 activity of LLC cells treated with curcumin.

Chapter 8
Structure of Polyacylated Anthocyanins and Their UV Protective Effect

Kumi Yoshida, Kin-ichi Oyama and Tadao Kondo

Abstract: The most important biological function of anthocyanins, besides insect pollination, is the prevention of harmful UV radiation. Polyacylated anthocyanins, which have two or more aromatic acyl residues, such as the *p*-coumaroyl, caffeoyl, feruloyl, and *p*-hydroxybenzoyl moieties, are found primarily in blue flower petals, leaves, and fruits, and the mechanism of the development of the blue color is closely correlated with the molecular stacking of the hydrophobic interactions between the anthocyanidin chromophore and the aromatic rings. These aromatic acyl residues absorb UV light around 260–320 nm in the UV-B region, which overlaps the absorption bands of nucleic DNAs. Therefore, polyacylated anthocyanins possess a stronger protecting effect against UV light than simple anthocyanins that do not contain the aromatic acyl moieties. In this review, we describe the chemical structure of polyacylated anthocyanins and the color development in flower petals. We also discuss their molecular stackings: the intramolecular stacking of aromatic acyl residues to anthocyanidin and the newly reported intermolecular stacking of chromophores. Finally, we review the UV photoreaction of the *E*, *Z*-isomerization of acyl residues, the distinctive quenching system that is due to the molecular associations, and the UV protective effect of the polyacylated anthocyanins.

Keywords: polyacylated anthocyanin, circular dichroism, intermolecular association, intramolecular association, UV protection.

8.1 Introduction

All organisms on the Earth are dependent on photosynthesis by solar energy for sustaining their life. However, ultraviolet rays (UV light) in sunlight can also cause damage to the living things. Although most of the UV-B radiation is absorbed by the ozone layer, the recent depletion of the ozone layer has generated a serious problem for all living

Recent Advances in Polyphenol Research, First Edition. Edited by Kumi Yoshida, Véronique Cheynier and Stéphane Quideau.

organisms on the Earth (Tevini, 1994; Madronich *et al.*, 1998; Bassman, 2004). One consequence of the UV light exposure to plants is the damage of various photosynthesis organs, including the damage to the photosystem II (Barber & Andersson, 1992; Takahashi & Badger, 2011), the hyperoxidation of the lipid membrane, the decreasing and inactivation of the plant hormones, and DNA damage. As a defense against these mechanisms, plants have developed not only a DNA repair mechanism but also the ability to synthesize UV absorbing compounds. In higher plants, flavonoid compounds are induced by UV irradiation and accumulated in high concentrations in the epidermal cells of flowers, leaves, and fruits.

Major plant pigments are chlorophyll, carotenoid, betalain, and anthocyanin (Davis, 2004; Brouillard *et al.*, 2010). Anthocyanin is responsible for wide range colors from red orange, red, purple, and blue in almost all vegetable tissues. In flowers and fruits, the color is used as a communication tool between plants, insects, and animals (primarily birds) for pollination and seed delivery (Harborne & Smith, 1978; Wilbert *et al.*, 1997; Perez-Barrales et al., 2006; Miller *et al.*, 2011). For this purpose, flowers keep their color stable while blooming; however, the isolated pigments are usually unstable and quickly decolorize in a dilute aqueous solution (Fig. 8.1).

The pigments exist in the vacuole of the colored cells in a high concentration up to 10^{-2} M. In the vacuole, simple anthocyanins that do not contain any aromatic acyl residues exhibit a self-associated structure at this high concentration. When a copigment exists, the molecular stacking structure of the self-association changes to copigmentation to produce the association of anthocyanin and copigment. In both cases, the molecular interactions via the hydrophobic interactions between the aromatic rings result in the prevention of a hydration reaction followed by a stable coloration in the living tissues.

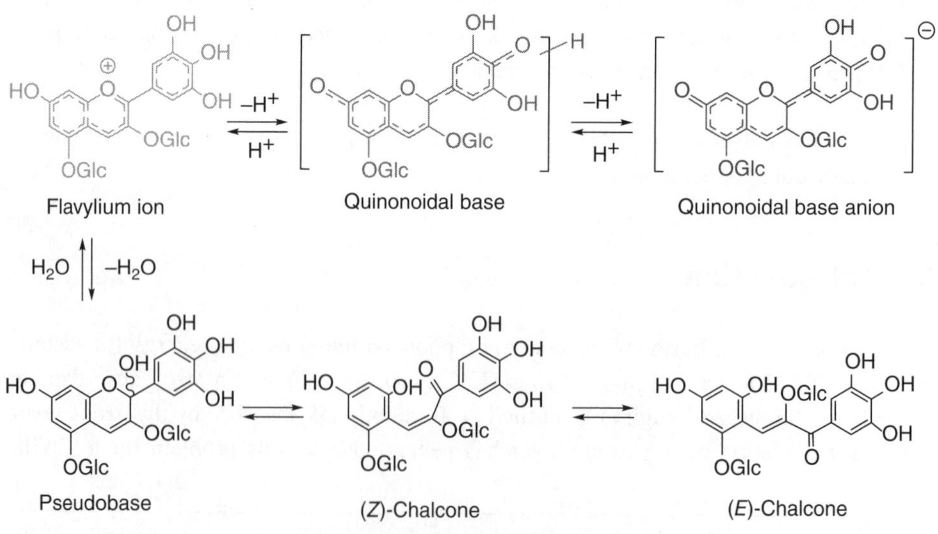

Fig. 8.1 Structure of anthocyanin forms, Flavylium ion-quinonoidal base-pseudobase-(*Z*)-chalcone-(*E*)-chalcone formed at various pH values. (*See insert for color representation of the figure*)

Currently, more than 20% of the naturally occurring anthocyanins are known to be polyacylated. Some of them possess more than two aromatic acyl moieties, and they are called polyacylated anthocyanins. The first finding of polyacylated anthocyanins was in the 1970s, and afterward, the isolation number increased to more than 100 (Andersen & Jordheim, 2006). The acyl residues of polyacylated anthocyanins are cinnamoyl and benzoyl derivatives, and the plural aromatic acyl moieties behave as an electron donor and are associated with the chromophore. Such a molecular interaction is called intramolecular copigmentation, and it produces a very stable form in diluted aqueous solutions (Goto *et al.*, 1982; Brouillard, 1988; Yoshida *et al.*, 1992, 2000; Brouillard & Dangles, 1994). The α,β-double bond of the cinnamic derivative is easily isomerized by solar UV radiation *in vitro* to give an *E,Z*-equilibrium mixture; however, almost all of the cinnamic derivatives in the naturally occurring polyacylated anthocyanins contain an *E*-configuration that does not include the *Z* (Yoshida *et al.*, 2009, 2012). In polyacylated anthocyanins, the absorbed light energy may transfer to the chromophore via the charge-transfer complex followed by the conversion to heat in a sub-picosecond (sub-ps) (Ferreira da Silva *et al.*, 2012). However, there are only a few chemical studies on the UV-B protection of the simple anthocyanin (Moreira *et al.*, 2003; Rodrigues *et al.*, 2009; Costa *et al.*, 2015; Gago *et al.*, 2015), and there are even fewer reports on acylated anthocyanins (Mori, 2005; Mori *et al.*, 2005; Ferreira da Silva *et al.*, 2012).

In this review, we describe the recent chemical studies related to the polyacylated anthocyanins, including the methodology for searching for the polyacylated anthocyanins and the determination of their chemical structure. We also discuss the elucidation of the spatial structure of the polyacylated anthocyanin and the molecular stacking structure of the polyacylated anthocyanins using nuclear magnetic resonance spectroscopy (NMR), circular dichroism (CD), and theoretical calculations (Fig. 8.2). In the final section, we describe the light protection effect of the polyacylated anthocyanins, and primarily, we discuss the DNA protective effect against solar radiation.

8.2 Occurrence and structure of polyacylated anthocyanins in blue flowers

Polyacylated anthocyanins were primarily found in blue flowers in early studies, but a considerable number of such pigments have now been isolated from purple colored flowers, fruits, seeds, roots, and stems (Goto & Kondo, 1991; Harborne & Williams, 2000; Andersen & Jordheim, 2006; Honda & Saito, 2012). The pigments are bearing one anthocyanidin, glycoside bearing one chromophore, plural number of sugars and aromatic acids, sometime additionally dicarboxylic acid in the structure. The chromophore is pelargonidin, cyanidin, peonidin, delphinidin, petunidin, or malvidin. As sugars, there are glucose, galactose, rhamnose, xylose, arabinose, and/or glucuronic acid; as aromatic acids, there are *p*-coumaric, caffeic, ferulic, sinapic, and/or *p*-hydroxybenzoic acid; as dicarboxylic acids, there are oxalic acid, malonic acid, tartaric acid, and/or malic acid. All of the pigments have two or more sugars acylated with plural aromatic acids, and the molecular weight of the polyacylated anthocyanins is at least 1000 Da (gentiodelphin from *Gentiana triflora*: 1114), and it may be over 2000 Da (ternatinD1 from *Clitoria ternatea*: 2108). Since 2005,

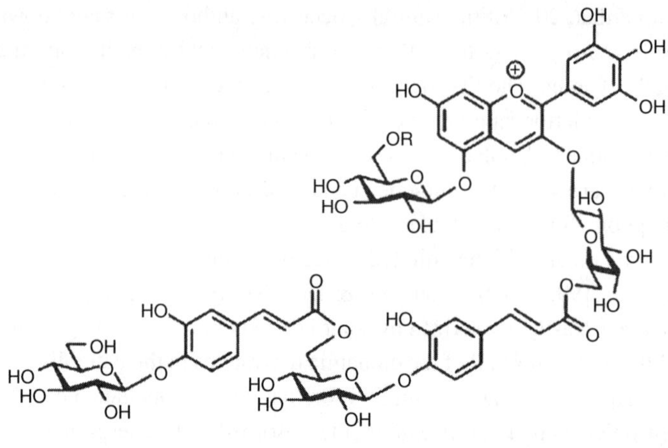

R = malonyl: Phacelianin (**1**)
R = H: Demalonylphacelianin (**2**)

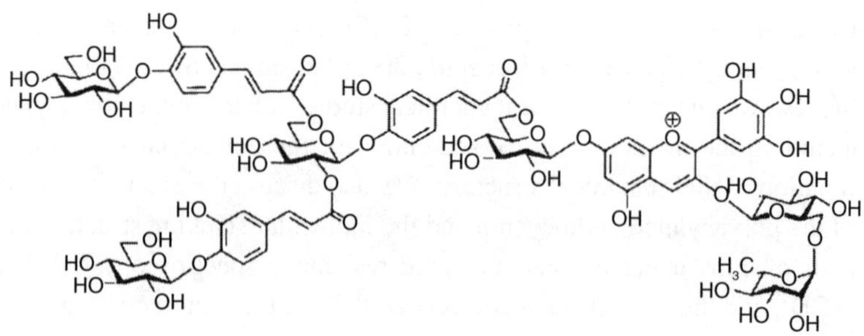

Tecophilin (**3**)

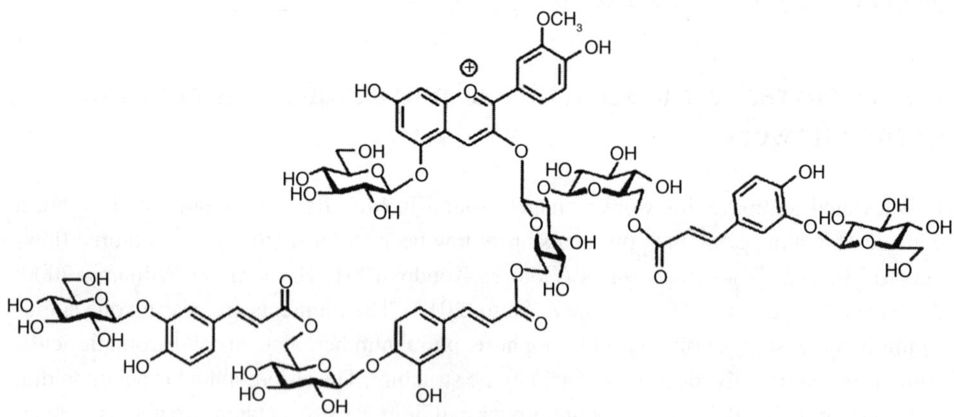

Heavenly blue anthocyanin (**4a**)

Fig. 8.2 The structures of polyacylated anthocyanins reviewed in this chapter: Phacelianin (**1**), demalonylphacelianin (**2**), tecophilin (**3**), heavenly blue anthocyanin (**4a**), gentiodelphin (**5a**), *cis*-gentiodelphin (**5b**), alatanin C (**6**), malonylshisonin (**7a**), and *cis*-malonylshisonin (**7b**).

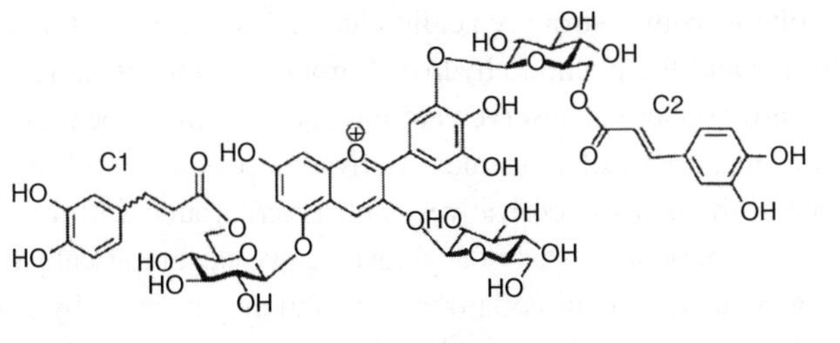

C1: *E*, Gentiodelphin (**5a**)
C1: *Z*, *cis*-Gentiodelphin (**5b**)

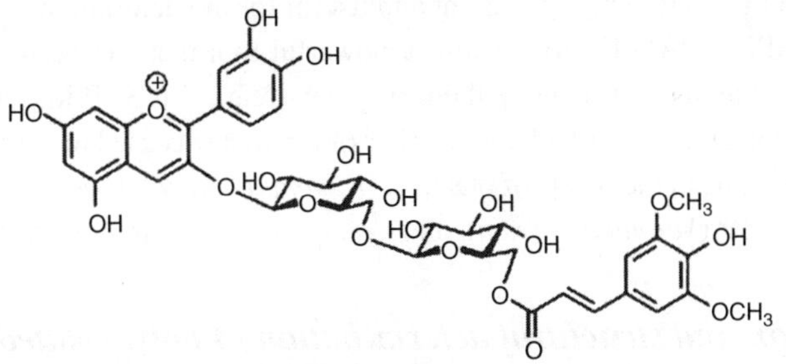

Alatanin C (**6**)

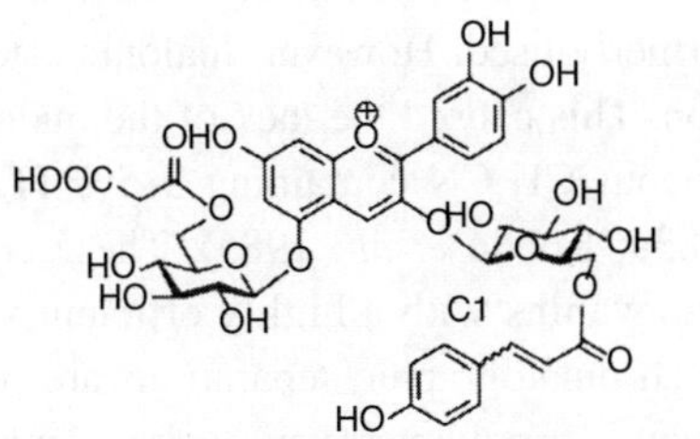

C1: *E*, Malonylshisonin (**MS,7a**)
C1: *Z*, *cis*-Malonylshisonin (**MS,7b**)

Fig. 8.2 (Continued)

new polyacylated anthocyanins from the leaves of *Gynura bicolor* (Shimizu *et al.*, 2010), the flowers of *Malcolmia maritima* (Tatsuzawa *et al.*, 2008), *Leschenaultia* cv. Violet Lena (Saito *et al.*, 2007), *Phacelia campanularia* (Mori *et al.*, 2006), and *Tecophilaea cyanocrocus* (Mori *et al.*, 2014) have been reported. We review the methodology of the search, isolation, structural elucidation, and the mechanism of the blue color development, particularly focusing on our recent research.

8.2.1 Searching for polyacylated anthocyanins

For the determination of polyacylated anthocyanins in plants, a simple detection method to elucidate the number of aromatic acid residues and the structure without isolation is necessary. The shape and intensity of the UV-Vis absorption spectrum that is obtained by HPLC-DAD analysis of the crude plant extracts provide the structural information, and the number of acyl components can be estimated. Usually, the HPLC analysis is performed using a reversed-phase column with a strongly acidic eluent. Therefore, all the anthocyanins take flavylium ion forms, and the phenolic hydroxyl group is protonated. Then, the absorption maximum of the chromophore is observed around 500–530 nm, depending on the structure, and the absorption maximum of cinnamoyl derivative residues is observed around 280–330 nm, while that of the benzoyl derivatives is observed around 260 nm. From the ratio of the absorbance of chromophore to that of acyl residue, we can empirically estimate the number; diacylated pigments have a ratio of around 1.0, and they increase by 0.5 with each additional acylation (Idaka *et al.*, 1987a; Yoshida, 1992). Therefore, the HPLC-DAD analysis of plant extracts with a 50% acetonitrile (CH_3CN) aqueous solution that contains 5% trifluoroacetic acid (TFA) is very convenient and useful for the determination of polyacylated anthocyanins. HPLC-DAD-ESIMS is also a powerful tool that can perform an exhaustive search for anthocyanins and identify them in plants (e.g., Wu & Prior, 2005). Using this searching protocol, an extract of a large number of flowers was used to clarify the new polyacylated anthocyanins in the petals of phacelia (*P. campanularia*), blue daze (*Evolvulus pilosus*), veronica flower (*Veronica persica*), and cyanocrocus (*T. cyanocrocus* var. Leichtlinii).

8.2.2 Isolation and structural determination of polyacylated anthocyanins

For the isolation of anthocyanins, a 1.0% concentrated hydrochloric acid (HCl)/methanol (MeOH) solution was formerly used. However, malonic ester is labile and easily hydrolyzed during the extraction. This caused the lack of the malonyl residue in reports before 1970. We chose 50% aqueous CH_3CN containing 3–5% TFA to replace the HCl/MeOH solution (Goto *et al.*, 1983a; Kondo *et al.*, 1985). These conditions are very useful for extracting the labile anthocyanins with a highly efficient yield. For purification of the pigment, several column chromatographic separations are performed. A crude extract is usually purified using an acrylate-polymer resin, such as Amberlite XAD-7; the absorption of the anthocyanin by this gel is very high, and the water-soluble impurities, such as sugars, amino acids, nucleic acids, and salts, can be removed via elution with acidic water (Goto *et al.*, 1982). Afterward, ODS-HPLC (Wilkinson *et al.*, 1977; Williams *et al.*, 1978; Akavia and Strack, 1980; Strack *et al.*, 1980), which is eluted with the appropriate acidic solution, is applied to produce a pure anthocyanin as an acidic salt.

The recent structure determination of polyacylated anthocyanins has been carried out with the combination of MS and NMR. The first MS spectrum of polyacylated anthocyanin was a measurement of gentiodelphin with a molecular ion peak at 1114 [M + 1] using FD-MS (Goto *et al.*, 1982). Currently, MS technology is advanced, and many instruments, such as FAB, ESI-MS, and MALDI-TOF, are available. NMR using a high field magnet (>500 MHz ^{1}H NMR) is another useful tool. Usually, one-dimensional (1D)- and two-dimensional (2D)-NMR spectra are recorded in 5–10% TFA-d_1 in methanol-d_4 or >10% TFA-d_1-dimethylsulfoxide-d_6 (DMSO-d_6) to keep the anthocyanin in the flavylium form (Nilsson, 1973; Brouillard, 1982). The chromophore and sugar connections are determined using HMBC and/or NOE correlation, and the position of esterification is elucidated via HMBC correlation between the carbonyl carbon and the sugar proton. The assignment of the sugar protons can be obtained using 1D and 2D TOCSY experiments, and the NMR measurement of the pertrifluoroacetate pigment is useful for the determination of the acyl position via the acyl shift effect. A degradation reaction is still essential. A higher magnetic field has enabled the measurement and assignment of the signals of the blue quinonoid form (Cheminat & Brouillard, 1986; Kondo *et al.*, 1992).

8.2.2.1 Structural determination of phacelianin and tecophilin

Using a general isolation procedure, phacelianin (170 mg) was purified from 300 g of fresh petals (Fig. 8.2). In the ^{1}H-NMR spectrum of phacelianin (**1**), the signals for one delphinidin, two (*E*)-caffeoyl residues, and four sugars were observed (Mori *et al.*, 2006). The molecular ion peak at $m/z = 1361$ in the FAB-MS confirmed the existence of malonate. The existence of the malonyl group can be determined from the fragment ion peak ($m/z = 1275$) minus 86 from the molecular ion, and it can also be determined from the methylene signals in the ^{1}H-NMR and the carbonyl signals in the ^{13}C-NMR. In this case, the alkaline methanolysis of **1** under argon atmosphere was performed because the NMR experiment could not differentiate between 4′-*O*-glucosyl-(*E*)-caffeate and 3′-*O*-glucosyl-(*E*)-caffeate. The reaction products were delphin and methyl 4′-*O*-glucosyl-(*E*)-caffeate ($J\alpha\beta = 16$ Hz), instead of methyl 3′-*O*-glucosyl-(*E*)-caffeate; it was then determined that the sugar is attached at the 4-OH of the caffeoyl residue. The proton signals of the delphinidin could be easily assigned based on the H-4, which is a characteristic low field signal. The 1D and 2D TOCSY revealed that all the four sugars were β-glucopyranosides ($^3J_{1,2} = 7.5$–10 Hz). The methylene proton at the 6-position of the three glucoses shifts by 0.3–0.8 ppm downfield, indicating that the three glucoses are acylated with the exclusion of the ○-glucose. A strong NOE observation between the ▲-1 and H-4, ●-1 and H-5, △-1 and H-5 of caffeoyl 1(C1), and ○-1 and H-5 of caffeoyl 2 (C2) determined the linkage positions of all of the sugars (Fig. 8.3). The HMBC spectrum showed a cross-peak between the carboxyl of C1 and ▲-6 and the carboxyl of C2 and △-6, indicating that the structure of **1** contains two 4-glucosylcaffeic acid blocks that link serially and that esterify to produce the hydroxyl of ▲-6. Therefore, the malonyl position is the hydroxyl of ●-6. The structure of **1** is determined to be 3-*O*-(6-*O*-(4′-*O*-(6-*O*-(4′-*O*-β-D-glucopyranosyl-(*E*)-caffeoyl)-β-D-glucopyranosyl)-(*E*)-caffeoyl)-β-D-glucopyranosyl)-5-*O*-(6-*O*-malonyl-β-D-glucopyranosyl)delphinidin. The long-distance NOE observation between the H-α of C1 and ●-2 and the △-1, H-2 and 6′ shows that the stereostructure of **1** may assume the conformation of both the C1 and C2 residue stacking onto the nucleus from the same front surface that is coincident with a negative exciton-type Cotton effect in CD (Fig. 8.3; Mori *et al.*, 2006).

In addition to **1**, a small amount of pigment (**2**, 24 mg) was obtained with the same absorption spectrum in the 3D-detection HPLC. There was a peak in the FAB-MS at $m/z = 1275$, indicating that a loss of the malonyl group occurred and that **2** was demalonylphacelianin. The ^{1}H-NMR of **2** was very similar to **1**; however, the signal of CH_2 of ▲-6 in **2** was observed more upfield compared with that of **1**. Thus, this pigment was 5-*O*-demalonylphacelianin (**2**, Fig. 8.2). The polyacylated anthocyanin from the flower petals of the Blue daze (*E. pilosus*, Toki *et al.*, 1994) was identical to the phacelianin (**1**) HPLC retention time, and it was also identical in the ^{1}H NMR. Therefore, the previously reported structure of the blue daze pigment should be revised to that of **1** (Toki *et al.*, 1994, 1998).

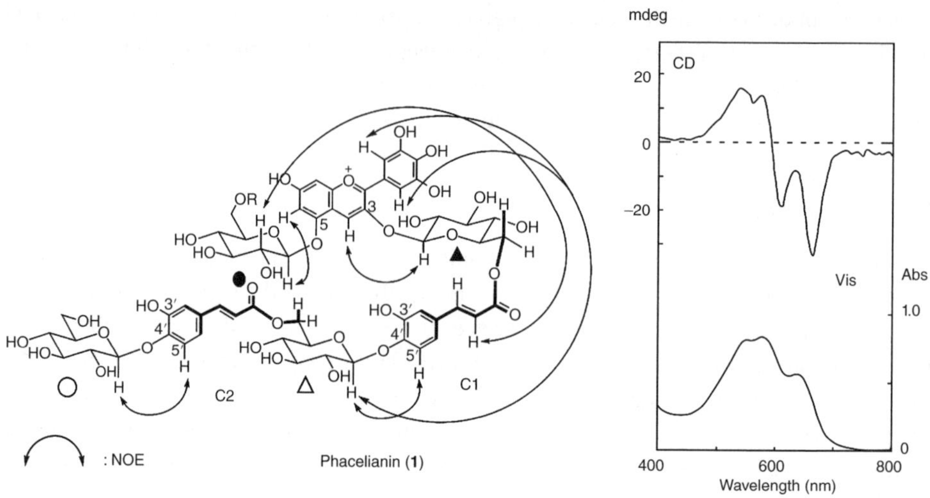

Fig. 8.3 Long-range NOEs in the NMR spectrum of phacelianin (**1**) and Vis-UV spectra and CD spectra of petal of *Phacelia campanularia*.

From the blue petals (12 g) of *T. cyanocrocus* var. Leichtlinii, 20 mg of the polyacylated anthocyanin was isolated and named tecophilin (**3**) (Mori *et al.*, 2014). The molecular weight was determined to be 1745 for the flavylium ion from the molecular ion at $m/z = 1767$ $[M\text{-}H+Na]^+$ using MALDI-TOF-MS. The ^1H NMR determined that **3** was composed of delphinidin, three (*E*)-caffeic acids, and six sugars. Compound **3** contained five β-glucosyl residues, and the remaining one was rhamnoside. The proton signals that were attributed to ■-2, ■-6, and ●-6 were shifted downfield by 0.5–1.5 ppm, indicating that the corresponding OHs were acylated. Using NOESY and HMBC, all of the connections of the components were determined. The structure of tecophilin (**2**) was determined to be 3-*O*-(6-*O*-α-L-rhamnopyranosyl-β-D-glucopyranosyl)-7-*O*-(6-*O*-(4′-*O*-(2-*O*-(4′-*O*-β-D-gluco-pyranosyl-(*E*)-caffeoyl)-6-*O*-(4′-*O*-β-D-glucopyranosyl-(*E*)-caffeoyl)-β-D-glucopyranosyl)-(*E*)-caffeoyl)-β-D-glucopyranosyl)delphinidin (Fig. 8.4; Mori *et al.*, 2014).

8.3 Molecular associations of polyacylated anthocyanins in blue flower petals

As described earlier, the stability of polyacylated anthocyanins is due to the prevention of the hydration reaction via intramolecular stacking of the aromatic acyl residues onto anthocyanidin chromophore. To clarify this phenomenon, studies on the 3D-structure using X-ray crystallographic analysis are essential; however, there are few successful examples of crystallization of anthocyanin. Only two have been reported: commelinin (Kondo *et al.*, 1992; Shiono *et al.*, 2008) and protocyanin (Shiono *et al.*, 2005). There have been few X-ray crystallographic analysis studies of polyacylated anthocyanins because no single crystal has been obtained until now. Therefore, all the 3-D structure

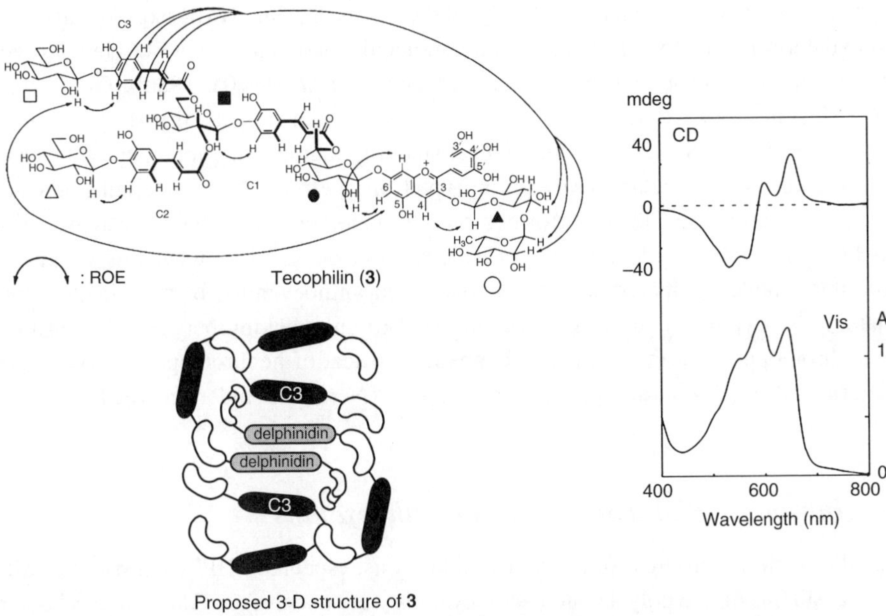

Fig. 8.4 Long-range NOEs of tecophilin (**3**), Vis-UV spectra and CD spectra of petal of *Tecophilaea cyanocrocus* and proposed 3-D structure of **3**.

studies and stacking conformational studies have been performed using NMR, CD, and molecular modeling calculations (Goto & Kondo, 1991; Yoshida *et al.*, 2009). In the NMR studies, the upfield shift of the protons of the chromophore and the acyl residues and the long-range NOEs are evidence of the intramolecular stacking (Yoshida *et al.*, 1991; Fernandes *et al.*, 2015). In CD, the absence of the exciton-type Cotton effect around the Vis absorption maximum is evidence of the intramolecular stacking. Simple anthocyanins self-associate and produce an exciton-type Cotton effect due to the close interaction of the chromophores. However, in the case of the polyacylated anthocyanins, the chromophore is covered by aromatic acyl residues, and no direct interaction of the anthocyanidins themselves exists. Then, the pigments are stabilized as a monomer molecule (Goto & Kondo, 1991; Yoshida *et al.*, 2009, 2012). However, some recently isolated polyacylated anthocyanins showed a strong exciton-type Cotton effect, indicating the existence of both intramolecular stacking of the aromatic residues to the chromophore via self-association and intermolecular stacking of the anthocyanidin nucleus (Yoshida *et al.*, 1991; Mori *et al.*, 2006, 2014).

8.3.1 Intermolecular associations of anthocyanins

In the early 20th century, the intermolecular association of colorless compound to anthocyanin was called copigmentation (Robinson & Robinson, 1931). At that time, the major force between the pigment and the copigment was thought to be hydrogen bond. The similar association of anthocyanin with itself in a highly concentrated solution was called self-association (Asen *et al.*, 1972). In the 1980s, CD and NMR experiments were used to determine that the mechanism of the molecular association was not via hydrogen bond but due to hydrophobic interactions (Hoshino *et al.*, 1980a, 1980b; Goto *et al.*, 1986; Hoshino, 1992; Houbiers *et al.*, 1998). As a result, the anthocyanidin 3,5-diglucosides stack vertically in a chiral stacking manner (the left-handed orientation was indicated by the negative exciton-type Cotton effect). More detailed studies on the self-association have been performed on the monomer–dimer–polymer aggregation model (Leydet *et al.*, 2012; Gavara *et al.*, 2013). Monoacylated anthocyanins are known to be more stable compared with the nonacylated anthocyanins. In petanin, a monoacylated anthocyanin, it forms a dimer via head-to-tail stacking between the chromophore–chromophore and the coumarate–coumarate, instead of the chromophore–coumarate, as determined from NOE analysis and calculations (Nerdal & Andersen, 1991).

8.3.2 Intramolecular associations of anthocyanins

In parallel with the progress in instrumental analysis, especially HPLC, NMR, and MS, chemical studies on the polyacylated anthocyanins were undertaken. The structural determination of the monoacylated anthocyanins, such as awobanin, shisonin, and violanin (Goto *et al.*, 1978) was performed, and the structures of the polyacylated anthocyanins were also reported (Goto *et al.*, 1982, 1983b, 1984; Idaka *et al.*, 1987a, 1987b; Kondo *et al.*, 1987). Goto *et al.* found the upfield shift of some the aromatic signals and the long-range NOEs in the ^{1}H NMR of the acidic CD_3OD samples of the polyacylated anthocyanins. These studies inspired the idea of the intramolecular stacking of the aromatic residues to the anthocyanidin chromophore. Until their proposal, intramolecular copigmentation was considered to be caused by hydrogen bond. However, Goto *et al.* proposed intramolecular stacking of aromatic acyl residues to the anthocyanidin chromophore with hydrophobic interaction such as self-association. Furthermore, it was reported that charge-transfer complexation might contribute to the copigmentation of anthocyanins (Ferreira da Silva *et al.*, 2005).

The petal pigment of the *Ipomoea tricolor* is a triacylated anthocyanin containing three caffeoyl residues and six glucosyl moieties (HBA, **4a**). Compound **4a** gives a simple bell shape in the absorption spectrum in the visible region and no exciton-type Cotton effect, suggesting that molecule **4** exists as a monomer (Goto *et al.*, 1986). The ^{1}H NMR of **4a** in CD_3OD/DCl exhibited a flavylium ion with δ values of all three caffeic acids that are shifted to the higher field by 0.5–0.75 ppm compared with methyl caffeate, thus, exhibiting a strong anisotropic effect. The long-distance NOE between the caffeoyl protons and the anomeric protons of the sugars that are attached directly to the chromophore indicates that the peonidin nucleus is intramolecularly covered by two or three caffeic acids on the surface to protect the hydration reaction (Kondo *et al.*, 1987; Goto and Kondo, 1991).

Gentiodelphin (**5**) is a diacylated anthocyanin with two caffeoyl residues and three glucosyl moieties (Goto *et al.*, 1982). The computer-assisted conformational analysis of **5** using NMR data (coupling constant of sugars, NOE network among residues) revealed that even in acidic methanol, one caffeate residue strongly stacks at the back surface of the

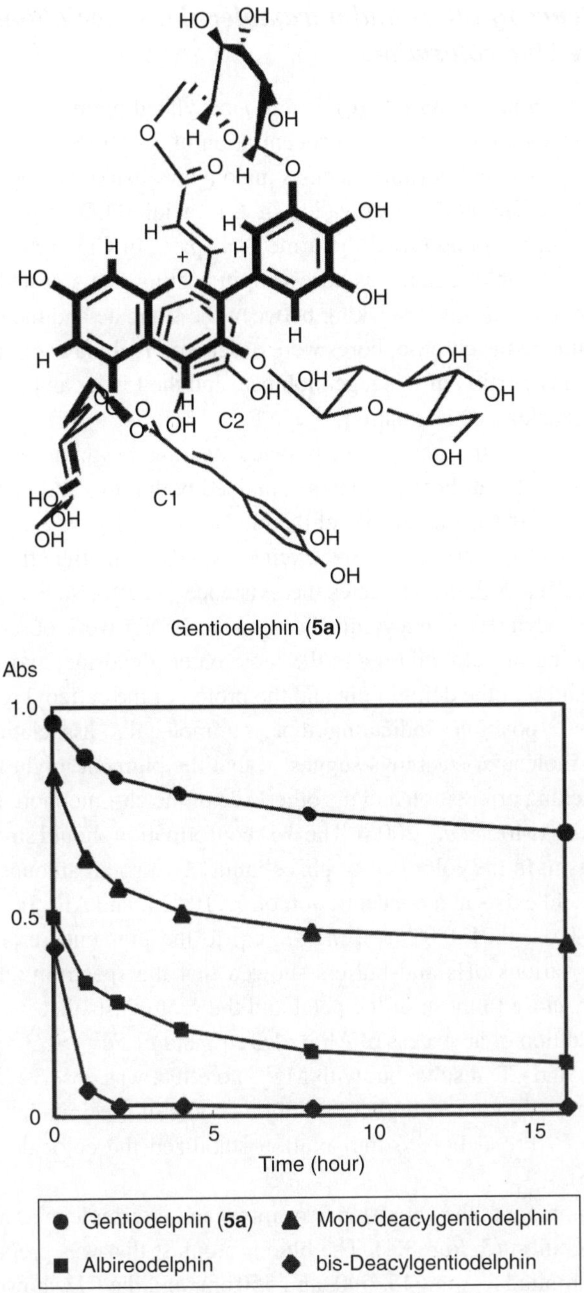

Fig. 8.5 The optimized 3-D structure of gentiodelphin (**5a**) obtained by calculation using NMR data and the comparison of the stability of **5a** and deacylated pigments.

chromophore plane and the other is free (Fig. 8.5). However, in aqueous solution, both of the caffeoyl residues may stack at the front and at the back of the surface, respectively. The pigment at a concentration of $5 \times 10\,M^{-5}$ and at a pH of 6.5 is stable due to a sandwich-type intramolecular association (Yoshida *et al.*, 1992).

8.3.3 Coexistence of inter- and intramolecular associations involved in the blue coloration

Purple yam anthocyanin, Alatanin C (**6**), is a monoacylated pigment; but, it is very stable in a pH 6.0 aqueous solution even at a concentration of 5×10^{-5} M (Yoshida *et al.*, 1991). However, monoacylated anthocyanin, such as malonylawobanin, is quickly decolorized in the same conditions. The NMR observation of **6** in acidic D_2O showed a long-distance NOE between the chromophore and the armed sinapate. In CD, a positive exciton-type Cotton effect in the visible region was observed, indicating that a nested-type association might exist. Both intramolecular stacking between the sinapate and the chromophore and self-association among the chromophores were observed (Yoshida *et al.*, 1991). This might be because the linker portion of **6** is a gentiobiose, and the longer and flexible sugar chain might enable the stacking conformation.

A recent detailed spectral analysis of polyacylated anthocyanins, phacelianin (**1**) and tecophilin (**3**), clarified that these pigments contained both intra- and intermolecular associations (Mori *et al.*, 2006). VIS spectra of the blue petals of **1** gave a λ_{max} at 636, 578, and 546 nm, and CD showed a strong negative exciton-type Cotton effect that corresponded to the visible peaks. This strongly indicates the existence of self-association in the chromophores of the polyacylated anthocyanin. Long-range NOEs were observed between the protons of the B-ring of delphinidin and the most outer glucoside, and the proton of the sugar at the 5-position of the delphinidin and the proton of the caffeoyl residue attached to the sugar of the 3-position, indicating that intramolecular associations also existed (Fig. 8.4). These molecular stackings suggested that the pigment might be dimer with an anticlockwise stacking orientation and the other side of the chromophore might be covered by caffeoyl moieties (Mori *et al.*, 2006). The 3-D conformation should strongly confirm the coloring mechanism. In the color cell of phacelianin (**1**), a small amount of demalonylate and chlorogenic acid exists at a concentration of 5×10^{-4} M, and Al^{3+}, Fe^{3+}, and Mg^{2+} were detected at 0.01–0.02, 0.01–0.02, and 0.2–0.5 eq. to the pigment, respectively. The Vis measurements at various pHs and buffers showed that the spectrum and the CD at pH 5.5–6.0 were very close to those of the petal and the protoplasts that were prepared from the petals. The addition of an excess of Al^{3+} (>1/3 eq.) and/or Fe^{3+} (>1/3 eq.) to the solution gave different Vis and CD results, but with Mg^{2+}, no effect was observed. The reproduced color solution returned to the natural color after 24 hours of precipitation. The addition of a small amount of $Al^{3+}$and Fe^{3+} to the solution stabilized the color during the 24 hours (Mori *et al.*, 2006).

The same phenomenon, the coexistence of intra- and intermolecular associations, was also found in tecophilin (**3**, Fig. 8.4). The blue protoplast that was prepared from a cyanocrocus petal exhibited λ_{max} at 645, 590, and 550 nm, and the CD showed a strong positive exciton-type Cotton effect that corresponded to the visible peaks. This indicates that the stacking orientation of the anthocyanidin chromophore was contrary to **1**: the orientation was clockwise in **3**. In **3**, a long-range NOE was observed between the protons at the 5-H position of the chromophore and the outermost sugar. NOEs between the protons of the rutinoside at the 3-position and the outer caffeoyl residue were also observed.

Compound **3** also exhibits a similar stacking conformation, but the stacking orientation of anthocyanidin was different. The concentration of **3** in the blue cells was quantified using HPLC to be approximately 25 mM, and some ions were detected (K^+, Mg^{2+}, Na^+, and Ca^{2+}). Repeated experiments for the same blue color as the petals clarified that compound **3** (5 mM) at a pH of 5.3 without the metal ions exhibited the same color and spectra in VIS and CD.

There are at least two types of blue color development in the polyacylated anthocyanins. One is a monomer pigment that exhibits the self-contained intramolecular copigmentation found in HBA (**4a**) and gentiodelphin (**5a**). The other is a cooperated-type pigment that might exist as a dimer with an intramolecular association between the aromatic residues and chromophore and with chiral self-association. It is not clear whether they are nested or unnested types. The metal ions are not primarily responsible for the blue coloration, but some might affect their stability.

8.4 UV protection of polyacylated anthocyanins from solar radiation

Polyacylated anthocyanins have a large absorption both in the visible and in the UV-B band region, and the absorption coefficient (ε) is 6.0 to $9.0 \times 10^3\,M^{-1}\,cm^{-1}$ near 280 nm and 1.0 to $3.5 \times 10^4\,M^{-1}\,cm^{-1}$ at the λ_{vismax} at the vacuolar pH (pH of ~5.5) of the plant. Therefore, polyacylated anthocyanins might be expected to exhibit a UV protecting effect over simple anthocyanins. There are several studies on the energy-transfer mechanism from excited anthocyanin and anthocyanin with copigment to the bulk system. Picosecond time-resolved fluorescence analysis has revealed that the excited singlet-state proton transfer from the flavylium ion of simple anthocyanins to water occurs on the picosecond timescale, and the excited quinonoidal base decay occurs on a sub-nanosecond timescale (Lima *et al.*, 1998). A more detailed study was reported that used nanosecond-laser flash photolysis (Maçanita *et al.*, 2002) for determination of rate constants of proton transfer in the ground state. An excited anthocyanin-copigment charge-transfer complex dissipates the absorbed energy into heat on a sub-picosecond timescale by internal conversion to the ground state. The efficient dissipation of the absorbed energy from the excess solar irradiation to the water results in protection of the plant tissue (Quina *et al.*, 2009; Rodrigues *et al.*, 2009). Regarding the UV-B dissipation capacity of polyacylated anthocyanin, the intramolecular copigmentation should be more efficient compared with the intermolecular copigmentation. A fluorescence study (Ferreira da Silva *et al.*, 2012) using cyanidin 3-6′-*O*-*p*-coumaroylsophoroside-5-glucoside in an acidic aqueous solution compared with a mixture of cyanin and *p*-coumaric acid demonstrated that the acylated anthocyanin is more efficient in dissipating the UV-B than the copigmentation. Despite the acidic conditions, the excited acylated anthocyanin decay (sub-ps) was too fast to avoid the other reaction events. The ability of the acylated anthocyanin to consume UV-B is of great advantage to plants.

8.4.1 **E,Z-*isomerization of cinnamoyl derivative residues in polyacylated anthocyanins***

The major acyl residues of polyacylated anthocyanin based on aromatic acyl groups include the cinnamoyl derivative, for example, *p*-coumaroyl, caffeoyl, feruloyl, and sinapoyl moieties. The addition of UV radiation to these acids *in vitro* produces an *E,Z*-equilibrium mixture. However, almost all of the naturally occurring acyl portions in the polyacylated anthocyanins exhibit the *E*-configuration instead of the *Z*-configuration. The first report of the *Z*-configured acyl moiety in anthocyanin was in the isolation of minor monoacylated pigments of purple *Perilla* leaves (Yoshida *et al.*, 1990). After that, several pigments with the *Z*-cinnamic acyl residue were reported (Kondo *et al.*, 1991; Hosokawa et al., 1995; Toki *et al.*, 1998; Yoshida *et al.*, 2003a; Tatsuzawa *et al.*, 2008, 2014; Zheng *et al.*, 2011; Matera *et al.*, 2012). However, the ratio of *Z*:*E* in the tissues obtained via HPLC analysis of the extracts has been very low at less than 5%. When the solution of each monoacylated anthocyanin was irradiated by UV, the *Z*:*E* ratio increased and sometimes reached 50% (Yoshida *et al.*, 1990). However, regarding the polyacylated anthocyanins, the *Z*-configured cinnamoyl derivatives have not been reported to our knowledge. Therefore, the question of why cinnamoyl derivatives in polyacylated anthocyanins in living cells do not isomerize to the *Z*-forms under strong solar radiation is attracting attention because of the biological functions of the anthocyanins.

Gentiodelphin (**5a**) and HBA (**4a**) produced *Z*-caffeoylated polyacylated anthocyanins via artificial UV irradiation. In both cases, not all the caffeoyl residues, but only the specific one(s), were isomerized, and the polarity of the solvent affected the ratio of the isomers. When **5a** was irradiated in the 0.5% TFA-MeOH solution, only the caffeoyl residue (C1) that was attached to the sugar of the 5-OH position of the anthocyanidin chromophore was isomerized (**5b**), and the residue (C2) at the 3′-OH position was not. The ratio of **5b** to **5a** was 10:90 in the 0.5% TFA-MeOH, but the isomer was not observed in the mixture of 0.5% aqueous TFA and 0.1 M phosphate buffer at a pH of 7.0. When HBA (**4a**) was irradiated in the 0.5% TFA-MeOH solution, two (C2 and C3) of the three caffeoyl residues were isomerized to the *Z*-form, and the C3 did not change (Fig. 8.6). The ratio of the isomers upon irradiation with UV at 280 nm for **4d**: **4c**: **4b**: **4a** at the equilibrium state was 5:12:20:63 in the 0.5% TFA-MeOH, 0:5:2:93 in the 0.5% aqueous TFA, and 0:3:1:96 in the buffer solution at a pH of 7.5. In acidic methanol, the equilibrium state was obtained from all of the isomers with the same isomer ratio within 10 minutes. The *Z,E*-ratio of the polyacylated anthocyanin was low compared with that of the monoacylated anthocyanin, malonylshisonin (**MS, 7a**). The ratio *cis*-MS (**7b**) to **7a** upon UV irradiation in 0.5% TFA-MeOH was 50:50. In the 0.5% aqueous TFA, the ratio was 20:80, and in the buffer solution at pH 7.0, the ratio was 20:80.

The reason why the isomerization reaction is suppressed in the polyacylated anthocyanins is explained by the intramolecular stacking of the aromatic acyl residues with the anthocyanidin chromophore. In gentiodelphin (**5a**), the nonisomerized caffeoyl residue in an acidic methanol solution stacks on the anthocyanidin nucleus stronger according to the upfield shift of the protons in the NMR studies. Under the neutral conditions, both C1 and C2 may stack. The chemical shift of the α,β protons of C1 in gentiodelphin (**5a**) moved

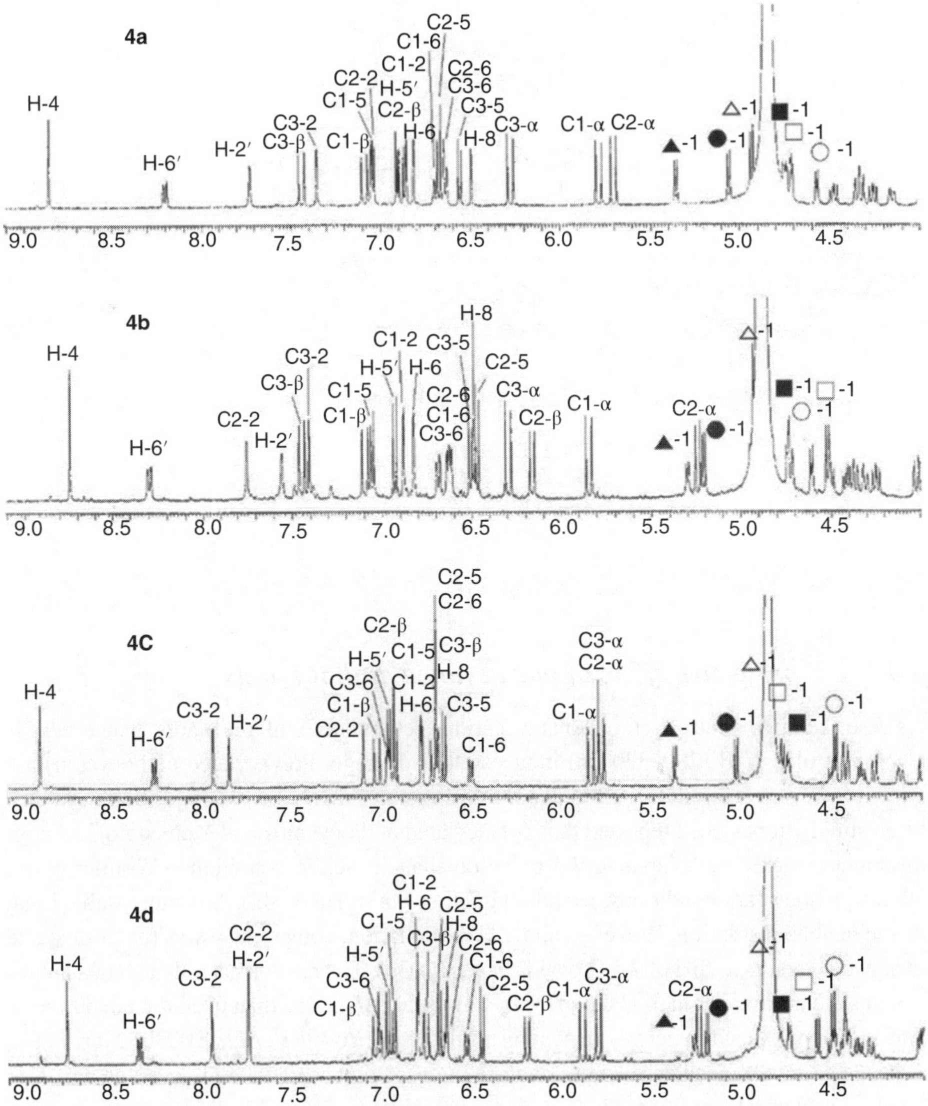

Fig. 8.6 Structure of UVB irradiation products of HBA, and their ^{1}H NMR spectra (500 MHz) in the low field region recorded in 5% TFA*d*-CD_3OD (500 MHZ).

toward a higher field, nearly identical to that of C2. The difference in 3-D conformation of **5a** between the acidic solution and the neutral solution could explain the suppression of the formation of the *Z*-isomers. In HBA (**4a**), the innermost caffeoyl residue (C1) never isomerized, and the ratio of the *Z*-isomer of C3 was the highest. The chemical shift of the α,β protons of C1 and C2 was toward the higher field compared with that of C3, indicating that the intramolecular stacking of C1 and C2 may be stronger than that of C3. Thus, C3 may easily isomerize.

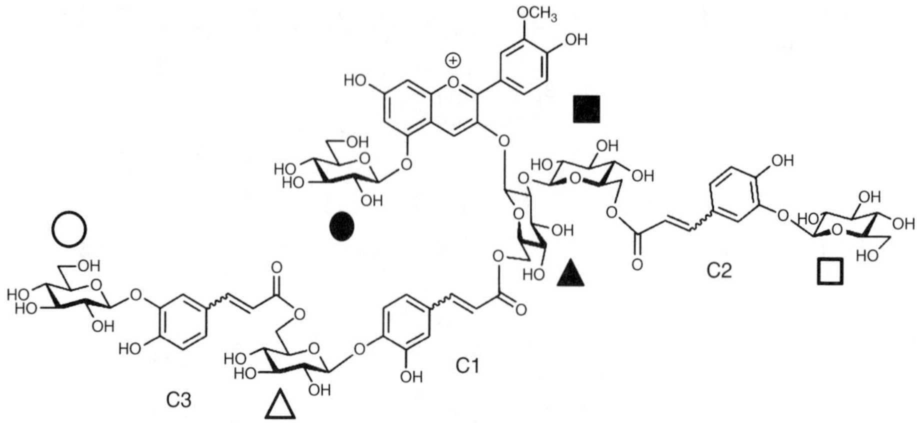

C1: *E*, C2: E, C3: *E*, HBA (**4a**)
C1: *E*, C2: *Z*, C3: *E*, **4b**
C1: *E*, C2: *E*, C3: *Z*, **4c**
C1: *E*, C2: *Z*, C3: *Z*, **4d**

Fig. 8.6 (Continued)

8.4.2 UV protective effect of polyacylated anthocyanins

As described in the introduction, solar radiation is essential for all the plants. However, it is also harmful; UV-B (280–320 nm) induces DNA damage. Polyacylated anthocyanins are known to be stable in dilute neutral aqueous solutions compared with simple anthocyanins. Interestingly, it has been reported that polyacylated anthocyanins under physiological conditions are more stable upon UV irradiation than in acidic conditions (Yoshida *et al.*, 2003b). Under dark conditions, tricaffeoyl anthocyanin, HBA (**4a**), was most stable in an acidic methanol solution. However, under UV irradiation, compound **4** was the most stable in aqueous media at a pH of 7.5 (Yoshida *et al.*, 2003b). The reason that polyacylated anthocyanins are more tolerant to UV under physiological conditions than in acidic conditions is due to a sophisticated energy quenching mechanism (Yoshida *et al.*, 2003b; Mori *et al.*, 2005). The intramolecular stacking conformation of polyacylated anthocyanins becomes stronger in neutral conditions than in acidic organic solvent, and in such a conformation, these pigments emit a strong fluorescence upon UV irradiation. For example, upon irradiation with UV-B, gentiodelphin (**5a**) emits fluorescence at a λ_{max} of 670 nm, and HBA (**4a**) emits fluorescence at λ_{max} of 652 nm. The intensity of the fluorescence of these pigments decreased with a decrease in the pH of the solution. Deacylated pigments in aqueous solutions at a pH of 7.5 emit very little fluorescence. These results indicate that the light energy absorbed with caffeoyl residues might be transferred to the anthocyanidin nucleus to emit a red fluorescence. These mechanisms may work efficiently via intramolecular stacking of the acyl residues onto the anthocyanidin nucleus. This could be reason why the polyacylated anthocyanins do not isomerize to the Z-form and are rather tolerant to UV-B radiation.

The stability and UV-resistance activity of the polyacylated anthocyanins must play an important biological role in protecting petal tissues from solar radiation,

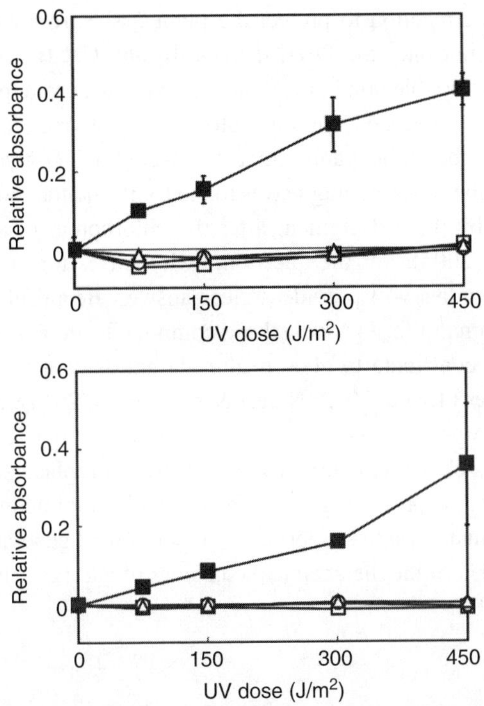

Fig. 8.7 Prevention of DNA lesions induced by broad UV-B with HBA solution screen (5 mM). ■: Control; □: pH 7.5; ○: 0.5% TFA aq. △: 0.5% TFA-MeOH. Upper: CPDs; lower: 6-4PPs.

especially the DNA protecting effect. The screening effects of HBA against UV-B induced damage were assessed using an epidermal cell mimicking experimental apparatus. A DNA solution was irradiated with UV-B through an HBA filter, and subsequently, the DNA damage was quantified as the amount of DNA lesions, cyclobutane pyrimidine dimers (CPDs), and (6-4)photoproducts (6-4PPs) using an enzyme-linked immunosorbent assay (ELISA, Fig. 8.7). The screening effect of the HBA against the monochromatic UV-B at 280 or 310 nm and broad UV-B was confirmed. The DNA damage was remarkably reduced to 5 mM in the HBA (**4a**). The effect of the protection was due to the absorption of UV-B radiation by the caffeoyl residue because the mixture of methyl caffeate with tris-deacyl HBA also showed the same effect. However, the stability during a long radiation may be different. HBA should be more stable, and the screening effect retains for longer time compared with methyl caffeate (Mori *et al.*, 2005).

8.5 Conclusions

Anthocyanin is stabilized and exhibits various colors via self-association, copigmentation, and metal complexation. All the mechanisms are caused by charge-transfer complex formation of the chromophore itself and/or the chromophore and intra- and/or intermolecular copigments. The anthocyanin complex dissipates fast and efficiently the excited energy into heat without a side reaction to protect the plant tissue from excess solar radiation. *In vitro* HBA inhibits not only the DNA damage by the UV-B irradiation but also the isomerization of the α,β-double bond of caffeate close to the chromophore, and the outside bonds are only partially isomerized. Autumn color turning of green leaves to red is responsible for production of the anthocyanin, almost a cyanidin 3-*O*-glycoside, and it allows nutrient retrieval from the senescencing leaves to the trunk and the roots to protect the cells from photo-oxidation by the red pigment after the chlorophyll disappears (Field *et al.*, 2001; Keskitalo *et al.*, 2005). For the question, "why leaves turn red in autumn" (Field *et al.*, 2001), further studies will provide a clear answer. Beautiful flowers enhance the spiritual richness of human life. In foods, these compounds are not only used to promote appetite, but they also contribute to good health via antioxidant activities by preventing lifestyle-related diseases (Tsuda, 2012; Nunez & Magnuson, 2014). Now, the compounds have attracted attention as potential functional materials in organic dye-sensitized solar cells (O'Regan & Grätzel, 1991; Cherepy *et al.*, 1997) to replace silicon cells. In food, polyacylated anthocyanin is promising as a safe and stable color additive and is promising for good health maintenance via the suppression of peroxidant generation. For its practical use, more research to determine the chemical character of anthocyanin is necessary.

References

Akavia, N. and Strack, D. (1980) High performance liquid chromatography of anthocyanidins as a new approach to study flower pigment genetics. *Zeitschrift für Naturforschung*, **35C**, 16–19.

Andersen, Ø.M. and Jordheim, M. (2006) The anthocyanins, in *Flavonoids, Chemistry, Biochemistry and Applications* (eds Ø.M. Andersen and K.R. Markham), CRC Press, Boca Raton, pp. 471–551.

Asen, S., Stewart, R.N. and Norris, K.H. (1972) Co-pigmentation of anthocyanins in plant tissues and its effect on color. *Phytochemistry*, **11**, 1139–1144.

Barber, J. and Andersson, B. (1992) Too much of a good thing: light can be bad for photosynthesis. *Trends in Biochemical Sciences*, **17**, 61–66.

Bassman, H.J. (2004) Ecosystem consequences of enhanced solar ultraviolet radiation: secondary plant metabolites as mediators of multiple trophic interactions in terrestrial plant communities. *Photochemistry and Photobiology*, **79**, 382–398.

Brouillard, R. (1982) Chemical structure of anthocyanins, in *Anthocyanins as Food Colors* (ed P. Markakis), Academic Press, New York, pp. 1–40.

Brouillard, R. (1988) Flavonoids and flower colour, in *The Flavonoids Advances in Research since 1980* (ed J.B. Harborne), Chapman & Hall, London, pp. 525–538.

Brouillard, R. and Dangles, O. (1994) Flavonoids and flower colour, in *The Flavonoids Advances in Research since 1986* (ed J.B. Harborne), Chapman & Hall, London, pp. 565–588.

Brouillard, R., Chassaing, S., Isorez, G., Kueny-Stotz, M. and Figueireido, P. (2010) The visible flavonoids or anthocyanins: from research to applications, in *Research Advances in Polyphenol Research*, vol. **2** (eds C. Santos-Buelga, M.T. Escriba-Bailon and V. Lattanzio), Wiley-Blackwell, Oxford, pp. 1–22.

Cheminat, A. and Brouillard, R. (1986) PMR investigation of 3-*O*-(β-D-glucosyl)malvidin structural transformations in aqueous solutions. *Tetrahedron Letters*, **27**, 4457–4460.

Cherepy, N.J., Smestad, G.P., Grätzel, M. and Zhang, J.Z. (1997) Ultrafast electron injection: implications for a photoelectrochemical cell utilizing an anthocyanin dye-sensitized TiO2 nanocrystalline electrode. *Journal of Physical Chemistry B*, **101**, 9342–9351.

Costa, D., Galvao, M., Adelino, M. *et al.* (2015) Photochemistry of the hemiketal form of anthocyanins and its potential role in plant protection from UV-B radiation. *Tetrahedron*, **71**, 3157–3162.

Davis, M.K. (2004) An introduction to plant pigments in biology and commerce, in *Plant Pigments and Their Manipulation*, Annual Plant Reviews 14 (ed K.M. Davis), Blackwell Publishing & CRC Press, Oxford, pp. 1–22.

Fernandes, A., Bras, N.F., Mateus, N. and Freitas, V. (2015) A study of anthocyanin self-association by NMR spectroscopy. *New Journal of Chemistry*, **39**, 2602–2611.

Ferreira da Silva, P., Lima, C.J., Freitas, A.A., Shimizu, K., Maçanita, L.A. and Quina, H.F. (2005) Charge-transfer complexation as a general phenomenon in the copigmentation of anthocyanins. *Journal of Physical Chemistry A*, **109**, 7329–7338.

Ferreira da Silva, P., Paulo, L., Barbafine, A., Elisei, F., Quina, H.F. and Maçanita, L.A. (2012) Photoprotection and the photophysics of acylated anthocyanins. *Chemistry A European Journal*, **18**, 3736–3744.

Field, T.S., Lee, D.W. and Holbrook, N.M. (2001) Why leaves turn red in autumn. The role of anthocyanins in senescing leaves of red-osier dogwood. *Plant Physiology*, **127**, 566–574.

Gago, S., Basílio, N., Moro, J.A. and Pina, F. (2015) Flavylium based dual photochromism: addressing *cis–trans* isomerization and ring opening-closure by different light inputs. *Chemical Communications*, **51**, 7349–7351.

Gavara, R., Petrov, V., Quintas, A. and Pina, F. (2013) Circular dichroism of anthocyanidin 3-glucoside self-aggregates. *Phytochemistry*, **88**, 92–98.

Goto, T. and Kondo, T. (1991) Structure and molecular stacking of anthocyanins-flower color variation. *Angewandte Chemie International Edition in English*, **30**, 17–33.

Goto, T., Takase, S. and Kondo, T. (1978) PMR spectra of natural acylated anthocyanins. Determination of stereostructure of awobanin, shisonin, and violanin. *Tetrahedron Letters*, **19**, 2413–2416.

Goto, T., Kondo, T., Tamura, H., Imagawa, H., Iino, A. and Takeda, K. (1982) Structure of gentiodelphin, an acylated anthocyanin isolated from *Gentiana makinoi*, that is stable in dilute aqueous solution. *Tetrahedron Letters*, **23**, 3695–3698.

Goto, T., Kondo, T., Tamura, H. and Kawahori, K. (1983a) Structure of platyconin, a diacylated anthocyanin isolated from the Chinese bell-flower *Platycodon grandiflorum. Tetrahedron Letters*, **24**, 2181–2184.

Goto, T., Kondo, T., Tamura, H. and Takase, S. (1983b) Structure of malonylawobanin, the real anthocyanin present in blue-colored flower petals of *Commelina communis. Tetrahedron Letters*, **24**, 4863–4866.

Goto, T., Kondo, T., Kawai, T. and Tamura, H. (1984) Structure of cinerarin, a tetra-acylated anthocyanin isolated from the blue garden cineraria, *Senecio cruentus. Tetrahedron Letters*, **25**, 6021–6024.

Goto, T., Tamura, H., Kawai, T., Hoshino, T., Harada, N. and Kondo, T. (1986) Chemistry of metalloanthocyanins. *Annals of the New York Academy of Sciences*, **471**, 155–173.

Harborne, B.J. and Smith, M.D. (1978) Correlations between anthocyanin chemistry and pollination ecology in the polemoniaceae. *Biochemical Systematics and Ecology*, **6**, 127–130.

Harborne, J.B. and Williams, C.A. (2000) Advances in flavonoid research since 1992. *Phytochemistry*, **55**, 481–504.

Honda, T. and Saito, N. (2012) Recent progress in the chemistry of polyacylated anthocyanins as flower color pigments. *Heterocycles*, **56**, 633–692.

Hoshino, T. (1992) Anthocyanin self-aggregates. Part 6. An approximate estimate of self-association constants and the self-stacking conformation of malvin quinonoidal bases studied by proton NMR. *Phytochemistry*, **31**, 647–653.

Hoshino, T., Matsumoto, U. and Goto, T. (1980a) The stabilizing effect of the acyl group on the co-pigmentation of acylated anthocyanins with *C*-glucosylflavones. *Phytochemistry*, **19**, 663–667.

Hoshino, T., Matsumoto, U. and Goto, T. (1980b) Evidence of the self-association of anthocyanins. I. Circular dichroism of cyanin anhydrobase. *Tetrahedron Letters*, **21**, 1751–1754.

Hosokawa, K., Fukunaga, Y., Fukushi, E. and Kawabata, J. (1995) Seven acylated anthocyanins in the blue flowers of Hyacinthus orientalis. *Phytochemistry*, **38**, 1293–1298.

Houbiers, C., Lima, J.C., Macanita, A.L. and Santos, H. (1998) Color stabilization of malvidin 3-glucoside: self-aggregation of the flavylium cation and copigmentation with the Z-chalcone form. *Journal of Physical Chemistry B*, **102**, 3578–3585.

Idaka, E., Yamakita, H., Ogawa, T., Kondo, T., Yamamoto, M. and Goto, T. (1987a) Structure of three diacylated anthocyanins from red cabbage, *Brassica oleracea*. *Chemistry Letters*, 1213–1216.

Idaka, E., Ogawa, T., Kondo, T. and Goto, T. (1987b) Isolation of highly acylated anthocyanins from commelinaceae plants, *Zebrina pendula*, *Rhoeo spathacea* and *Setcreasea purpurea*. *Agricultural and Biological Chemistry*, **51**, 2215–2220.

Keskitalo, J., Bergquist, G., Gardestroem, P. and Jansson, S. (2005) A cellular timetable of autumn senescence. *Plant Physiology*, **139**, 1635–1648.

Kondo, T., Nakane, Y., Tamura, H., Goto, T. and Eugster, C.H. (1985) Structure of monardaein, a bis-malonylated anthocyanin isolated from golden balm, *Monarda didyma*. *Tetrahedron Letters*, **26**, 5879–5882.

Kondo, T., Kawai, T., Tamura, H. and Goto, T. (1987) Structure determination of heavenly blue anthocyanin, a complex monomeric anthocyanin from the morning glory *Ipomoea tricolor*, by means of the negative NOE method. *Tetrahedron Letters*, **28**, 2273–2276.

Kondo, T., Yoshida, K., Yoshikane, M. and Goto, T. (1991) Mechanism for color development in purple flower of *Commelina communis*. *Agricultural and Biological Chemistry*, **55**, 2919–2921.

Kondo, T., Yoshida, K., Nakagawa, A., Kawai, T., Tamura, H. and Goto, T. (1992) Structural basis of blue-color development in flower petals from *Commelina communis*. *Nature*, **358**, 515–518.

Leydet, Y., Gavara, R., Petrov, V. *et al.* (2012) The effect of self-aggregation on the determination of the kinetic and thermodynamic constants of the network of chemical reactions in 3-glucoside anthocyanins. *Phytochemistry*, **83**, 125–135.

Lima, J.C., Abreu, I., Brouillard, R. and Macanita, A. (1998) Kinetics of ultra-fast excited state proton transfer form 7-hydroxy-4-methylflavylium chloride to water. *Chemical Physics Letters*, **298**, 189–195.

Maçanita, A.L., Moreira, P.F., Jr, Lima, J.C., Quina, F.H., Yihwa, C. and Vautier-Giongo, C. (2002) Proton transfer in anthocyanins and related flavylium salts. Determination of ground-state rate constants with nanosecond laser flash photolysis. *Journal of Physical Chemistry A*, **106**, 1248–1255.

Madronich, S., McKenzie, R.L., Bjorn, L.O. and Caldwell, M.M. (1998) Changes in biologically active ultraviolet radiation reaching the Earth's surface. *Journal of Photochemistry and Photobiology*, **46**, 5–19.

Matera, R., Gabbanini, S., De Nicola, G.R., Iori, R., Petrillo, G. and Valgimigli, L. (2012) Identification and analysis of isothiocyanates and new acylated anthocyanins in the juice of *Raphanus sativus* cv. Sango sprouts. *Food Chemistry*, **133**, 563–572.

Miller, R., Owens, J.S. and Rorslett, B. (2011) Plants and colour: flowers and pollinations. *Optics and Laser Technology*, **43**, 282–294.

Moreira, P.F., Jr, Giestas, L., Yihwa, C. *et al.* (2003) Ground- and excited-state proton transfer in anthocyanins: from weak acids to superphotoacids. *Journal of Physical Chemistry A*, **107**, 4203–4210.

Mori, M. (2005) Chemical studies on UV-protecting activities of polyacylated anthocyanins. Doctoral Thesis. Nagoya University.

Mori, M., Yoshida, K., Ishigaki, Y. *et al.* (2005) UV-B protecting effect of a polyacylated anthocyanin, HBA, in flower petals of the blue morning glory, *Ipomoea tricolor*, cv. Heavenly blue. *Bioorganic and Medicinal Chemistry*, **13**, 2015–2020.

Mori, M., Kondo, T., Toki, K. and Yoshida, K. (2006) Structure of anthocyanin from the blue petals of *Phacelia campanularia* and its blue flower color development. *Phytochemistry*, **67**, 622–629.

Mori, M., Ito, D., Miki, N., Kondo, T. and Yoshida, K. (2014) Structure of tecophilin, a tri-caffeoylanthocyanin from the blue petals of *Tecophilaea cyanocrocus*, and the mechanism of blue color development. *Tetrahedron*, **70**, 8657–8664.

Nerdal, W. and Andersen, Ø.M. (1991) Evidence for self-association of the anthocyanin petanin in acidified, methanolic solution using two-dimensional nuclear Overhauser enhancement NMR experiments and distance geometry calculations. *Phytochemical Analysis*, **2**, 263–270.

Nilsson, E. (1973) Flavylium compounds. 8. Application of proton NMR spectroscopy to the analysis of anthocyanin pigments. *Chemica Scripta*, **4**, 49–55.

Nunez, M.F. and Magnuson, B.A. (2014) Anthocyanins in health and disease prevention, in *Anthocyanins in Health and Disease* (eds T.C. Wallace and M.M. Giusti), CRC Press, Boca Raton, pp. 1–11.

O'Regan, B. and Gratzel, M. (1991) A low-cost, high-efficiency solar cell based on dye-sensitized colloidal TiO_2 films. *Nature*, **353**, 737–740.

Perez-Barrales, R., Vargas, P. and Arroyo, J. (2006) New evidence for the Darwinian hypothesis of heterostyly: breeding systems and pollinators in Narcissus sect. Apodanthi. *New Phytologist*, **171**, 553–567.

Quina, H.F., Moreira, P.F., Jr, Vautier-Giongo, C. *et al.* (2009) Photochemistry of anthocyanins and their biological role in plant tissues. *Pure and Applied Chemistry*, **81**, 1687–1694.

Robinson, M.G. and Robinson, R. (1931) A survey of anthocyanins. 1. *Biochemistry Journal*, **25**, 1687–1705.

Rodrigues, F.R., Silva, F.P., Shimizu, K. *et al.* (2009) Ultrafast internal conversion in a model anthocyanin–polyphenol complex: implications for the biological role of anthocyanins in vegetative tissues of plants. *Chemistry A European Journal*, **15**, 1397–1402.

Saito, N., Tatsuzawa, F., Yazaki, Y., Shigihara, A. and Honda, T. (2007) 7-Polyacylated delphinidin 3,7-diglucosides from the blue flowers of *Leschenaultia* cv. Violet Lena. *Phytochemistry*, **68**, 673–679.

Shimizu, Y., Imada, T., Huili, Z., Tanaka, R., Ohno, T. and Shimomura, K. (2010) Identification of novel poly-acylated anthocyanins from *Gynura bicolor* leaves and their antioxidative activity. *Food Science and Technology Research*, **16**, 479–486.

Shiono, M., Matsugaki, N. and Takeda, K. (2005) Structure of the blue cornflower pigment. *Nature*, **436**, 791.

Shiono, M., Matsugaki, N. and Takeda, K. (2008) Structure of commelinin, a blue complex pigment from the blue flowers of *Commelina communis*. *Proceedings of the Japan Academy, Series B*, **84**, 452–456.

Strack, D., Akavia, N. and Reznik, H. (1980) High performance liquid chromatographic identification of anthocyanins. *Zeitschrift für Naturforschung*, **35C**, 533–538.

Takahashi, S. and Badger, M.R. (2011) Photoprotection in plants: a new light on photosystem II damage. *Trends in Plant Science*, **16**, 53–60.

Tatsuzawa, F., Saito, N., Toki, K., Shinoda, K., Shigihara, A. and Honda, T. (2008) Triacylated cyanidin 3-(3^X-glucosylsambubioside)-5-glucosides from the flowers of *Malcolmia maritima*. *Phytochemistry*, **69**, 1029–1036.

Tatsuzawa, F., Toki, K., Ohtani, Y. *et al.* (2014) Floral pigments from the blue flowers of Nemophila menziesii "insignis blue" and the purple flower of its variants. *Journal of the Japanese Society for Horticultural Science*, **83**, 259–266.

Tevini, M. (1994) UV-B radiation in the biosphere, in *Stratosheric Ozone Depletion* (ed M. Tevini), Springer-Verlag, Berlin, pp. 37–62.

Toki, K., Saito, N., Kawano, K., Lu, T.S., Shigihara, A. and Honda, T. (1994) An acylated delphinidin glycoside in the blue flowers of *Evolvulus pilosus*. *Phytochemistry*, **36**, 609–612.

Toki, K., Saito, N. and Honda, T. (1998) Acylated anthocyanins from the blue-purple flowers of *Triteleia bridgesii*. *Phytochemistry*, **48**, 729–732.

Tsuda, T. (2012) Dietary anthocyanin-rich plants: biochemical basis and recent progress in health benefits studies. *Molecular Nutrition & Food Research*, **56**, 159–170.

Wilbert, S.M., Schemske, D.W. and Bradshaw, H.D., Jr (1997) Floral anthocyanins from two monkeyflower species with different pollinators. *Biochemical Systematics and Ecology*, **25**, 437–443.

Wilkinson, M., Sweeny, J.G. and Iacobucci, G.A. (1977) High-pressure liquid chromatography of anthocyanins. *Journal of Chromatography*, **132**, 349–351.

Williams, M., Hrazdina, G., Wilkinson, M.M., Sweeny, J.G. and Jacobucci, G.A. (1978) High-pressure liquid chromatographic separation of 3-glucosides, 3,5-diglucosides, 3-(6-*O*-*p*-coumaryl) glucosides and 3-(6-*O*-*p*-coumarylglucoside)-5-glucosides of anthocyanidins. *Journal of Chromatography*, **155**, 389–398.

Wu, X. and Prior, L.R. (2005) Identification and characterization of anthocyanins by high-performance liquid chromatography-electrospray ionization-tandem mass spectrometry on common foods in the United States: vegetables, nuts, and grains. *Journal of Agricultural and Food Chemistry*, **53**, 3101–3113.

Yoshida, K. (1992) Molecular stacking of anthocyanins with fine structure recognition. Doctoral thesis. Nagoya University.

Yoshida, K., Kondo, T., Kameda, K. and Goto, T. (1990) Structure of anthocyanins isolated from purple leaves of *Perilla ocimoides* L. var. *crispa* Benth and their isomerization by irradiation of light. *Agricultural and Biological Chemistry*, **54**, 1745–1751.

Yoshida, K., Kondo, T. and Goto, T. (1991) Unusually stable monoacylated anthocyanin from purple yam *Dioscorea alata*. *Tetrahedron Letters*, **32**, 5579–5580.

Yoshida, K., Kondo, T. and Goto, T. (1992) Intramolecular stacking conformation of gentiodelphin, a diacylated anthocyanin from *Gentiana makinoi*. *Tetrahedron*, **48**, 4313–4326.

Yoshida, K., Toyama, Y., Kameda, K. and Kondo, T. (2000) Contribution of each caffeoyl residue of the pigment molecule of gentiodelphin to blue color development. *Phytochemistry*, **54**, 85–92.

Yoshida, K., Okuno, R., Kameda, K., Mori, M. and Kondo, T. (2003a) Influence of *E*,*Z*-isomerization and stability of acylated anthocyanins under the UV irradiation. *Biochemical Engineering Journal*, **14**, 163–169.

Yoshida, K., Mori, M., Kawachi, M., Okuno, R., Kameda, K. and Kondo, T. (2003b) A UV-B resistant polyacylated anthocyanin, HBA, from blue petals of morning glory. *Tetrahedron Letters*, **44**, 7875–7880.

Yoshida, K., Mori, M. and Kondo, T. (2009) Blue flower color development by anthocyanins: from chemical structure to cell physiology. *Natural Product Reports*, **26**, 884–915.

Yoshida, K., Oyama, K. and Kondo, T. (2012) Chemistry of flavonoids in color development, in *Recent Advances in Polyphenol Research*, vol. **3** (eds V. Cheynier, P. Sarni-Manchado and S. Quideau), Wiley-Blackwell Publishing, Chichester, pp. 99–129.

Zheng, J., Ding, C., Wang, L. *et al.* (2011) Anthocyanins composition and antioxidant activity of wild *Lycium ruthenicum* Murr. From Qinghai-Tibet Plateau. *Food Chemistry*, **126**, 859–865.

Chapter 9

The Involvement of Anthocyanin-Rich Foods in Retinal Damage

Kenjirou Ogawa and Hideaki Hara

Abstract: Several retinal diseases, such as age-related macular degeneration (AMD) and diabetic retinopathy, are partly caused by vascular endothelial growth factor (VEGF)-induced angiogenesis. Recently, anti-VEGF drugs, such as pegaptanib, ranibizumab, and aflibercept, have been used as clinical treatments. However, current anti-VEGF drugs do not have sufficient efficacy to be used as effective therapies for the aforementioned retinal diseases; therefore, the development of more effective novel drugs is needed. Dietary supplements have been shown to be efficacious in the prophylactic treatment of retinal diseases. We previously showed that anthocyanin-rich foods such as bilberry, purple rice, and maqui berry possess preventative effects when tested in various retinal disease models. In this chapter, we discuss the effects of anthocyanin-rich foods in retinal disease models and their potential for clinical applications.

Keywords: anthocyanin, retinal diseases, bilberry, purple rice, maqui berry.

9.1 Introduction

The posterior eye diseases such as glaucoma, diabetic retinopathy, retinitis pigmentosa, and age-related macular degeneration (AMD), interfere with our quality of life. Posterior eye diseases are caused by many kinds of stress factors such as oxidative stress, endoplasmic reticulum (ER) stress, aging, genetic mutation, hyperglycemia, angiogenesis, smoking, and various light-related damage (Fig. 9.1). Some antioxidants such as polyphenols have been reported to possess preventive effects against posterior eye disease (Kubota *et al.*, 2009; Chen *et al.*, 2014).

Anthocyanins are the natural pigments present in almost all plants, which exhibit red, blue, and purple colors. The fruits and vegetables accumulate high levels of anthocyanins in order to protect themselves from ultraviolet light; similarly, exposure to a large amount

Recent Advances in Polyphenol Research, First Edition. Edited by Kumi Yoshida, Véronique Cheynier and Stéphane Quideau.

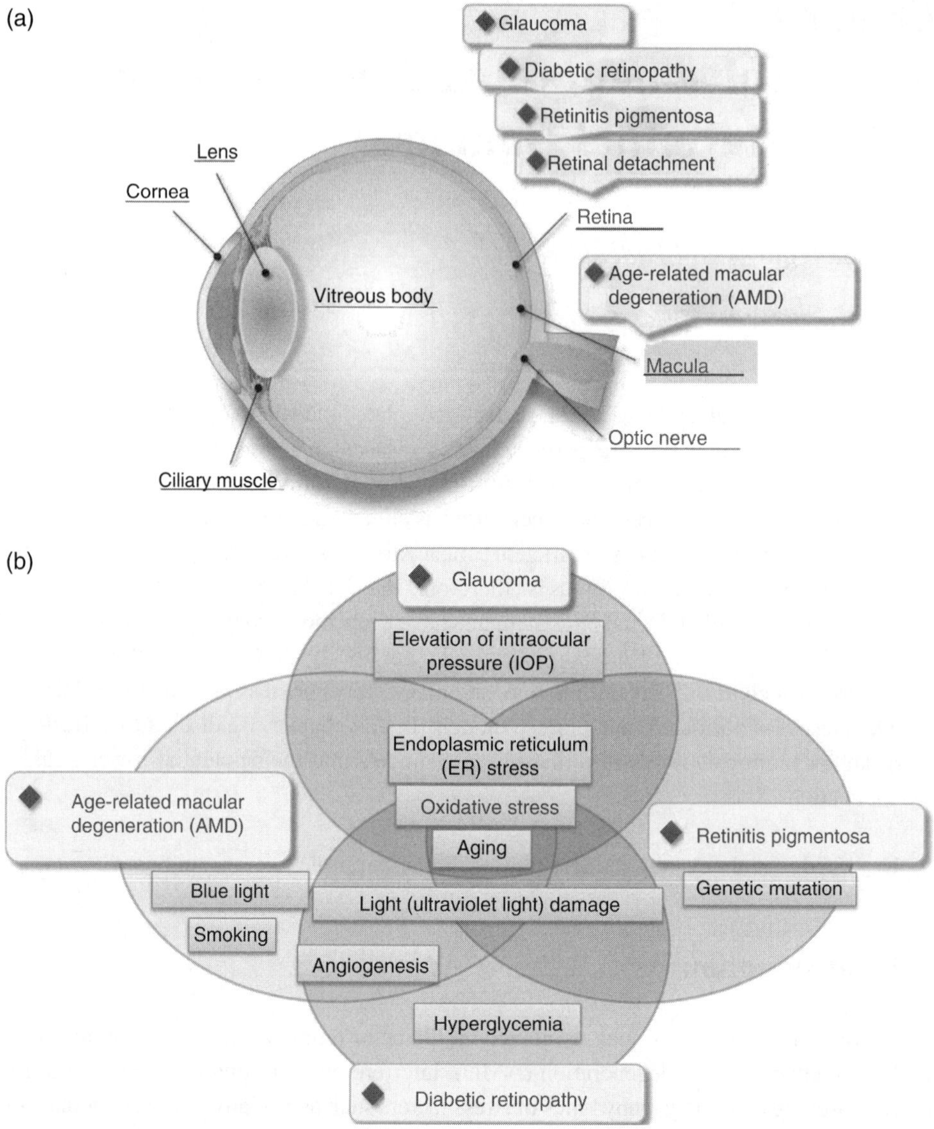

Fig. 9.1 Image showing posterior eye diseases and their common causes. (a) Image of eye structure and each posterior eye disease. (b) Mapping common causes of posterior eye diseases. (*See insert for color representation of the figure*)

of ultraviolet light stimulates further anthocyanin synthesis. The fruits and vegetables that contain high levels of anthocyanins include grape, blueberry, purple potato, and eggplant. Anthocyanins are a type of polyphenols known to possess a large antioxidative effect and high levels of safety, and therefore, they have been used to induce health benefits through health foods and dietary supplements. Examples include blueberry, bilberry, black currant, and purple rice, which are all used for maintaining eye health in Japan. We previously reported that bilberry, purple rice, and maqui berry anthocyanins maintained visual function by preventing retinal damage (Matsunaga *et al.*, 2009, 2010a, 2010b; Tanaka *et al.*, 2011, 2012, 2013; Ogawa *et al.*, 2013, 2014). Herein, we review some of our previous studies on the anthocyanins and anthocyanin-containing foods and their effect on eye health.

9.2 Anthocyanin-rich foods for eye health

The purpose of our study was to investigate the beneficial effects of anthocyanin-rich foods such as bilberry, purple rice, and maqui berry in maintaining eye health (Fig. 9.2).

Bilberry (*Vaccinium myrtillus* L.), a member of the Ericaceae family, grows in the forests of Northern Europe. Bilberry contains 15 different anthocyanins composed of a combination of five different anthocyanidins (delphinidin, cyanidin, malvidin, petunidin, and peonidin) with one of the three sugars (glucose, galactose, and arabinose). Bilberry extract possesses potent antioxidant properties (Ogawa *et al.*, 2008), stabilizes collagen fibers, promotes collagen biosynthesis (Morazzoni & Bombardelli, 1996), inhibits platelet aggregation (Morazzoni & Megistretti, 1990), and improves vascular tone, blood flow, and capillary action (Lietti *et al.*, 1976; Colantunoni *et al.*, 1991). Recently, the effects of bilberry in combatting insulin resistance and hyperglycemia were reported (Takikawa *et al.*, 2010). Hence, bilberry has been utilized as a popular edible aid or supplement.

Purple rice (*Oryza sativa* L. (Gramineae)) is widely cultivated in South East Asia, China, and Japan. Purple rice grain contains four types of anthocyanins: cyanidin-3-glucoside, cyanidin-3-rhamnoglucoside, peonidin-3-glucoside, and malvidin-3-galactoside. Previously, it was reported that purple rice possesses antioxidative (Ichikawa *et al.*, 2001), anti-inflammatory (Hu *et al.*, 2003), and hepatoprotective effects (Hou *et al.*, 2010), as well as demonstrates α-glucosidase inhibitory effects (Yao *et al.*, 2010).

Maqui berry (*Aristotelia chilensis* (Molina) Stuntz) is a plant of the Elaeocarpaceae family and is cultivated in Patagonia, Chile. Since ancient times, Mapuche Indians have used maqui berry as a food, juice, alcoholic drink, and medicine. Maqui berry contains eight types of antioxidative anthocyanins, including delphinidin-3-sambioside-5-glucoside, delphinidin-3,5-diglucoside, cyanidin-3-sambioside-5-glucoside, and cyanidin-3,5-glucoside (Escribano-Bailon *et al.*, 2006). Therefore, maqui berry possesses a large antioxidative

Fig. 9.2 Anthocyanin-rich foods used in our experiments. (*See insert for color representation of the figure*)

effect (Ruiz *et al.*, 2010). Furthermore, the anti-inflammatory (Schreckinger *et al.*, 2010), antidiabetic (Rojo *et al.*, 2012), antiatherogenic, and cardioprotective (Cespedes *et al.*, 2008) effects of maqui berry have already been published.

However, the detailed mechanism underpinning the improvements in visual function remains unclear. Thus, we investigated the effect of bilberry, purple rice, and maqui berry in maintaining eye health by using *in vivo* and *in vitro* studies.

9.3 Experimental models to mimic eye diseases and the effect of anthocyanin-rich foods

The eye diseases targeted in our study were glaucoma, AMD, and diabetic retinopathy. First, we established experimental models to replicate each eye disease; subsequently, we investigated the protective effects of anthocyanin-rich bilberry extract (BBE), purple rice extract (PRE), and maqui berry extract (MBE) in each model. BBE containing mostly carbohydrate, small amounts of organic acids, and an average of 36% total anthocyanins, including 15 different anthocyanins composed of 5 anthocyanidins (delphinidin, cyanidin, malvidin, petunidin, and peonidin) and 3 sugars (glucose, galactose, and arabinose) was provided by Wakasa Seikatsu Co., Ltd (Kyoto, Japan). PRE containing 3.1% of anthocyanins (cyanidin-3-glucoside and peonidin-3-glucoside), other phenolic compounds, including flavones (apigenin, diosmetin, and luteolin), flavonols (quercetin and isorhamnetin) and chlorogenic acids, and small amount of carotenoids (Pereira-Caro *et al.*, 2013), and MBE containing mostly carbohydrates and 35.4% of anthocyanins (delphinidin-3-glucoside, delphinidin-3-sambubioside, delphinidin-3-sambubioside-5-glucoside, delphinidin-3,5-diglucoside, cyanidin-3-glucoside, cyanidin-3-sambubioside, cyanidin-3-sambubioside-5-glucoside, and cyanidin-3,5-diglucoside) were provided by Oryza Oil & Fat Chemical Company, Ltd (Aichi, Japan).

9.3.1 3-(4-Morpholinyl) sydnonimine hydrochloride (SIN-1)-induced and N*-methyl-*D*-aspartate receptor (NMDA)-induced retinal ganglion cell damage models to mimic glaucoma* in vitro *and* in vivo

Glaucoma is caused by damage to the retinal ganglion cells due to ocular hypertension and oxidative stress. Thus, the key for prevention of glaucoma is to protect the retinal ganglion cells from damage. In previous *in vitro* studies, a high concentration of SIN-1 was found to induce murine retinal ganglion cell 5 (RGC-5) death (Eissa *et al.*, 2006) while NMDA-receptor-mediated neurotoxicity was reported to depend in part on the generation of nitric oxide and superoxide anions, which as mentioned earlier react to form peroxynitrite (Bonfoco *et al.*, 1995). Treatment with SIN-1 at 0.5 mM induced death in approximately 40% of RGC-5; however, administration of BBE at 10 μg/ml significantly inhibited SIN-1-induced RGC-5 cell death (~90% of the remaining cells were viable) (Matsunaga *et al.*, 2009). Anthocyanidins (cyanidin, delphinidin, and malvidin) in BBE also significantly inhibited the SIN-1-induced RGC-5 cell death. Furthermore, BBE, cyanidin, delphinidin, and malvidin significantly inhibited the SIN-1-induced reactive oxygen species (ROS) production in RGC-5 cells. This was determined using chloromethyl-2′,7′-dichlorodihydrofluorescein diacetate (CM-H_2DCFDA) fluorescent staining and lipid peroxidation in mouse forebrain homogenate and thiobarbituric acid reactive substance production (TBARS) method (Matsunaga *et al.*, 2009).

According to previous *in vivo* studies, intravitreal injection of NMDA induces RGC damage in the mouse retina, while the formation of superoxide, nitric oxide, and peroxynitrite plays a key role in NMDA-induced neurotoxicity in the rat retina (Matsunaga *et al.*, 2009). In our study, intravitreal injection of NMDA at 5 nmol/eye significantly decreased both the cell count in ganglion cell layer (GCL) and the thickness of inner plexiform layer (IPL) in the mouse retina (Fig. 9.3a). However, intravitreal injection of BBE at 100 μg/eye (coinjected with NMDA at 5 nmol/eye) significantly inhibited the reduction of GCL number and IPL thickness. The GCL cell counts in NMDA-injected mice and NMDA plus BBE (at 100 μg/eye)-injected mice were 66.5 ± 5.5 and 86.8 ± 4.4, respectively (Fig. 9.3b).

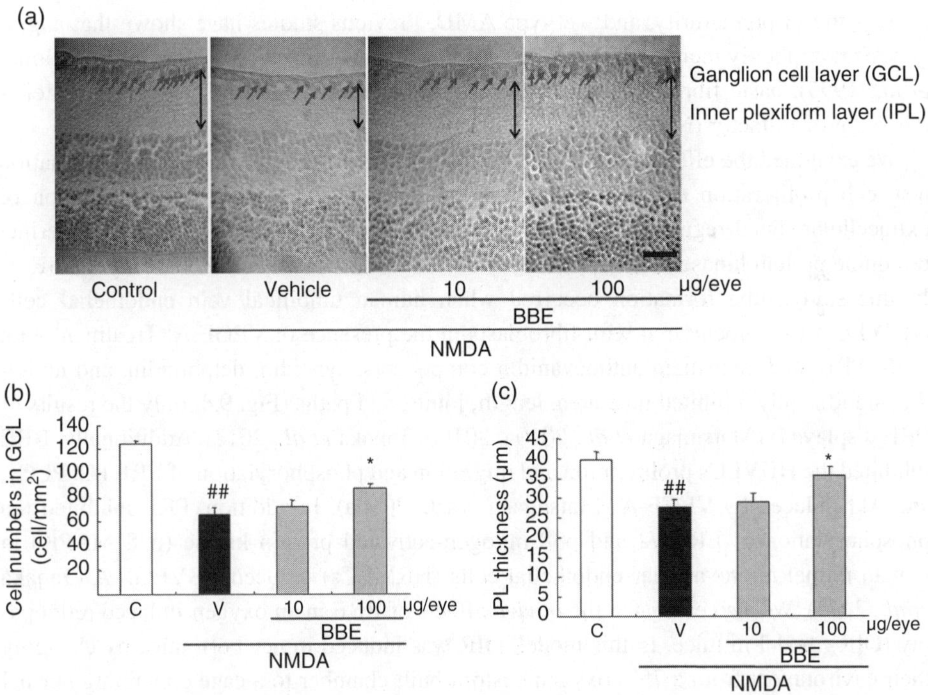

Fig. 9.3 Effect of bilberry extract (BBE) on *N*-methyl-D-aspartate receptor (NMDA)-induced retinal damage in mice. Hematoxylin and eosin staining of sections (thickness 5 μm) obtained from mice 7 days after intravitreal injection of NMDA. (a) Representative photographs show nontreated control retina, NMDA-injected vehicle-treated retina, retina treated with NMDA injection plus BBE at 10 μg/eye, and retina treated with NMDA injection plus BBE at 100 μg/eye. Horizontal bar represents 25 μm. Vertical bar and arrows indicate inner plexiform layer (IPL) and retinal ganglion cells (RGC), respectively. (b and c) Cell numbers in RGC and thickness of IPL. Data are shown as mean ± SEM (n = 6–11). C, control; V, vehicle; ##$p < 0.01$ versus control; *$p < 0.05$ versus vehicle. These results were cited in Matsunaga *et al.* (2009). (*See insert for color representation of the figure*)

The values obtained for IPL thicknesses in NMDA-injected mice and NMDA plus BBE (at 100 μg/eye)-injected mice were 27.5 ± 2.0 and $35.8 \pm 2.2\,\mu m$, respectively (Fig. 9.3c). Furthermore, the intravitreal injection of BBE at 100 μg/eye inhibited the NMDA-induced increase in (TDT)-mediated dUTP-digoxigenin nick end labeling (TUNEL)-positive cells, which are apoptotic cells in the retina (Matsunaga *et al.*, 2009).

These results showed that BBE and its main anthocyanidins possess a neuroprotective effect through inhibition of ROS production. BBE may therefore have potential as an agent to be used in the prevention of retinal degenerative diseases such as glaucoma.

9.3.2 Vascular endothelial growth factor (VEGF)-induced angiogenesis models that mimic diabetic retinopathy in vitro *and* in vivo

Angiogenesis is the process by which new blood vessels are formed from pre-existing ones. In childhood, angiogenesis is necessary for the development of blood vessels in the retina. However, abnormal angiogenesis, which can be caused by hyperglycemia, vessel infarct, and hypoxia, subsequently causes retinal diseases such as diabetic retinopathy, retinopathy of prematurity, and wet-type AMD. Previous studies have shown that angiogenesis is explicitly increased in response to several growth factors, such as VEGF (Tamar *et al.*, 1995), basic fibroblast growth factor (Gospodarowicz *et al.*, 1983), and platelet-derived growth factor (Napoleon *et al.*, 1992).

We examined the effects of BBE and PRE in a range of *in vitro* studies: tube formation test, cell proliferation test, and cell migration assay. We measured phosphorylation of extracellular signal-regulated kinase 1/2 (ERK1/2), phospholipase Cγ (PLCγ), and serine/threonine protein kinase family protein kinase B (Akt), which were induced by VEGF-A. In this study, tube formation occurred when human umbilical vein endothelial cells (HUVECs) were cocultured with fibroblasts in the presence of VEGF-A. Treatment with BBE, PRE, and their main anthocyanidin components, cyanidin, delphinidin, and malvidin, significantly inhibited tube area, length, joints, and paths (Fig. 9.4, only the results of PRE displayed) (Matsunaga *et al.*, 2010a, 2010b; Tanaka *et al.*, 2012). Additionally, BBE inhibited the HUVECs proliferation and migration and phosphorylation of ERK1/2, PLCγ, and Akt induced by VEGF-A (Matsunaga *et al.*, 2010a). In addition, PRE inhibited the phosphorylation of ERK1/2 and p38 mitogen-activated protein kinase (p38 MAPK) in human retinal microvascular endothelial cells (HRMECs) induced by VEGF-A (Tanaka *et al.*, 2012). We also evaluated the *in vivo* effects of BBE in an oxygen-induced retinopathy (OIR) model in mice. In this model, OIR was induced in newborn mice by changing their environment from a 75% oxygen custom-built chamber to a cage containing normal room air. Intravitreal administration of BBE significantly inhibited the area of the neovascular tufts (Fig. 9.5) (Matsunaga *et al.*, 2010a).

We suggested that BBE and PRE containing anthocyanins inhibited angiogenesis both *in vitro* and *in vivo*, and their effect may be partly due to reductions in vascular endothelial cell proliferation and migration through inhibition of the phosphorylation of both ERK1/2 and Akt. Our findings indicated that the anthocyanins contained in BBE and PRE may have the potential to prevent retinal angiogenetic diseases such as diabetic retinopathy.

9.3.3 Light-induced retinal damage models to mimic AMD in vitro *and* in vivo

AMD is known to be one of the leading causes of blindness in the United States. There are two types of AMD: (i) wet-type AMD, which is caused mainly by choroidal neovascularization, and (ii) dry-type AMD caused by accumulation of drusen, which may be due to failure of waste product disposal by the eye; these waste products are produced by degradation of photoreceptor cells following light exposure in the retinal pigment epithelial layer of the macular area. Thus, the prevention of retinal photoreceptor degradation is a key factor in the prevention of AMD.

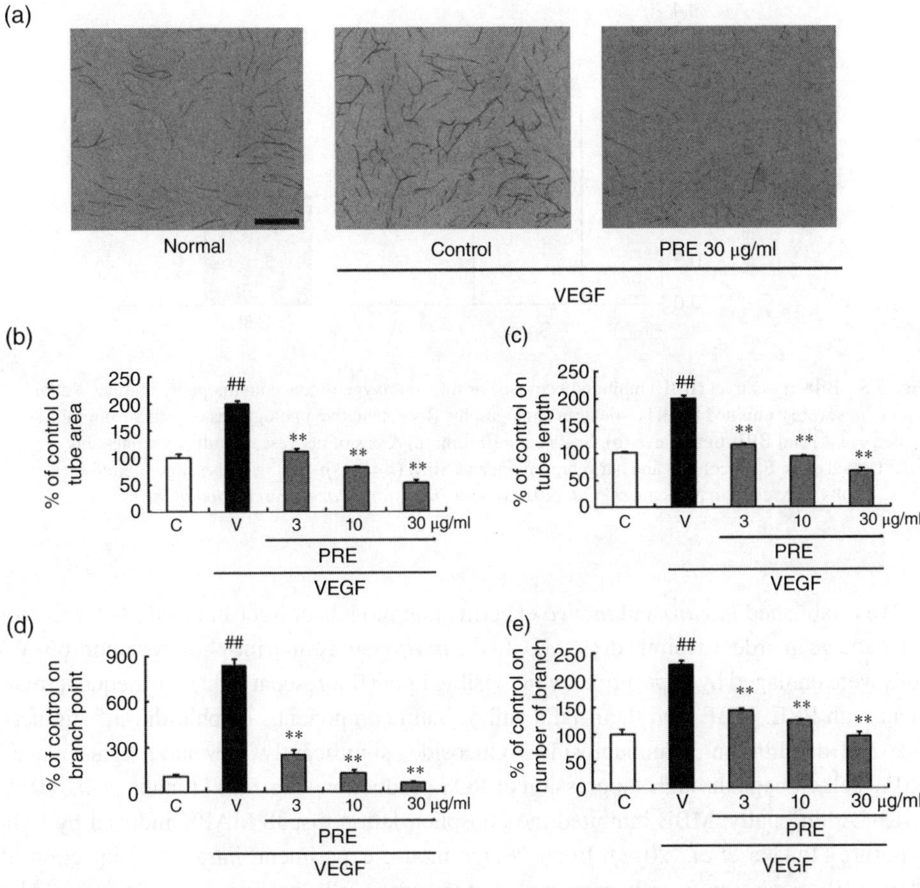

Fig. 9.4 Purple rice extract (PRE) inhibited tube formation induced by vascular endothelial growth factor (VEGF) A. (a) Representative photographs of tube formation. Scale bar = 1 mm. Human umbilical vein endothelial cells (HUVECs) were cocultured with human fibroblasts, as described in the section on methods, and incubated for 11 days with or without the indicated concentrations of PRE, with the concomitant addition of VEGF-A (10 ng/ml). Tube formation was observed in five randomly chosen fields, and tube area (b), length (c), joints (d), and paths (e) were measured using an Angiogenesis Image Analyzer. Data are shown as mean ± SEM (n = 3). C, control; V, vehicle; $^{##}p < 0.01$ versus control (Student's t-test); $^{**}p < 0.01$ versus vehicle (Dunnett's multiple-comparison test). These results were cited in Tanaka *et al.* (2012).

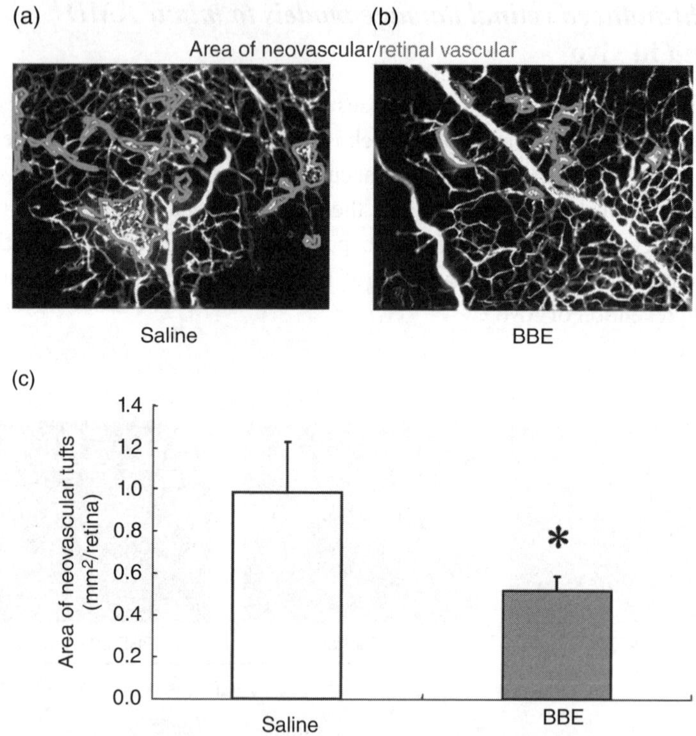

Fig. 9.5 Bilberry extract (BBE) inhibited neovascular tufts on oxygen-induced retinopathy in mice. Retinal flat mounts were examined by FITC–dextran angiography. Representative photographs of retina from saline-treated eye (a) and BBE-treated eye (b). Scale bar = 100 μm. (c) Areas of neovascular tufts in saline- and BBE-treated eyes. Each column and bar represent mean ± SEM ($n = 9$). *$p < 0.05$ versus saline (paired *t*-test). These results were cited in Matsunaga *et al.* (2010a). (*See insert for color representation of the figure*)

We established *in vitro* and *in vivo* experimental models of light-induced photoreceptor cell damage in order to mimic dry AMD. In the *in vitro* study, murine photoreceptor (661W) cells were damaged by exposure to white visible light (fluorescent light). Subsequent treatment with PRE, MBE, and their main anthocyanin components, delphinidin-3,5-diglucoside and delphinidin-3-sambubioside-5-glucoside, significantly prevented light-induced 661W cell damage through suppression of ROS production (Fig. 9.6) (Tanaka *et al.*, 2011, 2013). Additionally, MBE inhibited the phosphorylation of p38 MAPK induced by light exposure (Tanaka *et al.*, 2013). In an *in vivo* mouse experiment, intravitreal injection of PRE at 10 μg/eye significantly protected photoreceptor cells from degeneration induced by exposure to light, as revealed by histological analysis (Fig. 9.7) (Tanaka *et al.*, 2011). These results suggested that anthocyanins might be part of active components of PRE and MBE against visible-light-induced retinal damage.

High-energy light, such as ultraviolet and blue light of the sunlight, is related to the pathogenesis of age-related macular degeneration and retinitis pigmentosa (Margrain *et al.*, 2004; Glazer-Hockstein & Dunaief, 2006). A previous report suggested that retinal damage is inversely proportional to the wavelength (379–559 nm) of light in a rat *in vivo* model

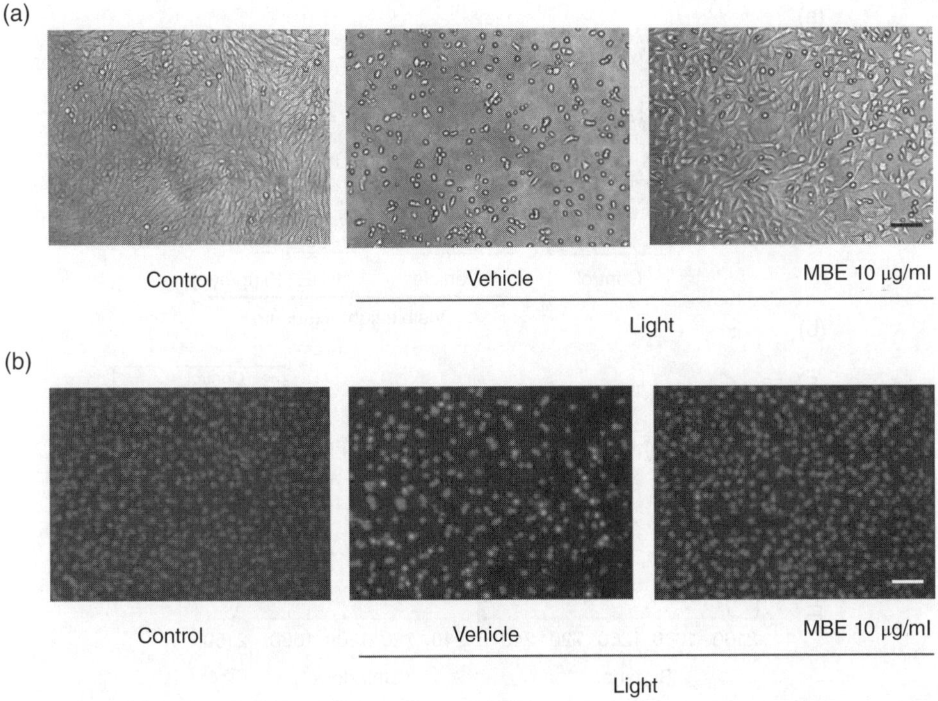

Fig. 9.6 Effects of maqui berry extract (MBE) on visible-light-induced 661W cell damage. (a) Representative microscopy at 24 hours after light exposure. (b) Representative fluorescence microscopy of Hoechst 33342 (blue color, cell nuclear stained) and propidium iodide (PI; red color, dead cell stained) fluorescence staining at 24 hours after light exposure. Each scale bar represents 250 μm. These results were cited in Tanaka *et al.* (2013). (*See insert for color representation of the figure*)

(van Norren & Schellekens, 1990). Thus, we established blue-light-emitting diode (LED) and ultraviolet A (UVA)-light-induced murine photoreceptor 661W cell damage in an *in vitro* model, and then we investigated the effect of BBE and its main anthocyanin components. In this model, BBE and its main anthocyanidins, cyanidin, delphinidin, and malvidin, significantly prevented UVA light– or blue LED light–induced photoreceptor cell damage through inhibition of ROS production and inhibition of phosphorylation of p38 MAPK, c-Jun-N-terminal kinase (JNK), and nuclear factor-kappa B (NF-kB) induced by light exposure (Ogawa *et al.*, 2013, 2014).

The main cause of light-induced retinal damage is ROS generated by light exposure. Therefore, BBE, PRE, and MBE containing high levels of antioxidants, not only anthocyanins but also other polyphenols, such as quercetin, isorhamnetin, and apigenin, may inhibit the occurrence of ROS production and could be useful in the prevention of dry AMD.

9.4 Conclusions

In summary, several eye diseases such as glaucoma, diabetic retinopathy, AMD, and retinitis pigmentosa, are caused by oxidative stress, angiogenesis, hyperglycemia, genetic mutation, aging, smoking, and light damage. Thus, it may be possible to prevent eye diseases

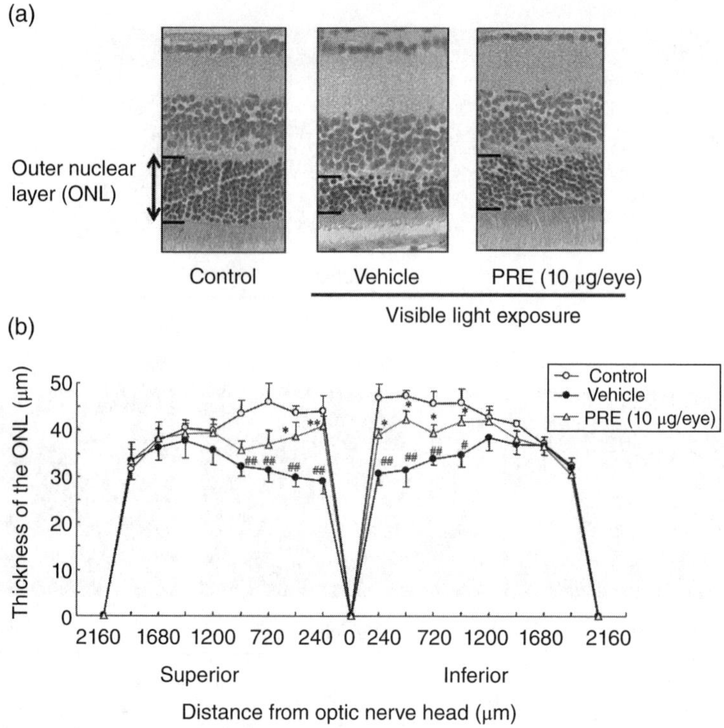

Fig. 9.7 Effects of purple rice extract (PRE) on retinal damage induced by exposure to light in mice. Hematoxylin and eosin staining of sections (thickness 5 lm) obtained from mice 5 days after light exposure. (a) Representative photographs show nontreated control retina, light-exposed vehicle-treated retina, light-exposed retina treated with PRE at 10 μg/eye. (b) Measurement of the thickness in the outer nuclear layer (ONL) 5 days after light exposure. Data are shown as mean ± SEM ($n=5$ or 6). $^{\#}p<0.05$ and $^{\#\#}p<0.01$ versus control. $^{*}p<0.05$ and $^{**}p<0.01$ versus vehicle. The scale bar represents 25 μm. These results were cited in Tanaka *et al.* (2011).

through reduction of these processes. Anthocyanins, which are a type of polyphenols, possess potent antioxidative, anti-inflammatory, and α-glucosidase inhibitory activities. In these studies, we have shown that extracts from anthocyanin-rich foods prevented angiogenesis, retinal ganglion cell damage, and light-induced retinal damage and therefore could have substantial benefits in maintaining eye health. Several anthocyanins having a characteristic structure exhibit different effects, and the mechanisms are not clear enough yet. There is a need to elucidate the mechanism of action of each anthocyanin and structure–function relationships in the prevention of eye diseases by anthocyanins in the future.

The pharmacokinetics of anthocyanins is an important issue in oral consumption of anthocyanin-rich foods. Some previous reports showed the metabolism of anthocyanins and their degradation by intestinal microflora *in vitro* (Keppler & Humpf, 2005; Hidalgo *et al.*, 2012). On the other hand, part of the anthocyanins are absorbed directly into the blood from the small intestine under their native form after oral consumption of bilberry or grape juice in clinical studies (Stalmach *et al.*, 2012; Sakakibara *et al.*, 2014). Anthocyanins of blackcurrant administered by intraperitoneal and/or intravenous injections in rat reached the ocular tissues, such as aqueous humor, cornea, choroid, ciliary body, iris, vitreous, and retina, and the amount of anthocyanins in the retina was higher than in the blood (Matsumoto *et al.*, 2006). Furthermore, it was reported in pig that ingested anthocyanins after dietary administration of blueberry accumulate in eye tissue and have potential passage through the blood–brain barrier (Kalt *et al.*, 2008). The purpose of our *in vivo* studies was investigation of the direct action to ocular tissue of anthocyanins by intravitreal injection of extracts containing high levels of anthocyanins. In our results, extracts containing anthocyanins showed the protective effect in some experimental *in vivo* model. These facts showed the potential of dietary supplements of anthocyanin-rich extract affecting the retina. However, there is no report suggesting the result of detection of anthocyanins on human ocular tissue after consumption of anthocyanin-rich foods. In the future, there is a need to perform clinical trials to reveal the pharmacokinetics and action mechanism of anthocyanin-rich foods.

References

Bonfoco, E., Krainc, D., Ankarcrona, M., Nicotera, P. and Lipton, S.A. (1995) Apoptosis and necrosis: two distinct events induced, respectively, by mild and intense insults with *N*-methyl-d-aspartate or nitric oxide/superoxide in cortical cell cultures. *Proceedings of the National Academy of Sciences of the United States of America*, **92**, 7162–7166.

Cespedes, C., El-Hafidi, M., Pavon, N. and Alarcon, J. (2008) Antioxidant and cardioprotective activities of phenolic extracts from fruits of Chilean blackberry *Aristotelia chilensis* (Elaeocarpaceae), Maqui. *Food Chemistry*, **108**, 820–829.

Chen, R., Hollborn, M., Grosche, A. *et al.* (2014) Effects of the vegetable polyphenols epigallocatechin-3-gallate, luteolin, apigenin, myricetin, quercetin, and cyanidin in primary cultures of human retinal pigment epithelial cells. *Molecular Vision*, **3**, 242–258.

Colantunoni, A., Bertuglia, S., Magistretti, M.J. and Donato, L. (1991) Effects of *Vaccinium myrtillus* anthocyanosides on arterial vasomotion. *Arzneimittel-Forschung*, **41**, 905–909.

Eissa, L.A., Smith, S.B. and El-Sherbeny, A. (2006) Regulation of taurine transporter activity in cultured rat retinal ganglion cells and rat retinal Müller cells. *Saudi Pharmaceutical Journal*, **14**, 16–26.

Escribano-Bailon, M.T., Alcalde-Eon, C., Munoz, O., Rivas-Gonzalo, J. and Santos-Buelga, C. (2006) Anthocyanins in berries of maqui (*Aristotelia chilensis* (Mol.) Stuntz). *Phytochemical Analysis*, **17**, 8–14.

Glazer-Hockstein, C. and Dunaief, J.L. (2006) Could blue light-blocking lenses decrease the risk of age-related macular degeneration? *Retina*, **26**, 1–4.

Gospodarowicz, D., Cheng, J. and Lirette, M. (1983) Bovine brain fibroblast growth factors: comparison of their abilities to support the proliferation of human and bovine vascular endothelial cells. *The Journal of Cell Biology*, **97**, 1677–1685.

Hidalgo, M., Oruna-Concha, M.J., Kolida, S. *et al.* (2012) Metabolism of anthocyanins by human gut microflora and their influence on gut bacterial growth. *Journal of Agricultural and Food Chemistry*, **60**, 3882–3890.

Hou, Z., Qin, P. and Ren, G. (2010) Effect of anthocyanin-rich extract from black rice (*Oryza sativa* L. Japonica) on chronically alcohol-induced liver damage in rats. *Journal of Agricultural and Food Chemistry*, **58**, 3191–3196.

Hu, C., Zawistowski, J., Ling, W. and Kitts, D.D. (2003) Black rice (*Oryza sativa* L. indica) pigmented fraction suppresses both reactive oxygen species and nitric oxide in chemical and biological model systems. *Journal of Agricultural and Food Chemistry*, **51**, 5271–5277.

Ichikawa, H., Ichiyanagi, T., Xu, B., Yoshii, Y., Nakajima, M. and Konishi, T. (2001) Antioxidant activity of anthocyanin extract from purple black rice. *Journal of Medicinal Food*, **4**, 211–218.

Kalt, W., Blumberg, J.B., McDonald, J.E. *et al.* (2008) Identification of anthocyanins in the liver, eye, and brain of blueberry-fed pigs. *Journal of Agricultural and Food Chemistry*, **56** (3), 705–712.

Keppler, K. and Humpf, H.U. (2005) Metabolism of anthocyanins and their phenolic degradation products by the intestinal microflora. *Bioorganic & Medicinal Chemistry*, **13**, 5195–5205.

Kubota, S., Kurihara, T., Mochimaru, H. *et al.* (2009) Prevention of ocular inflammation in endotoxin-induced uveitis with resveratrol by inhibiting oxidative damage and nuclear factor-kB activation. *Investigative Ophthalmology & Visual Science*, **50** (7), 3512–3519.

Lietti, A., Cristoni, A. and Piccci, M. (1976) Studies on *Vaccinium myrtillus* anthocyanosides. *Arzneimittel-Forschung*, **26**, 829–832.

Margrain, T.H., Boulton, M., Marshall, J. and Sliney, D.H. (2004) Do blue light filters confer protection against age-related macular degeneration? *Progress in Retinal and Eye Research*, **23**, 523–531.

Matsumoto, H., Nakamura, Y., Iida, H. and Ohguro, H. (2006) Comparative assessment of distribution of blackcurrant anthocyanins in rabbit and ocular tissues. *Experimental Eye Research*, **83**, 348–356.

Matsunaga, N., Imai, S., Inokuchi, Y., Shimazawa, M., Yokota, S. and Hara, H. (2009) Bilberry and its main constituents have neuroprotective effects against retinal neuronal damage *in vitro* and *in vivo*. *Molecular Nutrition & Food Research*, **53**, 869–877.

Matsunaga, N., Chikaraishi, Y., Shimazawa, M., Yokota, S. and Hara, H. (2010a) *Vaccinium myrtillus* (bilberry) extracts reduce angiogenesis *in vitro* and *in vivo*. *eCAM*, **7** (1), 47–56.

Matsunaga, N., Tsuruma, K., Shimazawa, M., Yokota, S. and Hara, H. (2010b) Inhibitory actions of bilberry anthocyanins on angiogenesis. *Phytotherapy Research*, **24**, S42–S47.

Morazzoni, P. and Bombardelli, E. (1996) *Vaccinium myrtillus*. *Fitoterapia*, **67**, 3–29.

Morazzoni, P.V. and Megistretti, M.J. (1990) Activity of myrtocyan®, an anthocyanoside complex from *Vaccinium myrtillus* (VMA), on platelet aggregation and adhesiveness. *Fitoterapia*, **61**, 13–21.

Napoleon, F., Keith, H., Lyn, J. and David, W.L. (1992) Molecular and biological properties of the vascular endothelial growth factor family proteins. *Endocrine Reviews*, **13**, 18–32.

van Norren, D. and Schellekens, P. (1990) Blue light hazard in rat. *Vision Research*, **30**, 1517–1520.

Ogawa, K., Sakakibara, H., Iwata, R. *et al.* (2008) Anthocyanin composition and antioxidant activity of the crowberry (*Empetrum nigrum*) and other berries. *Journal of Agricultural and Food Chemistry*, **56**, 4457–4462.

Ogawa, K., Tsuruma, K., Tanaka, J. *et al.* (2013) The protective effects of bilberry and lingonberry extracts against UV light-induced retinal photoreceptor cell damage *in vitro*. *Journal of Agricultural and Food Chemistry*, **61**, 10345–10353.

Ogawa, K., Kuse, Y., Tsuruma, K., Kobayashi, S., Shimazawa, M. and Hara, H. (2014) Protective effects of bilberry and lingonberry extracts against blue light-emitting diode light-induced retinal photoreceptor cell damage *in vitro*. *BMC Complementary and Alternative Medicine*, **14**, 120.

Pereira-Caro, G., Cros, G., Yokota, T. and Crozier, A. (2013) Phytochemical profiles of black, red, brown, and white rice from the Camargue region of France. *Journal of Agricultural and Food Chemistry*, **61** (33), 7976–7986.

Rojo, L.E., Ribnicky, D., Logendra, S. *et al.* (2012) *In vitro* and *in vivo* anti-diabetic effects of anthocyanins from Maqui Berry (*Aristotelia chilensis*). *Food Chemistry*, **131**, 387–396.

Ruiz, A., Hermosín-Gutiérrez, I., Mardones, C. *et al.* (2010) Polyphenols and antioxidant activity of calafate (*Berberis microphylla*) fruits and other native berries from Southern Chile. *Journal of Agricultural and Food Chemistry*, **58**, 6081–6089.

Sakakibara, H., Ichikawa, Y., Tajima, S. *et al.* (2014) Practical application of flavonoid-poor menu meals to the study of the bioavailability of bilberry anthocyanins in human subjects. *Bioscience, Biotechnology, and Biochemistry*, **78**, 1748–1752.

Schreckinger, M., Wang, J., Yousef, G., Lila, M. and de Mejia, E. (2010) Antioxidant capacity and *in vitro* inhibition of adipogenesis and inflammation by phenolic extracts of *Vaccinium floribundum* and *Aristotelia chilensis*. *Journal of Agricultural and Food Chemistry*, **58**, 8966–8976.

Stalmach, A., Edwards, C.A., Wightman, J.D. and Crozier, A. (2012) Gastrointestinal stability and bioavailability of (poly)phenolic compounds following ingestion of Concord grape juice by humans. *Molecular Nutrition & Food Research*, **56** (3), 497–509.

Takikawa, M., Inoue, S., Horio, F. and Tsuda, T. (2010) Dietary anthocyanin-rich bilberry extract ameliorates hyperglycemia and insulin sensitivity via activation of AMP-activated protein kinase in diabetic mice. *Journal of Nutrition*, **140** (3), 527–533.

Tamar, A., Itzhak, H., Ahuva, I., Jacob, P., Jonathan, S. and Eli, K. (1995) Vascular endothelial growth factor acts as a survival factor for newly formed retinal vessels and has implications for retinopathy of prematurity. *Nature Medicine*, **1**, 1025–1028.

Tanaka, J., Nakanishi, T., Ogawa, K. *et al.* (2011) Purple rice extract and anthocyanidins of the constituents protect against light-induced retinal damage *in vitro* and *in vivo*. *Journal of Agricultural and Food Chemistry*, **59**, 528–536.

Tanaka, J., Nakamura, S., Tsuruma, K., Shimazawa, M., Shimoda, H. and Hara, H. (2012) Purple rice (*Oryza sativa* L.) extract and its constituents inhibit VEGF-induced angiogenesis. *Phytotherapy Research*, **26**, 214–222.

Tanaka, J., Kadekaru, T., Ogawa, K., Hitoe, S., Shimoda, H. and Hara, H. (2013) Maqui berry (*Aristotelia chilensis*) and the constituent delphinidin glycoside inhibit photoreceptor cell death induced by visible light. *Food Chemistry*, **139**, 129–137.

Yao, Y., Sang, W., Zhou, M. and Ren, G. (2010) Antioxidant and α-glucosidase inhibitory activity of colored grains in China. *Journal of Agricultural and Food Chemistry*, **58**, 770–774.

Chapter 10
Prevention and Treatment of Diabetes Using Polyphenols via Activation of AMP-Activated Protein Kinase and Stimulation of Glucagon-like Peptide-1 Secretion

Takanori Tsuda

Abstract: There is a pressing need for a countermeasure to diabetes, and methods for prevention and treatment based on dietary factors are highly anticipated. AMP-activated protein kinase and glucagon-like peptide-1 are considered to be the most important molecular targets for these purposes. In this chapter, we introduce the health benefits of polyphenols based on our research, focusing on the prevention and suppression of diabetes. We particularly focus on anthocyanins, curcumin, and a derivative of caffeoylquinic acid. Finally, we discuss the challenges surrounding studies on polyphenol health benefits and future prospects.

Keywords: AMP-activated protein kinase, antidiabetes, anthocyanins, caffeoylquinic acid, curcumin, glucagon-like peptide-1

10.1 Introduction

Owing to the high background of interest in the health benefits of polyphenol-rich edible plants, the health benefits of polyphenols are currently being actively researched. So far, great advances have been made in studies focusing on the presence and mechanism of action of polyphenols that may contribute to health maintenance and disease prevention in humans, as well as their mechanism of action. Various food and dietary supplements have been developed based on these findings and are in current use. Initially, research on the health benefit of polyphenols was centered on their antioxidant- and oxidative-stress-repressing effects. Presently, it has been found that polyphenols have various other

Recent Advances in Polyphenol Research, First Edition. Edited by Kumi Yoshida, Véronique Cheynier and Stéphane Quideau.

health benefits, and research is now centered on these so-called beyond antioxidant functions (Barrajón-Catalán *et al.*, 2014).

The increase in incidence of diabetes has become a global issue. Recently, there is a marked increase of diabetes especially in Asian regions (Scully, 2012). There is a pressing need for a countermeasure of diabetes, and methods for prevention and treatment based on dietary factors are anticipated.

Thus far, various studies have been conducted on dietary factors, including polyphenols, in relation to the prevention and treatment of diabetes. Some drugs are being used due to their ability to improve insulin sensitivity and glucose homeostasis. For example, thiazolidinediones are a group of synthetic peroxisome proliferator-activated receptor (PPAR) agonists that are used as antidiabetic drugs, with effects targeting adipocyte differentiation and adipocyte gene activation (Yamauchi *et al.*, 2001; Lehrke & Lazar, 2005). Another drug, metformin, improves hyperglycemia without stimulating insulin secretion (Zhou *et al.*, 2001; Hawley *et al.*, 2002). Administration of metformin induces AMP-activated protein kinase (AMPK) activation, which is suggested as a potential target for the treatment of type 2 diabetes (Zhou *et al.*, 2001; Hawley *et al.*, 2002). Although some drugs have been used as therapeutic agents for diabetes, there is little evidence that polyphenols themselves directly modulate insulin sensitivity or glucose-dependent insulin secretion. AMPK and glucagon-like peptide-1 (GLP-1) are considered to be the most significant molecular targets for these purposes. These molecules are important targets of various drugs; in particular, medications that increase the effect of GLP-1 are highly effective in many instances for treating patients with diabetes. Drawing from our research, in the following sections, we first describe the diet-derived factors involved in the activation of AMPK, followed by the diet-derived factors involved in the regulation of GLP-1 function in relation to meals and the promotion of GLP-1 secretion.

10.2 Activation of AMPK and metabolic change

10.2.1 Activation of AMPK

AMPK is a serine/threonine kinase expressed in most eukaryotic cells, from yeast and plants to mammals (Hardie, 2007). AMPK is activated by a decrease in intracellular energy level (i.e., an increase in the ratio of AMP/ATP). AMPK is activated by the phosphorylation of the Thr172 residue in the catalytic region of its α-subunit by the upstream factor AMPK kinase. LKB1 and the calcium signaling pathway are involved as upstream factors of AMPK (Hardie, 2008). The activation of AMPK stimulates the energy production system, as well as suppresses the synthesis of glucose and fat. Glucose transporter 4 (Glut4) is an insulin-sensitive Glucose transporter and is involved in regulating insulin-stimulated glucose uptake into skeletal muscle and adipose tissue. In type 2 diabetes, insulin resistance leads to defects in the muscle glucose transport system (Shepherd & Kahn, 1999). The glucose transport system can also be regulated via AMPK in skeletal muscle and adipose tissue. In skeletal muscles, the activation of AMPK induces the insulin-independent translocation of intracellular Glut4 to the cell membrane, increases the intracellular uptake

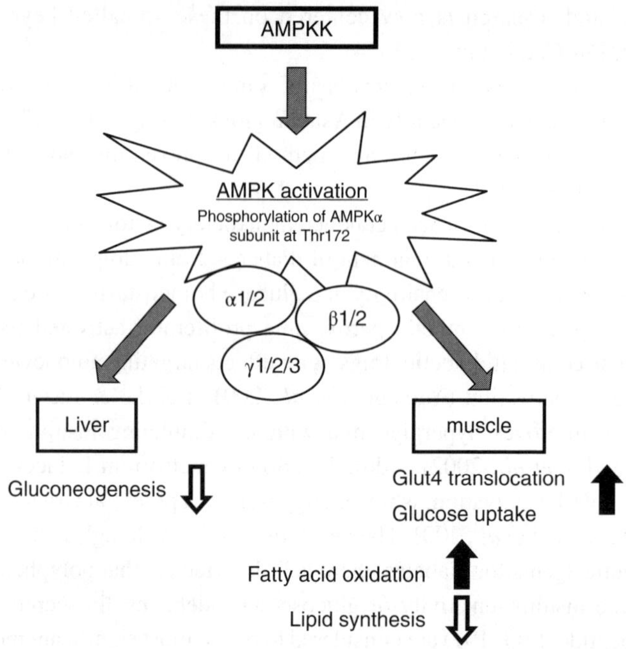

Fig. 10.1 Activation of AMPK leads to various metabolic changes. AMPKK, AMPK kinase.

of glucose, and promotes glucose usage. In the liver, the activation of AMPK suppresses gluconeogenesis and glucose output independently of insulin. cAMP-response element-binding protein-regulated transcription coactivator 2 (CRTC2) has emerged as a critical regulator of gluconeogenesis in mice (Foretz *et al.*, 2005). AMPK activation reduces the gluconeogenic program by promoting CRTC2 phosphorylation and inhibiting its accumulation in the nucleus (Koo *et al.*, 2005). Further, in lipid metabolism, the AMPK activation phosphorylates and thus suppresses the activity of acetyl-CoA carboxylase (ACC), which is the rate-limiting enzyme of lipid synthesis. This leads to the decrease of intracellular malonyl-CoA, which is the product of ACC. Malonyl-CoA inhibits carnitine acyltransferase-I, which is necessary for the uptake of long-chain fatty acids into mitochondria; thus, a decrease of intracellular malonyl-CoA concentration halts the allosteric inhibition of this enzyme (Minokoshi *et al.*, 2002). This results in the uptake of fatty acids into mitochondria, promoting β-oxidation (Fig. 10.1). From the studies performed over one decade, it has become clear that adipocytokines such as leptin and adiponectin activate AMPK (Minokoshi *et al.*, 2002; Yamauchi *et al.*, 2002). In this way, activated AMPK came to attract attention as the signaling molecule that prevents obesity and diabetes.

10.2.2 Dietary factors that exert diabetes-preventing and -suppressing effects through the activation of AMPK

Dietary factors including polyphenols have been reported to activate AMPK (Hardie, 2011). For example, the administration of epicatechin-3-gallate (1 μM) inhibits glucose production via activation of AMPK in isolated hepatocytes (Collins *et al.*, 2007).

Theaflavins, a major group of polyphenols in black tea, also attenuate hepatic lipid accumulation through activating AMPK in HepG2 cells (Lin *et al.*, 2007). Resveratrol improved health and survival of mice on a high-calorie diet accompanying activation of AMPK (Baur *et al.*, 2006). Curcumin, which gives turmeric its yellow color, stimulates glucose uptake through AMPK-p38 mitogen-activated protein kinase pathways in L6 myotube cells (Kim *et al.*, 2010). Quercetin, a flavonol, exerts inhibition of adipogenesis by activating the AMPK signal pathway in 3T3-L1 preadipocytes (Ahn *et al.*, 2008). Here, we present the recent reports of our group.

10.2.2.1 Blueberry (bilberry)

Anthocyanins are flavonoid plant pigments that exist as glycosides in plants. Anthocyanins are present in many plants consumed as staples, such as grains, tubers, vegetables, legumes, and fruits. Berries in particular are known to be rich in anthocyanins. Blueberries contain various types of anthocyanins, and anthocyanins from bilberry (BBE), the wild variant of blueberry, show a diabetes-suppressing effect. Its mechanism of action can be explained by the activation of AMPK (Takikawa *et al.*, 2010) (Fig. 10.2). Dietary BBE leads to the activation of AMPK in skeletal muscles, white adipose tissue, and the liver in a type 2

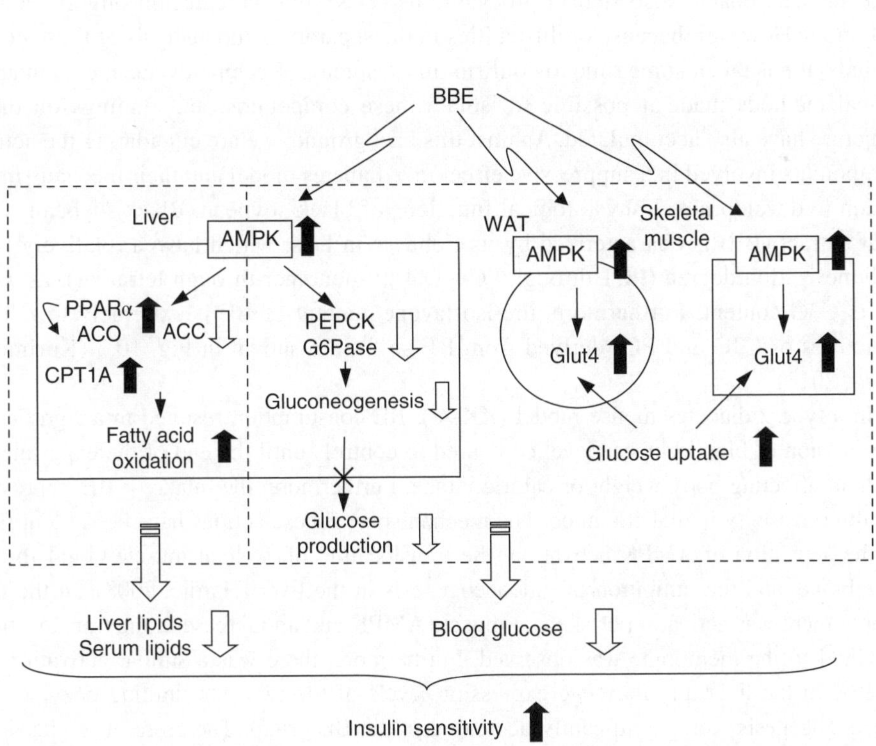

Fig. 10.2 Proposed mechanism for amelioration of hyperglycemia and insulin sensitivity by dietary bilberry extract, BBE. ACC, acetyl-CoA carboxylase; ACO, acyl-CoA oxidase; CPT1A, carnitine palmitoyltransferase-1A; G6Pase, glucose-6-phosphatase; PEPCK, phosphoenolpyruvate carboxykinase. Takikawa *et al.* (2010). Reproduced from with permission from the American Society for Nutrition.

diabetes mouse model (KK-A^y). The activation of AMPK caused by this intake of BBE leads to an increase in Glut4 expression in skeletal muscle and in white adipose tissues and to the suppression of gluconeogenesis in the liver. These findings are supported by the fact that BBE intake suppresses the gene expression level of the rate-limiting enzyme for gluconeogenesis, as well as glucose synthesis function after a pyruvic acid load. Furthermore, in lipid metabolism, fat synthesis is suppressed and fat usage is stimulated through the phosphorylation of ACC. These effects are understandable, especially when considered together with the finding that intake of BBE by the mouse model of diabetes resulted in suppressed fatty liver. These results show that the intake of BBE suppresses the rise in blood sugar value and increases sensitivity to insulin (Takikawa *et al.*, 2010). Whether the presence of multiple types of anthocyanin molecules, as in BBE, is relevant to antidiabetes effects via AMPK activation has yet to be elucidated from the analysis of the degradation and metabolic products of anthocyanin.

10.2.2.2 Black soybean

Black soybean is known to contain the anthocyanin cyanidin 3-glucoside (C3G) as the pigment component of its seed coat. Other than this anthocyanin, it has been suspected that the seed coat is also rich in procyanidins (PCs), that is, catechin oligomers and polymers. However, because of difficulties in the separation and analysis of these compounds, it has taken some time to confirm this suspicion. Recent advancements in analytical methods made it possible to isolate these compounds, and findings on their structure have also accumulated. Against this background, we are elucidating the active compounds involved in a suppressive effect in a diabetes model and their mechanism of action to develop novel physiological functions of black soybean. Black soybean seed coat extract (BE) is characterized by its richness in PCs, which have a relatively low polymerization degree (PC1 through PC4-1, 4-2; monomer through tetramer), as well as its C3G content. Furthermore, the isoflavone content of BE is extremely low. The structures of C3G and PCs derived from BE so far are shown in Fig. 10.3 (Kurimoto *et al.*, 2013).

In a type 2 diabetes mouse model (KK-A^y), BE consumption resulted in a significant suppression of blood glucose level compared to controls until the end of the experiment, without affecting body weight or caloric intake. Furthermore, the intake of BE improved insulin sensitivity in diabetic mice. The mechanisms of these actions have been identified as the activation of AMPK activation, the translocation of skeletal muscle Glut4 to the membrane, and the inhibition of gluconeogenesis in the liver (Hardie, 2008). In the BE group, there was activation of skeletal muscle AMPK and an increase in the translocation of Glut4 to the membrane was observed. Furthermore, there was a similar activation of AMPK in the liver and the gene expression levels of the two rate-limiting enzymes of gluconeogenesis were significantly decreased in the BE group. Therefore, it is plausible that the consumption of BE activates AMPK and stimulates Glut4 translocation to the membrane in the skeletal muscles, while decreasing the expression of the rate-limiting enzymes of gluconeogenesis via the AMPK activation in the liver, resulting in the suppression of increasing blood glucose levels.

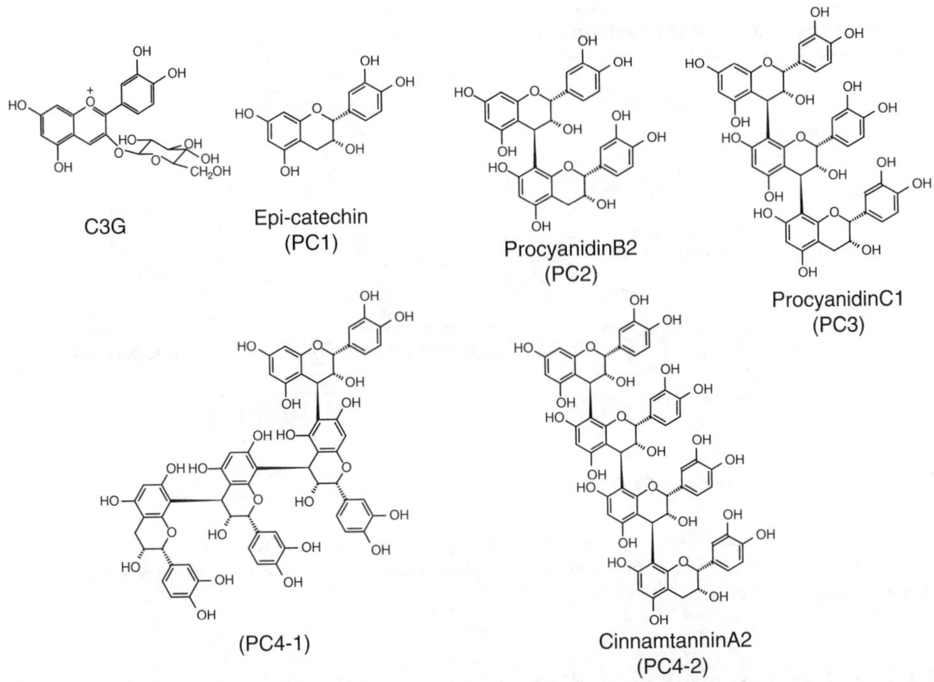

Fig. 10.3 Chemical structures of cyanidin 3-glucoside (C3G) and procyanidins (PCs) contained in dietary black soybean seed coat (BE). Kurimoto *et al.* (2013). Reproduced with permission from the American Chemical Society.

We have investigated the AMPK-activating effect of the components in BE and the change in the glucose uptake into cells induced by these components, using myotubes. Our findings indicated that C3G activates AMPK, and the activation of AMPK by PC increased with the degree of polymerization. The translocation of Glut4 to the cell membrane was similarly promoted during this process. The relative amount of cellular glucose uptake was parallel to the activation of AMPK and the translocation of Glut4 to membrane, increasing significantly. Interestingly, the simultaneous administration of PC dimer or trimer and C3G has resulted in significant increase of AMPK activation, compared to the administration of each compounds on their own. Furthermore, in animal studies, serum glucose concentration of the group given the fraction with C3G removed showed a markedly weakened suppression effect on the rise in blood glucose compared to the group given the fraction that contained C3G (BE).

In summary, the dietary BE ameliorates hyperglycemia and insulin sensitivity in type 2 diabetic mice. The AMPK activation and the resulting increase of cellular glucose uptake, which follows the membrane translocation of Glut4 in the skeletal muscle, and the suppression of gluconeogenesis in the liver are involved in the mechanisms of these actions. It is considered that among the components in BE, PCs are involved in these actions besides C3G and that the simultaneous intake of these two components is most effective (Kurimoto *et al.*, 2013) (Fig. 10.4).

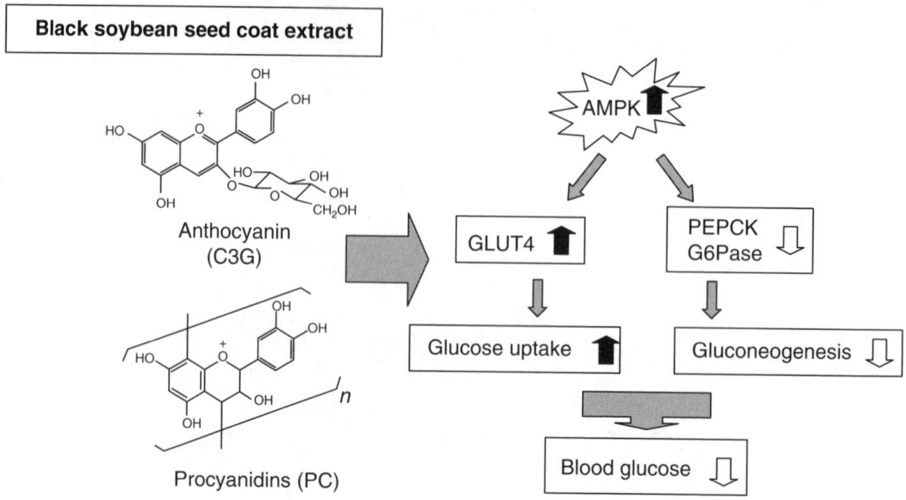

Fig. 10.4 Proposed mechanism for amelioration of hyperglycemia and insulin sensitivity by dietary black soybean seed coat (BE) in mice. Kurimoto *et al.* (2013). Reproduced with permission from the American Chemical Society.

10.3 GLP-1 action and diabetes prevention/suppression

10.3.1 GLP-1 action

Similarly to AMPK, GLP-1 is an important molecular target in light of the diabetes-preventing and -suppressing effects of polyphenols. Japanese people typically have low insulin secretion and thus are prone to diabetes with the slight increase in obesity caused by the Western diet and lifestyle. In type 2 diabetes, it is of importance to resolve the insulin resistance as well as the impairment in insulin secretion. As novel type 2 diabetes treatment strategies, incretin and related medications are bringing great benefits. "Incretin" is a collective term for peptide hormones released from the digestive tract with the food intake and acting on the β cells of the pancreas to stimulate insulin release. Glucose-dependent insulinotropic polypeptide (GIP) and GLP-1 are the two incretins known so far (Fig. 10.5). Because the sensitivity of GLP-1 in the β cells of the pancreas is not reduced in type 2 diabetic patients, GLP-1 is considered to be effective for the prevention and treatment of type 2 diabetes (Lovshin & Drucker, 2009).

GLP-1 is secreted from enteroendocrine L cells in the lower small intestine and the large intestine. GLP-1 enhances insulin secretion depending on the blood glucose concentration (Fig. 10.5). Therefore, the risks of pancreatic β cell exhaustion, induction of hypoglycemia, and weight gain that are associated with conventional insulin secretagogues are low. In addition, many antidiabetes actions have been reported, such as the promotion of β cell proliferation and protection of the cells, as well as effects on organs other than the pancreas (Baggio & Drucker, 2007). This background has resulted in the recent development of GLP-1 related drugs, which are being put to clinical use. GLP-1 is inactivated by dipeptidyl peptidase-4 (DPP-4), a peptidase located in the blood. The half-life of GLP-1 in the blood

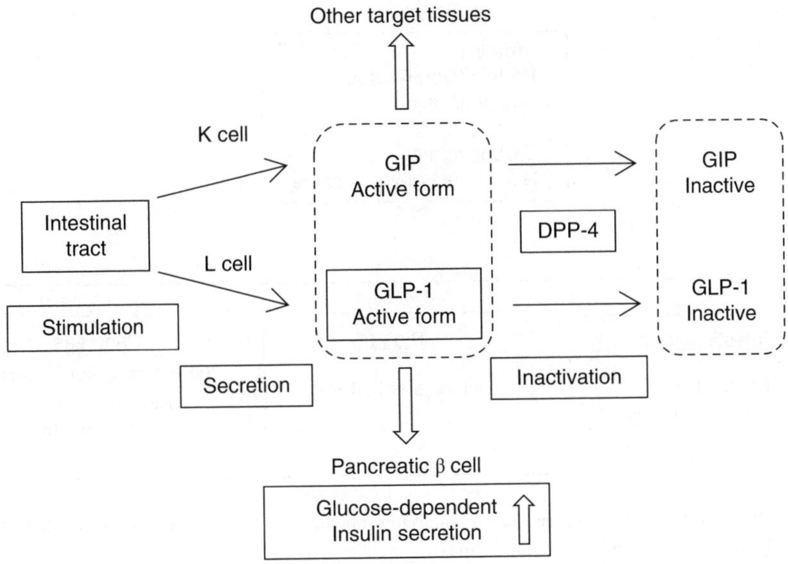

Fig. 10.5 Overview of actions of GLP-1 and GIP. Tsuda (2015). Reproduced with permission of John Wiley & Sons.

is therefore short (about 2 minutes); this prompted the development of a dipeptidyl peptidase-4 (DPP-4) inhibitor, which raises the active GLP-1 concentration in the blood by suppressing DPP-4 activity (Fig. 10.5). Meanwhile, GLP-1 receptor agonist has a degradation-resistant GLP-1-like structure and binds to GLP-1 receptor. However, GLP-1 receptor agonist, being a peptide, cannot be taken orally and requires administration by hypodermic injection (Herman *et al.*, 2006). As this is not ideal, the preferable way for dietary factors to elevate GLP-1 actions is by stimulating the endogenous secretion of GLP-1.

There is generally considered to be a close relationship between incretin and diet. Besides directly stimulating GLP-1 release in the intestinal tract, it is possible that dietary factors exert such effects as increasing incretin sensitivity, functioning as GLP-1 receptor agonists, and inhibiting DPP-4 in the blood (Fig. 10.6). Considering the treatment of diabetes, it is very likely that the efficacies of the incretin-related drugs currently in use are influenced by diet content. That is, it is likely that diet content may increase or decrease the efficacy of incretin-related drugs. The interactions between drugs and polyphenol intake must be thoroughly investigated when dealing with the meals of diabetic patients under treatment.

10.3.2 Dietary factors that promote GLP-1 secretion

As already mentioned, the best strategy to increase GLP-1 action by dietary factors is considered to be the stimulation of the secretion of intrinsic GLP-1. Besides glucose, fatty acids and glutamines have been reported as nutrients with GLP-1 secretion stimulating function (Hirasawa *et al.*, 2005; Edfalk *et al.*, 2008; Samocha-Bonet *et al.*, 2011; Clemmensen *et al.*, 2013). Further, it has been reported that bile acid promotes GLP-1 secretion via G-protein-coupled receptor (Parker *et al.*, 2012) and that gustatory receptor

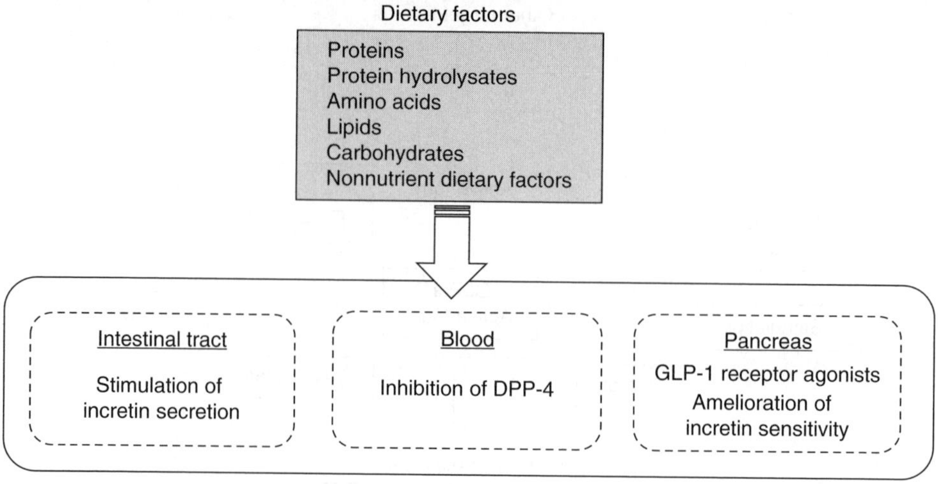

Fig. 10.6 Relationship between dietary factors (nutrients, nonnutrients) and incretin action. Tsuda (2015). Reproduced with permission of John Wiley & Sons.

may participate in the secretion (Jang *et al.*, 2007).Whether these findings apply to human as well or not must be verified. Although they are not polyphenols, Hira and coworkers have reported the hydrolysate of zein, a maize-derived protein (Hira *et al.*, 2009; Mochida *et al.*, 2010; Higuchi *et al.*, 2013), and palatinose (Hira *et al.*, 2011) as dietary factors that function as GLP-1 secretagogue. Other reports have also examined the stimulation of GLP-1 by polyphenols. For example, resveratrol reduces blood glucose level via the stimulation of GLP-1 secretion (Dao *et al.*, 2011). Large amount of procyanidins are contained in grape seeds, and it was found to stimulate GLP-1 secretion by a Spanish research group (González-Abuín *et al.*, 2014a). It has also been reported that the treatment of STC-1 cells with procyanidins resulted in depolarization at low concentration (0.05 mg/l) and hyperpolarization at high concentration (50 mg/l) (González-Abuín *et al.*, 2014b). González-Abuín and coworkers (2014c) also found that procyanidins prevent the decrease of GLP-1 production induced by the cafeteria diet. Stimulation of GLP-1 secretion is not significantly enhanced by α-glycosyl-isoquercitrin; however, its simultaneous administration with fructooligosaccharides enhanced GLP-1 secretion (Phuwamongkolwiwat *et al.*, 2014). There is one report that found an effect of green tea extract in type 2 diabetes patients (Liu *et al.*, 2014). Blood GLP-1 concentration was significantly increased by consumption of green tea extract (1500 mg green tea extract/day, 856.8 mg epigallocatechin gallate per day) after 16 weeks.

In the following sections, we show our recent findings on the stimulation of GLP-1 secretion by curcumin, an extract of dietary sweet potato leaves that contains the derivative of caffeoylquinic acid (CQA) as polyphenol and anthocyanins.

10.3.2.1 Curcumin

Our group has investigated the dietary factors that function as GLP-1 secretagogues at the cellular level and found curcumin to be one such factor (Takikawa *et al.*, 2013). The diabetes-preventing and -suppressing effect of curcumin has been recently reported in a study

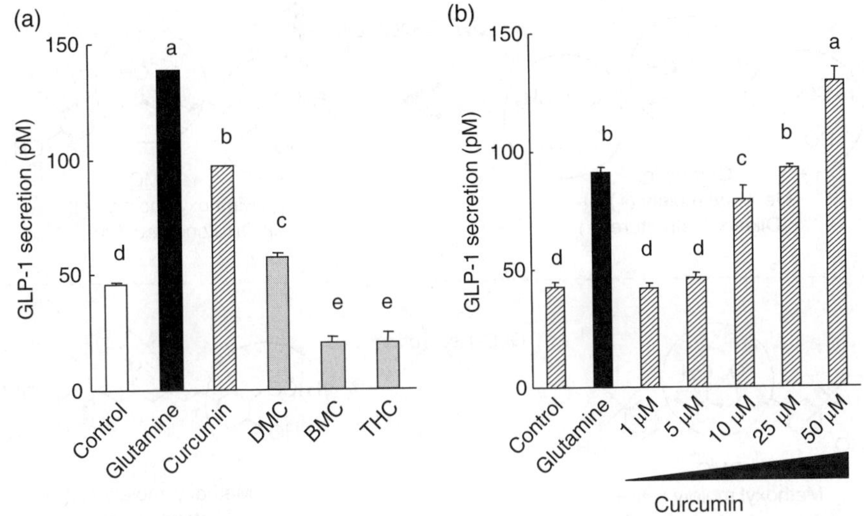

Fig. 10.7 Stimulation of GLP-1 secretion by curcuminoids (25 µM) (a), and the dose response of curcumin (1–50 µM) on GLP-1 secretion (b) in GLUTag cells. Glutamine (10 mM) was used as a positive control. Values are means ± SEM, $n = 3$. Values without a common letter significantly differ at $P < 0.05$. Takikawa *et al.* (2010). Reproduced with permission of Elsevier.

involving human subjects (Chuengsamarn *et al.*, 2012). Here, we present our results on the GLP-1-secretion-promoting function of curcumin using enteroendocrine cells.

Curcumin displayed a strong GLP-1-releasing effect, stimulating the secretion significantly at 10 µM (Fig. 10.7). Turmeric contains dimethoxycurcumin (DMC), which is missing one methoxy group, and bisdemethoxycurcumin (BMC), which does not contain any methoxy groups. It also contains curcumin derivatives and tetrahydrocurcumin (THC), which lacks the β-diketone structure. We examined the structure–activity relationship of these curcumin derivatives and found that for a curcumin derivative to have GLP-1-releasing activity, it must possess a β-diketone structure and at least one methoxy group on the aromatic ring. The most powerful structure was curcumin, which possesses two methoxy groups in addition to having β-diketone structure. However, neither BMC, which is missing the methoxy groups, nor THC, which is missing the β-diketone structure, promoted GLP-1 secretion significantly (Fig. 10.8). The bioavailability of curcumin is low, detected mostly as conjugates in blood. However, considering that curcumin directly stimulates the secretion of GLP-1 by intestinal surface endocrine cells, factors such as its solubility in the intestine become more important than its metabolism, absorption, and concentration in the blood. GLP-1 secretion is induced via the activation of G-protein-coupled receptors (GPCR) in intestinal L cells (Janssen & Depoortere, 2013). Curcumin can interact with possible GPCR candidates and stimulate GLP-1 secretion; however, its exact molecular target remains unclear. Studies to elucidate this target are currently underway.

Next, we investigated curcumin mechanism of action as GLP-1 secretagogue. It is considered that the secretion of GLP-1 involves an increase of intracellular Ca^{2+} level. The elevation of intracellular Ca^{2+} level is achieved by the influx of exogenous Ca^{2+} or the

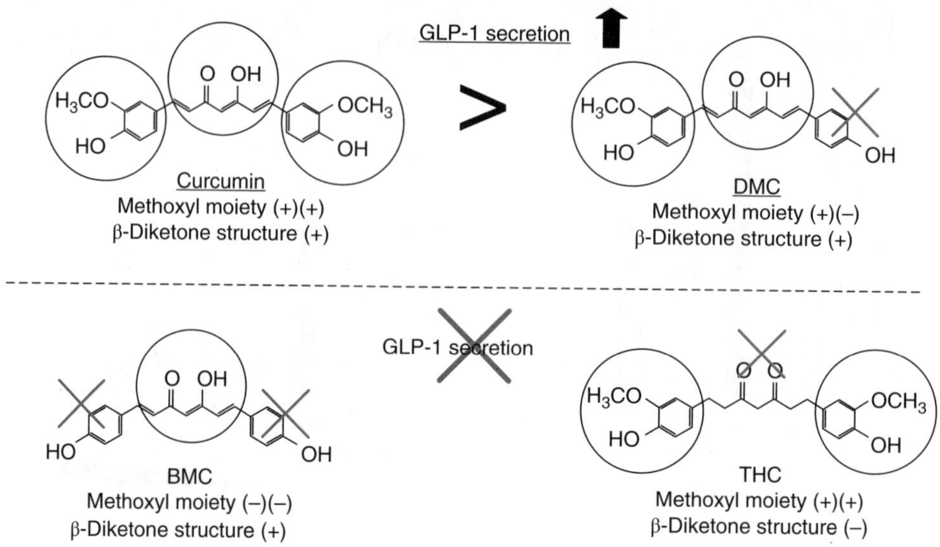

Fig. 10.8 Structure–activity relationship between curcumin derivatives and stimulation of GLP-1 secretion (GLUTag cells). Tsuda (2015). Reproduced with permission of John Wiley & Sons.

mobilization of endogenous Ca^{2+} via ryanodine receptor (RyR) or inositol 1,4,5-triphosphate receptor (IP_3R) on the endoplasmic reticulum. Our study showed that curcumin administration raises the intracellular Ca^{2+} level. Curcumin's GLP-1-secretion-promoting effect was significantly inhibited by Ca^{2+} channel blockers and by regulating the release of endogenous Ca^{2+} storage.

There has been considerable advancement in the elucidation of the molecular mechanism of the insulin-secretion-promoting effect of glucose stimulation on pancreatic β cells. Yet many factors, such as the role of the K_{ATP} channel in GLP-1 secretion by enteroendocrine cells, are still unknown, as they differ from the case of insulin secretion by pancreatic β cells. In the case of the insulin-secretion mechanism, the involvement of elevated intracellular Ca^{2+} and subsequent activation of Ca^{2+}/calmodulin-dependent kinase II (CaMK II) have been reported (Lee *et al.*, 2009; Dixit *et al.*, 2013). Thus, it is possible that CaMK II is also involved in the secretion of GLP-1. We therefore examined the involvement of CaMK II in the GLP-1 secretion induced by curcumin. As a result, treatment with the CaMK II inhibitor completely blocked the curcumin-induced GLP-1 secretion. The protein analysis of CaMK II at that moment revealed that the administration of curcumin has markedly increased the phosphorylation (activation) of CaMK II in a time- and concentration-dependent manner. These results indicate that curcumin stimulates GLP-1 secretion by increasing the intracellular Ca^{2+} level and activating CaMK II (Fig. 10.9).

Promotion of GLP-1 secretion can be achieved through the involvement of PKA and PKC, without the elevation of intracellular Ca^{2+} (Iakoubov *et al.*, 2007). The involvement of extracellular signal-regulated kinase (ERK)1/2 has also been reported (Reimer, 2006; Sangle *et al.*, 2012). In addition, in several cases, it is accompanied by the increase

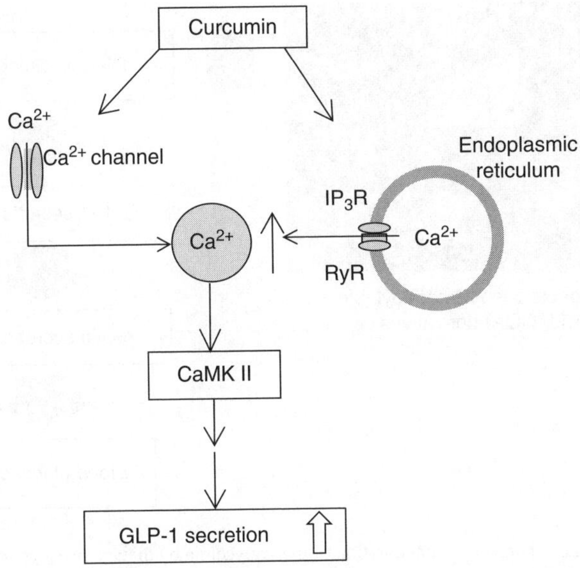

Fig. 10.9 Proposed mechanism for stimulation of GLP-1 secretion by curcumin. Tsuda (2015). Reproduced with permission of John Wiley & Sons.

of intracellular cAMP. However, these pathways are not involved in the promotion of GLP-1 secretion by curcumin.

10.3.2.2 Edible young leaves of sweet potato (culinary sweet potato leaves)

The sweet potato variant "Suioh" (*Ipomoea batatas* L.) was developed by the Japanese Ministry of Agriculture, Forestry and Fisheries through selective breeding so that the leaves would be edible as vegetables. Suioh leaves contain CQA derivative, and there have been numerous health benefits reported so far regarding its consumption (Kurata *et al.*, 2007; Nagai *et al.*, 2011). Dietary Suioh leaf extract (SP) significantly suppresses the increase in blood glucose level in type 2 diabetes model mice. The possible mechanisms of this action include α-glucosidase inhibition in the intestinal tract and upregulation of adiponectin in white adipose tissues. However, as the inhibitory action of α-glucosidase by SP is quite weak and SP does not induce the expression of adiponectin, the stimulation of GLP-1 secretion was proposed as an alternative mechanism of action by SP. The administration of SP strongly stimulated GLP-1 secretion by enteroendocrine cells. In addition, GLP-1 secretion was also significantly stimulated by 3,4-diCQA and 3,4,5-triCQA, which are components of SP. When blood glucose concentration was monitored in rats intraperitoneally administered glucose after they were given SP, a significant increase in blood insulin concentration was accompanied by a significant decrease in blood glucose concentration, compared to the control group (no SP administration + intraperitoneal glucose). In the same experiment, both active and total GLP-1 concentrations in the blood were significantly increased after SP administration compared to the control. The transition of blood glucose and insulin concentrations after the administration of insulin secretagogue sulfonylurea

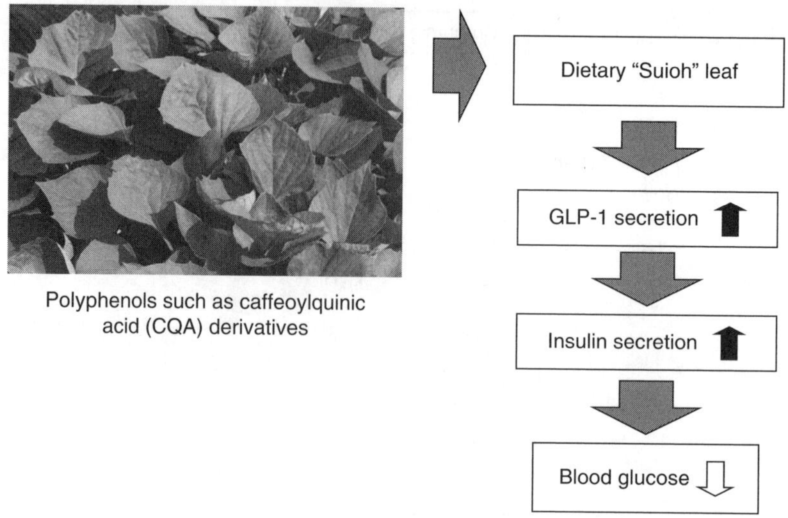

Fig. 10.10 Proposed mechanism for amelioration of hyperglycemia by dietary sweet potato leaves, Suioh, *in vivo*. Nagamine *et al.* (2014). Reproduced with permission from the Royal Society of Chemistry. (*See insert for color representation of the figure*)

and SP preparations has also been studied in rats. The result revealed that the administration of sulfonylurea causes hypoglycemia due to the strong induction of insulin secretion even without the administration of glucose, whereas the administration of SP alone did not cause hypoglycemia. From these results, it has become evident that the mechanism for prevention and suppression of diabetes by SP involves the stimulation of GLP-1 secretion, and in the case of high blood glucose, the promotion of insulin secretion, causing normalization of blood glucose concentration (Nagamine *et al.*, 2014) (Fig. 10.10). Unlike insulin secretagogues, SP does not induce hypoglycemia, as its action of blood glucose reduction is mediated via its promotion of GLP-1 secretion.

These studies demonstrate that the elucidation of the mechanisms of GLP-1-secretion-stimulating action of polyphenols and their diabetes-preventing and -suppressing effects may advance the development of novel diabetes-preventing foods and medication. Future prospects include the validation of such effects in humans and the elucidation of the primary target of dietary factor action on GLP-1 secretion.

10.3.2.3 Delphinidin 3-rutinoside (D3R)

A recent epidemiological study has indicated that higher consumption of anthocyanins and anthocyanin-rich fruit is associated with lower risk of type 2 diabetes (Wedick *et al.*, 2012). The molecular mechanism underlying this effect can be explained by the activation of AMPK (Takikawa *et al.*, 2010; Kurimoto *et al.*, 2013). To date, six principal types of anthocyanidins have been identified, and many anthocyanins can be derived from these six types through structural modifications, such as attachment of various numbers and types of sugars. For this reason, anthocyanins may have various antidiabetic effects through mechanisms other than activation of AMPK, such as modification of GLP-1. Here, we show that

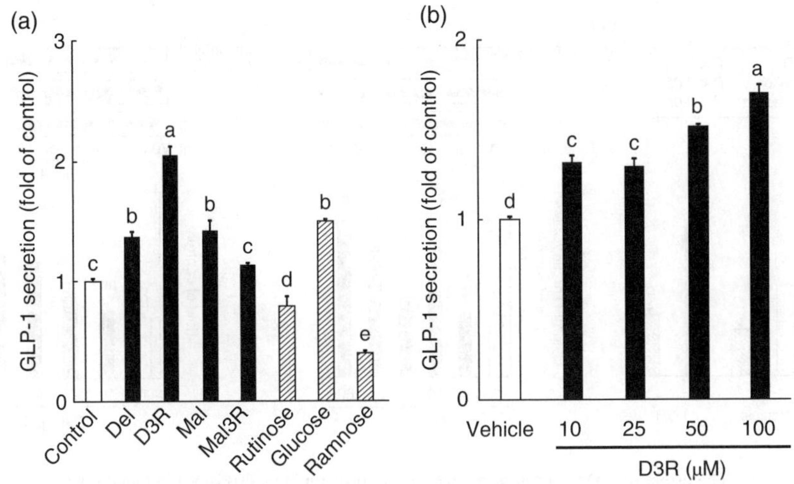

Fig. 10.11 GLP-1 secretion in the medium of GLUTag cells treated with various anthocyanins (a) and/or varying concentrations of D3R (b). (a) The final concentration of anthocyanins and sugars are 100 μM. Values are expressed as the means ± SEM, $n = 3–9$. Values without a common letter differ significantly at $P < 0.05$ (Kato *et al.*, 2015). Mal3R, malvidin 3-rutinoside. Kato *et al.* (2015). Used under CC-BY-4.0 http://creativecommons.org/licenses/by/4.0/.

D3R significantly increased secretion of GLP-1 in an enteroendocrine L cell model (GLUTag cells) and demonstrated its structure–activity relationship using anthocyanin derivatives. Moreover, this increase was found to involve the IP_3R-mediated intracellular Ca^{2+}-CaMKII pathway.

The screening showed that three anthocyanins, D3R (which contains three B-ring hydroxy moieties and rutinose), delphinidin (Del, the aglycone of D3R), and malvidin (Mal, which has two methoxy moieties on the B-ring), significantly increased in GLP-1 secretion in GLUTag cells (Fig. 10.11A). Notably, the D3R-induced GLP-1 secretion was markedly higher compared with the secretion induced by Del and Mal treatment. In addition, we confirmed that treatment of GLUTag cells with 10 μM D3R nevertheless significantly stimulates GLP-1 secretion, and rutinose itself did not enhance GLP-1 secretion (Fig. 10.11A and B). From these screening results, D3R was selected for further studies of the GLP-1 secretion mechanism.

To clarify how D3R induces GLP-1 secretion in *in vitro* cells, two molecular mechanisms are possible: (i) elevation of cytosolic Ca^{2+} and results in activation of CaMKII, which is involved in GLP-1 secretion (Takikawa *et al.*, 2013); and (ii) stimulation of the cAMP/protein kinase (PK) A, PKCζ, or ERK pathways (Reimer, 2006; Iakoubov *et al.*, 2007; Lauffer *et al.*, 2009; Parker *et al.*, 2012; Sangle *et al.*, 2012). In this study, we demonstrated that stimulation of GLP-1 secretion by D3R is completely abolished by the preadministration of an intracellular Ca^{2+} chelator. Furthermore, D3R-stimulated GLP-1 secretion was significantly inhibited by preadministration of an IP_3 receptor antagonist, but not by an RyR antagonist or L-type Ca^{2+} channel blocker. These results indicate that D3R stimulates intracellular Ca^{2+} mobilization via IP_3R. We have already demonstrated that curcumin-stimulated GLP-1 secretion is regulated by the Ca^{2+}-CaMKII pathway

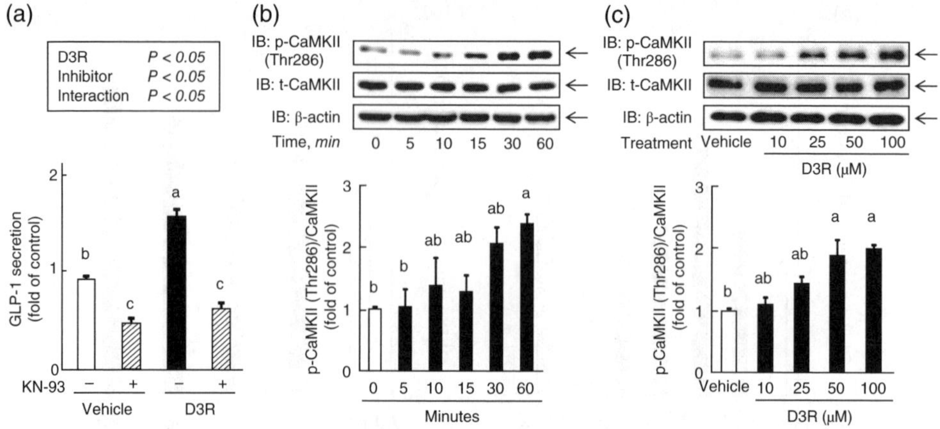

Fig. 10.12 (a) The effect of CaMKII inhibitor (KN-93, 10 μM) on D3R (100 μM)-induced GLP-1 secretion. (b and c) Immunoblot analysis of the phosphorylated CaMKII, total CaMKII, and β-actin protein treated with D3R duration (b) and dose (c) in GLUTag cells. Values are expressed as the means ± SEM, $n = 3$. Values without a common letter significantly differ at $P < 0.05$. Kato *et al.* (2015). Used under CC-BY-4.0 http://creativecommons.org/licenses/by/4.0/.

(Takikawa *et al.*, 2013). As shown in Fig. 10.12, pretreatment of GLUTag cells with KN-93 (a CaMKII inhibitor) abolished D3R-stimulated GLP-1 secretion. In addition, D3R enhanced the phosphorylation of CaMKII in GLUTag cells in both a time- and a concentration-dependent manner, indicating that the D3R-induced GLP-1 secretion is regulated via Ca^{2+}-CaMKII pathway in GLUTag cells. However, the exact exocytosis mechanism of GLP-1 from L cells remains unclear. A recent study indicated that vesicle-associated membrane protein 2 (VAMP2), which is involved in vesicular transport and membrane fusion, has an essential role in GLP-1 exocytosis (Li *et al.*, 2014). One report indicated that binding of CaMKII to syntaxin1a, which interacts with VAMP2, is an important process in the regulation of exocytosis (Ohyama *et al.*, 2002). These results suggest that the D3R-induced stimulation of CaMKII modulates VAMP2-mediated GLP-1 exocytosis.

GPR agonists stimulate GLP-1 secretion in intestinal L cells (Hirasawa *et al.*, 2005; Edfalk *et al.*, 2008; Lan *et al.*, 2012; Parker *et al.*, 2012). The pretreatment of GLUTag cells with a GPR40/120 antagonist resulted in a significant decrease of D3R-stimulated GLP-1 secretion. However, GPR119- or TGR5-related Gαs subunit antagonists did not affect D3R-induced GLP-1 secretion, suggesting that GPR40/120–intracellular Ca^{2+} mobilization via IP3R is involved in this process. Additionally, D3R-stimulated GLP-1 secretion is not regulated by the GPR119- or TGR5-related cAMP/PKA, PKCζ, or ERK pathways. Figure 10.13 shows the proposed mechanism for stimulation of GLP-1 secretion by D3R in GLUTag cells.

10.4 Future issues and prospects

This chapter describes the health benefit of polyphenols from the point of view of diabetes prevention and suppression, as well as their mechanisms of action. Finally, we present four issues that remain concerning the research of polyphenol health benefits.

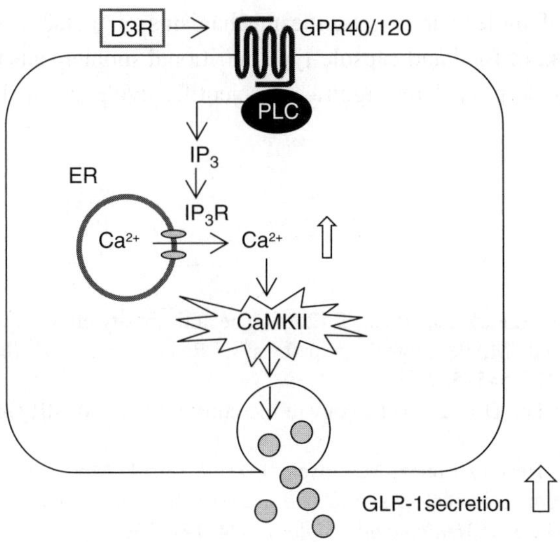

Fig. 10.13 Proposed mechanism for stimulation of GLP-1 secretion by D3R in GLUTag cells. ER, endoplasmic reticulum. Kato *et al.* (2015). Used under CC-BY-4.0 http://creativecommons.org/licenses/by/4.0/.

The first issue is the determination of the involvement of the various degradation products and metabolites derived from the intake of polyphenols in their manifested health benefits. For instance, anthocyanins are absorbed directly in the form of glycosides. However, chemical degradation products and metabolites produced by intestinal flora derived from anthocyanins are considered to play an important role in the manifestation of these health benefits. A study of anthocyanin bioavailability in humans using ^{13}C-labeled compounds has revealed that degradation products and metabolites derived from the parent anthocyanin can be isolated for quite a long time in the body, urine, and so on (Czank *et al.*, 2013). It follows that in explaining the involvement of polyphenols with health benefits, the enterobacterial flora and their metabolites must be considered and the most effective polyphenol structure verified.

Secondly, the signaling molecules that interact with polyphenols and their signaling pathways are still being elucidated through technological innovations. Further continuation of the detailed elucidation of polyphenol mechanisms of action in their health benefit is necessary.

The third issue is that there are many unknowns in the interaction of polyphenols with other components. Polyphenols are ingested simultaneously with various nutritional components. It is thus necessary to determine whether the presence of other components is necessary for polyphenols to manifest their effects.

The fourth issue is that there is insufficient validation of polyphenol health benefits in humans, despite a recent number of studies, which has great relevance for the industrial application of polyphenol health benefits.

The research on polyphenols would certainly continue to be attractive in health benefit studies as well as in other fields. We anticipate that polyphenol health benefit studies will continue to advance the elucidation of the mechanism of actions at the molecular level, the identification of molecular targets, the mechanisms of metabolism and absorption, and the development of food and capsule-type nutritional supplements that provide health benefits that are underpinned by significant scientific evidence, including randomized controlled trials.

References

Ahn, J., Lee, H., Kim, S., Park, J. and Ha, T. (2008) The anti-obesity effect of quercetin is mediated by the AMPK and MAPK signaling pathways. *Biochemical and Biophysical Research Communications*, **373**, 545–549.

Baggio, L.L. and Drucker, D.J. (2007) Biology of incretins: GLP-1 and GIP. *Gastroenterology*, **132**, 2131–2157.

Barrajón-Catalán, E., Herranz-López, M., Joven, J. *et al.* (2014) Molecular promiscuity of plant polyphenols in the management of age-related diseases: far beyond their antioxidant properties. *Advances in Experimental Medicine and Biology*, **824**, 141–159.

Baur, J.A., Pearson, K.J., Price, N.L. *et al.* (2006) Resveratrol improves health and survival of mice on a high-calorie diet. *Nature*, **444**, 337–342.

Chuengsamarn, S., Rattanamongkolgulz, S., Luechapudiporn, R., Phisalaphong, C. and Jirawatnotai, S. (2012) Curcumin extract for prevention of type 2 diabetes. *Diabetes Care*, **35**, 2121–2127.

Clemmensen, C., Smajilovic, S., Smith, E.P. *et al.* (2013) *Endocrinology*, **154**, 3978–3983.

Collins, Q.F., Liu, H.Y., Pi, J., Liu, Z., Quon, M.J. and Cao, W. (2007) Epigallocatechin-3-gallate (EGCG), a green tea polyphenol, suppresses hepatic gluconeogenesis through 59-AMP-activated protein kinase. *Journal of Biological Chemistry*, **282**, 30143–30149.

Czank, C., Cassidy, A., Zhang, Q. *et al.* (2013) Human metabolism and elimination of the anthocyanin, cyanidin-3-glucoside: a ^{13}C-tracer study. *American Journal of Clinical Nutrition*, **97**, 995–1003.

Dao, T.M., Waget, A., Klopp, P. *et al.* (2011) Resveratrol increases glucose induced GLP-1 secretion in mice: a mechanism which contributes to the glycemic control. *PLoS One*, **6**, e20700.

Dixit, S.S., Wang, T., Manzano, E.J. *et al.* (2013) Effects of CaMKII-mediated phosphorylation of ryanodine receptor type 2 on islet calcium handling, insulin secretion, and glucose tolerance. *PLoS One*, **8**, e58655.

Edfalk, S., Steneberg, P. and Edlund, H. (2008) Gpr40 is expressed in enteroendocrine cells and mediates free fatty acid stimulation of incretin secretion. *Diabetes*, **57**, 2280–2287.

Foretz, M., Ancellin, N., Andreelli, F. *et al.* (2005) Short-term overexpression of a constitutively active form of AMP-activated protein kinase in the liver leads to mild hypoglycemia and fatty liver. *Diabetes*, **54**, 1331–1339.

González-Abuín, N., Martínez-Micaelo, N., Margalef, M. *et al.* (2014a) A grape seed extract increases active glucagon-like peptide-1 levels after an oral glucose load in rats. *Food & Function*, **5**, 2357–2364.

González-Abuín, N., Martínez-Micaelo, N., Blay, M., Green, B.D., Pinent, M. and Ardévol, A. (2014b) Grape-seed procyanidins modulate cellular membrane potential and nutrient-induced GLP-1 secretion in STC-1 cells. *American Journal of Physiology. Cell Physiology*, **306**, C485–C492.

González-Abuín, N., Martínez-Micaelo, N., Blay, M., Ardévol, A. and Pinent, M. (2014c) Grape-seed procyanidins prevent the cafeteria-diet-induced decrease of glucagon-like peptide-1 production. *Journal of Agricultural and Food Chemistry*, **62**, 1066–1072.

Hardie, D.G. (2007) AMP-activated/SNF1 protein kinases: conserved guardians of cellular energy. *Nature Reviews Molecular Cell Biology*, **8**, 774–785.

Hardie, D.G. (2008) Role of AMP-activated protein kinase in the metabolic syndrome and in heart disease. *FEBS Letters*, **582**, 81–89.

Hardie, D.G. (2011) Sensing of energy and nutrients by AMP-activated protein kinase. *American Journal of Clinical Nutrition*, **93**, 891S–896S.

Hawley, S.A., Gadalla, A.E., Olsen, G.S. and Hardie, D.G. (2002) The anti-diabetic drug metformin activates the AMP-activated protein kinase cascade via an adenine nucleotide-independent mechanism. *Diabetes*, **51**, 2420–2425.

Herman, G.A., Bergman, A., Stevens, C. *et al.* (2006) Effect of single oral doses of sitagliptin, a dipeptidyl peptidase-4 inhibitor, on incretin and plasma glucose levels after an oral glucose tolerance test in patients with type 2 diabetes. *Journal of Clinical Endocrinology & Metabolism*, **91**, 4612–4619.

Higuchi, N., Hira, T., Yamada, N. and Hara, H. (2013) Oral administration of corn zein hydrolysate stimulates GLP-1 and GIP secretion and improves glucose tolerance in male normal rats and Goto-Kakizaki rats. *Endocrinology*, **154**, 3089–3098.

Hira, T., Mochida, T., Miyashita, K. and Hara, H. (2009) GLP-1 secretion is enhanced directly in the ileum but indirectly in the duodenum by a newly identified potent stimulator, zein hydrolysate, in rats. *American Journal of Physiology-Gastrointestinal and Liver Physiology*, **297**, G663–G671.

Hira, T., Muramatsu, M., Okuno, M. and Hara, H. (2011) GLP-1 secretion in response to oral and luminal palatinose (isomaltulose) in rats. *Journal of Nutritional Science and Vitaminology*, **57**, 30–35.

Hirasawa, A., Tsumaya, K., Awaji, T. *et al.* (2005) Free fatty acids regulate gut incretin glucagon-like peptide-1 secretion through GPR120. *Nature Medicine*, **11**, 90–94.

Iakoubov, R., Izzo, A., Yeung, A., Whiteside, C.I. and Brubaker, P.L. (2007) Protein kinase Czeta is required for oleic acid-induced secretion of glucagon-like peptide-1 by intestinal endocrine L cells. *Endocrinology*, **148**, 1089–1098.

Jang, H.J., Kokrashvili, Z., Theodorakis, M.J. *et al.* (2007) Gut-expressed gustducin and taste receptors regulate secretion of glucagon-like peptide-1. *Proceedings of the National Academy of Sciences of the United States of America*, **104**, 15069–15074.

Janssen, S. and Depoortere, I. (2013) Nutrient sensing in the gut: new roads to therapeutics? *Trends in Endocrinology & Metabolism*, **24**, 92–100.

Kato, M., Tani, T., Terahara, N. *et al.* (2015) The anthocyanin delphinidin 3-rutinoside stimulates glucagon-like peptide-1 secretion in murine GLUTag Cell Line via the Ca^{2+}/Calmodulin-dependent kinase II Pathway. *PLoS ONE*, **10**, e0126157.

Kim, J.H., Park, J.M., Kim, E.K. *et al.* (2010) Curcumin stimulates glucose uptake through AMPK-p38 MAPK pathways in L6 myotube cells. *Journal of Cellular Physiology*, **223**, 771–778.

Koo, S.H., Flechner, L., Qi, L. *et al.* (2005) The CREB coactivator TORC2 is a key regulator of fasting glucose metabolism. *Nature*, **437**, 1109–1111.

Kurata, R., Adachi, M., Yamakawa, O. and Yoshimoto, M. (2007) Growth suppression of human cancer cells by polyphenolics from sweet potato (*Ipomoea batatas* L.) leaves. *Journal of Agricultural and Food Chemistry*, **55**, 185–190.

Kurimoto, Y., Shibayama, Y., Inoue, S. *et al.* (2013) Black soybean seed coat extract ameliorates hyperglycemia and insulin sensitivity via the activation of AMP-activated protein kinase in diabetic mice. *Journal of Agricultural and Food Chemistry*, **61**, 5558–5564.

Lan, H., Lin, H.V., Wang, C.F. *et al.* (2012) Agonists at GPR119 mediate secretion of GLP-1 from mouse enteroendocrine cells through glucose-independent pathways. *British Journal of Pharmacology*, **165**, 2799–2807.

Lauffer, L.M., Iakoubov, R. and Brubaker, P.L. (2009) GPR119 is essential for oleoylethanolamide-induced glucagon-like peptide-1 secretion from the intestinal enteroendocrine L-cell. *Diabetes*, **58**, 1058–1066.

Lee, S.J., Kim, H.E., Choi, S.E. *et al.* (2009) Involvement of Ca^{2+}/calmodulin kinase II (CaMK II) in genistein-induced potentiation of leucine/glutamine-stimulated insulin secretion. *Molecules and Cells*, **28**, 167–174.

Lehrke, M. and Lazar, M.A. (2005) The many faces of PPARγ. *Cell*, **123**, 993–999.

Li, S.K., Zhu, D., Gaisano, H.Y. and Brubaker, P.L. (2014) Role of vesicle-associated membrane protein 2 in exocytosis of glucagon-like peptide-1 from the murine intestinal L cell. *Diabetologia*, **57**, 809–818.

Lin, C.L., Huang, H.C. and Lin, J.K. (2007) Theaflavins attenuate hepatic lipid accumulation through activating AMPK in human HepG2 cells. *Journal of Lipid Research*, **48**, 2334–2343.

Liu, C.Y., Huang, C.J., Huang, L.H., Chen, I.J., Chiu, J.P. and Hsu, C.H. (2014) Effects of green tea extract on insulin resistance and glucagon-like peptide 1 in patients with type 2 diabetes and lipid abnormalities: a randomized, double-blinded, and placebo-controlled trial. *PLoS One*, **9**, e91163.

Lovshin, J.A. and Drucker, D.J. (2009) Incretin-based therapies for type 2 diabetes mellitus. *Nature Reviews Endocrinology*, **5**, 262–269.

Minokoshi, Y., Kim, Y.B., Peroni, O.D. *et al.* (2002) Leptin stimulates fatty-acid oxidation by activating AMP-activated protein kinase. *Nature*, **415**, 339–343.

Mochida, T., Hira, T. and Hara, H. (2010) The corn protein, zein hydrolysate, administered into the ileum attenuates hyperglycemia via its dual action on glucagon-like peptide-1 secretion and dipeptidyl peptidase-IV activity in rats. *Endocrinology*, **151**, 3095–3104.

Nagai, M., Tani, M., Kishimoto, Y. *et al.* (2011) Sweet potato (*Ipomoea batatas* L.) leaves suppressed oxidation of low density lipoprotein (LDL) *in vitro* and in human subjects. *Journal of Clinical and Biochemical Nutrition*, **48**, 203–208.

Nagamine, R., Ueno, S., Tsubata, M. *et al.* (2014) Dietary sweet potato (*Ipomoea batatas* L.) leaf extract attenuates hyperglycaemia by enhancing the secretion of glucagon-like peptide-1 (GLP-1). *Food & Function*, **5**, 2309–2316.

Ohyama, A., Hosaka, K., Komiya, Y. *et al.* (2002) Regulation of exocytosis through Ca^{2+}/ATP-dependent binding of autophosphorylated Ca^{2+}/calmodulin-activated protein kinase II to syntaxin 1A. *Journal of Neurosciences*, **22**, 3342–3351.

Parker, H.E., Wallis, K., Le Roux, C.W., Wong, K.Y., Reimann, F. and Gribble, F.M. (2012) Molecular mechanisms underlying bile acid-stimulated glucagon-like peptide-1 secretion. *British Journal of Pharmacology*, **165**, 414–423.

Phuwamongkolwiwat, P., Hira, T. and Hara, H. (2014) A nondigestible saccharide, fructooligosaccharide, increases the promotive effect of a flavonoid, α-glycosyl-isoquercitrin, on glucagon-like peptide 1 (GLP-1) secretion in rat intestine and enteroendocrine cells. *Molecular Nutrition & Food Research*, **58**, 1581–1584.

Reimer, R.A. (2006) Meat hydrolysate and essential amino acid-induced glucagon-like peptide-1 secretion, in the human NCI-H716 enteroendocrine cell line, is regulated by extracellular signal-regulated kinase1/2 and p38 mitogen-activated protein kinases. *Journal of Endocrinology*, **191**, 159–170.

Samocha-Bonet, D., Wong, O., Synnott, E.L. *et al.* (2011) Glutamine reduces postprandial glycemia and augments the glucagon-like peptide-1 response in type 2 diabetes patients. *Journal of Nutrition*, **141**, 1233–1238.

Sangle, G.V., Lauffer, L.M., Grieco, A., Trivedi, S., Iakoubov, R. and Brubaker, P.L. (2012) Novel biological action of the dipeptidylpeptidase-IV inhibitor, sitagliptin, as a glucagon-like peptide-1 secretagogue. *Endocrinology*, **153**, 564–573.

Scully, T. (2012) Diabetes in numbers. *Nature*, **485**, S2–S3.

Shepherd, P.R. and Kahn, B.B. (1999) Glucose transporters and insulin action: implications for insulin resistance and diabetes mellitus. *New England Journal of Medicine*, **341**, 248–257.

Takikawa, M., Inoue, S., Horio, F. and Tsuda, T. (2010) Dietary anthocyanin-rich bilberry extract ameliorates hyperglycemia and insulin sensitivity via activation of AMP-activated protein kinase in diabetic mice. *Journal of Nutrition*, **140**, 527–533.

Takikawa, M., Kurimoto, Y. and Tsuda, T. (2013) Curcumin stimulates glucagon-like peptide-1 secretion in GLUTag cells via Ca^{2+}/calmodulin-dependent kinase II activation. *Biochemical and Biophysical Research Communications*, **435**, 165–170.

Tsuda, T. (2015) Possible abilities of dietary factors to prevent and treat diabetes via the stimulation of glucagon-like peptide-1 secretion. *Molecular Nutrition & Food Research*, **59**, 1264–1273.

Wedick, N.M., Pan, A., Cassidy, A. *et al.* (2012) Dietary flavonoid intakes and risk of type 2 diabetes in US men and women. *American Journal of Clinical Nutrition*, **95**, 925–933.
Yamauchi, T., Kamon, J., Waki, H. *et al.* (2001) The mechanisms by which both heterozygous peroxisome proliferator-activated receptorγ (PPARγ) deficiency and PPARγ agonist improve insulin resistance. *Journal of Biological Chemistry*, **276**, 41245–41254.
Yamauchi, T., Kamon, J., Minokoshi, Y. *et al.* (2002) Adiponectin stimulates glucose utilization and fatty-acid oxidation by activating AMP-activated protein kinase. *Nature Medicine*, **8**, 1288–1295.
Zhou, G., Myers, R., Li, Y. *et al.* (2001) Role of AMP-activated protein kinase in mechanism of metformin action. *Journal of Clinical Investigation*, **108**, 1167–1174.

Chapter 11

Beneficial Vascular Responses to Proanthocyanidins: Critical Assessment of Plant-Based Test Materials and Insight into the Signaling Pathways

Herbert Kolodziej

Abstract: A large body of evidence has demonstrated beneficial effects of polyphenol-rich diets and herbal medicines on the vascular system. However, polyphenols encompass diverse compounds, and conclusions are limited by poor extract compositional data and lack of bioassay-guided fractionation aimed at identifying the active constituents. Research on the vasorelaxant potential of chemically defined proanthocyanidins is hitherto confined. A few studies focused only on purified fractions or well-characterized compounds, supporting unambiguously the beneficial effects of proanthocyanidins on endothelial dysfunction. The proanthocyanidin-containing samples evoked endothelium-dependent relaxation in arteries through multiple mechanisms, including the NO-cGMP pathway and, to a lesser extent, activation of the endothelium-derived hyperpolarizing factors (EDHF)-signaling cascade in a redox-sensitive manner. EDHF events are primarily associated with the modulation of Ca^{2+}-dependent and -independent K^+ channel functions. In addition, the vascular tone appears to be regulated to various extents by the intracellular Ca^{2+} concentration. This chapter presents a brief overview of the signaling events in relaxation, summarizes the literature on the vasorelaxant effects of proanthocyanidins in isolated arteries, and provides a critical assessment of plant-based test materials.

Keywords: proanthocyanidins, vasorelaxation, nitric oxide, eNOS, EDHF, hyperpolarization, signaling pathways, K^+ channels

---

Recent Advances in Polyphenol Research, First Edition. Edited by Kumi Yoshida, Véronique Cheynier and Stéphane Quideau.

11.1 Introduction

Cardiovascular diseases (CVDs) remain a leading cause of morbidity and mortality worldwide and are responsible for increased costs of medical treatment (Fig. 11.1). Although death rates from CVDs are decreasing in many developed countries due to current therapeutic concepts that commonly effectively target the various pathological conditions, developing and underdeveloped nations show increasing numbers attributable, at least in part, to greater longevity and lifestyle changes. For example, the demographic and nutritional transition in sub-Saharan Africa is reported to cause a dramatic increase in cardiovascular disease related deaths (Seedat & Rosenthal, 2006; Mensah, 2008). An estimated 17.3 million people died from CVDs in 2008 representing 30% of all global deaths. Of these, about 7.3 million were due to coronary heart disease and 6.2 million were due to stroke, with high blood pressure representing a major risk factor. Conspicuously, most deaths take place in low- and middle-income countries, where people have less access to effective and equitable health care services and do not have the benefit of prevention programs. According to WHO reports, CVD deaths will increase to 23.3 million by 2030, a rather frightening scenario.

CVDs comprise a group of disorders of the heart and blood vessels leading to atherosclerosis, heart attack, stroke, hypertension, hyperlipidemia, obesity, and diabetes. It should be noted that atherosclerosis is the leading cause of death in the Western World. The major reasons from the perspective of lifestyle include tobacco use, harmful consumption of alcohol, an unhealthy diet, physical inactivity, and genetic disposition. Accordingly, a healthy diet, for example, fruits and vegetables, regular physical activity to maintain a wholesome body weight, and the avoidance of excessive tobacco and alcohol uses are considered to have favorable effects on the pathogenesis of CVDs. These disorders engage a variety of cell types, including endothelial cells, vascular smooth muscle cells (VSMC), leukocytes, and platelets. Although a complex array of contributing factors have been implicated in these pathological conditions, many are associated with vascular endothelial dysfunction. Indeed, a functional endothelium is recognized to play a vital role in vascular homeostasis and vascular tone regulation (Furchgott & Zawadzki, 1980), preventing unwanted vascular

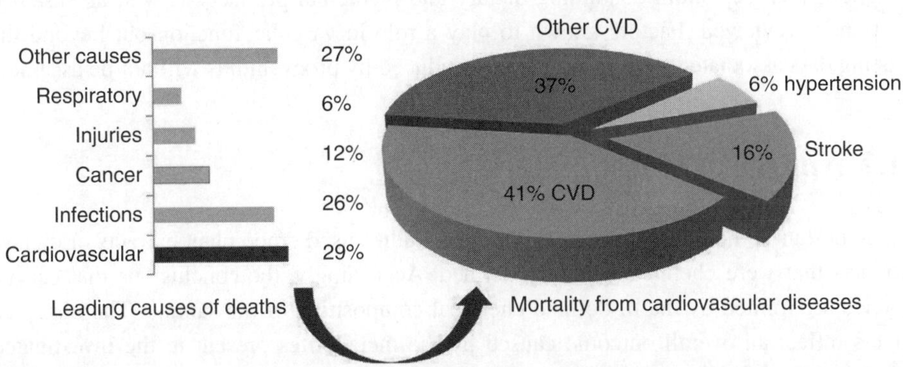

Fig. 11.1 Global fatalities and cardiovascular diseases. Data from WHO (2008).

changes in healthy vessels that lead to cardiovascular pathologies. Moreover, endothelial dysfunction is an early key event in these pathogeneses that occurs before structural changes in the vessel wall become evident.

Numerous epidemiological studies have indicated an inverse association between the regular consumption of diets rich in foods of plant origin in Mediterranean countries and risk for developing CVDs. The amount of published work on polyphenols related to vascular health benefits is immense. Because of space limitations, the reader is referred for a more complete coverage to a few selected key reviews (Stoclet *et al.*, 2004; Rasmussen *et al.*, 2005; Clifford & Brown, 2006; Schini-Kerth *et al.*, 2010a, 2010b, 2011; Visioli, *et al.*, 2011). The protective effect of plant-based diets has been related, at least in part, to their high level of bioactive polyphenols. Independent support for this finding is provided by experimental and clinical investigations demonstrating that polyphenol-rich natural products are able to improve endothelial dysfunction and related pathological conditions. An excellent account of general current knowledge can be found in some recent reviews (Manach *et al.*, 2005a; Chong *et al.*, 2010; Habauzit & Morand, 2012; Del Rio *et al.*, 2013; Rodriguez-Mateos *et al.*, 2014). However, plant polyphenols including flavonoids and tannins vary widely in their chemical structures and even within subclasses there is a broad structural diversity. Even though various *in vitro* studies have reported beneficial effects of diverse classes of plant polyphenols on the vascular system, the characterization of individual active compounds remains largely elusive. A more detailed assessment of health-promoting effects of specific polyphenols is needed.

Although widely distributed in plants and common constituents of herbal medicines and the human diet, research on proanthocyanidins in the context of vascular disease is limited. The objective of this chapter is to record recent advances in the perceived role of proanthocyanidins as vasoprotective plant polyphenols and place these findings in a context of mechanisms by which these polyphenols may confer cardiovascular protection. It will first present a critical evaluation of proanthocyanidin-containing test materials, then introduce the endothelium and briefly the underlying signaling pathways including the phosphatidylinositol-3 kinase (PI3K)/Akt pathway, the endothelial nitric oxide synthase (eNOS)/nitric oxide (NO)/cyclic guanosyl monophosphate (cGMP)- and the endothelium-derived hyperpolarizing factor (EDHF)-signaling pathway before current knowledge on the mode of actions of procyanidin-containing dietary and medicinal products as well as enriched fractions is reviewed. Factors known to play a role in vascular function but beyond the existing data associated with vasorelaxation induced by procyanidins will not be assessed.

11.2 Appraisal of test materials

It is important to note that the vast majority of studies used crude plant extracts or dietary products that were chemically poorly defined. Accordingly, the conclusions that can be reached are limited by the ill-defined chemical composition of the sample. The observed effects reflect an overall outcome caused by the metabolites present in the investigated extracts or products. Interactions between phenolic compounds may have important consequences for the activation of signaling cascades and hence efficacy of the plant product.

The type and quantity of polyphenols in foodstuffs and plant products are of paramount importance to the biological effect, and additive, synergistic, and even antagonistic effects of co-occurring polyphenols should be considered. The variability in extract composition and the procedure employed in product preparation may also be critical when comparing different studies carried out on the same plant-based items. Quantification of total phenolic content of complex mixtures related to activity makes only sense if the proportions of the various polyphenols are taken into account. These notions imply that studies on biological effects of polyphenols require detailed information about the origin of the test material, extraction/product preparation, and chemical composition.

Although some proanthocyanidin-enriched materials provided useful data for vasoprotective effects, chemically well-defined samples represent more convincing probes. To date, these are confined to a few examples including 2,3-*cis*- (Kaufeld *et al.*, 2013), 2,3-*trans*-procyanidins (Kaufeld *et al.*, 2014) as well as the trimer epicatechin-[4β,8]-epicatechin-[4β,8]-epicatechin (procyanidin C1) (Byun, 2012; Byun *et al.*, 2012a, 2014). A more detailed assessment of effects of individual proanthocyanidins and their metabolites may significantly contribute to a better understanding of their impact on pathologic conditions. An account of key questions involved in accounting for a causal relationship between intake of procyanidins and health benefits in humans was recently addressed (Schroeter *et al.*, 2010). In this context, a brief note on the bioavailability of proanthocyanidins appears appropriate. Current information is still confined and conflicting, with indications of a limited absorption of low-molecular-weight candidates (≤trimers) and a key role of the bacterial intestinal flora in the production of more absorbable, biologically active metabolites (Rasmussen *et al.*, 2005; Manach *et al.*, 2005a; Appeldoorn *et al.*, 2009b; Borges *et al.*, 2013; Del Rio *et al.*, 2013; Guerrero *et al.*, 2013). On the contrary, the detection of higher oligomers in urea-extracted rat plasma to overcome the binding to proteins indicates how important the analytical approach may be (Shoji *et al.*, 2006). Although the absorption of higher oligomeric proanthocyanidins appears questionable, low but sustained proanthocyanidin plasma levels reached by long-term dietary exposures may well mediate beneficial effects. Supporting evidence for this conjecture may be obtained from long-term clinical studies using dietary-relevant doses of defined proanthocyanidin mixtures. Also worthy to consider is possible tissue accumulation of potentially beneficial compounds, exemplified by tea polyphenols (Henning *et al.*, 2006). Such "arrested" metabolic pools are not included in measured circulating forms. Finally, proanthocyanidins larger than trimers proved to be most effective in some experimental models. Clearly, the human body is the ideal site on which to demonstrate the ultimate evidence rather than *in vitro* studies. However, while the absorbability of proanthocyanidins is commonly judged under physiological conditions, the bioavailability might differ in a diseased status. Information on this point is hitherto not available.

11.2.1 Analytical challenges of proanthocyanidin composition

The complexity of proanthocyanidin extract composition and the close structural similarity of constituent oligomers (Fig. 11.2) represent severe complicating factors in their isolation and purification. Application of combined chromatographic methods facilitates successful

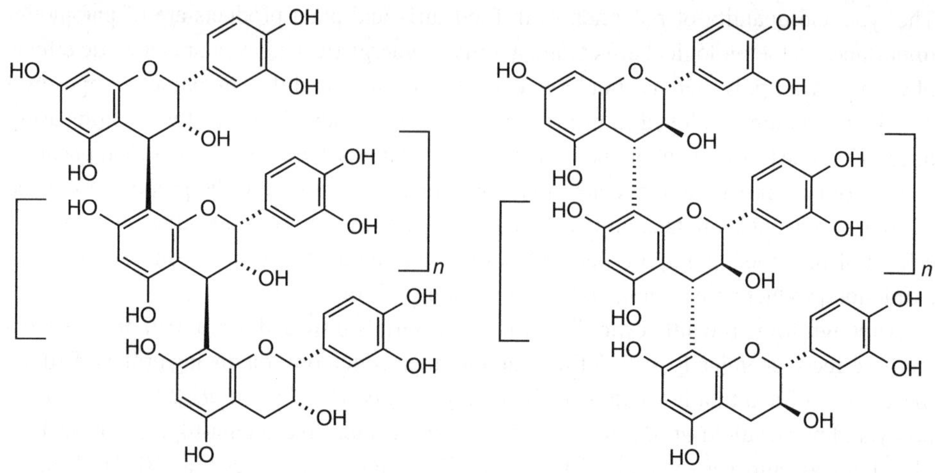

Fig. 11.2 Representative structures of 2,3-*cis*- (left) and 2,3-*trans*-procyanidins (right).

purification of enriched fractions and individual proanthocyanidins. On the other hand, problems in the structure elucidation of proanthocyanidins include assessment of bonding positions, assignment of the configuration at the points of bonding, characterization of constituent flavanyl units, and definition of their sequential order in oligomers. Clearly, comprehensive structure elucidation of proanthocyanidins present in test materials appears extremely difficult if not impossible, but the quantity, the degree of polymerization (DP), and the proportion of the different oligomers and the constituent flavanyl units should be given. For this, size elution chromatography, HPLC analysis, chemical methods, and mass spectrometry are useful tools.

11.2.2 Chemical data on proanthocyanidin-containing materials

Although widely distributed in nature, only a limited number of proanthocyanidin-containing natural products and plant extracts have hitherto been reported to provide cardiovascular protection. Conspicuously, experiments were commonly conducted on chemically poorly defined extracts. Reasonably well-characterized procyanidin-rich extracts that have been shown to induce vasorelaxations in aortic rings of experimental animals include extracts of maritime pine bark (Pycnogenol), hawthorn (WS 1442), willow, and *Nelia meyeri*, along with some apple- and grape-derived products. Table 11.1 provides an overview of proanthocyanidin-containing plant sources studied for vasorelaxant activity with reported chemical data on proanthocyanidin profiles. It should be noted that quantification of polyphenols is based on different methods and identification of proanthocyanidins remains largely disputable.

French maritime pine bark (*Pinus pinaster*, syn.: *Pinus maritima*) extracts contain a complex mixture of polyphenols, with a substantial proportion of procyanidins. Pycnogenol is a standardized water extract obtained from a monoculture with consistent quality of the raw material (Rohdewald, 2002). Besides a range of phenolic acids, the extract contains taxifolin,

Table 11.1 Data on the chemical composition of procyanidin samples.

Plant source	Product	DP	Quantitative data	Constituent units	References
Combretum celastroides	Extract	OPC			Manga *et al.* (2012)
Crataegus monogyna *Crataegus laevigata*	WS 1442	2–8	17.3–20.1% OPC 18.75% OPC		Idris-Khodja *et al.* (2012a), Anselm *et al.* (2009), and Brixius *et al.* (2006)
	Extract	OPC	5.1% OPC		Kim *et al.* (2000)
	Fruits	OPC			Chen *et al.* (1998)
	Juice	OPC	80% OPC		Corder *et al.* (2004)
Croton celtidifolius	Extract	OPC	43.3% OPC	C, EC, GC	DalBó *et al.* (2008a, 2008b)
Diospyros kaki	Extract	mDP 6.2		C, EC, EGC ECG, EGCG	Kawakami *et al.* (2011)
Elaeis guineensis	Extract	OPC	Phenols 10.7%		Ndiaye *et al.* (2010)
Fragaria vesca	Extract	OPC	Phenols 9.3%		Mudnic *et al.* (2009)
Guazuma ulmifolia	Extract	2–4		EC, EGC	Magos *et al.* (2008)
Hymenocardia acida	Extract	mDP 15	70% OPC	C, EC, GC, EGC	Manga *et al.* (2013)
Malus pumila	Extract	2–15	65% OPC	C, EC	Byun *et al.* (2012b) and Matsui *et al.* (2009)
Maytenus ilicifolia	Extract	OPC			Rattmann *et al.* (2006)
Musanga cecropioides	Extract	OPC			Dongmo *et al.* (2002)
Nelia meyeri	Extract	2–6	100% OPC	E	Kaufeld *et al.* (2013)
Parkia biglobosa	Extract	OPC	Phenols 10%		Tokoudagba *et al.* (2010)
Pinus maritima	Extract	2–12	70% OPC	C, E	Fitzpatrick *et al.* (1998) and Rohdewald (2002)
Rubus ursinus × *idaeus*	Extract	OPC	4.4% OPC	C, EC	Furuuchi *et al.* (2012)
Salix spp.	Extract	2–6	100% OPC	C	Kaufeld *et al.* (2014)
Theobroma cacao	Extract	2–10			Karim *et al.* (2000)
Vaccinium vitis-idaea	Juice	OPC	Phenols 0.6%; 0.2% OPC		Kivimäki *et al.* (2011)
Vigna angularis	Extract	OPC	Phenols 9.3%; 9% OPC		Mukai and Sata (2009)
Vitis vinifera	Juice	OPC	Phenols 2.3%		Alhosin *et al.* (2013) and Anselm *et al.* (2007)
	Extract	2–7	85% OPC	C, EC, G	Aldini *et al.* (2003) and Facino *et al.* (1999)
	Extract	2–8		C, EC, ECG	Alvarez *et al.* (2012)
	Extract	2–4	OPC		Appeldoorn *et al.* (2009a)
	Extract	mDP 2.3	92% OPC		Edirisinghe *et al.* (2008)
	Extract	2–5	Phenols 0.8%	G,C, EC, ECG	Fitzpatrick *et al.* (2000, 2002)
	Extract	2–13	80% OPC		Quiñones *et al.* (2013)
	Extract	OPC		C, EC	Rodriguez-Rodriguez *et al.* (2012)
Vitis labrusca		OPC	Phenols 5.5%		Madeira *et al.* (2009)
	Red wine polyphenols	2–7	Phenols 16%		Idris-Khodja *et al.* (2012b, 2013), Dal-Ros *et al.* (2011, 2012), Auger *et al.* (2010), Ndiaye *et al.* (2003, 2004, 2005), and Oak *et al.* (2003)
	Trimer	3		EC	Byun (2012) and Byun *et al.* (2012a, 2014)

C, catechin; DP, degree of polymerization; EC, epicatechin; EGC, epigallocatechin; EGCG, epigallocatechin-3-O-gallate; mDP, mean DP; OPC, oligomeric procyanidins.

catechin, epicatechin, and about 70% procyanidins with a DP between 2 and 12 (Fitzpatrick *et al.*, 1998; Rohdewald, 2002). A considerable portion thereof is composed of pentamers to heptamers, as concluded from size separation (Packer *et al.*, 1999). To date, the dimers epicatechin-[4ß,8]-catechin, catechin-[4α,8]-catechin and their [4,6]-linked analogs were identified, and tentatively also the trimer, epicatechin-[4ß,8]-epicatechin-[4ß,8]-catechin.

Hawthorn (*Crataegus* spp.) extracts have a long tradition in the treatment of cardiovascular diseases and are reputed to contain a wealth of procyanidins. Their structures have been elucidated up to the tetrameric level (Kolodziej *et al.*, 1984), while higher oligomers were tentatively characterized by mass spectrometric analysis (Zumdick *et al.*, 2009). Epicatechin constitutes a prominent flavanyl constituent unit of these oligomers and to a lesser degree its diastereomeric analog, catechin (ratio of about 8–10 : 1). Hawthorn extracts are obtained from various plant parts, including leaves and flowers (European Pharmacopoeia, *Crataegus monogyna*/*Crataegus laevigata*, or other European species including *Crataegus pentagyna*/*Crataegus nigra*/*Crataegus azarolus*), or berries. While many preparations are poorly defined, some standardized extracts used in the treatment of cardiovascular disorders are well characterized. For example, the hydroalcoholic extract WS 1442 (manufacturer: Dr. Willmar Schwabe GmbH & Co. KG, Germany) is adjusted to a content of 18.75% oligomeric procyanidins (OPC) and a minimum of 6% flavonoids expressed as hyperoside (Koch & Malek, 2011).

Earlier work on grape seed extracts indicated the presence of procyanidins in the range of DP2 to DP9, with catechin, epicatechin, and epicatechin-3-*O*-gallate constituent units. Analysis of grape skin extracts revealed a more complex composition with (epi)gallocatechin units in the oligomer structure. These proanthocyanidin profiles are essentially the same in grape-based products studied for their vasorelaxant activities. Definition of the OPC fractions commonly included DPs, which were between 2 and 5, and the type of terminal and extender units. Particularly noteworthy are the successful fractionation of a commercial grape seed extract and the characterization of subfractions with bioassay at each step, providing insight into structural features of vasorelaxant-active grape proanthocyanidins (Fitzpatrick *et al.*, 2002). A different picture emerges from examining chemical data on a grape juice product (Anselm *et al.*, 2007; Alhosin *et al.*, 2013). The information is strictly limited to the total amount of polyphenols and the content of proanthocyanidins and anthocyanins, also known to be vasoactive, thus rendering interpretations difficult. A similar note applies to a commercial red wine polyphenol extract, but with specifications for dimers as a minimum of information about the OPC composition (Auger *et al.*, 2002).

Within the large group of proanthocyanidin-rich dietary products (Gu *et al.*, 2004), apples represent a source that is reasonably well-studied for vasorelaxation (Matsui *et al.*, 2009; Byun *et al.*, 2012b). The polyphenol profile of an apple extract was well characterized, with about 65% procyanidins and about 12% (epi)catechin and other phenolic components. HPLC and LC-MS analyses facilitated estimation of the procyanidins ranging from dimers to hexamers and pooled oligomers (>heptamers).

Although the procyanidin profile of *Theobroma cacao* was well characterized (Porter *et al.*, 1991; Kelm *et al.*, 2006) and chocolate consumption suggested having potential beneficial effects on vascular health (Fisher *et al.*, 2003; Flammer *et al.*, 2012; Del Rio *et al.*, 2013), cacao polyphenols including procyanidins remain largely undefined in processed dietary products. However, a number of well-designed studies provided evidence that vascular improvements correlated with flavan-3-ol-rich cacao intake, though the high content of sugars and saturated fatty acids in chocolate products has questioned their dietary benefits (Ceballos & Villarreal, 2011).

The emerging picture from analyzing chemical data of the remaining herbal samples revealed that DPs and constituent flavanyl units are given, but the proportions of the diverse oligomers are missing. For example, bioassay-guided fractionation identified a proanthocyanidin-rich fraction containing minor quantities of prodelphinidins as the most effective subfraction of *Hymenocardia acida* (Manga *et al.*, 2013). Size-exclusion analysis and thiolytic studies of this fraction suggested a mean DP of 15 with no further details on the proanthocyanidin composition. Analysis of the most effective fraction of *Guazuma ulmifolia* indicated the presence of predominantly trimeric and tetrameric procyanidins, which, however, remained structurally undefined (Magos *et al.*, 2008). Similarly, the presence of unidentified proanthocyanidins was shown in active fractions of *Maytenus ilicifolia* (Rattmann *et al.*, 2006).

The complex composition of crude extracts and the lack of a precise description of the proanthocyanidin components in test materials recently prompted a more systematic approach to the assessment of the vasorelaxant potential of procyanidins (Kaufeld *et al.*, 2013, 2014). Selection of highly purified and chemically well-defined 2,3-*cis*- and 2,3-*trans*-procyanidin samples obtained from *N. meyeri* and *Salix* spp., respectively, permitted for the first time a limited structure–activity relationship within the group of homogeneous procyanidins. Their preliminary characterization was facilitated by HPLC analysis and mass spectrometry, indicating the presence of dimers to hexamers in the proportions dimers : trimers : tetramers : pooled pentamers/hexamers 1 : 1.6 : 2.3 : 7.5 and 1 : 1.4 : 2.1 : 6.4, respectively. Structural assessment of low-molecular-weight procyanidins was achieved by a combination of NMR analysis, electronic circular dichroism (ECD) measurements, and chemical methods.

11.3 Endothelial dysfunction

The endothelium is the active inner monolayer of blood vessels, forming an interface between circulating blood in the lumen and the adjacent vascular smooth muscle cells (VSMCs). It has a central role in the regulation of blood flow through continuous modulation of vascular tone. In this regard, endothelial cells release a variety of relaxing (NO, EDHF, PGI_2) and contracting factors (endothelin, thromboxanes, reactive oxygen species (ROS), angiotensin II). Endothelial dysfunction is a key event in the development of CVDs and includes an impairment of endothelium-dependent vasorelaxation caused by a reduced release/production or enhanced inactivation of vasodilators (Fig. 11.3). The dampened vasorelaxation was shown to be associated with an increased oxidative stress caused by an enhanced expression and activity of NADPH oxidase, a major source of superoxide anion, or other enzymatic ROS producing systems (Sarr *et al.*, 2006; Dal-Ros *et al.*, 2011). The shift of endothelial actions toward reduced vasodilation, with a proinflammatory and prothrombic state as associated phenomenon, is implicated in a broad spectrum of CVDs such

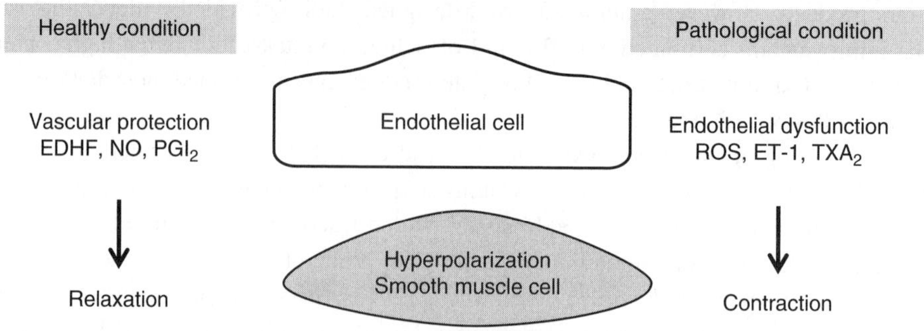

Fig. 11.3 Simplified illustration of the regulation of vascular tone.

as hypertension, atherosclerosis, coronary artery disease, congestive heart failure, chronic renal failure, and diabetes. The pathophysiologic mechanisms related to endothelial dysfunction have been reviewed in detail elsewhere (Endemann & Schiffrin, 2004; Feletou *et al.*, 2010, 2012; Garland *et al.*, 2011).

11.4 *In vitro* test systems

Isolated arterial rings of experimental animals (rat, rabbit, and swine) were commonly employed as useful models to study the vasorelaxant response to proanthocyanidins. Hearts are obtained from local slaughter houses. The descending coronary artery is excised and dissected free from connective tissue. Care has to be taken to avoid damage of the endothelium. Endothelium-denuded rings are prepared by gently rubbing the intimal surface with a pair of tweezers. The arterial rings are suspended in an organ bath containing oxygenated Krebs–Henseleit buffer and contractions are recorded isometrically. Once a stable baseline tone is reached, the tissue rings are precontracted (KCl, the thromboxane A2 mimetic U46619, or norepinephrine). To indicate the functional integrity of the endothelium, the responsiveness is proven with a known vasodilator (acetylcholine or bradykinin). Inhibitors are added before the final stable contraction plateau is reached, followed by the addition of the procyanidin sample in various concentrations for constructing a dose–response curve. The superior role of nitric oxide (NO) effects over endothelium-dependent hyperpolarizing factor (EDHF)-mediated responses in large conducting vessels imposes a serious limitation to these preparations. It is generally accepted that EDHF contributions to vasorelaxation become more evident as the internal diameter of vessels decreases. Thus, when studying the mechanism of action of procyanidins, it is vital to include small resistance arteries, with a vascular bed as perfect system. This methodology was employed in a few studies on procyanidin-containing materials (Duarte *et al.*, 2004; Kwak *et al.*, 2009).

A useful parameter in characterizing vasorelaxant activity is the EC_{50} value, that is, the sample concentration causing a 50% relaxation in precontracted vessels. However, the complex composition of polyphenolic test materials (Table 11.1) limits the conclusions from reported data and calls for a generally accepted proper control. An interesting result emerged from recent studies (Kaufeld *et al.*, 2013, 2014). With EC_{50} values of about

1.8 μg/ml, the 2,3-*cis*- and 2,3-*trans*-procyanidin samples composed of oligomers up to hexamers in comparable proportions exhibited similar relaxant potentials under the same *in vitro* experimental conditions. It will be of interest to assess the relaxing activity of other structural variants to gain insight into structure–activity relationships as a basis for recommendations of dietary and medicinal-plant-based products to prevent CVDs.

11.5 Vasorelaxant mechanisms

Diverse signaling pathways mediate vasorelaxation in small and conductance vessels. The relevant mechanisms of action activated by plant-based products are closely associated with endothelial NO production, contribution of eicosanoid products, modulation of K^+ channel functions, and regulation of cytosolic Ca^{2+} homeostasis (McNeill & Jurgens, 2006; Schmitt & Dirsch, 2009). Proanthocyanidins evoke vasorelaxation apparently through multiple mechanisms including predominantly modulation of the NO- and EDHF-signaling pathways as well as Ca^{2+}-signaling events. However, many studies examined more or less the contributions of the various pathways, thus providing an incomplete picture of how samples induce vasorelaxation. The ability of inhibitors to affect the vasorelaxant response evoked by the test material provided evidence for the role of signaling molecules. For convenience, Table 11.2 lists important blocking agents and their targets used in studies on proanthocyanidins.

11.5.1 Endothelium-dependent vasorelaxation

The endothelium plays an important role in the control of vascular-tone-mediating vasorelaxation by release of endothelium-derived relaxing factors (NO, PGI_2), activation of muscarinic receptors, and endothelium-dependent hyperpolarizations (Feletou *et al.*, 2010). Accordingly, all studies compared responses in endothelium-intact vessel preparations to those in endothelium-denuded preparations to demonstrate endothelium-dependency or otherwise. Evidence of a functional endothelium was shown by recording, for example, the acetylcholine-evoked vasorelaxant response. There are no reports on procyanidins that addressed endothelium-independent mechanisms by denuding the endothelium before any experiments, except for the effect of a *Crataegus* fruit extract on high K^+-induced contraction in rat mesenteric arteries (Chen *et al.*, 1998). The relaxant effect of this extract was attributed to a direct action on VSMCs through inhibition of Ca^{2+} influx via voltage-sensitive Ca^{2+} channels.

11.5.2 eNOS-NO-cGMP signaling pathway

eNOS catalyzes oxidation of L-arginine in the presence of molecular oxygen and various cofactors, resulting in the production of NO and L-citrulline. The endothelium-derived mediator permeates biological membranes and diffuses to adjacent VSMCs. Stimulation of soluble guanylyl cyclase (sGC), the primary target of NO, catalyzes conversion of guanosine 5-triphosphate (GTP) to cGMP (Fig. 11.4). Activation of cGMP-dependent protein

Table 11.2 Overview of important inhibitors and their targets.

Inhibitor	Target
N^G-nitro-L-arginine methyl ester (L-NAME)	eNOS
N^G-nitro-L-arginine (L-NMMA)	
1*H*-[1,2,4]oxadiazolo[4,3-*a*]quinoxalin-1-one (ODQ)	Soluble guanylyl cyclase (sGC)
Wortmannin	PI3 kinase/Akt
Thapsigargin	Sarcoplasmatic/endoplasmic reticulum Ca^{2+}-ATPase (SERCA)
2-Aminoethyl diphenylborinate (2-APB)	Store-operated Ca^{2+} channels (SOCCs)
Nifedipine, verapamil	Ca^{2+} channels
Tetraethylammonium chloride (TEA)	K^+ channels
Apamin (apa)	SK_{Ca} (small conductance Ca^{2+}-sensitive K^+ channel)
Charybdotoxin (CTX)	BK_{Ca}/IK_{Ca} (large and intermediate conductance Ca^{2+}-sensitive K^+ channel, respectively)
Iberiotoxin	BK_{Ca} (K_{Ca} channels)
TRAM 34	IK_{Ca} (K_{Ca} channels)
4-Aminopyridine (4-AP)	Voltage gated K^+ channels ($K\nu$)
Glibenclamide	ATP-sensitive K^+ channels (K_{ATP})
$BaCl_2$	Inwardly rectifying K^+ channels (K_{IR})
Ouabain	Na^+/K^+-ATPase
Atropine	Muscarinic receptors
Carbenoxolone	Gap junctions
ICI 182,780	Estrogen receptor (ER)
Mn(III)tetrakis(1-methyl-4-pyridyl)porphyrin (MnTMPy)	Cell-permeable mimic of superoxide dismutase (SOD): O_2^-
Allopurinol	Xanthine oxidase (XOD)
Apocynin	NADPH oxidase
Rotenone	Mitochondrial complex I
N-acetylcysteine	ROS scavenger
Indomethacin	COX components
4-Amino-5-(4-chlorophenyl)-7-(dimethylethyl)-pyrazolo[3-4-*d*]pyrimidine (PP2)	Src kinases

kinases induces vasorelaxation by the reduction of intracellular Ca^{2+} levels ($[Ca^{2+}]_i$) in VSMCs (Bredt *et al.*, 1992; Schlossmann *et al.*, 2000).

11.5.2.1 Key role of the NO-cGMP signaling pathway

All papers focused on the role of NO in mediating the vasorelaxant response to procyanidin-containing products. Abolished or reduced vasodilatory responses upon treatment with an inhibitor of eNOS indicated that NO production significantly contributed to vascular relaxation. Independent support of the activation of eNOS as NO-generating enzyme was repeatedly demonstrated by Western blotting through expression of phosphorylated eNOS (p-eNOS) in polyphenol-treated human umbilical vein endothelial cells (HUVEC) (Ndiaye *et al.*, 2005; Edirisinghe *et al.*, 2008; Anselm *et al.*, 2009; Madeira *et al.*, 2009; Kaufeld *et al.*, 2013, 2014) or rat aortic endothelial cells (RAEC) (Byun *et al.*, 2012a). Direct evidence of increased NO synthesis by the endothelium of animal vessel segments was also provided by electron paramagnetic resonance spectroscopy and NO-selective microsensors (Andriambeloson *et al.*, 1997; Stoclet *et al.*, 1999; Taubert *et al.*, 2002; Auger *et al.*, 2010a).

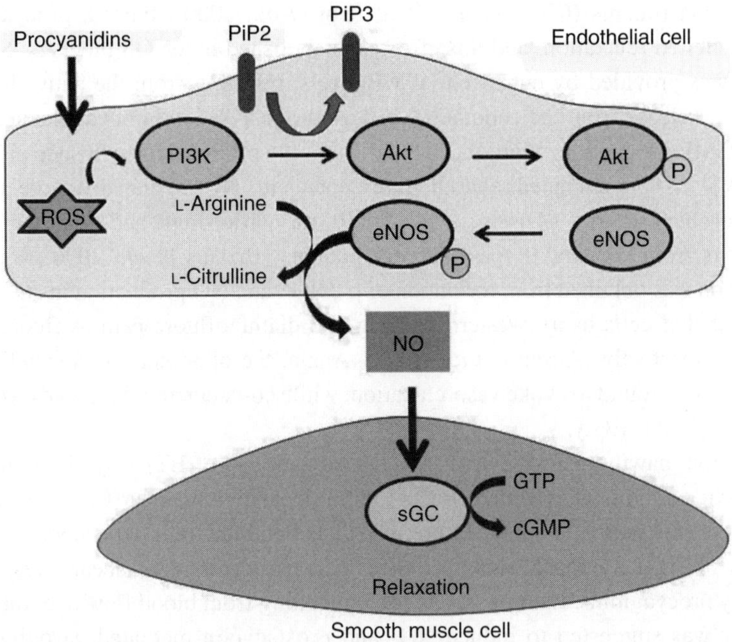

Fig. 11.4 Schematic illustration of the endothelium-dependent NO-signaling pathway and its role in the response to procyanidins. Endothelial production of NO is triggered via a redox-sensitive activation of the phosphaditylinositol-3-kinase (PI3K)/Akt pathway promoting eNOS catalysis, that is, conversion of L-arginine to L-citrulline and NO. Stimulation of the NO-sensitive soluble guanosyl cyclase (sGC) generates cyclic guanosyl monophosphate (cGMP) from guanosine triphosphate (GTP). Relaxation of the smooth muscle cell is due to cyclic GMP-dependent protein kinases. PIP2, phosphatidylinositol biphosphate; PIP3, phosphatidylinositol triphosphate; ROS, reactive oxygen species.

In some instances, NO-mediated activation of sGC was confirmed by the measurement of increased cGMP levels including that induced by procyanidin C1 (Byun *et al.*, 2014). The key role of the NO-cGMP pathway is best documented by the following examples.

Accumulating evidence suggested a variety of health-promoting effects of the commercially available maritime pine bark extracts on the development of cardiovascular diseases (Iravani & Zolfaghari, 2011). The extracts induced endothelium-dependent relaxations in rat and mouse aortic rings and rat mesenteric vascular beds, attributed to enhanced NO production resulting from activation of eNOS (Kwak *et al.*, 2009). Supporting evidence was obtained from Western blot analysis, demonstrating phosphorylation of eNOS at Ser^{1177}, inhibition of NO release in the presence of the NOS inhibitor L-NAME, and decreased vasorelaxant responses following pretreatment with the NO-sensitive sGC inhibitor 1*H*-[1,2,4]oxadiazolo[4,3-*a*]quinoxalin-1-one (ODQ). In addition, aortic rings of eNOS knockout mice and endothelium-denuded preparations failed to induce vasorelaxation. Fractionation procedures provided evidence that oligomers with DPs from 4 to 7 showed the highest vasorelaxant activity. However, studies on endothelial hyperpolarization-dependent and -independent pathways are missing.

A number of studies provided evidence that hawthorn extracts may evoke an endothelium-dependent vasorelaxation in isolated rat aortic and mesenteric rings, with procyanidins as the active constituents (Chen *et al.*, 1998; Kim *et al.*, 2000). Indeed, procyanidin-rich fractions elicited relaxation, and this effect was abolished by eNOS blockers. Supporting evidence was provided by enhanced cGMP levels, resulting from the activation of sGC through NO release from the endothelial cell. Again, eNOS inhibitors attenuated the augmented cGMP production. More recent studies on the procyanidin-rich special Crataegus extract WS 1442 confirmed endothelium-dependent NO-mediated vasorelaxation in porcine coronary arteries (Anselm *et al.*, 2009), rat aortic rings, and human mammarian preparations from coronary bypass surgery patients (Brixius *et al.*, 2006). Independent confirmation was obtained from formation of p-eNOS and NO production in cultured coronary endothelial cells using Western blotting and diaminofluorescein as fluorescent NO indicator, respectively (Anselm *et al.*, 2009). Again, the oligomeric procyanidin fraction of WS 1442 was found to evoke vasorelaxation, while co-occurring flavonoids were devoid of this biological activity.

In addition, hawthorn fruit extracts decreased endothelin (ET)-1 synthesis in cultured bovine aortic endothelial cells in a concentration-dependent manner (Corder *et al.*, 2004). Excess ET-1 production is a key feature of reduced endothelium-dependent vasodilation. Inhibition of ET-1 synthesis was attributed to a pseudo laminar shear stress response induced by procyanidins. Laminar shear stress resulting from blood flow over the endothelium layer was suggested to induce NO- and prostacyclin-mediated vasodilation with concurrent reduction of ET-1 synthesis. Thus, decreased blood flow may cause diminished endothelium-dependent vasodilation. Conspicuously, oligomers comprised of 3–5 flavanyl constituent units were most effective, suggesting a specific, yet unknown, mechanism of action.

Although various epidemiological studies have demonstrated that a moderate wine consumption and intake of other grape-derived products are associated with reduced risk of cardiovascular diseases, experimental evidence of endothelium-dependent relaxation was first provided in rat aortic rings via the NO-cGMP signaling pathway (Fitzpatrick *et al.*, 1993). Subsequent work confirmed the endothelium-dependent NO-mediated relaxation of grape-derived products in several conductance and resistance arteries of experimental animals and humans (Schini-Kerth *et al.*, 2010a, 2011). The vasorelaxant potential was shown to correlate with the phenolic content (Burns *et al.*, 2000). Notably, trimeric to pentameric procyanidins exhibited marked relaxant activities (Andriambeloson *et al.*, 1998; Fitzpatrick *et al.*, 2000, 2002).

Investigations with traditional herbal medicines for the treatment of vascular diseases indicated that some procyanidin-rich extracts showed endothelium-dependent relaxations in coronary artery rings (Table 11.3). Recent reports include extracts of *H. acida* (Euphorbiaceae) and *Combretum celastroides* subsp. *laxiflorum* (Combretaceae) used for the treatment of hypertension in Congolese traditional medicine (Manga *et al.*, 2012, 2013). Methanolic extracts produced a concentration-dependent vasorelaxation on rat aortic rings. In experiments with L-NAME and ODQ, the extract-induced relaxations were markedly attenuated, while pretreatment with indomethacin failed to affect the relaxant response. These findings suggested that the modulation of the vascular tone was via the NO/cGMP pathway, while participation of prostanoids can be excluded. A closer look into EDHF-mediated responses is missing, rendering the mode of action undefined.

Table 11.3 Vasorelaxant mechanisms studied in procyanidin-containing plant products.

			eNOS	sGC					
Plant	Extract/product	Model	NO	cGMP	PI3/Akt	K^+/K_{Ca}	$[Ca^{2+}]_i$	EC50	References
	EC-[4β,8]-EC-[4β,8]-EC	Rat aortic endothelial cells	+	+	+	+	+		Byun (2012)
		Rat thoracic aortic rings							Byun *et al.* (2012a, 2014)
Combretum celastroides	Leaves	Rat aortic	+	+				3.9 µg/ml	Manga *et al.* (2012)
Crataegus monogyna	WS 1442	Rat mesenteric artery	+			+			Idris-Khodja *et al.* (2012a)
Crataegus laevigata	Leaves and flowers								
		Porcine coronary artery	+	+	+			2.2 µg/ml	Anselm *et al.* (2009)
		Rat aortic	+					7.4 µg/ml	Brixius *et al.* (2006)
		Human mammary aorta						14.6 µg/ml	
	Extract	Rat aortic	+	+		+		27.5 µg/ml	Kim *et al.* (2000)
	OPC							1.5 µg/ml	
	Fruits	Rat mesenteric artery	+	+		+	+	220 µg/ml	Chen *et al.* (1998)
Croton celtidifolius	Bark	Rat mesenteric arterial bed	+	+		+			DalBó *et al.* (2008a)
		Rat thoracic aorta	+	+					DalBó *et al.* (2008b)
Diospyros kaki	Leaves, aqueous	Rat thoracic aorta	+					0.8 µg/ml	Kawakami *et al.* (2011)
	OPC								
Elaeis guineensis	Leaves	Porcine coronary artery	+		+	+		>50 µg/ml	Ndiaye *et al.* (2010)
Fragaria vesca	Leaves	Rat aortic	+						Mudnic *et al.* (2009)
Guazuma ulmifolia	Bark, acetone	Rat aortic	+					35 ng/ml	Magos *et al.* (2008)
	OPC								
Hymenocardia acida	Bark	Rat aortic	+	+					Manga *et al.* (2013)
Malus pumila	OPC	Rat aortic endothelial cells	+	+		+	+		Byun *et al.* (2012b)
	Extract	Rat aortic	+	+		+		15 µg/ml	Matsui *et al.* (2009)
	OPC							5 µg/ml	
Maytenus ilicifolia	Leaves	Rat aortic	+	+		+		200 µg/ml	Rattmann *et al.* (2006)
	OPC							4.3 µg/ml	
Musanga cecropioides	Leaves	Rat aortic	+						Dongmo *et al.* (2002)
N. meyeri	Leaves	Porcine coronary artery	+	+	+	+		1.9 µg/ml	Kaufeld *et al.* (2013)
	OPC								
Parkia biglobosa	Leaves	Porcine coronary artery	+			+			Tokoudagba *et al.* (2010)

(*Continued*)

Table 11.3 (Continued)

Plant	Extract/product	Model	eNOS	sGC					
			NO	cGMP	PI3/Akt	K^+/K_{Ca}	$[Ca^{2+}]_i$	EC50	References
Pinus maritima	Bark	Rat aortic	+					3 µg/ml	Fitzpatrick *et al.* (1998)
		Rat aorta and mesenteric beds	+	+					Kwak *et al.* (2009)
Rubus ursinus × idaeus	Boysenberry seeds	Hypertensive rats	+	+					Furuuchi *et al.* (2012)
Salix spp.	Bark	Porcine coronary artery	+	+	+	+	+	1.7 µg/ml	Kaufeld *et al.* (2014)
	OPC								
Vaccinium vitis-idaea	Juice	Rat mesenteric aorta	+			+			Kivimäki *et al.* (2011)
Vigna angularis	Fruits	Hypertensive rats	+						Mukai and Sato (2009)
	Aqueous EtOH								
Vitis vinifera	Grape juice	Porcine endothelial cells	+	+	+				Alhosin *et al.* (2013)
	Grape seeds	Human mammary aorta	+	+					Aldini *et al.* (2003)
	Grape juice	Porcine coronary artery	+		+	+			Anselm *et al.* (2007)
	Grape seeds	Human endothelial cells	+						Appeldoorn *et al.* (2009a)
	Grape seeds	Rabbit aortic	+		+				Edirisinghe *et al.* (2008)
	Grape seeds	Rat aortic	+	+					Fitzpatrick *et al.* (2002)
	Grape seeds	Rat aortic	+						Mendes *et al.* (2003)
	Grape seeds	Hypertensive rats							Quiñones *et al.* (2013, 2014)
	Grape pomace	Rat aortic and mesenteric artery	+						Rodriguez-Rodriguez *et al.* (2012)
	Grape seeds	Hypertensive mice	+						Wang *et al.* (2013)
Vitis labrusca	Grape skin	Porcine coronary artery	+	+		+			Madeira *et al.* (2009)
	RWP	Rat mesenteric artery	+			+			Idris-Khodja *et al.* (2013)
	RWP	Rat mesenteric artery	+			+			Idris-Khodja *et al.* (2012b)
	RWP	Rat mesenteric artery	+						Dal-Ros *et al.* (2011, 2012)
	RWP	Endothelial cells	+		+				Auger *et al.* (2010)
	RWP	Porcine coronary artery	+	+	+				Ndiaye *et al.* (2005)
	RWP	Porcine coronary artery			+				Ndiaye *et al.* (2004)
	RWP	Porcine coronary artery	+			+			Ndiaye *et al.* (2003)

+, studied component; $[Ca^{2+}]_i$, intracellular Ca^{2+} concentration; EC, epicatechin; eNOS, endothelial nitric oxide synthase; K^+/K_{Ca} channel function; NO, nitric oxide; OPC, oligomeric procyanidins; PI3K, phosphatidylinositol-3-kinase; RWP, red wine polyphenols.

Bioassay-guided fractionation identified a procyanidin-rich fraction containing minor quantities of prodelphinidins as the most effective subfraction of *H. acida*. Size-exclusion analysis and thiolytic studies of this fraction suggested a mean DP of 15, rendering their bioavailability ambiguous. No such information was given for *C. celastroides* subsp. *laxiflorum*, except for indicating the presence of saponins, terpenoids, and tannins as concluded from a TLC screening. Accordingly, the bioactive metabolites still remain to be identified.

Similar findings were reported for *G. ulmifolia* (Sterculiaceae), traditionally used as an antihypertensive in Latin America (Magos *et al.*, 2008), and *M. ilicifolia* (Celastraceae) with a broad spectrum of uses in South American folk medicine (Rattmann *et al.*, 2006). In isolated aortic rings of rats, extracts of both plants evoked vasorelaxation. The relaxant response was abolished by endothelium denudation or attenuated by pretreatment with L-NAME, but not by indomethacin or atropine. The structures of the oligomeric procyanidins detected in the most effective fractions remain ill-defined.

In an interesting study, efforts were made to explore the mechanism leading to activation of eNOS in rat mesenteric and thoracic aorta rings by a procyanidin-rich fraction from *Croton celtidifolius*, with catechin and epicatechin constituent flavanyl units (DalBò *et al.*, 2008a, 2008b). The relaxant effect was significantly reduced by L-NAME and ODQ, but remained unaffected by indomethacin and atropine. In addition, activation of eNOS in small resistance arteries was shown to be mediated by modulation of K_{Ca} channel functions, while increased $[Ca^{2+}]_i$ through stimulation of phospholipase C and Ca^{2+} entry via opening of transmembrane Ca^{2+} channels apparently enhanced eNOS activity in thoracic aorta (*vide infra*).

Other proanthocyanidin-containing extracts tested for their ability to evoke vasorelaxation included enriched fractions from *Diospyros kaki* (Kawakami *et al.*, 2011) and *Parkia biglobosa* (Tokoudagba *et al.*, 2010), apple procyanidins (Matsui *et al.*, 2009; Byun *et al.*, 2012b), leave extracts from *Elaeis guineensis* (Ndiaye *et al.*, 2010), *Fragaria vesca* (Mudnic *et al.*, 2009), and *Musanga cecropioides* (Dongmo *et al.*, 2002), a seed extract from *Rubus ursinus* × *idaeus* (Furuuchi *et al.*, 2012), and a fruit preparation from *Vigna angularis* (Mukai & Sato, 2009). With the exception of a few reports, merely the role of NO was studied, leaving many questions related to the underlying mechanism unresolved (Table 11.3).

Beneficial effects of other dietary proanthocyanidins on cardiovascular diseases followed from experimental *in vitro* and animal studies. For example, rats were fed for 8 weeks on cranberry (*Vaccinium oxycoccus*), lingonberry (*Vaccinium vitis-idaea*), and blackcurrant (*Ribes nigrum*) fruit juices (Kivimäki *et al.*, 2011). The given amounts of fruit polyphenols, including simple phenolic acids, anthocyanins, flavonoids, flavan-3-ols, and procyanidins, indicated a 4–5 fold higher concentration of procyanidins in lingonberry compared with the other berry juices. Although the vasorelaxant responses to all juice samples were strongly dependent on NO production, conspicuous differences in effectiveness were noted. Results showed that only lingonberry treatment normalized the impaired endothelium-dependent relaxation of mesenteric arteries from spontaneously hypertensive rats. A key difficulty in drawing clear conclusions is the complex product composition, rendering an association between distinct polyphenols and vasoprotection difficult.

However, on the basis of compositional data, the lingonberry procyanidins may be assumed to represent the major active principles.

Along that line, bioassay-guided studies and investigations with purified fractions including 2,3-*cis*- (Kaufeld *et al.*, 2013), 2,3-*trans*-procyanidins (Kaufeld *et al.*, 2014), and, most notably, procyanidin C1 (Byun, 2012; Byun *et al.*, 2012a, 2014) provided convincing evidence that procyanidins played an important role in the NO-regulated control of vascular tone. In porcine coronary or rat thoracic aortic rings, the procyanidins produced an endothelium-dependent relaxation that was significantly reduced by L-NAME and ODQ, while the trimer produced a similar marked inhibitable increase in NO formation in endothelial cultures.

In contrast to the key role of NO in endothelium-dependent proanthocyanidin-induced vasorelaxation, membrane receptors did not play any role in studies performed so far. Preincubations of aortic sections with a range of antagonists suggested that muscarinic, histamine, substance P receptors, and ß-adrenoceptors did not affect the relaxations to procyanidins (Kim *et al.*, 2000).

11.5.2.2 Activation of eNOS via the phosphatidylinositol-3-kinase (PI3K)/Akt pathway

The PI3K/Akt pathway (Fig. 11.4) is an important cell survival mechanism in response to various stimuli (Fulton *et al.*, 1999). In particular, a key role in the activation of eNOS is played by this signaling pathway, leading to an increased formation of vasodilating NO (Dimmeler *et al.*, 1999; O'Neill & Abel, 2005). Activation of PI3K leads via phosphatidylinositol-3,4,5-triphosphate (PIP3) to phosphorylation of Akt, which, in turn, promotes phosphorylation of eNOS at Ser^{1177} (activator site) and dephosphorylation at Thr^{495} (inhibitor site). It should also be noted that, besides the PI3K/Akt pathway, activation of eNOS may also be achieved by an increase in cytosolic $[Ca^{2+}]_i$ in endothelial cells (see below) (Monbouli & Vanhoutte, 1999) and in response to shear stress induced, for example, by blood flow (Corder *et al.*, 2004). Upregulation and rapid activation of eNOS lead to increased and more sustained formation of NO.

It is, therefore, evident that activation of eNOS is intimately linked to the PI3K/Akt pathway. Indeed, Western blotting proved that procyanidins induced phosphorylation of the upstream effector Akt on Thr^{308} and Ser^{473} in cultured endothelial cells (Anselm *et al.*, 2009; Byun *et al.*, 2012a; Kaufeld *et al.*, 2013, 2014). This Akt modification is known to mediate eNOS-dependent NO formation. The relaxations of aortic rings were considerably attenuated by wortmannin and provided independent evidence of the involvement of the PI3K/Akt pathway.

Estrogens have been shown to trigger the PI3K/Akt/eNOS pathway (Chambliss & Shaul, 2002). Indeed, studies on red wine extracts suggested the involvement of the estrogen receptor (ER) alpha in intact aortic rings, but not in preparations from ER knockout mice. Furthermore, the use of the selective ER inhibitor fulvestrant and proper siRNA abolished phosphorylation of eNOS and NO production in a human endothelial cell line (Chalopin *et al.*, 2010). However, grape-derived products are complex mixtures of polyphenols, with anthocyanidins and procyanidins as potent vasodilators. The observation that ER alpha did play a role in the endothelial production of NO only in response to some red wine polyphenols and in certain types of endothelial cells (Padilla *et al.*, 2005; Kane *et al.*, 2009) emphasizes the need for meticulous chemical profiling to identify the ER-activating substances. Unpublished work from the author's group precludes activation of ER by highly purified procyanidin fractions.

11.5.2.3 Role of reactive oxygen species and redox-sensitive kinases

Continued studies of the PI3K/Akt and NO signaling pathways have indicated a key role for an intracellular redox-sensitive mode of action in eNOS activation triggered by grape polyphenols (Ndiaye *et al.*, 2003, 2004, 2005; Anselm *et al.*, 2007; Madeira *et al.*, 2009) and the hawthorn extract WS 1442 (Anselm *et al.*, 2009). Indeed, moderate intracellular ROS formation in response to procyanidins, which, in turn, stimulates the PI3K/Akt pathway leading to rapid eNOS activation, was recently confirmed for 2,3-*cis*- and 2,3-*trans*-procyanidins (Kaufeld *et al.*, 2013, 2014). The relaxant response to the procyanidins in isolated artery rings was abolished by the membrane-permeant mimic of superoxide dismutase (SOD), Mn(III)tetrakis(1-methyl-4-pyridyl)porphyrin (MnTMPyP). Since SOD catalyzes the dismutation of $O_2^{\cdot-}$ into hydrogen peroxide and oxygen, the effect may be attributable, at least in part, to decreased $O_2^{\cdot-}$ levels. Confirmatory evidence for an intracellular ROS production was obtained in HUVEC cultures and cultured coronary artery endothelial cells using 2′,7′-dichlorofluorescein diacetate (Kaufeld *et al.*, 2013, 2014) and dihydroethidine, respectively (Ndiaye *et al.*, 2003; Anselm *et al.*, 2007; Madeira *et al.*, 2009). The nonfluorescent redox-sensitive probes penetrate the cell membrane readily and are oxidized to highly fluorescent products by ROS/$O_2^{\cdot-}$. The procyanidins concentration-dependently increased ROS production.

Several endothelial cell enzymes such as xanthine oxidase (XOD), NADPH oxidase or the mitochondrial respiratory chain are capable of generating ROS including $O_2^{\cdot-}$. Indeed, experiments performed in the presence of allopurinol (nonselective inhibitor of XOD), apocynin (inhibitor of NADPH oxidase), or the mitochondrial electron transport chain inhibitor rotenone affected the relaxant response to procyanidins (Kaufeld *et al.*, 2013, 2014). However, conspicuous differences in the origin of $O_2^{\cdot-}$ were noted with respect to 2,3-*cis*- and 2,3-*trans*-procyanidins in that potential sources apparently included NADPH oxidases/XOD in the first instance and the mitochondrial respiratory chain in the latter. This recent finding may explain conflicting data resulting from studies on polyphenolic mixtures and the discussion about potential sources of ROS (Kim *et al.*, 2014).

In the absence of decisive information about ROS sources, autoxidation of polyphenols involving a two-step oxidation process of *ortho*-hydroxy groups to produce *ortho*-quinones via a semiquinone intermediate with concomitant $O_2^{\cdot-}$ generation was plausibly suggested (Schini-Kerth *et al.*, 2010a). Methylation of the hydroxy functionalities of grape seed extract polyphenols with a major proportion of procyanidins and EGCG resulted in complete loss of its relaxant ability, giving credence to this conjecture (Edirisinghe *et al.*, 2008; Auger *et al.*, 2010b). Interestingly, selective methylation of hydroxy groups in EGCG indicated that in particular the presence of an unsubstituted functionality at C-3 of the galloyl residue and, to some extent, at C-3 and C-4 on the B-ring is crucial for the EGCG-induced redox-sensitive activation of eNOS (Kurita *et al.*, 2013). However, free hydroxy groups may also be essential trigger functions of $O_2^{\cdot-}$ production by enzymatic sources. Recalling notable procyanidin-dependent differences in sources of ROS, further studies are needed to clarify their origin in response to individual procyanidins and other polyphenols.

It is also worth mentioning that polyphenol-induced pro-oxidant responses have been reported to be limited to the endothelium, while ROS production was inhibited in VSMCs and in platelets (Oak *et al.*, 2003; Vitseva *et al.*, 2005; Rattmann *et al.*, 2012). This very special endothelial cell behavior may reflect sensing properties for polyphenols, thereby providing a clue for their vasoprotective effects. Clearly, more studies are needed to get insight into this emerging phenomenon with a focus on the role of individual polyphenolic compounds.

A large body of evidence indicates that ROS may trigger the activation of redox-sensitive kinases (Irani, 2000). For example, the pro-oxidant status in endothelial cells can provoke the activation of sarcoma oncogene homolog (Src) kinase by phosphorylation, which subsequently leads to a PI3K-dependent activation of Akt, required for successive eNOS stimulation via phosphorylation at Ser^{1177} to enhance its activity in response to polyphenols (Madeira *et al.*, 2009). In this context, it appears appropriate to note that EGCG-induced vasorelaxation and NO production in rat mesenteric vascular beds and bovine aortic endothelial cells were attenuated by the Src family kinase inhibitor 4-amino-5-(4-chlorophenyl)-7-(dimethylethyl)pyrazolo[3-4-*d*]pyrimidine (PP2) (Kim *et al.*, 2007). Of the known members of the Src family, Fyn was identified as essential activator required for the EGCG-stimulated phosphorylation of Akt and eNOS. It may therefore be concluded that members of the Src and mitogen-activated protein (MAP) kinase families are also required in the procyanidin-mediated signaling pathway leading to activation of Akt and eNOS. Supporting evidence of this conjecture was provided by the inhibition of *Crataegus* WS 1442 extract-induced relaxation in porcine coronary artery by PP2 (Anselm *et al.*, 2009) and demonstration of the redox-sensitive upregulation of eNOS by purple grape juice (Anselm *et al.*, 2007; Alhosin *et al.*, 2013) and red wine polyphenols (Madeira *et al.*, 2009) involving the upstream redox-sensitive kinases Src, p38 MAPK and c-JUN N-terminal kinases (JNK). Furthermore, Akt-dependent inactivation of FoxO transcription factors, which downregulate eNOS expression, was demonstrated. This intriguing finding was based on a set of experiments using modulators of ROS, kinase inhibitors, determination of the cellular formation of ROS and NO, and Western Blot analysis. Conspicuously, relaxation in response to red wine polyphenols was not affected by inhibitors of p38 MAPK and ERK 1/2 in other experiments using porcine artery rings, while their phosphorylation could be demonstrated in endothelial cell cultures (Ndiaye *et al.*, 2004). A rationale for this may be that ROS production is caused by the culture conditions, but clarification is needed. Whether members of the Src and MAP kinase family are activated by procyanidins remains elusive at this point, since relaxation induced by purified procyanidin fractions was not significantly affected in the presence of the Src inhibitor PP2 (unpublished results). On the other hand, red wine polyphenols have been shown to inhibit vascular growth factor expression, which is strongly expressed in atherosclerotic plaques, by preventing the redox-sensitive activation of the p38 MAPK pathway in VSMCs (Oak *et al.*, 2003). A plethora of questions remains at this point in view of the hitherto limited studies with purified procyanidins on effectors upstream of Akt and the diverse interrelated MAP cascades.

For example, do different polyphenols trigger individual redox-sensitive kinases, and if so, is there a regulatory principle? Further, excess ROS (oxidative stress) produced, for example, by increased activity of vascular NADPH oxidase has been implicated in certain pathological conditions. The Janus-faced role of ROS and the signaling pathways regulated by ROS are still not completely understood (Thannickal & Fanburg, 2000).

11.5.3 Eicosanoid-mediated vasorelaxation

Eicosanoids generated by lipoxygenase and cytochrome P450 monooxygenase pathways are recognized as endothelium-derived relaxing factors (Chawengsub *et al.*, 2009; Feletou *et al.*, 2010). The mechanisms by which these metabolites mediate vasorelaxation include the activation of K_{Ca} channels resulting in hyperpolarization, closure of voltage-gated Ca^{2+} channels, and thus relaxation by inhibition of Ca^{2+} influx into VSMCs.

Cyclooxygenase-dependent mechanisms related to procyanidin research are hitherto confined to a few reports. Grape seed procyanidins dose-dependently relaxed human internal mammary aortic rings precontracted with norepinephrine. This effect was significantly attenuated by the COX inhibitor indomethacin, as assessed by measuring the 6-ketoprotaglandin-$F_1\alpha$ (Aldini *et al.*, 2003). In addition, concurrent work demonstrated that the commercially available grape seed procyanidin sample stimulated release of prostacyclin (PGI_2), known to improve infarcted heart function. Decreased leukotriene–prostacyclin ratios were reported in human aortic endothelial cells and in humans, following treatment with chocolate procyanidins (Schramm *et al.*, 2001). Favorably altered eicosanoid synthesis was demonstrated for procyanidin-rich chocolate in a short-term trial by monitoring plasma epicatechin, leukotriene, and prostacyclin concentrations of the volunteers. The fact that the induced changes in eicosanoid plasma levels were due to chocolate procyanidins was supported by similar results using endothelial cell cultures. Additional investigations are required to better characterize the molecular mechanism involved in eicosanoid-mediated vascular relaxation in response to procyanidins.

11.5.4 Endothelium-derived hyperpolarizing signaling cascade

Numerous lines of evidence have shown that endothelium-dependent vasorelaxations are also mediated by the induction of endothelium-derived hyperpolarization (Garland *et al.*, 2011; Feletou *et al.*, 2012). EDHF signaling is a powerful vasorelaxant system that becomes more important as the internal diameter of the vessel decreases (Luksha *et al.*, 2009) or when the NO production is markedly reduced (Feletou *et al.*, 2012). EDHF-mediated responses are not sensitive to inhibitors of eNOS and cyclooxygenases and are caused by the release of a variety of structurally very distinct factors or even a nonchemical signal and may function simultaneously with the NO-signaling pathway in vessels. Activation of endothelial Ca^{2+}-dependent K^+ channels (IK_{Ca}: intermediate conductance Ca^{2+}-sensitive K^+ channels; SK_{Ca}: small conductance Ca^{2+}-sensitive K^+ channels) induces endothelial hyperpolarization associated with K^+ efflux (Fig. 11.5). Increases in extracellular K^+ evoke hyperpolarization of VSMCs and relaxation by activating inwardly rectifying K^+ channels (K_{IR}) and Na^+/K^+-ATPase associated with reduced cytosolic Ca^{2+} levels. A decrease in

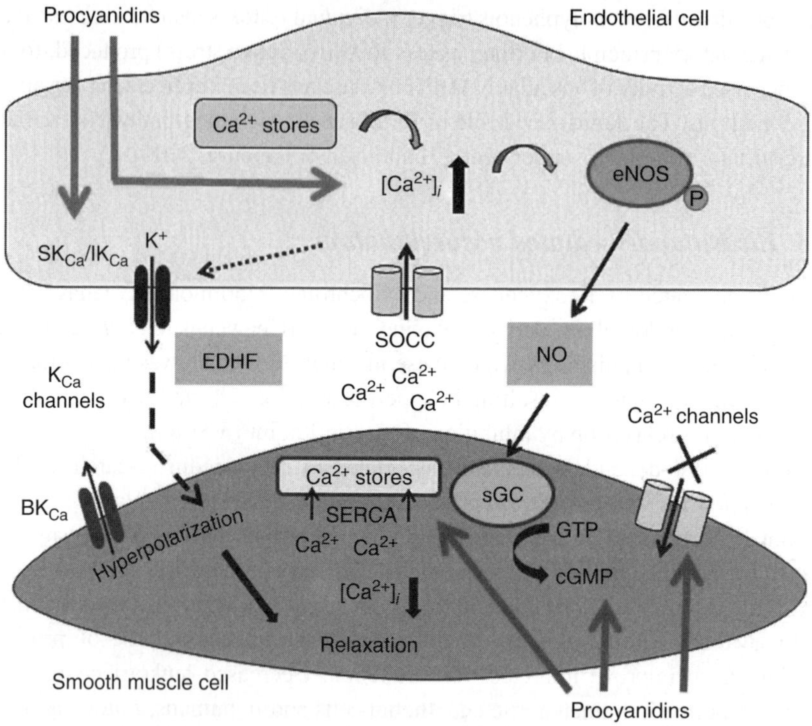

Fig. 11.5 Schematic illustration of the endothelium-dependent hyperpolarization factor (EDHF)-mediated relaxation and modulation of intracellular Ca^{2+} concentrations ($[Ca^{2+}]_i$): current view on responses to procyanidins. Ca^{2+} release from intracellular stores and Ca^{2+} influx from extracellular sources via store-operated calcium channels (SOCCs) upon store depletion stimulate the opening of endothelial Ca^{2+}-dependent K^+ channels (K_{Ca}). Activation of K_{Ca} hyperpolarizes the endothelial cell and the effluxing K^+ can spread the effect to the smooth muscle. Increased endothelial $[Ca^{2+}]_i$ may enhance eNOS activation. A decrease of cytosolic Ca^{2+} concentration in the smooth muscle cell involves cGMP-dependent protein kinases (NO signaling), closure of Ca^{2+} channels, and refilling of intracellular stores (SERCA).

$[Ca^{2+}]_i$ is mediated by multiple mechanisms, including activation of cGMP-dependent kinases (NO signaling), refilling of intracellular Ca^{2+} stores by the sarcoplasmic reticulum Ca^{2+}-ATPase (SERCA), and closure of voltage-gated Ca^{2+} channels (Feletou *et al.*, 2012).

The first evidence for a contribution of EDHF to endothelium-dependent relaxations in response to polyphenols was provided by experiments with red wine polyphenols using isolated porcine coronary rings (Ndiaye *et al.*, 2003). Subsequent research indicated EDHF-mediated responses to a bark extract of *Eucommia ulmoides* (Kwan *et al.*, 2004), a Brazilian red wine (De Moura *et al.*, 2004), purple grape juice (Anselm *et al.*, 2007), and red wine polyphenols (Dal-Ros *et al.*, 2009). It is, therefore, evident that procyanidins and structurally related variants may exert relaxation via EDHF pathways. Indeed, the available data provide compelling evidence indicating that modulation of K^+ channel functions and Ca^{2+} -signaling events contribute to the vasorelaxation evoked by procyanidins, but leave the participation of other EDHF components unanswered.

11.5.4.1 Modulation of K^+ channel functions

Exploration of the functional significance of K_{Ca} channels has been greatly aided by the use of highly purified 2,3-*cis*- (Kaufeld *et al.*, 2013) and 2,3-*trans*-procyanidin fractions (Kaufeld *et al.*, 2014). Relaxations evoked in precontracted porcine aortic rings in response to these samples were only moderately affected in the presence of the selective K_{Ca} inhibitors apamin, TRAM 34, iberiotoxin, and the nonselective blocker charybdotoxin (CTX) (Table 11.2). However, when K_{Ca} blockers were incubated in combination with the eNOS inhibitor L-NAME, vasorelaxation was abolished, differing from only inhibitory effects in experiments with L-NAME alone. This finding suggests that NO signaling outweighs EDHF-mediated responses under the experimental conditions, which, however, become obvious upon eNOS inhibition. The observation is consistent with different contributions of NO and EDHF in large and small conduit arteries and reduced NO bioavailability (Luksha *et al.*, 2009; Feletou *et al.*, 2012).

Interestingly, procyanidin C1 was recently reported to modulate multiple K^+ channel functions. Using RAECs, addition of the trimer exerted a time-dependent hyperpolarization measured by means of the membrane potential-dependent fluorescence probe bis-(1,3-dibutyl-barbituric acid) trimethine oxonol [DiBAC$_4$(3)] (Byun, 2012; Byun *et al.*, 2012a). Conspicuously, DiBAC$_4$(3) fluorescence remained unaffected by the analogous dimer procyanidin B2 and the monomer epicatechin, suggesting differences in the action of individual procyanidins. On the other hand, effects may also depend on the type of cells. For example, procyanidin C1 but not B2 inhibited proliferation of endothelial cells (Garcia-Conesa *et al.*, 2009; Byun *et al.*, 2012a), while only B2 was effective on VSMCs (Cai *et al.*, 2011). Hyperpolarization was significantly inhibited in the presence of tetraethylammonium (TEA) and modestly by the BK_{Ca} blocker iberiotoxin, while inhibitory effects were not observed in experiments with inhibitors of $K\nu$, K_{IR}, and K_{ATP} channels (Table 11.2). Since BK_{Ca} channels are poorly expressed in most endothelial cells except for proliferating cultures and SK_{Ca}/IK_{Ca} are preferentially located in these cells, experiments with inhibitors of K_{Ca} constitutively expressed in endothelial cells may be of more interest. Also worthy of mentioning is that procyanidin C1 stimulated Ca^{2+} entry across the plasma membrane in response to depletion of intracellular stores (see Section 11.5.4.2) (Byun *et al.*, 2012a). Subsequent work verified the previous finding of procyanidin C1-induced relaxation in thoracic aortic rings via endothelial NO production involving both the PI3K/Akt pathway and hyperpolarization (Byun *et al.*, 2014).

Studies on hawthorn extracts using isolated arterial ring preparations produced conflicting data, which may be attributed to differences in their preparations and the plant materials used. While the BK_{Ca} channel inhibitor iberiotoxin slightly, but significantly, reduced the relaxation in response to a fruit extract (Chen *et al.*, 1998), apamin and CTX (Table 11.2) did not affect the relaxed vascular tone induced by an extract (*Crataegus oxyacantha*/*C. monogyna*) and a procyanidin fraction (Kim *et al.*, 2000). On the other hand, glibenclamide, a selective inhibitor of K_{ATP} channels, was invariably ineffective, suggesting that K_{ATP} channels were not involved. The fact that K^+ channels play a role in the hawthorn-induced relaxant effect follows from the pronounced inhibition of relaxation observed in the presence of the nonselective K^+ channel blocker TEA. The current data suggest that vasorelaxation elicited by *Crataegus* extracts and their procyanidins may be due to a certain extent via modulation of K^+ channel functions, but appropriate experiments still need to be conducted to identify the various types of activated K^+ channels.

Experiments with rat aorta rings showed that, besides an NO-cGMP-dependent event, voltage-sensitive and K_{Ca} channels may play a role in the evoked response to procyanidin-containing *M. ilicifolia* extracts, as concluded from experiments with the inhibitors 4-aminopyridine (4-AP) and TEA, respectively (Rattmann *et al.*, 2006). Glibenclamide again failed to affect the vasodilatory activity of the samples. Similarly, the relaxation elicited by *C. celtidifolius* in thoracic aorta rings was attenuated in the presence of TEA and in experiments with the combination CTX plus apamin (DalBò *et al.*, 2008a, 2008b). Comparable inhibitory effects were demonstrated for a procyanidin-enriched apple extract with regard to the employment of 4-AP and TEA (Matsui *et al.*, 2009). From additional experiments with the inhibitors apamin, glibenclamide and $BaCl_2$ (Table 11.2) causing reduced relaxation, the authors concluded that the apple extract evoked relaxation by the combined activation of K^+ channels (BK_{Ca}, SK_{Ca}, K_{ATP}, $K\nu$, K_{IR}). Further examples of contributions of endothelial K_{Ca} channels to relaxation include some procyanidin-containing fruit juices, as concluded from affected responses in the presence of apamin and TRAM 34 (Kivimäki *et al.*, 2011). These observations agree with EDHF-mediated hyperpolarization via multiple K_{Ca} channel activation.

Even though the questions of the precise molecular mechanism of proanthocyanidin-activated K^+ channels and how these components are interrelated in the relaxant response are still to be resolved, accumulating evidence indicates that procyanidins exert their vasorelaxant effects by the simultaneous modulation of such ion channels. An explanation for seemingly conflicting data may be seen in the different extract compositions, and systematic approaches are required to address this.

11.5.4.2 Ca^{2+} signaling events and modulation of Ca^{2+} channel functions

Many endothelial functions including vasorelaxation depend to various extents on changes in $[Ca^{2+}]_i$ concentration. For example, endothelium-dependent hyperpolarization is closely related with an increase in endothelial $[Ca^{2+}]_i$, which is due to either release of Ca^{2+} from the endoplasmic reticulum or Ca^{2+} influx from extracellular sources or both (Fig. 11.5). Notably, depletion of intracellular Ca^{2+} stores activates Ca^{2+} influx by opening of store-operated calcium channels (SOCCs) in the plasma membrane, while refilling blocks transmembrane Ca^{2+} channels (Pareki & Putney, 2005; Smyth *et al.*, 2010). The Ca^{2+} elevating action amplifies Ca^{2+}-dependent eNOS activation and NO production (Nelson *et al.*, 1995; Ignarro *et al.*, 1999; Feletou & Vanhoutte, 2006). On the other hand, a decrease in $[Ca^{2+}]_i$ is a key event involved in VSMC relaxation in response to vasodilators (*vide supra*). There is a serious gap in our understanding of the effects of procyanidins on Ca^{2+} signaling events. Indeed, little attention has been paid to the impact of procyanidins on Ca^{2+} homeostasis.

A few studies including procyanidin-rich hawthorn (Chen *et al.*, 1998) and apple extracts (Byun *et al.*, 2012b) demonstrated that Ca^{2+}-free solutions inhibited the induced relaxation, signifying the importance of transmembrane Ca^{2+} influx from extracellular sources. A question yet to be answered is the transmembrane Ca^{2+} entry mechanism mediated either via store-operated or receptor-operated Ca^{2+} channels. Investigation of a crude bark extract of *C. celtidifolius* also showed that relaxation disappeared in the absence of extracellular Ca^{2+} in the medium and was significantly decreased by the sulfhydryl alkylating agent *N*-ethylmaleimide and the phospholipase C (PLC) blocker neomycin (DalBò *et al.*, 2008b). The authors concluded that the activation of eNOS is due to increased $[Ca^{2+}]_i$ derived from PLC activation and an *N*-ethylmaleimide-sensitive Ca^{2+} associated pathway. However, the extract preparation does not necessarily relate the effect to procyanidins.

More recently, chemically defined fractions of 2,3-*cis*- (Kaufeld *et al.*, 2013) and 2,3-*trans*-procyanidins (Kaufeld *et al.*, 2014) were tested for their ability to mediate vasorelaxation by affecting $[Ca^{2+}]_i$. Thapsigargin (SERCA inhibitor) and 2-APB (inhibitor of store-operated calcium channels; SOCCs) abolished the vasorelaxant response, while ryanodine (inhibitor of Ca^{2+} release from ryanodine-sensitive stores) failed to affect the relaxation. This indicates that the nonrefilling of intracellular stores in VSMCs by the SERCA inhibitor thapsigargin leads to depletion of these Ca^{2+} stores and, in turn, to Ca^{2+} influx via SOCCS associated with an increase in vascular tone.

The fact that procyanidins may modulate $[Ca^{2+}]_i$ was also verified in RAECs and thoracic aortic rings (Byun, 2012; Byun *et al.*, 2012a, 2014). Using confocal fluorescence microscopy and the Ca^{2+} indicator fluo-4 acetoxymethyl ester (Fluo-4 AM), addition of procyanidin C1 caused a marked increase in endothelial cytosolic Ca^{2+} levels but only when using a Ca^{2+}-containing buffer. The nonfluorescent probe is cleaved inside living cells to highly fluorescent Fluo-4 upon binding Ca^{2+}. Conversely, addition of the store-operated Ca^{2+} entry pathway inhibitor 2-APB significantly diminished cytosolic Ca^{2+} concentrations. This finding suggests procyanidin C1-evoked opening of membrane-associated Ca^{2+} channels via the store-operated Ca^{2+} entry pathway.

Although the scope of this work is rather limited, there is strong evidence that proanthocyanidins are able to modulate $[Ca^{2+}]_i$ by regulating Ca^{2+} influx across the plasma membrane and release from intracellular stores in both endothelial cells and VSMCs.

11.6 Bioavailability and metabolic transformation: the missing link in the evidence to action in the body

Although not in the focus of this overview, it appears appropriate to give a brief note on the significance of bioavailability and metabolism of proanthocyanidins in the context of human health. While isolated arteries constitute a particularly versatile model for assessing the vasorelaxant mode of action of proanthocyanidins and other polyphenols as demonstrated earlier, the biological effects *in vivo* strongly depend on their bioavailability. Intestinal absorption, distribution in body tissues, and bioconversion are important efficacy-influencing parameters. The intestine and its colonic microflora play a crucial role in these processes. The fact that proanthocyanidins are vasoactive compounds clearly follows from numerous organ bath experiments using precontracted arterial rings, an experimental design that excludes metabolic transformations. However, *in vitro* models do not reflect *in vivo* conditions, thus studies with animals and humans are needed to provide conclusive verification of these health effects. Indeed, feeding studies have indicated that oligomeric and polymeric proanthocyanidins are poorly absorbed. For example, the low levels of intact procyanidin dimers reaching nanomolar concentrations in the circulatory system cannot provide convincing evidence for the benefit *in vivo* (Manach *et al.*, 2005b), and absorption rates of higher oligomers and polymers are considerably lower, if absorbed at all. Although the measurements of plasma levels have not considered binding of proanthocyanidins to proteins and tissue accumulation, it remains to be determined if these quantities can contribute to more effective levels. The detection of proanthocyanidins from dimers up to pentamers in urea-extracted plasma of rats shows that absorption of high-molecular-weight species may occur (Shoji *et al.*, 2006). The feeding experiment was, however, performed under nonphysiological conditions.

Current evidence for poor absorption of proanthocyanidins supports the crucial role of microbial metabolism for the production of more absorbable metabolites and their impact on biological effects (Selma *et al.*, 2009; Williamson & Clifford, 2010). Nonabsorbed proanthocyanidins reach the large intestine where microbial transformation may occur, yielding a diversity of bioavailable phenolic acids and some lactones. The metabolism by intestinal microorganisms can lead to a fairly complex metabolic picture. Conspicuously, conversion of oligomers to monomeric flavan-3-ols was reported to be a minor catabolic pathway in the microbial degradation, while the main routes suggested cleavage of the heterocyclic C-ring (Del Rio *et al.*, 2013). Their microbial origin was verified by treatment of procyanidins with rat cecal contents or human fecal slurries (Gonthier *et al.*, 2003; Stoupi *et al.*, 2010). The spectrum of microbial catabolites has essentially been studied with some dimeric procyanidins, while the fate of higher oligomeric forms is largely unknown. Interestingly, the rate of microbial metabolites decreased as the degree of polymerization increased (Gonthier *et al.*, 2003). Notably, this fraction constitutes a substantial portion of proanthocyanidin-rich diets with supposed beneficial health effects. From this, it is evident that there are still some gaps in our understanding of the microbial metabolism of proanthocyanidins.

More importantly, the microbial metabolites of proanthocyanidins have not yet been studied for biological activity. Therefore, the vasorelaxant potential of metabolites present in the circulatory system should be evaluated and at the *in vivo* concentrations reached in plasma. For this, isolated arteries may well serve as a convenient probe to identify bioactive metabolites, followed by studies with animals and humans. If, and in contrast to the current view, none of the microbial metabolites should exhibit adequate bioefficacy, vasorelaxation of proanthocyanidins may originate, at least to some extent, from modulatory effects on components of signaling cascades. The concentrations required to activate initial signaling events may be much lower, thereby allowing modulation of intracellular downstream signaling pathways required for a biological response. Further research is warranted to clarify the impact of microbial degradation metabolites on health benefits.

11.7 Conclusions

Polyphenols are abundantly present in foodstuffs of plant origin and herbal medicines. There is strong evidence that regular consumption of a polyphenol-rich diet has beneficial effects on cardiovascular diseases. However, these secondary metabolites represent a chemically heterogeneous group and most studies have used chemically poorly defined test materials. There is an urgent need for taking into account the extract composition in more detail to correlate health-beneficial effects with the responsible active polyphenol(s). At present, uncertainty remains regarding which distinct polyphenols of a complex cocktail mediate the beneficial actions and what are their contributions to the overall effect. This notion applies in particular to the structurally complex proanthocyanidins to get insight into structure–activity relationships. Although research on vasoprotective effects of proanthocyanidins is rather limited, emerging evidence supports at least an important *in vitro* role in the improvement of endothelial dysfunction in large arteries through predominantly the eNOS/NO/cGMP pathway and, to a lesser extent, the EDHF-signaling pathway. Regulation of the intracellular Ca^{2+} level in endothelial cells and VSMCs associated with hyperpolarization is apparently a further contributing component of procyanidins to vasorelaxation. More work with purified proanthocyanidins, particularly in smaller resistant vessels, is required to address the role of EDHF and the gaps in our understanding of orchestrating K^+ and Ca^{2+} channel functions. Another key issue concerns Janus-faced ROS, apparently important activators of redox-sensitive kinases in the signaling pathways preventing and counteracting endothelial dysfunction. Much more is to be gleaned about the bioavailability and metabolism of proanthocyanidins for a better understanding of their mode of actions *in vivo*.

References

Aldini, G., Carini, M., Piccoli, A. *et al.* (2003) Procyanidins from grape seeds protect endothelial cells from peroxynitrite damage and enhance endothelium-dependent relaxation in human artery: new evidences for cardio-protection. *Life Sciences*, **73**, 2883–2898.

Alhosin, M., Anselm, E., Rashid, S. *et al.* (2013) Redox-sensitive up-regulation of eNOS by purple grape juice in endothelial cells: role of PI3-kinase/Akt, p38 MAPK, JNK, FoxO1 and FoxO3a. *PlosOne*, **8**, e57883.

Alvarez, E., Rodiño-Janeiro, B.K., Jerez, M. *et al.* (2012) Procyanidins from grape pomace are suitable inhibitors of human endothelial NADPH oxidase. *Journal of Cellular Biochemistry*, **113**, 1386–1396.

Andriambeloson, E., Kleschyov, A.L., Muller, B. *et al.* (1997) Nitric oxide production and endothelium-dependent vasorelaxation induced by wine polyphenols in rat aorta. *British Journal of Pharmacology*, **120**, 1053–1058.

Andriambeloson, E., Magnier, C., Haan-Archipoff, G. *et al.* (1998) Natural dietary polyphenolic compounds cause endothelium-dependent vasorelaxation in rat thoracic aorta. *The Journal of Nutrition*, **128**, 2324–2333.

Anselm, E., Chataigneau, M., Ndiaye, M. *et al.* (2007) Grape juice causes endothelium-dependent vasorelaxation via a redox-sensitive Src- and Akt-dependent activation of eNOS. *Cardiovascular Research*, **73**, 404–413.

Anselm, E., Socorro, V.F.M., Dal-Ros, S. *et al.* (2009) Crataegus special extract WS 1442 causes endothelium-dependent relaxation via a redox-sensitive Src- and Akt-dependent activation of endothelial NO synthase but not via activation of estrogen receptors. *Journal of Cardiovascular Pharmacology*, **53**, 253–260.

Appeldoorn, M.M., Venema, D.P., Peters, T.H.F. *et al.* (2009a) Some phenolic compounds increase the nitric oxide level in endothelial cells *in vitro*. *Journal of Agricultural and Food Chemistry*, **57**, 7693–7699.

Appeldoorn, M.M., Vincken, J.P., Aura, A.M. *et al.* (2009b) Procyanidin dimers are metabolized by human microbiota with 2-(3,4-dihydroxyphenyl)acetic acid and 5-(3,4-dihydroxyphenyl)-γ-valerolactone as the major metabolites. *Journal of Agricultural and Food Chemistry*, **57**, 1084–1092.

Auger, C., Caporiccio, B., Landrault, N. *et al.* (2002) Red wine phenolic compounds reduce plasma lipids and apolipoprotein B and prevent early aortic atherosclerosis in hypercholesterolemic golden Syrian hamsters (Mesocricetus auratus). *Journal of Nutrition*, **132**, 1207–1213.

Auger, C., Chaabi, M., Anselm, E. *et al.* (2010a) The red wine extract-induced activation of endothelial nitric oxide synthase is mediated by a great variety of polyphenolic compounds. *Molecular Nutrition & Food Research*, **54**, S171–S183.

Auger, C., Kim, J.H., Chabert, P. *et al.* (2010b) The EGCg-induced redox-sensitive activation of endothelial nitric oxide synthase and relaxation are critically dependent on hydroxyl moieties. *Biochemical and Biophysical Research Communications*, **393**, 162–167.

Borges, G., Lean, M.E.J., Roberts, S.A. *et al.* (2013) Bioavailabilty of dietary polyphenols: a study with ileostomists to discriminate between absorption in small and large intestine. *Food & Function*, **4**, 754–762.

Bredt, D.S., Ferris, C.D. and Snyder, S.H. (1992) Nitric oxide synthase regulatory site: phosphorylation by cyclic AMP-dependent protein kinase, protein kinase C, and calcium/calmodulin protein kinase, identification of flavin and calmodulin binding sites. *The Journal of Biological Chemistry*, **267**, 10976–10981.

Brixius, K., Willms, S., Napp, A. *et al.* (2006) Crataegus special extract WS® 1442 induces an endothelium-dependent, NO-mediated vasorelaxation via eNOS-phosphorylation at serine 1177. *Cardiovascular Drugs and Therapy*, **20**, 177–184.

Burns, J., Gardner, P.T., O'Neil, J. *et al.* (2000) Relationship among antioxidant activity, vasodilation capacity, and phenolic content of red wines. *Journal of Agricultural and Food Chemistry*, **48**, 220–230.

Byun, M.W. (2012) Effect of procyanidin C1 on nitric oxide production and hyperpolarization through Ca^{2+}-dependent pathway in endothelial cells. *Journal of Medicinal Food*, **15**, 1032–1037.

Byun, E.B., Ishikawa, T., Suyama, A. *et al.* (2012a) A procyanidin trimer, C1, promotes NO production in rat aortic endothelial cells via both hyperpolarization and PI3K/Akt pathways. *European Journal of Pharmacology*, **692**, 52–60.

Byun, E.B., Korematsu, S., Ishikawa, T. *et al.* (2012b) Apple procyanidins induce hyperpolarization of rat aorta endothelial cells via activation of K^+ channels. *Journal of Nutritional Biochemistry*, **23**, 278–286.

Byun, E.B., Sung, N.Y., Yang, M.S. *et al.* (2014) Procyanidin C1 causes vasorelaxation through activation of the endothelial NO/cGMP pathway in thoracic aortic rings. *Journal of Medicinal Food*, **17**, 742–748.

Cai, Q., Li, B.Y., Gao, H.Q. *et al.* (2011) Grape seed procyanidin B2 inhibits human aortic smooth muscle cell proliferation and migration induced by advanced glycation end products. *Bioscience, Biotechnology, and Biochemistry*, **75**, 1692–1697.

Ceballos, G. and Villarreal, F. (2011) The beneficial vascular effects of cacao flavanols: having your cake and eating it too. *Journal of Applied Physiology*, **111**, 1544–1545.

Chalopin, M., Tesse, A., Martinez, M.C. *et al.* (2010) Estrogen receptor alpha as a key target of red wine polyphenol action on the endothelium. *PlosOne*, **5**, e8554.

Chambliss, K.L. and Shaul, P.W. (2002) Estrogen modulation of endothelial nitric oxide synthase. *Endocrine Reviews*, **23**, 665–686.

Chawengsub, Y., Gauthier, K.M. and Campbell, W.B. (2009) Role of arachidonic acid lipoxygenase metabolites in the regulation of vascular tone. *American Journal of Physiology—Heart and Circulatory Physiology*, **297**, H495–H507.

Chen, Z.Y., Zhang, Z.S., Kwan, K.Y. *et al.* (1998) Endothelium-dependent relaxation induced by hawthorn extract in rat mesenteric artery. *Life Sciences*, **63**, 1983–1991.

Chong, M.F., Macdonald, R. and Lovegrove, J.A. (2010) Fruit polyphenols and CVD risk: a review of human intervention studies. *British Journal of Nutrition*, **104**, S28–S39.

Clifford, M. and Brown, J.E. (2006) Dietary flavonoids and health—broadening the perspective, in Flavonoids: Chemistry, Biochemistry and Applications (eds Ø.M. Andersen and K.R. Markham), CRC Taylor & Francis, Boca Rotan, pp. 319–370.

Corder, R., Warburton, R.C., Khan, N.Q. *et al.* (2004) The procyanidin-induced pseudo laminar shear stress response: a new concept for the reversal of endothelial dysfunction. *Clinical Science*, **107**, S13–S17.

DalBó, S., Moreira, E.G., Brandão, F.C. *et al.* (2008a) Mechanisms underlying the vasorelaxant effect induced by proanthocyanidin-rich fraction from *Croton celtidifolius* in rat small resistance arteries. *Journal of Pharmacological Sciences*, **106**, 234–241.

DalBó, S., Goulart, S., Horst, H. *et al.* (2008b) Activation of endothelial nitric oxide synthase by proanthocyanidin-rich fraction from *Croton celtidifolius* (Euphorbiaceae): involvement of extracellular calcium influx in rat thoracic aorta. *Journal of Pharmacological Sciences*, **107**, 181–189.

Dal-Ros, S., Bronner, C., Schott, C. *et al.* (2009) Angiotensin II-induced hypertension is associated with a selective inhibition of endothelium-derived hyperpolarization factor-mediated responses in the rat mesenteric artery. *The Journal of Pharmacology and Experimental Therapeutics*, **328**, 478–486.

Dal-Ros, S., Zoll, J., Lang, A.L. *et al.* (2011) Chronic intake of red wine polyphenols by young rats prevents aging-induced endothelial dysfunction and decline in physical performance: role of NADPH oxidase. *Biochemistry and Biophysics Research Communications*, **404**, 743–749.

Dal-Ros, S., Bronner, C., Auger, C. *et al.* (2012) Red wine polyphenols improve an established aging-related endothelial dysfunction in the mesenteric artery of middle-aged rats: role of oxidative stress. *Biochemical and Biophysical Research Communications*, **419**, 381–387.

De Moura, R.S., Miranda, D.Z., Pinto, A.C. *et al.* (2004) Mechanism of the endothelium-dependent vasodilation and the antihypertensive effect of a Brazilian red wine. *Journal of Cardiovascular Pharmacology*, **44**, 302–309.

Del Rio, D., Rodriguez-Mateos, A., Spencer, J.P. *et al.* (2013) Dietary (poly)phenolics in human health: structures, bioavailability, and evidence of protective effects against chronic diseases. *Antioxidants and Redox Signaling*, **18**, 1818–1896.

Dimmeler, S., Fleming, I., Fisslthaler, B. *et al.* (1999) Activation of nitric oxide synthase in endothelial cells by Akt-dependent phosphorylation. *Nature*, **399**, 601–605.

Dongmo, A.B., Kamanyi, A., Franck, U. *et al.* (2002) Vasodilating properties of extracts from the leaves of *Musanga cecropioides* (R. Brown). *Phytotherapy Research*, **16**, S6–S9.

Duarte, J., Andriambeloson, E., Diebolt, M. *et al.* (2004) Wine polyphenols stimulate superoxide anion production to promote calcium signaling and endothelial-dependent vasodilatation. *Physiological Research*, **53**, 595–602.

Edirisinghe, I., Burton-Freeman, B. and Tissa, K.C. (2008) Mechanism of the endothelium-dependent relaxation evoked by a grape seed extract. *Clinical Science*, **114**, 331–337.

Endemann, D.H. and Schiffrin, E.L. (2004) Endothelial dysfunction. *Journal of the American Society of Nephrology*, **15**, 1983–1992.

Facino, R.M., Carini, M., Aldini, G. *et al.* (1999) Diet enriched with procyanidins enhances antioxidant activity and recuces myocardial post-ischaemic damage in rats. *Life Sciences*, **64**, 627–642.

Feletou, M. and Vanhoutte, P.M. (2006) Endothelium-derived hyperpolarizing factor: where are we now? *Arteriosclerosis, Thrombosis, and Vascular Biology*, **26**, 1215–1225.

Feletou, M., Köhler, R. and Vanhoutte, P.M. (2010) Endothelium-derived vasoactive factors and hypertension: possible roles in pathogenesis and as treatment targets. *Current Hypertensive Reports*, **12**, 267–275.

Feletou, M., Köhler, R. and Vanhoutte, P.M. (2012) Nitric oxide: orchestrator of endothelium-dependent responses. *Annals of Medicines*, **44**, 694–716.

Fisher, N.D.L., Hughes, M., Gerhard-Herman, M. *et al.* (2003) Flavanol-rich cacao induces nitric-oxide-dependent vasodilation in healthy humans. *Journal of Hypertension*, **21**, 2281–2286.

Fitzpatrick, D.F., Hirschfield, S.L. and Coffey, R.G. (1993) Endothelium-dependent vasorelaxing activity of wine and other grape products. *American Journal of Physiology*, **265**, H774–H778.
Fitzpatrick, D.F., Bing, B. and Rohdewald, P. (1998) Endothelium-dependent vascular effects of pycnogenol. *Journal of Cardiovascular Pharmacology*, **32**, 509–515.
Fitzpatrick, D.F., Fleming, R.C., Bing, B. *et al.* (2000) Isolation and characterization of endothelium-dependent vasorelaxing compounds from grape seeds. *Journal of Agricultural and Food Chemistry*, **48**, 6384–6390.
Fitzpatrick, D.F., Bing, B., Maggi, D.A. *et al.* (2002) Vasodilating procyanidins derived from grape seeds. *Annals of the New York Academy of Sciences*, **957**, 78–89.
Flammer, A.J., Sudano, I., Wolfrum, M. *et al.* (2012) Cardiovascular effects of flavanol-rich chocolate in patients with heart failure. *European Heart Journal*, **33**, 2172–2180.
Fulton, D., Gratton, J.P., McCabe, T.J. *et al.* (1999) Regulation of endothelium-derived nitric oxide production by the protein kinase Akt. *Nature*, **399**, 597–601.
Furchgott, R.F. and Zawadzki, J.V. (1980) The obligatory role of the endothelial cells in the relaxation of arterial smooth muscle by acetylcholine. *Nature*, **288**, 373–376.
Furuuchi, R., Sakai, H., Hirokawa, M. *et al.* (2012) Antihypertensive effect of boysenberry seed polyphenols on spontaneously hypertensive rats and identification of orally absorbable proanthocyanidins with vasorelaxant activity. *Bioscience, Biotechnology, and Biochemistry*, **76**, 1694–1701.
Garcia-Conesa, M.T., Tribolo, S., Guyot, S. *et al.* (2009) Oligomeric procyanidins inhibit cell migration and modulate the expression of migration and proliferation associated genes in human umbilical vascular endothelial cells. *Molecular Nutrition and Food Research*, **53**, 266–276.
Garland, C.J., Hiley, C.R. and Dora, K.A. (2011) EDHF: spreading the influence of the endothelium. *British Journal of Pharmacology*, **164**, 839–852.
Gonthier, M.P., Donovan, J.L., Texier, O. *et al.* (2003) Metabolism of dietary procyanidins in rats. *Free Radical Biology and Medicine*, **35**, 837–844.
Gu, L., Kelm, M.A., Hammerstone, J.F. *et al.* (2004) Concentrations of proanthocyanidins in common foods and estimations of normal consumption. *The Journal of Nutrition*, **134**, 613–617.
Guerrero, L., Margalef, M., Pons, Z. *et al.* (2013) Serum metabolites of proanthocyanidin-administered rats decrease lipid synthesis in HepG2 cells. *Journal of Nutritional Biochemistry*, **24**, 2092–2099.
Habauzit, V. and Morand, C. (2012) Evidence for a protective effect of polyphenols-containing foods on cardiovascular health: an update for clinicians. *Therapeutic Advances in Chronic Disease*, **3**, 87–106.
Henning, S.M., Aronson, W., Niu, Y. *et al.* (2006) Tea polyphenols and theaflavins are present in prostate tissue of humans and mice after green and black tea consumption. *The Journal of Nutrition*, **136**, 1839–1843.
Idris-Khodja, N., Auger, C., Koch, E. *et al.* (2012a) Crataegus special extract prevents aging-related endothelial dysfunction. *Phytomedicine*, **19**, 699–706.
Idris-Khodja, N., Chataigneau, T., Auger, C. *et al.* (2012b) Grape-derived polyphenols improve aging-related endothelial dysfunction in rat mesenteric artery: role of oxidative stress and the angiotensin system. *PlosOne*, **7**, e32039.
Idris-Khodja, N., Di Marco, P., Mona Farhat, M. *et al.* (2013) Grape-derived polyphenols prevent doxorubicin-induced blunted EDH-mediated relaxations in the rat mesenteric artery: role of ROS and angiotensin II. *Evidence-Based Complementary and Alternative Medicine*, **2013**, 516017.
Ignarro, L.J., Cirino, G., Casini, A. *et al.* (1999) Nitric oxide as a signaling molecule in the vascular system: an overview. *Journal of Cardiovascular Pharmacology*, **34**, 879–886.
Irani, K. (2000) Oxidant signaling in vascular cell growth, death, and survival: a review of the roles of reactive oxygen species in smooth muscle and endothelial mitogenic and apoptotic signaling. *Circulation Research*, **87**, 179–183.
Iravani, S. and Zolfaghari, B. (2011) Pharmaceutical and nutraceutical effects of *Pinus pinaster* bark extract. *Research in Pharmaceutical Sciences*, **6**, 1–11.
Kane, M.O., Anselm, E., Rattmann, Y.D. *et al.* (2009) Role of gender and estrogen receptors in the rat aorta endothelium-dependent relaxation to red wine polyphenols. *Vascular Pharmacology*, **51**, 140–146.

Karim, M., McCarmick, K. and Kappagoda, C.T. (2000) Effects of cacao extracts on endothelium-dependent relaxation. *The Journal of Nutrition*, **130**, 2105S–2108S.

Kaufeld, A.M., Pertz, H.H. and Kolodziej, H. (2013) 2,3-*cis* procyanidins elicit endothelium-dependent relaxation in porcine coronary arteries via activation of the PI3/Akt kinase signaling pathway. *Journal of Agricultural and Food Chemistry*, **61**, 9609–9616.

Kaufeld, A.M., Pertz, H.H. and Kolodziej, H. (2014) A chemically defined 2,3-*trans* procyanidin fraction from willow bark causes redox-sensitive endothelium-dependent relaxation in porcine coronary arteries. *Journal of Natural Products*, **77**, 1607–1614.

Kawakami, K., Aketa, S., Sakai, H. *et al.* (2011) Antihypertensive and vasorelaxant effects of water-soluble proanthocyanidins from persimmon leaf tea in spontaneously hypertensive rats. *Bioscience, Biotechnology, and Biochemistry*, **75**, 1435–1439.

Kelm, M.A., Johnson, J.C., Robbins, R.J. *et al.* (2006) High-performance liquid chromatography separation and purification of cacao (*Theobroma cacao* L.) procyanidins according to degree of polymerization using a diol stationary phase. *Journal of Agricultural and Food Chemistry*, **54**, 1571–1576.

Kim, S.H., Kang, K.W., Kim, K.W. *et al.* (2000) Procyanidins in crataegus extract evoke endothelium-dependent vasorelaxation in rat aorta. *Life Sciences*, **67**, 121–131.

Kim, J., Formosa, G., Li, Y. *et al.* (2007) Epigallocatechin gallate, a green tea polyphenol, mediates NO-dependent vasodilation using signaling pathways in vascular endothelium requiring reactive oxygen species and Fyn. *The Journal of Biological Chemistry*, **282**, 13736–13745.

Kim, J.H., Auger, C. and Schini-Kerth, V.B. (2014) Activation of eNOS by polyphenol-rich products and polyphenolic compounds. *Current Pharmaceutical Design*, **20**, 3521–3529.

Kivimäki, A.S., Ehlers, P.I., Turpeinen, A.M. *et al.* (2011) Lingonberry juice improves endothelium-dependent vasodilatation of mesenteric arteries in spontaneously hypertensive rats in a long-term intervention. *Journal of Functional Foods*, **3**, 267–274.

Koch, E. and Malek, F.A. (2011) Standardize extracts from hawthorn leaves and flowers in the treatment of cardiovascular disorders—preclinical and clinical studies. *Planta Medica*, **77**, 1123–1128.

Kolodziej, H., Ferreira, D. and Roux, D.G. (1984) Direct access to [4,6]- and [4,8]-all-2,3-cis procyanidin derivatives from (-)-epicatechin: assessment of bonding positions in oligomeric analogues from *Crataegus oxyacantha* L. *Journal of Chemical Society Perkin Transaction I*, 343–350.

Kurita, I., Kim, J.H., Auger, C. *et al.* (2013) Hydroxylation of (-)-epigallocatechin-3-O-gallate at 3″, but not 4″, is essential for the PI3-kinase/Akt-dependent phosphorylation of endothelial NO synthase in endothelial cells and relaxation of coronary artery rings. *Food and Functions*, **4**, 249–257.

Kwak, C.J., Kubo, E., Fujii, K. *et al.* (2009) Antihypertensive effect of French maritime pine bark extract (Flavangenol): possible involvement of endothelial nitric oxide-dependent vasorelaxation. *Journal of Hypertension*, **27**, 92–101.

Kwan, C.Y., Zhang, W.B., Sim, S.M. *et al.* (2004) Vascular effects of Siberian ginseng (*Eleutherococcus senticosus*); endothelium-dependent NO- and EDHF-mediated relaxation depending on vessel size. *Naunyn-Schmiedeberg's Archives of Pharmacology*, **369**, 473–480.

Luksha, L., Agewall, S. and Kublickiene, K. (2009) Endothelium-derived hyperpolarizing factor in vascular physiology and cardiovascular disease. *Atherosclerosis*, **202**, 330–344.

Madeira, S.V., Auger, C., Anselm, E. *et al.* (2009) eNOS activation induced by a polyphenol-rich grape skin extract in porcine coronary arteries. *Journal of Vascular Research*, **46**, 406–416.

Magos, G.A., Mateos, J.C., Paez, E. *et al.* (2008) Hypotensive and vasorelaxant effects of the procyanidin fraction from *Guazuma ulmifolia* bark in normotensive and hypertensive rats. *Journal of Ethnopharmacology*, **117**, 58–68.

Manach, C., Mazur, A. and Scalbert, A. (2005a) Polyphenols and prevention of cardiovascular diseases. *Current Opinion in Lipidology*, **16**, 77–84.

Manach, C., Williamson, G., Morand, C. *et al.* (2005b) Bioavailability and bioefficacy of polyphenols in humans. I. Review of 97 bioavailability studies. *The American Journal of Clinical Nutrition*, **81**, 230S–242S.

Manga, F.N., Khattabi, C.E., Fontaine, J. *et al.* (2012) Vascular effects and antioxidant activity of two *Combretum* species from Democratic Republic of Congo. *Journal of Ethnopharmacology*, **142**, 194–200.
Manga, F.N., Khattabi, C.E., Fontaine, J. *et al.* (2013) Vasorelaxant and antihypertensive effects of methanolic extracts from *Hymenocardia acida*. *Journal of Ethnopharmacology*, **146**, 623–631.
Matsui, T., Korematsu, S., Byun, E.B. *et al.* (2009) Apple procyanidins induced vascular relaxation in isolated rat aorta through NO/cGMP pathway in combination with hyperpolarization by multiple K^+ channel activations. *Bioscience, Biotechnology, and Biochemistry*, **73**, 2246–2251.
McNeill, J.R. and Jurgens, T.M. (2006) A systematic review of mechanisms by which natural products of plant origin evoke vasodilatation. *Canadian Journal of Physiology and Pharmacology*, **84**, 803–821.
Mendes, A., Desgranges, C., Chèze, C. *et al.* (2003) Vasorelaxant effect of grape polyphenols in rat isolated aorta. Possible involvement of a purinergic pathway. *Fundamental and Clinical Pharmacology*, **17**, 673–681.
Mensah, G.A. (2008) Epidemiology of stroke and high blood pressure in Africa. *Heart*, **84**, 697–705.
Monbouli, J.V. and Vanhoutte, P.M. (1999) Endothelial dysfunction: from physiology to therapy. *Journal of Molecular and Cellular Cardiology*, **31**, 61–74.
Mudnic, I., Moduna, D., Brizica, I. *et al.* (2009) Cardiovascular effects *in vitro* of aqueous extract of wild strawberry (*Fragaria vesca*, L.) leaves. *Phytomedicine*, **16**, 462–469.
Mukai, Y. and Sato, S. (2009) Polyphenol-containing azuki bean (*Vigna angularis*) extract attenuates blood pressure elevation and modulates nitric oxide synthase and caveolin-1 expressions in rats with hypertension. *Nutrition, Metabolism, and Cardiovascular Diseases*, **19**, 491–497.
Ndiaye, M., Chataigneau, T., Andriantsitohaina, R. *et al.* (2003) Red wine polyphenols cause endothelium-dependent EDHF-mediated relaxations in porcine coronary arteries via a redox-sensitive mechanism. *Biochemical and Biophysical Research Communications*, **310**, 371–377.
Ndiaye, M., Chataigneau, T., Chataigneau, M. *et al.* (2004) Red wine polyphenols induce EDHF-mediated relaxations in porcine coronary arteries through the redox-sensitive activation of the PI3-kinase/Akt pathway. *British Journal of Pharmacology*, **142**, 1131–1136.
Ndiaye, M., Chataigneau, M., Lobysheva, I. *et al.* (2005) Red wine polyphenols induced endothelium-dependent NO-mediated relaxation is due to the redox-sensitive PI3-kinase/Akt-dependent phosphorylation of endothelial NO-synthase in the isolated porcine coronary artery. *The FASEB Journal*, **19**, 455–457.
Ndiaye, M., Anselm, E., Sénea, M. *et al.* (2010) Mechanisms underlying the endothelium-dependent vasodilatory effect of an aqueous extract of *Elaeis guineensis* Jacq. (Arecaceae) in porcine coronary artery rings. *African Journal of Traditional, Complementary and Alternative Medicine*, **7**, 118–124.
Nelson, M.T., Cheng, H., Rubart, M. *et al.* (1995) Relaxation of arterial smooth muscle by calcium sparks. *Science*, **270**, 633–637.
Oak, M.H., Chataigneau, M., Keravis, T. *et al.* (2003) Red wine polyphenolic compounds inhibit vascular endothelial growth factor expression in vascular smooth muscle cells by preventing the activation of the p38 mitogen-activated protein kinase pathway. *Arteriosclerosis, Thrombosis, and Vascular Biology*, **23**, 1001–1007.
O'Neill, B.T. and Abel, E.D. (2005) Akt1 in the cardiovascular system: friend or foe. *Journal of Clinical Investigation*, **115**, 2059–2064.
Packer, L., Rimbach, G. and Virgilj, F. (1999) Antioxidant activity and biological properties of a procyanidin-rich extract from pine (*Pinus maritima*) bark, pycnogenol. *Free Radical Biology and Medicine*, **27**, 704–724.
Padilla, E., Ruiz, E., Redondo, S. *et al.* (2005) Relationship between vasodilation capacity and phenolic content of Spanish wines. *European Journal of Pharmacology*, **517**, 84–91.
Pareki, A.B. and Putney, J.W. (2005) Store-operated calcium channels. *Physiological Reviews*, **85**, 757–810.
Porter, L.J., Ma, Z. and Chan, B.G. (1991) Cacao procyanidins: major flavanoids and identification of some minor metabolites. *Phytochemistry*, **30**, 1657–1663.

Quiñones, M., Guerrero, L., Suarez, M. *et al.* (2013) Low-molecular procyanidin rich grape seed extract exerts antihypertensive effect in males spontaneously hypertensive rats. *Food Research International*, **51**, 587–595.

Quiñones, M., Guerrero, L., Fernàndez-Vanillas, S. *et al.* (2014) Involvement of nitric oxide and prostacyclin in the antihypertensive effect of low-molecular-weight procyanidin rich grape seed extract in male spontaneously hypertensive rats. *Journal of Functional Foods*, **6**, 419–427.

Rasmussen, S.E., Fredeeriksen, H., Krogholm, K.S. *et al.* (2005) Dietary proanthocyanidins: occurrence, dietary intake, bioavailability, and protection against cardiovascular disease. *Molecular Nutrition and Food Research*, **49**, 169–174.

Rattmann, Y.D., Cipriani, T.R., Sassaki, G.L. *et al.* (2006) Nitric oxide-dependent vasorelaxation induced by extractive solutions and fractions of *Maytenus ilicifolia* Mart ex Reissek (Celastraceae) leaves. *Journal of Ethnopharmacology*, **104**, 328–335.

Rattmann, Y.D., Anselm, E., Kim, J.H. *et al.* (2012) Natural product extract of *Dicksonia sellowiana* induces endothelium-dependent relaxations by a redox-sensitive Src- and Akt-dependent activation of eNOS in porcine coronary arteries. *Journal of Vascular Research*, **49**, 284–298.

Rodriguez-Mateos, A., Heiss, C., Borges, G. *et al.* (2014) Berry (poly)phenols and cardiovascular health. *Journal of Agricultural and Food Chemistry*, **62**, 3842–3851.

Rodriguez-Rodriguez, R., Justo, M.L., Claro, C.M. *et al.* (2012) Endothelium-dependent vasodilator and antioxidant properties of a novel enzymatic extract of grape pomace from wine industrial waste. *Food Chemistry*, **135**, 1044–1051.

Rohdewald, P. (2002) A review of the French maritime pine bark extract (Pycnogenol), a herbal medication with a diverse clinical pharmacology. *International Journal of Clinical Pharmacology and Therapeutics*, **40**, 158–168.

Sarr, M., Chataigneau, T., Auger, C. *et al.* (2006) Red wine polyphenols prevent angiotensin II-induced hypertension and endothelial dysfunction in rats: role of NADPH oxidase. *Cardiovascular Research*, **71**, 794–802.

Schini-Kerth, V.B., Auger, C., Etienne-Selloum, N. *et al.* (2010a) Polyphenol-induced endothelium-dependent relaxations: role of NO and EDHF, in Advances in Pharmacology, vol. **60** (ed P.M. Vanhoutte), Academic Press, Burlington, pp. 133–175.

Schini-Kerth, V.B., Auger, C., Kim, J.H. *et al.* (2010b) Nutritional improvement of the endothelial control of vascular tone by polyphenols: role of NO and EDHF. *Pflügers Archive—European Journal of Physiology*, **459**, 853–862.

Schini-Kerth, V.B., Etienne-Selloum, N., Chataigneau, T. *et al.* (2011) Vascular protection by natural product-derived polyphenols: *in vitro* and *in vivo* evidence. *Planta Medica*, **77**, 1161–1167.

Schlossmann, J., Ammendola, A., Ashman, K. *et al.* (2000) Regulation of intracellular calcium by a signaling complex of IRAG, IP3 receptor and cGMP kinase Ibeta. *Nature*, **404**, 197–201.

Schmitt, C.A. and Dirsch, V.M. (2009) Modulation of endothelial nitric oxide by plant-derived products. *Nitric Oxide*, **21**, 77–91.

Schramm, D.D., Wang, J.F., Holt, R.R. *et al.* (2001) Chocolate procyanidins decrease the leukotriene-prostacyclin ratio in humans and human aortic endothelial cells. *The American Journal of Clinical Nutrition*, **73**, 36–40.

Schroeter, H., Heiss, C., Spencer, J.P.E. *et al.* (2010) Recommending flavanols and procyanidins for cardiovascular health: current knowledge and future needs. *Molecular Aspects of Medicine*, **31**, S46–S57.

Seedat, Y.K. and Rosenthal, T. (2006) Meeting the challenges of hypertension in sub-Saharan Africa. *Cardiovascular Journal of South Africa*, **17**, 47–48.

Selma, M.V., Espin, J.C. and Tomas-Barberan, F.A. (2009) Interaction between phenolics and gut microbiota: role in human health. *Journal of Agricultural and Food Chemistry*, **57**, 6485–6501.

Shoji, T., Masumoto, S., Moriichi, N. *et al.* (2006) Apple procyanidin oligomers absorption in rats after oral administration: analysis of procyanidins in plasma using the porter method and high-performance liquid chromatography/tandem mass spectrometry. *Journal of Agricultural and Food Chemistry*, **54**, 884–892.

Smyth, J.T., Hwang, S.Y., Tomita, T. *et al.* (2010) Activation and regulation of store-operated calcium entry. *Journal of Cellular and Molecular Medicine*, **14**, 2337–2349.
Stoclet, J.C., Kleschyov, A., Andriambesolon, E. *et al.* (1999) Endothelial NO release caused by red wine polyphenols. *Journal of Physiology and Pharmacology*, **50**, 535–540.
Stoclet, J.C., Chataigneau, T., Ndiaye, M. *et al.* (2004) Vascular protection by dietary polyphenols. *European Journal of Pharmacology*, **500**, 299–313.
Stoupi, S., Williamson, G., Drynan, J.W. *et al.* (2010) Procyanidin B2 catabolism by human fecal microflora. Partial characterization of 'dimeric' intermediates. *Archives of Biochemistry and Biophysics*, **501**, 73–78.
Taubert, D., Berkels, R., Klaus, W. *et al.* (2002) Nitric oxide formation and corresponding relaxation of porcine coronary arteries induced by plant phenols: essential structural features. *Journal of Cardiovascular Pharmacology*, **40**, 701–713.
Thannickal, V.J. and Fanburg, B.L. (2000) Reactive oxygen species in cell signaling. *American Journal of Physiology—Lung Cellular and Molecular Physiology*, **279**, L1005–L1028.
Tokoudagba, J.M., Auger, C., Bréant, L. *et al.* (2010) Procyanidin-rich fractions from *Parkia biglobosa* (Mimosaceae) leaves cause redox-sensitive endothelium-dependent relaxation involving NO and EDHF in porcine coronary artery. *Journal of Ethnopharmacology*, **132**, 246–250.
Visioli, F., De La Lastra, C.A., Andres-Lacueva, C. *et al.* (2011) Polyphenols and human health: a prospectus. *Critical Reviews in Food Science and Nutrition*, **51**, 524–546.
Vitseva, O., Varghese, S., Chakrabarti, S. *et al.* (2005) Grape seed and skin extracts inhibit platelet function and release of reactive oxygen intermediates. *Journal of Cardiovascular Pharmacology*, **46**, 445–451.
Wang, X., Huang, L., Yu, T. *et al.* (2013) Effects of oligomeric grape seed proanthocyanidins on heart, aorta, kidney in DOCA-salt mice: role of oxidative stress. *Phytotherapy Research*, **27**, 869–876.
WHO (2008) *Health Statistics and Health Information Systems*, http://www.who.int/healthinfo/statistics (accessed on July 7, 2016).
Williamson, G. and Clifford, M.N. (2010) Colonic metabolites of berry polyphenols: the missing link to biological activity? *British Journal of Nutrition*, **104**, S48–S66.
Zumdick, S., Petereit, F., Luftmann, H. *et al.* (2009) Preparative isolation of oligomeric procyanidins from Hawthorn (*Crataegus* spp.) *Pharmazie*, **64**, 286–288.

Chapter 12
Polyphenols for Brain and Cognitive Health

Katherine H. M. Cox and Andrew Scholey

Abstract: Polyphenols are known to influence a number of biological systems relevant to brain function. Converging evidence from epidemiology and randomized controlled trials suggests that polyphenol intake can both protect against age-related cognitive decline and enhance cognitive function (although dose and duration of intake appear critical). This research has included studies into the mood and cognitive effects of polyphenols within the diet and also on individual classes of polyphenols from sources including tea, cocoa, wine, and soy as well as nondietary botanical sources. These studies show that polyphenols can improve cognitive function and/or mood both acutely and chronically in populations with compromised cognitive function, and also in those who are cognitively intact. Recent research has revealed some of the biologically plausible systemic and central mechanisms contributing to these effects. These include benefits to cerebral blood flow and patterns of brain activation measured using functional imaging methodology. Taken together, this body of evidence suggests that polyphenols may be useful additions to lifestyle changes, which may help to offset dementia and cognitive ageing.

Keywords: brain and cognition, mood, cognitive aging, bioactive nutrients, tea, cocoa, wine, soy.

12.1 Introduction

Similarly to other biological processes, brain function is influenced by nutritional factors. It is known that dietary patterns can influence long-term cognitive function, for example, healthy diet at baseline start of study predicted better cognitive function some 25 years later (Zhu *et al.*, 2015). There are also "nonnutritional" elements of diet, which have the potential to affect neurocognitive health. Polyphenols are one example of bioactive dietary components.

Recent Advances in Polyphenol Research, First Edition. Edited by Kumi Yoshida,
Véronique Cheynier and Stéphane Quideau.

Cognitive aging, mental decline, dementia, and even the fluctuations in day-to-day function are influenced by diverse and interacting processes, which are themselves influenced by diet.

The brain is the most metabolically active organ in the human body. This makes it particularly vulnerable to oxidative stress as well as inflammatory processes. It is also richly vascularized and exquisitely susceptible to fluctuations in blood flow and vascular processes. Cerebral blood flow must be maintained and is adjusted by coupling to local energetic demands dictated by ongoing cognitive processes. This involves multiple biological systems, many of which are influenced by diet including, in some cases, polyphenols.

While the expression of dementia and age-associated cognitive impairment is usually associated with old age there is emerging evidence that the processes that underlie these conditions can begin earlier in life. Boosting or preserving cognitive performance in early or midlife may have the potential to delay these conditions. Cognitively intact adults (Small, 2004) and even children (Bloss *et al.*, 2008) who possess risk factors for later developing Alzheimer's disease (AD) display impaired cognitive function. Therefore, treatments that combat the process involved in AD may serve to protect them from developing dementia but may also enhance cognition throughout life.

A large prospective study by Singh-Manoux and colleagues (2012) found significant decline over 10 years on a range of cognitive tasks, including among those aged 45–49 years at baseline. Further, it has been shown that cognitive function may already be impaired relative to peers, almost a decade before a dementia diagnosis is reached (Amieva *et al.*, 2005). Therefore, interventions that are able to improve cognition in any age group may have the benefit of not only improving cognition at that time but also reducing the risk of cognitive decline and dementia.

From early adulthood, there is an often linear relationship between increasing age and decreasing performance on tasks relating to memory, reasoning, spatial abilities, and cognitive processing speed. Conversely, in healthy aging, there is a pattern for improved performance on test of concrete knowledge, such vocabulary, until about the age of 60 years, after which there is a slow and mild decline. About 1–2% of the normal elderly population develop AD, whereas 10–15% of individuals with mild cognitive impairment (MCI) progress to AD within a year (Petersen *et al.*, 1999). Slowing cognitive decline in those with relatively marked cognitive impairment could significantly impact on AD prevalence. A simulation by Vickland and colleagues (2010) suggested that an intervention that could delay the onset of dementia by just 2 years would reduce its 2040 prevalence by 16%. As the pathology of Alzheimer's involves a complex cascade of interacting, deleterious processes that begin long before signs of cognitive decline are apparent, such an intervention would need to have wide spread effects and target multiple aberrant pathways (Frautschy & Cole, 2010). The following sections detail the effects of polyphenols on cognitive function, with later sections focusing on the role of the flavonoid subclass of polyphenols.

12.2 Studies of total polyphenols and cognition

Valls-Pedret and colleagues (2012) showed a link between total polyphenol intake and better cognitive function among men aged 55–80 years and women aged 60–80 years, with high cardiovascular risk. The study used urinary polyphenol excretion as an objective measure of polyphenol intake, which has the obvious advantage of overcoming potential inaccuracies inherent in self-reported measures of dietary patterns. A continuous dose-related effect was found between higher polyphenol excretion and better performance on a task of immediate episodic memory.

In a large population-based study combining data across 23 countries, Beking and Vieira (2010) investigated the relationship between per capita dietary flavonoid intake and disability-adjusted life years (DALY) caused by dementia. DALY represents a combination of years of life lost due to premature death and years of life spent in disability. There was a significant inverse correlation, that is greater total per capital intake of flavonoids was associated with lower rate of DALY caused by dementia. The relationship between flavonoids and DALY did not appear to be influenced by socioeconomic and demographic factors including national income, education, and ratio of men to women. When flavonoid subclasses were examined individually, flavones showed the strongest effects, explaining an estimated 19% of the variance in DALY due to dementia. The relationship between dietary intake and dementia was not seen for antioxidant vitamins nor fruit and vegetables, suggesting that this was not simply an effect of healthy diet or increased antioxidant intake *per se*. This finding is consistent with a number of other large epidemiological studies linking higher flavonoid intake with lower risk of dementia or cognitive decline (Commenges *et al.*, 2000; Engelhart *et al.*, 2002; Letenneur *et al.*, 2007). Letenneur and colleagues (2007) reported that among 1640 community dwelling adults aged 65 years or older, higher flavonoid intake was associated with better MMSE score at baseline and a slower rate of decline over 10 years. For example, after controlling for many health and lifestyle factors, the average 10-year decrease in MMSE score was 2.1 (out of 30) points for those in the lowest quartile of flavonoid intake (0–10.39 mg/day) and just 1.2 points for those is the highest quartile intake (17.70–36.94 mg/day). In a sample of 5395 individuals aged 55 years or older and free from dementia, Engelhart and colleagues (2002) also found that higher MMSE scores were associated with greater baseline function; every point increase on the MMSE score was associated with 0.24 mg/day greater flavonoid intake. In this study, flavonoids did not appear to reduce the 6-year risk of developing dementia. However, among older participants, aged 65+ years, Commenges and colleagues (2000) found that participants who on average consumed more than 11.5 mg of flavonoids per day had a significantly reduced 5-year risk of developing dementia.

Nurk and colleagues (2009) examined the relationship between self-reported consumption of particular flavonoid rich foods (tea, wine, and chocolate) by older adults and their performance on a battery of cognitive tasks. Global cognitive function, episodic memory, access to semantic memory, perceptual speed, and visual spatial skills were included in the assessment. Performance was better on all tasks by consumers of the flavonoid-rich foods compared to those who reported no consumption of tea, wine, or chocolate. Flavonoid consumption was also associated with a reduced risk of poor performance, defined as being among the poorest performing 10%. The cognitive benefits afforded by tea, wine, and chocolate appeared to increase with increasing consumption up to 200 ml, 75–100 ml, and 10 g/day, respectively. Interestingly, in addition to there being a dose response, with higher does affording better cognitive protection, there was an increased benefit to consuming a greater number of these foods. This suggests that both increasing quantity and variety of polyphenol (type and source) intake may be beneficial to cognition.

It is the polyphenol content of foods in these aforementioned studies that is thought to lead to enhanced or protected cognitive function and underlying biological processes; however, pragmatically dietary recommendations are likely to be easier to understand and more readily implemented if they are based on food products rather than particular polyphenol requirements. The remainder of this chapter will discuss a selection of specific polyphenol-rich foods and their extracts that have demonstrated the potential for cognitive enhancement or protection. They have each shown biological and cognitive benefits in animal studies; however, human studies will be the focus here. Selected polyphenol fractions from nondietary sources will also be briefly discussed.

12.2.1 Tea

After water, tea is thought to be the most widely consumed beverage in the world. Humans have been drinking tea for thousands of years both for pleasure and because of its purported health benefits. Descriptions of the health benefits of tea in Chinese medicine date back to 3000 BC, but archaeological evidence suggests that tea drinking may have started as early as the Paleolithic period (Martin & Cooper, 2011).

Catechins, particularly epigallocatechin gallate (EGCG), epigallocatechin (EGC), epicatechin gallate (ECG), and epicatechin (EC), are the most abundant polyphenols in tea. Phenolic acids are also present, mainly as gallic acid along with flavonols such as kaempferol, myricetin, and quercetin. While all true teas (as opposed to herbal or fruit "teas," which are more accurately described as tisanes) are made from the leaves of the *Camellia sinensis* plant, the polyphenol content varies according to the type of tea, which is often categorized as green (unfermented), oolong (semifermented), or black (fermented). Green tea has the highest catechin level being particularly rich in EGCG, followed by oolong, then black tea. This difference is largely due to the effects of fermentation (Zuo *et al.*, 2002; Lin *et al.*, 2003).

There is a great deal of variation between studies in the estimated polyphenol content of different tea samples. However, on average, estimates suggest that if infused at a ratio of 1 g tea leaves per 100 ml of boiling water, a standard serving (200 ml) of green tea would provide approximately 140 mg of EGCG, 58 mg of EGC, 36 mg of ECG, and 16 mg of EC. By comparison, when prepared in the same way, a serving of black tea would provide just 19 mg EGCG, 16 mg of EGC, 11 mg of ECG, and 4 mg of EC (Bhagwat *et al.*, 2014).

Many epidemiological studies have examined the relationship between habitual tea intake and cognitive function and have supported its potential for cognitive enhancement and protect. Ng and colleagues (2008) examined the association between tea drinking and global cognitive function as measured by the mini mental state examination (MMSE), in 2501 Chinese men and women aged 55 years or older. After controlling for demographic variables, which may influence tea drinking choice, and for health promoting or confounding variables such as fish, fruit, and vegetable intake, physical activity, depression, and APOE4 status, there remained a highly significant association between tea intake and cognitive function. People who reported drinking tea more often were less likely to show cognitive impairment. Furthermore, at a 1–2 year follow-up, having been a frequent tea drinker was associated with a reduced risk of cognitive decline. This relationship with cognition was not seen for coffee drinking, making it unlikely that the caffeine content of tea played a role in cognitive protection. The cognitive enhancing and protective effects of tea were more pronounced for black/oolong tea than green tea; however, this was likely due to the small percentage of the study sample who reported frequently drinking green tea at all (only 7% drink at least 1 cup per day) and very small number who drank green tea exclusively ($N = 10/2501$). Nevertheless, green tea (with or without also drinking other forms of tea) was still associated with a significant reduction in the odds of cognitive impairment.

Kuriyama and colleagues (2008) found a similar effect in an older population: among 1003 participants aged 70+ years. The odds of cognitive impairment, defined at any of three MMSE cutoff levels (slight impairment <28, the standard MMSE impairment cutoff of <26, or severe impairment MMSE <24), reduced dramatically with increasing tea consumption. However, in this study, little or no association was seen between cognitive impairment and drinking black/oolong tea.

Feng and colleagues (2012) conducted a study of 7139 Chinese men and women categorized as being among the oldest-old (aged 80–115). Again here, it was reported that frequent or occasional tea consumption was associated with better cognitive function (assessed by verbal fluency) when compared to rarely or never drinking tea. There was a clear pattern for increasing frequency of tea drinking to be associated with better cognitive function. Additionally, having been a frequent tea drinker at the age of 60 years (average ~30 years prior to the study) appeared to have protective effects as it was also associated with better cognitive function at the time of assessment. However, as the authors reported a high level of consistency between previous and current tea drinking behaviors, it is difficult to separate protective and enhancing effects that tea may have had in this study. Interestingly, the tea drinkers did show a faster rate of decline over the 7-year observation period. The authors suggest that this may have been due to a previously reported phenomenon whereby increased cognitive reserve delays the onset cognitive decline but is then associated with faster progression once dementia has set in.

These studies each assessed the relationship between tea drinking cognition using only a single cognitive outcome measure. In order to further investigate if there were aspects of cognition that particularly benefited from tea consumption, Feng and colleagues (2010) used both the MMSE and a battery of tasks probing different cognitive domains in their assessment of elderly men and women who were free from cognitive impairment. Again it was found that habitual tea consumption was independently and linearly related to greater global cognitive function (measured with the MMSE). These benefits were seen for both black/oolong tea and green tea. In the cognitive test battery, tea particularly benefited memory, executive function, and information processing speed. This is particularly interesting as these are cognitive functions known to be vulnerable to aging.

The benefits of tea drinking have not just been observed in the elderly. In a naturalist cross-sectional study, Bryan and colleagues (2012) examined the relationship between beverage consumption in a cognitively demanding workplace (university) and work performance and mood. Among 95 academic and professional staff, drinking tea without added sugar or milk was related to reduced tiredness over the workday and better perceived work performance. These effects were lost when sugar or milk was added to the tea.

Interestingly, adding milk to tea when serving has been shown to reduce its antioxidant effects (Tewari *et al.*, 2000).

For a recent review of the acute neurocognitive effects of tea constituents, see Camfield *et al.* (2014). Three controlled, crossover studies have investigated the acute effects of polyphenol tea extracts on brain function and cognition. In the first, 12 healthy young participants completed four testing sessions across which they received four treatments: 250 or 500 ml of a milk-whey-based drink that did or did not contain a green tea extract (0.05%) (Borgwardt *et al.*, 2012). Following treatment administration, brain activation while completing working memory task was examined using functional magnetic resonance imaging (fMRI). Compared to control, the drinks containing green tea extract were associated with increased activation in a number of brain regions. There was also a pattern for greater activation with greater dose of green tea extract. However, there was no significant effect of treatment condition on task performance.

Wightman and colleagues (2012) administered placebo, 135, and 270 mg of an EGCG extract to 27 healthy young participants. Following a 45-minute absorption period, a battery of cognitive tasks aimed at inducing high or low frontal cortex activation were completed. Throughout each testing session, the cerebral hemodynamics were assessed using near-infrared spectroscopy (NIRS). It was found that while treatment did not change task outcomes, during task performance, relative to placebo, the lower dose of EGCG was associated with reduced cerebral blood blow in the frontal cortex. Although a reduction in cerebral blood flow can be interpreted as a negative effect, the authors suggest that it may have been the result of reduced demand on blood flow owing to improvement in some unexamined mechanism. This is plausible given that no change in task performance was observed.

In 31 healthy young adults (mean age 27 years), Scholey and colleagues (2012) compared electrical brain activity in both eyes by open and resting-state electroencephalogram (EEG) before and 120 minutes after administration of 300 mg a caffeine-free tea extract (94% EGCG) and placebo. The active treatment was associated with an overall increase in alpha, beta, and theta activities coupled with increased self-rated calmness and reduced stress. The authors speculated that these results reflect a more relaxed and attentive state following the tea extract, in keeping with tea's purported "refreshing" properties.

These three trials demonstrate that tea extracts can acutely influence human brain physiology, whereas the cognitive effects may be more fragile. The studies had relatively small sample sizes and healthy young participants who would likely have less potential for benefit than an older population. Further, the tasks selected in these trials were primary done so based on their known patterns of brain activation rather than their potential for improvement following treatment.

In a small pilot study conducted by De la Torre and colleagues (2014) examining the effects of EGCG in individuals with Down Syndrome with moderate intellectual disability, ECGC was able to produce significant improvements in immediate recognition compared to placebo, with trends toward improving social function and working memory. Parents of participants receiving ECGC also reported positive effects in day-to-day life.

De Bruin and colleagues (2011) showed that, in healthy young adults, black tea was associated with better control of attention compared to placebo. Participants performed faster and more accurately on a task in which they had to consciously focus their attention on visual or auditory stimuli or divide it between both. They were also more accurate on a task of switching attention. The polyphenol content of the tea was not specified.

Thus, there appears to be increasing evidence of cognitive and mood benefits from tea and its components. However, it is clear that the nature of the polyphenol fraction, as well as the dose and timing of administration, may differentially affect cognitive outcomes.

12.2.2 Cocoa

Similarly to tea, the cocoa bean has a long history of medicinal and culinary use and is rich in flavanols, particularly catechins and proanthocyanidins. The concentration of catechins in chocolate is greater than in tea, although it contains a smaller variety of catechins (catechin and epicatechin) (Arts *et al.*, 1999). The polyphenol content of cocoa beans, and therefore any chocolate products produced, varies dramatically according to where they were grown (Natsume *et al.*, 2000). The processes of fermenting and roasting cocoa beans, required for manufacturing cocoa products, also alter the polyphenol profile, reducing concentrations and promoting conversion to less bioavailable forms (Andres-Lacueva *et al.*, 2008; Hurst *et al.*, 2011). Finally, as polyphenols are found primarily in the cocoa solids, rather than cocoa fat, the percentage of the final chocolate product that is composed of nonfat coca solids determines its final polyphenol content (Miller *et al.*, 2006). In short, less processing and a greater cocoa solid content (e.g., cocoa powder or dark chocolate) produce chocolate products richer in polyphenols.

The effects of chocolate constituents on mood and cognition have been reviewed recently (Scholey & Owen, 2013). Consuming chocolate and cocoa flavanols produces significant neurocognitive changes including increased cerebral blood flow (Francis *et al.*, 2006), flow velocity (Sorond *et al.*, 2008), and altered brain activation (Francis *et al.*, 2006; Camfield *et al.*, 2012) during cognitive task performance. Recently, Brickman and colleagues (2014) demonstrated that cocoa flavanols were able to improve function of the dentate gyrus, an area of the brain known to be vulnerable to aging. A group of healthy older adults received either high (900 mg/day) or low (10 mg/day) doses of cocoa flavanols for 3 months, and the effects of treatment on visual memory task performance and cerebral blood volume (CBV) were examined. Changes in CBV reflect changes in metabolic activity and capillary density. It was found that, compared to the low-dose treatment, high-dose cocoa flavanols were associated with significantly increased CBV in the dentate gyrus and other hippocampal structures. Furthermore, increased CBV correlated with significant improvement in the memory task performance. Improvements in the dentate gyrus are especially interesting as it plays an important role in learning and memory and is linked to the susceptibility of these processes to aging.

The Cocoa, Cognition, and Aging (CoCoA) Study has reported that cocoa flavanols are able to improve cognitive function both in older adults free from memory complaints (Mastroiacovo *et al.*, 2015) and in those with mild cognitive impairment (MCI) (Desideri *et al.*, 2012). Participants consumed daily cocoa drinks containing high (993 mg), medium (520 mg), or low (45 mg) doses of cocoa flavanols (epicatechin, catechin, and procyanidins) for 8 weeks. In the cognitively intact cohort, supplementation with either the high- or intermediate-dose cocoa flavanols produced significant improvements in processing speed, executive function, working memory, verbal fluency, and a composite score of overall cognitive function, which were not seen in a low-flavanol group. MMSE scores did not significantly change, but this is unsurprising given the near ceiling measures at baseline (mean of >29/30 in each treatment group), which are to be expected in a cognitively healthy group. In participants with MCI, the high- and medium-flavanol treatments produced significantly improved processing speed, executive function, and working memory. The high-flavanol dose also improved verbal fluency.

Interestingly, while the antioxidant capacity of polyphenols is among the most well studied of their properties and most frequently credited for their beneficial effects, they appeared to be only a minor contributor in this study. In the MCI participants, changes in plasma isoprostane concentration, a marker of oxidative, accounted for only 7% of the improvement in cognitive function. Rather improvement in insulin resistance was likely the main mechanism of effect, as it explained an estimated 40% of the variability in cognitive improvement. Insulin resistance has been implicated in the etiology of these conditions age-related cognitive decline and dementia (Cholerton *et al.*, 2011). In the cognitively normal group, high- and intermediate-doses of flavanols were associated with significant reductions in blood pressure, fasting plasma insulin and glucose, insulin resistance, lipid peroxidation, triglycerides, and low-density lipoprotein (LDL) or "bad cholesterol" and increases in high-density lipoprotein (HDL) or "good cholesterol," total cholesterol, and the HDL-to-LDL ratio. Changes in insulin resistance, oxidative stress, and blood pressure explained 17, 4, and 3% of the variability in change in overall cognitive function, respectively.

Crews and colleagues (2008) administered cocoa and dark chocolate to healthy older adults for 6 weeks but failed to find a significant effect on any of their implemented cognitive tasks or hematological measures (cholesterol profile and inflammatory marker). Similarly, despite a longer-term benefit to mood, Pase and colleagues (2013) did not find significant cognitive effects of high (500 mg) or low (250 mg) levels of cocoa polyphenols in healthy middle-aged adults after either a single acute dose or 30 days of continuous supplementation.

In a double-blind, crossover study in healthy young adults, Scholey and colleagues (2010) compared the acute effects of cocoa flavanols at high (994 mg), intermediate (520 mg), and low (control, 40 mg) doses. These doses are equivalent to those that showed chronic effects in the (CoCoA) study. Importantly, in this study, all three treatments were matches for macro- and micronutrients, caffeine, and theobromine, allowing any effects to be confidently attributed to the variation in flavanol content. Before treatment administration, participants completed a single performance of the study's 10-minute mood and cognitive assessment battery. Following treatment administration, a 90-minute absorption period was observed and then six consecutive performances of the battery were undertaken. Compared to placebo, both doses of cocoa flavanols were associated with better accuracy on a mental subtraction task during the postdose battery, the intermediate dose at all six repetitions, and the high dose during the first four. During the third and fourth postdose cycles of the assessment battery, the higher cocoa flavanol dose was associated with significantly more errors on the serial subtraction task; however, the number of correct responses was not effects and during the same repetitions, the high-dose treatment improved performance of the Rapid Visual Information Processing task of vigilance. The low dose of cocoa flavanols significantly protected against the fatiguing effects of undertaking the battery at all except the third postdose assessment. This antifatiguing effect was not shown with the high dose of cocoa flavanols. The authors speculated that the dissimilarity between the effects of cocoa flavanols on serial 3 and serial 7 tasks may be due to the tasks engaging different cognitive processes. They also noted that the time at which the strongest and most widespread benefits of treatment were observed (~130 minutes postdose) roughly coincides with the changes in cerebral blood flow reported by Francis and colleagues (2006) following acute administration of 450 mg of cocoa flavanols. The mixture of results seen in this study suggests that cocoa flavanols may acutely benefit some cognitive processes more than others and the nature of exerted benefits may vary according to quantity and time since consumption.

Thus, cocoa flavanols appear to benefit aspects of cognitive performance. The effects are dependent on a number of variables with acute effects being more evident during cognitive fatigue and longer-term effects linked to changes in insulin sensitivity and central angiogenesis.

12.2.3 Wine and grapes

Drinking moderate amounts of wine, especially red wine, is a prominent feature of the Mediterranean diet, which has been suggested to reduce the risk of many negative health outcomes including cognitive decline, mild cognitive impairment, and Alzheimer's disease (Sofi *et al.*, 2008; Féart *et al.*, 2010).

The phenolic compounds in grapes and grape products such as wine and juice include anthocyanins, flavanols, flavonols, the stilbene resveratrol, and phenolic acids. These are most abundant in the skins, stems, and seeds of the grapes, which is why red wine is usually richer in polyphenols than white. Stilbenes are not prominent in human diet, and grapes and wine are the major sources of resveratrol along with some nuts and berries.

Data from the Canadian Study of Health and Aging, a population-based study in cognitively intact adults, showed a protective effect of wine drinking against the development of dementia. Compared to those who did not drink wine, people who reported drinking wine at least once per week at baseline were half as likely to have developed dementia when they were reassessed 5 years later (Lindsay *et al.*, 2002). Similarly, in the third Copenhagen City Heart Study (Truelsen *et al.*, 2002), it was found that weekly or monthly intake of wine was associated with a lower risk of dementia 15 years later, compared to never or hardly ever drinking wine. Greater total alcohol or beer intake had the opposite effect and was associated with increased risk of dementia, suggesting that wine's alcohol content did not contribute to its protective effects. Epidemiological studies have also reported an association between drinking wine and better current cognitive function (Nurk *et al.*, 2009; Valls-Pedret *et al.*, 2012).

In a small ($n = 12$), randomized placebo-controlled trial, Krikorian and colleagues (2010) reported that 12 weeks of daily supplementation with concord grape juice was able to significantly improve memory in participants with mild cognitive impairment. The same group later performed another trial of concord grape juice in 21 adults aged 68 years or older ($n=21$), with MCI (Krikorian *et al.*, 2012). In this trial, the intervention period was extended to 16 weeks. The polyphenolic content of the concord grape juice treatment was reported as 2091 mg/l as gallic acid equivalents, which comprised 46% anthocyanins, 29% tartaric esters of hydroxycinnamates, and 10% procyanidins. Dosage was stratified by weight in order to maintain a daily intake of 6.3–7.8 ml/kg. The grape juice treatment was associated with fewer false-positives on a recognition memory task. The authors suggest that this demonstrated a resistance to memory interference and superior inhibitory control of working memory, two processes known to deteriorate with aging (Robert *et al.*, 2009). In the subset of the trial's participants ($n=8$), fMRI was performed while completing a difficult working memory task. The grape juice treatment produced a significant increase in regional brain activity; however, the authors did not report whether this change in activation was related to behavioral changes. Hendrickson and colleagues (2008) failed to find any acute effects of the same grape juice on cognition in healthy adults aged 18–50 years. However, it should be noted that only a single and intentionally monotonous task of implicit memory was used, which may have limited the ability of this trial to detect cognitive effects in a healthy young population.

Witte and colleagues (2014) examined the effects of a combination of two wine polyphenols, resveratrol and quercetin, on memory performance, function and structure of the hippocampus, hematological measures, and vascular markers in a sample of 46 overweight but healthy participants aged 50–75. Participants received a daily dose of 200 mg of resveratrol plus 320 mg of quercetin or placebo for 26 weeks. At the end of the intervention, the polyphenol group should show significantly better memory retention than the placebo group. They also showed a significant improvement in delayed recall and recognition not seen in the placebo group; however; this was not sufficient to produce significant between-group differences. Neuroimaging measures revealed that the polyphenol treatment significantly enhanced functional connectivity between the hippocampus and a number of cortical regions. Of particular note was the improved functional connectivity to the medial prefrontal cortex because this correlated with significant improvement in memory retention scores. Plasma markers of long-term glucose control also significantly improved in the polyphenol group, and again this correlated with cognitive and functional connectivity changes. Finally, the polyphenol group showed significant decreases in leptin (hunger-signaling hormone) and body fat. These results suggest that improved glucose regulation, even within a nondiabetic cohort, by polyphenols may contribute to better cognition and brain function.

Cimrová and colleagues (2011) investigated the acute effects of a red wine polyphenol extract in healthy undergraduate students. Participants received 4 mg/kg body weight of the red wine polyphenols (containing, per gram, 480 mg proanthocyanidins, 61 mg total anthocyanins, 19 mg free anthocyanins, 38 mg catechin, 18 mg hydroxycinnamic acids, 14 mg flavonols, and 370 mg polymeric tannins), placebo, or no treatment. Assessments were carried out before and 2 hours after dose administration. The red wine polyphenols were found to significantly improve spatial memory, resulting in greater accuracy for remembering the precise location at which stimuli had previously been presented. In the period between stimulus presentation/removable and participant's response, during which the presentation location had to be held in memory, those who had received red wine polyphenols showed higher levels of activation in brain regions associated with spatial orientation and memorization.

Scholey and colleagues (2014) compared the effects of red wine alone and red wine enriched with 200 mg resveratrol on demanding cognitive tasks in an older cohort (mean age 70 years). The two interventions differentially affected performance, with the resveratrol group performing better on more-demanding tasks but less well on less-demanding tasks. However, the authors caution that the data were preliminary and the study did not include an inert control. Acute physiological effects have also been observed in healthy young people following ingestion of *trans*-resveratrol alone, without out other wine polyphenols. In a placebo-controlled crossover study (Kennedy *et al.*, 2010), 22 healthy adults aged 18–25 years received placebo, 250 mg *trans*-resveratrol, and 500 mg *trans*-resveratrol. Cerebral blood flow in the prefrontal cortex was assessed over the course of each session, which included predose cognitive tasks, dose administration, a 45-minute absorption period, and, finally, four repetitions of the cognitive tasks. Resveratrol was associated with a dose-dependent increase in cerebral blow flow in the prefrontal cortex during the postdose cognitive tasks. This was attributed to resveratrol's vasorelaxatory properties. However, the effects did not produce any significant changes in cognitive task performance at any timepoint.

The data on resveratrol are mixed. Changes to physiological processes are not always matched with cognitive benefits. It may be that further work comparing specific doses and dosing regimens may resolve the issue of whether there are real cognitive benefits from resveratrol and in which populations.

12.2.4 Soy

Soybean (*Glycine max*) has been cultivated for thousands of years. Today, soy foods such as edamame, tofu, and dairy alternatives are popular around the world including in Asian and vegetarian populations. In addition to their high protein content, soy foods are a major dietary source of isoflavones (genistin, daidzin, and glycitin). The average daily intake of isoflavones varies widely between regions. In Europe, intake is quiet low, as little as 0.5–0.9 mg/day (van Erp-Baart *et al.*, 2003) while estimates in Asian nations are usually in the range of 20–50 mg/day (Messina *et al.*, 2006). For reference, 100 g of tofu or one cup of soy milk each provides approximately 25 mg isoflavones (Messina *et al.*, 2006).

While possessing beneficial properties common to other polyphenol families, isoflavones are additionally of interest in the field of cognition because of their ability to bind to the ERβ estrogen receptors and act as nonsteroidal estrogens (Setchell, 1998). Although usually associated with reproductive function, estrogens exert influence on many bodily systems including the central nervous system (Maggi *et al.*, 2004). It is therefore not surprising that alterations in estrogen levels can impact upon these processes. Reduced estrogen levels among peri- or postmenopausal women have been associated with memory complaints (Devi *et al.*, 2005) and poorer cognitive performance particularly in the domains of verbal or working memory (Duff & Hampson, 2000; Shaywitz *et al.*, 2003; Fuh *et al.*, 2006), which may be alleviated by hormone replacement therapy. Cognitive benefits of estrogen treatment have also been reported in men awaiting gender reassignment

(Miles *et al.*, 1998), which suggests that estrogen may be able to enhance cognition even among males or a healthy younger population without altered hormone levels. However, estrogen supplementation is not a viable option for cognitive enhancement either long term or among healthy populations, not least of all because of the many reports of increased risk of serious negative health outcomes including breast cancer (Lacey *et al.*, 2002; Nelson *et al.*, 2002; Million Women Study Collaborators, 2003). Therefore, as phytoestrogens, isoflavones may afford the health and cognitive benefits associated with estrogen while avoiding the corresponding health risks. For example, a meta-analysis by Trock and colleagues (2006) found that soy intake did not increase breast cancer risk and may even be associated with a mild risk reduction.

In their randomized control trial, File and colleagues (2005) used a battery of cognitive tasks targeting specific cognitive abilities. Participants received either soy isoflavones (60 mg/day) or placebo for 6 weeks, and cognitive assessment was conducted before and after the treatment period. It was revealed that compared both to placebo and baseline, soy isoflavones produced significant improvements in mental flexibility and planning ability. This result did not differ according to time since menopause. Within other cognitive domains, the effect of treatment reached significance in only one of the many memory tests, that of short-term nonverbal memory. The authors suggested that isoflavones specially benefited cognitive processes linked to frontal lobe function.

Duffy and colleagues (2003) investigated the effects of a daily soy extract containing the equivalent of 60 mg of isoflavones for 12 weeks, on cognition in postmenopausal women. They found that compared to placebo, the soy extract was associated with significant benefits in tasks assessing sustained attention, recall, planning ability, and mental flexibility.

Fournier and colleagues (2007) trailed the effects of approximately 70 mg of isoflavones either in the form of soy milk or in a tablet for 16 weeks, in postmenopausal women aged 48–65 years. The isoflavone group did not outperform placebo on any measure of attention or memory.

There have been three double-blind, randomized control studies investigating the effects of varying doses of isoflavones on the cognitive function of postmenopausal women. Casini and colleagues (2006) conducted crossover trial in which participants received 60 mg of isoflavones or placebo daily for 6 months after which they observed a 1 month washout period before receiving the alternate treatment for a further 6 months. Compared to placebo, isoflavones significantly improved incidental learning, mental flexibility, and verbal working memory. Ho and colleagues (2007) used a higher dose of isoflavones, 80 mg/day in their parallel group trial, but found no significant effects of isoflavones over placebo on any of their cognitive outcomes. Kritz-Silverstein and colleagues (2003) used a dose of 110 mg isoflavone per day in their 6-month parallel group trial. They found that qualitatively, the pattern of improvement favored the isoflavone group on all tasks; however, only the effect on category fluency reached significance. The category fluency task is a measure of verbal fluency with a small demand on semantic memory. Over the intervention period, the performance of women in the isoflavone group improved by 23% compared to just a 3% improvement in the placebo group. When the sample was divided to examine effects in women aged 50–59 years and 60–74 years separately, additional benefits were observed in the younger group. Here, the effect of treatment on a measure of visuomotor tracking and attention reached significance. The time taken to complete the task decreased by almost 20% in the isoflavone group and showed no improvement in the placebo group.

Henderson and colleagues (2012) found that 2.5 years of daily soy isoflavone supplementation (52 mg of genistein, 36 mg of daidzein, and 3 mg glycitein per day) did not significantly improve global cognition in postmenopausal women. However, the neuropsychological battery revealed a significant benefit of treatment on visual memory especially during immediate recall. Results also suggested that the effects of isoflavone supplementation may vary according to the time since menopause, type of menopause (natural or surgical), and individual variation in the ability to metabolize isoflavones (non-, intermittent- or consistent-equol producers).

The majority of studies of soy or isoflavones and cognition have focused on postmenopausal women; however, a very few studies have included an examination of their effects in men. Gleason and colleagues (2009) investigated the effects of 6 months of soy isoflavone supplementation (100 mg/day, 85% daidzein and genistein) in cognitively healthy men and women aged over 60 years. Blood measures and a neuropsychological battery assessing a range of cognitive domains were completed at 1, 3, and 6 months after commencing treatment and again 2 months after treatment termination. Compared to placebo, isoflavones significantly improved performance on tests of visual spatial memory, visual motor function, and categorical verbal fluency. Two months after treatment cessation, the between-group differences on cognitive task performance were no longer evident. The investigators noted that, despite all having received the same dose, plasma levels of daidzein and genistein and their metabolite equol, within the active treatment group, showed substantial interindividual variation.

Thorp and colleagues (2009) completed a double-blind crossover study in which men aged 60–80 years (mean age 49 years) received soy isoflavones or placebo for 6 weeks before switching to the alternate treatment for additional 6 weeks. Tests of memory, executive function, and visual spatial processing were completed at the end of each treatment.

Compared to placebo, isoflavone supplementation improved all three outcomes on the novel spatial working memory test; participants were 17% faster and required 18% few trials to complete the task and committed 23% few errors in the process. This is a task that usually shows superior performance in women. It has been speculated that phytoestrogen supplementation in men may improve performance on tasks in which females usually perform better but be detrimental to male dominant tasks. Importantly, there were no detrimental effects of isoflavones on any tasks, including those that usually favor men.

As for the effects of soy isoflavones in younger populations, File and colleagues (2001) compared the effects a high- or low-soy diets on attention, memory and frontal lobe function, and mood in healthy male and female university students. The high-soy diet provided 100 mg of isoflavones per day while the low-soy diet contained just 0.5 mg. Importantly, the two diets were also matched on a number of potentially confounding qualities such as total energy intake, fats, and proteins. It was reported that at the end of the 10-week intervention period, the high-soy diet resulted in significantly better performance of immediate visual and verbal episodic memory and mental flexibility. There were gender differences observed in the effects of the high-soy diet on other tasks. In a task of planning ability among females, the high-soy diet produced an improvement in performance that was not seen in the low-soy group. In males, both groups improved equally. On the letter fluency task, which is associated with frontal lobe function, females improved under the high-soy diet, while males showed deterioration in performance.

These studies have shown that soy isoflavones may benefit a number of cognitive processes and that these effects are not limited to postmenopausal women. It has been speculated that those aspects of cognition that most benefit are the ones that traditionally show superior performance by women, possibly due to higher circulating levels of estrogens and those that relate to frontal lobe and hippocampal function, which may be because the Erβ estrogen receptors that isoflavones act upon are particularly prevalent in these areas (File *et al.*, 2005; Thorp *et al.*, 2009).

12.3 Pine bark

Not all polyphenols that have undergone scrutiny as potential cognitive enhancers are derived from dietary sources. Some have come to the attention of scientific enquiry through their long-term use in traditional medicine systems. One such example is Pycnogenol® outlined as follows.

Traditionally, pine bark has had many uses ranging from treating scurvy to promoting would healing. Today, two well-standardized pine bark extracts have received attention with regard to the potential for cognitive benefits. Pycnogenol (Horphag Research, Geneva, Switzerland) is produced from the bark of the French maritime pine (*pinus pinaster*). It contains a high content of procyanidins (65–75%) as well as phenolic acids, catechin, and taxifolin and has been indicated for the treatment of a wide variety of conditions (Rohdewald, 2004). Enzogenol® is derived from the bark of New Zealand grown *pinus radiata*. It contains approximately 84% proanthocyanidins, 1–2% taxifolin, and 8% other flavonoids, phenolic acids, and stillbenes. The cognitive effects of these extracts have been examined in several populations.

Trebatická and colleagues (2006) gave children aged 6–14 years, with diagnosed attention deficit/hyperactivity disorder, 1 mg/kg/day of Pycnogenol ($n = 44$) or placebo ($n = 17$). Assessment of their cognitive function and behavioral symptoms was carried out before treatment commenced, at the end of 1 month of treatment and again following a 1-month washout period from termination of treatment. One month of Pycnogenol treatment produced a significant improvement in visual–motoric coordination and concentration both compared to pretreatment assessment and compared to placebo. Teacher ratings of inattention and hyperactivity also significantly decreased in the active treatment group but not in the placebo group. These ratings returned to baseline 1 month after treatment was terminated.

Belcaro and colleagues (2014) examined the effects of Pycnogenol in a sample of young professionals. Participants were provided with a suggested, unenforced, health plan with or without the addition of 12 weeks of supplementation with 150 mg of Pycnogenol per day. Before and after the treatment period, a large assessment battery was completed with included objective and subjective measures of cognitive function. The results revealed a number of diverse benefits of Pycnogenol. At the 12-week assessment, the Pycnogenol group showed significantly superior general cognitive function, attention, mental performance, sustained attention, memory, and executive function.

Interestingly, the investigators included a number of items in the assessment battery for rating subjective impressions of well-being and performance of everyday tasks. They also included a number of tests of ability to carry out professional activities, such as remembering phone numbers, taking and sending messages, and finding addresses on a map. These were scored on the time, accuracy, and efficiency with which they were completed. This is valuable as it gives an indication as to whether supplement effects might translate to real-life benefit. Here, Pycnogenol also showed benefits and was related to greater self-rated ease and enjoyment when performing a number of activities including dealing with money or people, decision-making and problem-solving, and many aspects of mood and well-being. Participants who received Pycnogenol showed significant improvements on the professional multitasks, which the unsupplemented group did not. In this study, Pycnogenol treatment also produced a significant 30% decrease in plasma oxidative stress markers, which was unchanged in the unsupplemented group.

In a study of 155 perimenopausal women aged 45–55 years, Pycnogenol (200 mg/day for 6 months) was found to relieve all assessed symptoms of climacteric syndrome, including problems with memory and concentration. The effects were superior to placebo after just 1 month and remained so at the final 6-month follow-up (Yang *et al.*, 2007).

In a randomized control trial, Pipingas and colleagues (2008) examined the effects of daily supplementation with a different extract, Enzogenol (960 mg) plus vitamin C (120 mg) compared to vitamin C alone on cognitive function in healthy males aged 50–65. At the end of the treatment period, the Enzogenol group performed significantly faster than placebo on tasks of immediate recognition memory and spatial memory. Further, this represented a significant improvement from baseline.

In healthy older adults ($n=101$ mean age = 68 years), 3 months of daily supplementation with either Pycnogenol (150 mg) or placebo showed significant benefits of the active treatment on cognition (Ryan *et al.*, 2008). Specifically, scores on a cognitive factor denoted as "quality of memory" were significantly better at the 3-month timepoint in the Pycnogenol group compared to the placebo group. This improvement appeared to be driven by an improvement in spatial working memory. These improvements in cognition were accompanied by a significant reduction in markers of lipid peroxidation.

Similarly to age-related cognitive decline and dementia, the pathophysiology of traumatic brain injury (TBI) includes oxidative stress, inflammation, and vascular and metabolic alterations (Werner & Engelhard, 2007). Therefore, similarly to age-related cognitive decline and dementia, cognitive function in patients with TBI may benefit from the effects of polyphenols. Theadom and colleagues (2013) conducted a pilot study of Enzogenol for the improvement of cognitive function in patients with mild TBI. Participants were aged 21–64 and had sustained a mild TBI 3–12 months prior to enrolment that had resulted in persistent cognitive difficulties. They were randomized to receive 1000 mg Enzogenol per day or placebo for 6 weeks. Results showed that at the end of the treatment period, patients receiving Enzogenol reported significantly fewer cognitive failures than those receiving placebo.

Table 12.1 Controlled trials of the cognitive effects of polyphenols/polyphenol rich products.

Design	Dosage	Duration	Population	Cognitive effects	Reference
Tea					
Randomized, double-blind, placebo-controlled, parallel group trial	Average of 9 mg/kg body weight of EGCG, or placebo	3 months. Reassessed 3 months after treatment discontinuation	Individuals aged 14–29 years with Down syndrome. $N=29$	Significant beneficial effect on episodic memory compared to placebo, which was revered by treatment discontinuation. Trends toward improved working memory, psychomotor speed, and social functioning also reported	De la Torre *et al.* (2014)
A double-blind, placebo-controlled, crossover trial	250 or 500 ml milk whey drink with or without (control) 0.05% m/m green tea extract (47.5–52.5% m/m polyphenols). Administered via feeding tube	Single dose Time between treatment administration and cognitive assessment not specified	Healthy Caucasian right-handed males aged 21–28 years. $N=12$	None	Borgwardt *et al.* (2012)
Randomized, double-blind, placebo-controlled, crossover trial	135 or 270 mg of EGCG or placebo	Single dose Cognitive effects assessed starting 45 minutes after treatment administration	Health adults aged 18–30 years. $N=27$	None	Wightman *et al.* (2012)
Randomized, double-blind, placebo-controlled, crossover, trial	Study 1: 2 × 200 ml black tea (1040 mg tea solids), or placebo, consumed over 60 minutes	Study 1: 1 serving administered in each of two breaks between three assessments	Study 1: Volunteers' average age 30.7 years. $N=26$	Significantly benefits the ability to selectively focus and switch attention	De Bruin *et al.* (2011)
	Study 2: 3 × 200 ml black tea (1140 mg tea solids), or placebo, consumed over 90 minutes	Study 2: 1 serving administered in each of three breaks between four assessments	Study 2: Volunteers' average age 30.3 years. $N=32$		
	Polyphenol content not specified				

Cocoa					
Randomized, double-blind, placebo-controlled, parallel group trial	993, 520, or 48 mg (placebo) of cocoa flavanols. Cocoa flavanol and procyanidin content calculated as epicathechin + catechin + dimers + trimers + tetramers + pentamers – decamers	8 weeks	Older adults, aged 61–85 years, without cognitive dysfunction. $N=90$	Significant benefit of high and intermediate doses on Trail Making Task A and B. High dose resulted in significantly better verbal fluency compared to both placebo and intermediate dose	Mastroiacovo *et al.* (2015)
Experimenter-blind, controlled, crossover trial	35 g of dark chocolate containing 773 mg of cocoa flavanols or 35 g of calorically similar white chocolate (control) containing only trace amounts of flavanols	Single dose	Undergraduate students aged 18–25 years. $N=30$	Improved visual function (contrast sensitivity, detection of motion direction), spatial memory, and aspects of predictable but not unpredictable choice reaction time task	Field *et al.* (2011)
Randomized, double-blind, placebo-controlled, parallel group trial	Flavanoid- and proanthocyanin-rich dark chocolate bar (37 g, 397.30 mg proanthocyanins per gram) and cocoa drink (≈11 g cocoa, 357.41 mg proanthocyanins per gram) or calorically matched placebo, taken daily	6 weeks	Healthy adults aged 60+ years, free from a history of dementia or significant neurocognitive impairment. $N=90$	None	Crews *et al.* (2008)
Randomized, double-blind, controlled, crossover trial	172 mg (high) or 13 mg (low/control) flavanols, taken daily	5 days, cognition assessed 1.5 hours after final drink consumption	Females aged 18–30 years. $N=16$	None	Francis *et al.* (2006)
Randomized, double-blind, parallel group trial	993 mg (high), 520 mg (intermediate), or 48 mg (control) flavanols, taken once daily. Cocoa flavanol and procyanidin content calculated as epicathechin + catechin + dimers + trimers + tetramers + pentamers – decamers	8 weeks	Adults aged 64–82 years with mild cognitive impairment	High and intermediate doses resulted in significant improvements in speed of completing the Trail Making Test part A and B. High dose also improved verbal fluency	Desideri *et al.* (2012)
Randomized, double-blind, placebo-controlled, parallel group trial	≈500, ≈250, or ≈0 mg (placebo) of cocoa flavanol polyphenols, daily	30 days	Healthy adults aged 40–65 years. $N=61$	None	Camfield *et al.* (2012)

(*Continued*)

Table 12.1 (Continued)

Design	Dosage	Duration	Population	Cognitive effects	Reference
Randomized, double-blind, placebo-controlled, parallel group trial	500, 250, or 0 mg (placebo) of cocoa polyphenols	Single dose, cognitive assessments at 1, 2.5, and 4 hours after administration and 30 days	Healthy, adults aged 40–65 years. $N=72$	None	Pase *et al.* (2013)
Double-blind, placebo-controlled, counterbalanced, crossover trial	994 mg (high), 520 mg (intermediate), or 46 mg (control) of flavanols. Cocoa flavanol content calculated as epicatechin + catechin + dimers + trimers – decamers	Single dose. Cognitive effects assessed by six repeats of task battery beginning 90 minutes after dose administration	Healthy, university students aged 18–35 years. $N=30$	High and low doses produced significantly improvement on serial 3 subtraction task. High-dose treatment associated with more errors in serial 7 subtractions task at 2 postdose assessments. High dose produced faster responses on a rapid visual information processing task at 2 postdose assessments	Scholey *et al.* (2010)
Randomized, placebo-controlled, parallel group trial	900 mg cocoa flavanols + 138 mg epicatechin, or 10 mg cocoa flavanols + <2 mg epicatechin (control)	3 months	Healthy, sedentary older adults, aged 50–69 years. $N=37$	Significant beneficial effect of visual memory	Brickman *et al.* (2014)
Wine					
Randomized, double-blind, placebo-controlled, crossover trial	250 mg *trans*-resveratrol, 250 mg *trans*-resveratrol + 20 mg piperine or placebo	Single dose. Cognitive effects assessed 40 minutes after administration	Healthy adults aged 19–34 years. $N=23$	None	Wightman *et al.* (2014)
Randomized, double-blind, placebo-controlled, parallel group trial	Dose dependant on body weight; 58–68 kg = 444 ml/day, 69–82 kg = 532 ml/day, 83–95 kg = 621 ml/day, concord grape juice. Containing 2091 mg/l total polyphenols as gallic acid equivalents (prominently anthocyanins: 46%, tartaric acid esters of hydroxycinnamates: 29%, and procyanidins: 10%)	16 weeks	Individuals with MCI aged 68–90 years. $N=21$. Substudy $N=8$	Significant reduction in interference errors during recognition memory task	Krikorian *et al.* (2012)
Double-blind, placebo-controlled, crossover trial	10 ml/kg body weight. Concord Grape Juice containing 2100 mg/l total phenolics as gallic acid equiv. or energy-matched placebo	Single dose	Smokers aged 18–50 years, $N=35$	None	Hendrickson and Mattes (2008)

Randomized, double blind, placebo-controlled, parallel group trial	6–9 ml/kg body weight per day. Concord grape juice or nutritionally matched placebo. Polyphenol content not detailed	12 weeks	Older adults with early memory decline but no dementia. Mean age 78.2 (range not provided). $N=12$	Significant improvement in verbal list learning. Nonsignificant benefit on spatial and verbal memory	Krikorian *et al*. (2010)
Randomized, double-blind, placebo controlled, three-stage crossover trial	500 mg (high), 250 mg (low) of *trans*-resveratrol or placebo	Single dose	Healthy adults aged 18–25 years (mean 20.17). $N=22$	None	Kennedy *et al*. (2010)
Randomized, placebo-controlled, parallel group trial	Provinols™ 4 mg/kg body weight containing, (in milligram per gram of dry powder) 480 proanthocyanidins, 61 total anthocyanins, 19 free anthocyanins, 38 catechin, 18 hydroxycinnamic acids, 14 flavonols, and 370 polymeric tannins. No details of placebo or "control" provided	Single dose	Healthy undergraduate students. Age not specified. $N=30$	Significant beneficial effect on spatial memory (memory-guided saccade task)	Cimrová *et al*. (2011)
Randomized, double-blind, placebo-controlled, parallel group trial	Initially 500 mg daily, with dose escalation by 500 mg increments every 13 weeks up to 2000 mg/day	1 year	Patients with diagnosis of probable AD, aged >49 years. $N=104$	None (underpowdered for cognitive outcomes)	Turner *et al*. (2015)
Randomized, double-blind, placebo-controlled, matched-pairs trial	200 mg resveratrol + 320 mg quercetin, or placebo	26 weeks	Overweight adults aged 50–70 years. $N=46$	Beneficial effects on memory retention	Witte *et al*. (2014)
Randomized, double-blind, placebo-controlled, crossover trial	100 ml red wine enriched or not (control) with 200 mg resveratrol	Single dose Cognitive assessments performed 60 minutes after dose administration	Healthy adults aged 65–78 years. $N=16$	Significant benefits of more-demanding task but poorer performance on less-demanding task	Scholey *et al*. (2014)
Soy					
Randomized, blind, placebo-controlled, parallel group trial	50 g of soy protein isolate powder (54 mg total isoflavones) or 50 g whey protein (0 g isoflavones)	Single dose Cognitive assessments performed 1.75 hours after dose administration	Healthy college students, mean age 20.1 years. $N=50$	None	Vanata and Metzger (2007)

(*Continued*)

Table 12.1 (Continued)

Design	Dosage	Duration	Population	Cognitive effects	Reference
Randomized, double-blind, placebo controlled, parallel group trial	25 g soy isoflavone-rich soy protein containing 91 mg aglycone weight isoflavones (154 mg total isoflavone equivalents) composed of genistein (52 mg aglycone equivalents), daidzein (36 mg aglycone equivalents), and glycitein (3 mg aglycone equivalents), or milk-protein-matched placebo (0 mg isoflavones)	2.5 years	Healthy postmenopausal women aged 42–92 years (mean 61 years) $N = 313$	Significant benefit to visual memory	Henderson *et al.* (2012)
Randomized, double-blind, placebo-controlled, parallel group trial	20 g of soy protein (160 mg total isoflavones: genistein 64 mg, daidzein 63 mg, and glycitein 34 mg) or nutritionally matched placebo (20 g whole milk protein)	12 weeks	Postmenopausal women. Mean age 56 years. $N = 84$	None	Basaria *et al.* (2009)
Randomized, double-blind, placebo-controlled, crossover trial	Two capsules, twice a day containing 116 mg total isoflavones (68 mg was daidzein, 12 mg was genistein and 36 mg glycitin) or placebo	6 weeks	Healthy men aged 30–80 years (mean 49 years) $N = 34$	Significant beneficial effect on spatial working memory	Thorp *et al.* (2009)
Randomized, double-blind, placebo-controlled, parallel group trial	110 mg soy isoflavones or placebo	6 months	Postmenopausal women aged 54–75 years (mean of 60 years). $N = 51$	Significant benefit to category verbal fluency test. Younger women (≤59 years) also showed a significant benefit on visuomotor tracking and attention	Kritz-Silverstein *et al.* (2003)
Randomized, double-blind, placebo-controlled, parallel group trial	100 mg/day of purified glycosidic isoflavones (~85% daidzein and genistein) or placebo	6 months	Cognitively healthy adults aged 62–89 years. $N = 30$	Significant benefits in tasks of visual–spatial memory and construction, verbal fluency, and speeded dexterity. Placebo was associated with better performance on tasks of executive function	Gleason *et al.* (2009)
Randomized, double-blind, placebo-controlled, parallel groups	25.6 g of isoflavone-rich, soy protein containing 52 mg genistein, 41 mg daidzein, and 6 mg glycitein (aglycone weights) or placebo (25.6 g total milk protein) taken daily	12 months	Women aged 60–75 years. $N = 175$	None	Kreijkamp-Kaspers *et al.* (2004)

Randomized, parallel groups trial	High-soy diet providing 100 mg isoflavones per day or nutritionally matched low-soy diet providing 0.5 mg isoflavones per day	10 weeks	University students aged $N=17$	Improvements in visual and verbal memory and mental flexibility in males and females. In females, significant benefits were also seen in planning ability. High-soy diet associated with verbal fluency improvements in females but poorer performance in males	File *et al.* (2001)
Randomized, double-blind, placebo-controlled, parallel groups	60 mg isoflavones or placebo per day	6 weeks	Postmenopausal women aged 51–66 years. $N=50$	Significant benefits to mental flexibility, planning ability, and short-term nonverbal memory	File *et al.* (2005)
Randomized, double-blind, placebo-controlled, parallel group trial	80 mg soy isoflavones or placebo	6 months	Chinese women aged 55–75 years. $N=176$	None	Ho *et al.* (2007)
Randomized, double blind, placebo-controlled, parallel group trial	Participants received cow's milk and placebo supplement (control), soy milk (containing ~71.3 mg/day isoflavones; 30.9 mg daidzein, 37.4 mg genistein, and 3.6 mg glycitein) and placebo supplement or cow's milk and isoflavone supplement (containing 70 mg/day isoflavones; 30 mg daidzein, 33 mg genistein, and 7 mg glycitein)	16 weeks	Postmenopausal women aged 48–65 years (mean 56.1 years), not receiving hormone replacement therapy. $N=79$	None	Fournier *et al.* (2007)
Randomized, double-blind, placebo-controlled, crossover trial	One tablet daily containing 60 mg of isoflavones as aglycone	6 months	Postmenopausal women with a mean age of approximately 50 years. $N=77$	Significantly improved incidental learning and mental flexibility/ verbal working memory	Casini *et al.* (2006)
Randomized, double blind, placebo-controlled, parallel group trial	60 mg isoflavones equivalents per day, or placebo	12 weeks	Postmenopausal women aged 50–65 years, not receiving hormone replacement therapy. $N=33$	Significantly benefits to immediate and delayed episodic memory, mental flexibility, and attention	Duffy *et al.* (2003)

(Continued)

Table 12.1 (Continued)

Design	Dosage	Duration	Population	Cognitive effects	Reference
Pine bark					
Controlled, parallel group trial	150 mg/day Pycnogenol, in conjunction with a suggested, unenforced, health plan. Compared to health plan alone	12 weeks	Healthy professionals with high oxidative stress aged 35–55 years. $N=59$	Significant benefits on general cognitive function, attention, mental performance, sustained attention, memory, executive function, and professional tasks. Subjective cognitive benefits also reported	Belcaro *et al.* (2014)
Randomized, double-blind, placebo-controlled, matched-pairs trial	150 mg Pycnogenol or placebo per day	3 months	Adults aged 60–85 years. $N=101$	Significant benefit on quality of working memory, particularly spatial working memory	Ryan *et al.* (2008)
Randomized, double-blind, placebo-controlled, matched-parallel groups trial		Half sample: 6-week placebo run-in followed by 6-week treatment Half sample 12 weeks treatment. Reassessed following by 4 week placebo run out	Patients with mild TBI, aged 21–64 years. $N=56$	Significant improvement in self-reported cognitive failures at 6 weeks. Benefit estimated to increase until 11 weeks before stabilizing and remained after treatment cessation	Theadom *et al.* (2013)
Randomized, double-blind, placebo-controlled, parallel group trial	1 mg/kg/day or Pycnogenol® or placebo	1 month treatment and reassessed after 1 month washout	Children, aged 6–14 years, with ADHD. $N=44$	Significant reductions is teacher, but not parent, ratings of hyperactivity and inattention, which were reversed by treatment cessation	Trebatická *et al.* (2006)
Randomized, double-blind, controlled, parallel group trial	960 mg of Enzogenol® + 120 mg of vitamin C, or 120 mg of vitamin C (control)	5 weeks	Males aged 50–65 years. $N=42$	Improved performance on spatial working memory and immediate recognition memory	Pipingas *et al.* (2008)
Randomized, double-blind, placebo-controlled, parallel group trial	200 mg Pycnogenol or placebo	6 months	Perimenopausal women aged 45–55 years, $N=155$	Significant beneficial effect of frequency of memory/ concentration problems	Yang *et al.* (2007)

Randomized, blind, placebo-controlled, parallel group trial	50 g of soy protein isolate powder (54 mg total isoflavones) or 50 g whey protein (0 g isoflavones)	Single dose Cognitive assessments performed 1.75 hours after dose administration	Healthy college students, mean age 20.1 years. $N=50$	None	Vanata and Metzger (2007)
Randomized, double-blind, placebo-controlled, parallel groups trial	20 g of soy protein (160 mg total isoflavones: genistein 64 mg, daidzein 63 mg, and glycitein 34 mg) or nutritionally match placebo (20 g whole milk protein)	12 weeks	Postmenopausal women. Mean age 56 years. $N=84$	None	Basaria *et al.* (2009)
Randomized, double-blind, placebo-controlled, two-stage crossover trial	Two capsules, twice a day containing 116 mg total isoflavones (68 mg was daidzein, 12 mg was genistein and 36 mg glycitin) or placebo	6 weeks	Healthy men aged 30–80 years (mean 49 years) $N=34$	Significant beneficial effect on spatial working memory	Thorp *et al.* (2009)
Randomized, double-blind, placebo-controlled, parallel groups trial	110 mg soy isoflavones or placebo	6 months	Postmenopausal women age 54–75 years (mean of 60 years). $N=51$	Significant benefit to category verbal fluency test. Younger women (≤59 years) also showed a significant benefit on visuomotor tracking and attention	Kritz-Silverstein *et al.* (2003)
Randomized, double-blind, placebo controlled, parallel groups	25.6 g of isoflavone-rich, soy protein containing 52 mg genistein, 41 mg daidzein, and 6 mg glycitein (aglycone weights) or placebo (25.6 g total milk protein) taken daily	12 months	Women aged 60–75 years. $N=175$	None	Kreijkamp-Kaspers *et al.* (2004)
Randomized, parallel groups trial	High-soy diet providing 100 mg isoflavones per day or nutritionally matched low-soy diet providing 0.5 mg isoflavones per day	10 weeks	University students age $N=17$	Improvements in visual and verbal memory and mental flexibility in males and females. In females, significant benefits were also seen in planning ability. High-soy diet associated with verbal fluency improvements in females but poorer performance in males	File *et al.* (2001)
Randomized, double-blind, placebo controlled, parallel groups	60 mg isoflavones or placebo per day	6 weeks	Postmenopausal women aged 51–66 years. $N=50$	Significant benefits to mental flexibility, planning ability, and short-term nonverbal memory	File *et al.* (2005)

(Continued)

Table 12.1 (Continued)

Design	Dosage	Duration	Population	Cognitive effects	Reference
Randomized, double-blind, placebo controlled, parallel group trial	80 mg soy isoflavones or placebo	6 months	Chinese women aged 55–75 years. $N=176$	None	Ho *et al.* (2007)
Randomized, double blind, placebo-controlled, parallel group trial	Participants received cow's milk and placebo supplement (control), soy milk (containing ~71.3 mg/day isoflavones; 30.9 mg daidzein, 37.4 mg genistein, and 3.6 mg glycitein), and placebo supplement or cow's milk and isoflavone supplement (containing 70 mg/day isoflavones; 30 mg daidzein, 33 mg genistein and 7 mg glycitein)	16 weeks	Postmenopausal women aged 48–65 years (mean 56.1 years), not receiving hormone replacement therapy. $N=79$	None	Fournier *et al.* (2007)
Randomized, double-blind, placebo-controlled, crossover trial	One tablet daily containing 60 mg of isoflavones as aglycone	6 months	Postmenopausal women with a mean age of approximately 50 years. $N=77$	Significantly improved incidental learning and mental flexibility/ verbal working memory	Casini *et al.* (2006)
Randomized, double blind, placebo-controlled, parallel group trial	60 mg isoflavones equivalents per day, or placebo	12 weeks	Postmenopausal women aged 50–65 years, not receiving hormone replacement therapy. $N=33$	Significantly benefits to immediate and delayed episodic memory, mental flexibility, and attention	Duffy *et al.* (2003)

MCI, mild cognitive impairment; TBI, traumatic brain injury.

12.4 Discussion and conclusions

Overall, polyphenols show great promise for cognitive enhancement in a number of populations. Studies that have examined the underlying central mechanisms involved suggest that there may be differential effects of polyphenols on different neural systems, with the frontal lobes and hippocampus being particular targets. However, this does not preclude activation in other brain regions. Polyphenols may also affect various central and systemic processes, which impact on cognitive function including cerebral blood flow, insulin resistance, oxidative stress, inflammation, and cardiovascular function.

It should also be noted that several studies have reported polyphenol effects on processes, which are known to modulate neurocognitive effects in the absence of any direct mood or cognitive effects of the intervention. These studies serve to illuminate the fact that there is no simple correspondence between physiological and behavioral processes. They also raise the question as to whether polyphenol effects on behavior may be dependent on dose, duration, and population. Clearly, this area is ripe for further investigation (Table 12.1).

References

Amieva, H., Jacqmin-Gadda, H., Orgogozo, J.M. *et al.* (2005) The 9 year cognitive decline before dementia of the Alzheimer type: a prospective population-based study. *Brain*, **128**, 1093–1101.

Andres-Lacueva, C., Monagas, M., Khan, N. *et al.* (2008) Flavanol and flavonol contents of cocoa powder products: influence of the manufacturing process. *Journal of Agricultural and Food Chemistry*, **56**, 3111–3117.

Arts, I.C.W., Hollman, P.C.H. and Kromhout, D. (1999) Chocolate as a source of tea flavonoids. *The Lancet*, **354**, 488.

Basaria, S., Wisniewski, A., Dupree, K. *et al.* (2009) Effect of high-dose isoflavones on cognition, quality of life, androgens, and lipoprotein in post-menopausal women. *Journal of Endocrinological Investigation*, **32**, 150–155.

Beking, K. and Vieira, A. (2010) Flavonoid intake and disability-adjusted life years due to Alzheimer's and related dementias: a population-based study involving twenty-three developed countries. *Public Health Nutrition*, **13**, 1403–1409.

Belcaro, G., Luzzi, R., Dugall, M., Ippolito, E. and Saggino, A. (2014) Pycnogenol(R) improves cognitive function, attention, mental performance and specific professional skills in healthy professionals aged 35–55. *Journal of Neurosurgical Sciences*, **58**, 239–248.

Bhagwat, S., Haytowitz, D.B., and Holden, J.M. (2014) *USDA Database for the Flavonoid Content of Selected Foods*, Release 3.1. U.S. Department of Agriculture, Agricultural Research Service, http://www.ars.usda.gov/nutrientdata/flav (accessed July 7, 2016).

Bloss, C.S., Delis, D.C., Salmon, D.P. and Bondi, M.W. (2008) Decreased cognition in children with risk factors for Alzheimer's disease. *Biological Psychiatry*, **64**, 904–906.

Borgwardt, S., Hammann, F., Scheffler, K., Kreuter, M., Drewe, J. and Beglinger, C. (2012) Neural effects of green tea extract on dorsolateral prefrontal cortex. *European Journal of Clinical Nutrition*, **66**, 1187–1192.

Brickman, A.M., Khan, U.A., Provenzano, F.A. *et al.* (2014) Enhancing dentate gyrus function with dietary flavanols improves cognition in older adults. *Nature Neuroscience*, **17**, 1798–1803.

Bryan, J., Tuckey, M., Einöther, S.J.L., Garczarek, U., Garrick, A. and De Bruin, E.A. (2012) Relationships between tea and other beverage consumption to work performance and mood. *Appetite*, **58**, 339–346.

Camfield, D.A., Scholey, A., Pipingas, A. *et al.* (2012) Steady state visually evoked potential (SSVEP) topography changes associated with cocoa flavanol consumption. *Physiology and Behavior*, **105**, 948–957.

Camfield, D.A., Stough, C., Farrimond, J. and Scholey, A.B. (2014) Acute effects of tea constituents L-theanine, caffeine, and epigallocatechin gallate on cognitive function and mood: a systematic review and meta-analysis. *Nutrition Reviews*, **72**, 507–522.

Casini, M.L., Marelli, G., Papaleo, E., Ferrari, A., D'ambrosio, F. and Unfer, V. (2006) Psychological assessment of the effects of treatment with phytoestrogens on postmenopausal women: a randomized, double-blind, crossover, placebo-controlled study. *Fertility and Sterility*, **85**, 972–978.

Cholerton, B., Baker, L.D. and Craft, S. (2011) Insulin resistance and pathological brain ageing. *Diabetic Medicine*, **28**, 1463–1475.

Cimrová, B., Budáč, S., Melicherová, U., Jergelová, M. and Jagla, F. (2011) Electrophysiological evidence of the effect of natural polyphenols upon the human higher brain functions. *Neuroendocrinology Letters*, **32**, 464–468.

Commenges, D., Scotet, V., Renaud, S., Jacqmin-Gadda, H., Barberger-Gateau, P. and Dartigues, J.F. (2000) Intake of flavonoids and risk of dementia. *European Journal of Epidemiology*, **16**, 357–363.

Crews, W.D., Harrison, D.W. and Wright, J.W. (2008) A double-blind, placebo-controlled, randomized trial of the effects of dark chocolate and cocoa on variables associated with neuropsychological functioning and cardiovascular health: clinical findings from a sample of healthy, cognitively intact older adults. *The American Journal of Clinical Nutrition*, **87**, 872–880.

De Bruin, E.A., Rowson, M.J., Van Buren, L., Rycroft, J.A. and Owen, G.N. (2011) Black tea improves attention and self-reported alertness. *Appetite*, **56**, 235–240.

De La Torre, R., De Sola, S., Pons, M. *et al.* (2014) Epigallocatechin-3-gallate, a DYRK1A inhibitor, rescues cognitive deficits in Down syndrome mouse models and in humans. *Molecular Nutrition and Food Research*, **58**, 278–288.

Desideri, G., Kwik-Uribe, C., Grassi, D. *et al.* (2012) Benefits in cognitive function, blood pressure, and insulin resistance through cocoa flavanol consumption in elderly subjects with mild cognitive impairment: the Cocoa, Cognition, and Aging (CoCoA) study. *Hypertension*, **60**, 794–801.

Devi, G., Hahn, K., Massimi, S. and Zhivotovskaya, E. (2005) Prevalence of memory loss complaints and other symptoms associated with the menopause transition: a community survey. *Gender Medicine*, **2**, 255–264.

Duff, S.J. and Hampson, E. (2000) A beneficial effect of estrogen on working memory in postmenopausal women taking hormone replacement therapy. *Hormones and Behavior*, **38**, 262–276.

Duffy, R., Wiseman, H. and File, S.E. (2003) Improved cognitive function in postmenopausal women after 12 weeks of consumption of a soya extract containing isoflavones. *Pharmacology Biochemistry and Behavior*, **75**, 721–729.

Engelhart, M.J., Geerlings, M.I., Ruitenberg, A. *et al.* (2002) Dietary intake of antioxidants and risk of alzheimer disease. *Journal of the American Medical Association*, **287**, 3223–3229.

Féart, C., Samieri, C. and Barberger-Gateau, P. (2010) Mediterranean diet and cognitive function in older adults. *Current Opinion in Clinical Nutrition and Metabolic Care*, **13**, 14–18.

Feng, L., Gwee, X., Kua, E.H. and Ng, T.P. (2010) Cognitive function and tea consumption in community dwelling older Chinese in Singapore. *The Journal of Nutrition, Health & Aging*, **14**, 433–438.

Feng, L., Li, J., Ng, T.P., Lee, T.S., Kua, E.H. and Zeng, Y. (2012) Tea drinking and cognitive function in oldest-old Chinese. *The Journal of Jutrition, Health & Aging*, **16**, 754–758.

Field, D.T., Williams, C.M. and Butler, L.T. (2011) Consumption of cocoa flavanols results in an acute improvement in visual and cognitive functions. *Physiology and Behavior*, **103**, 255–260.

File, S.E., Jarrett, N., Fluck, E., Duffy, R., Casey, K. and Wiseman, H. (2001) Eating soya improves human memory. *Psychopharmacology*, **157**, 430–436.

File, S.E., Hartley, D.E., Elsabagh, S., Duffy, R. and Wiseman, H. (2005) Cognitive improvement after 6 weeks of soy supplements in postmenopausal women is limited to frontal lobe function. *Menopause*, **12**, 193–201.

Fournier, L.R., Ryan Borchers, T.A., Robison, L.M. *et al.* (2007) The effects of soy milk and isoflavone supplements on cognitive performance in healthy, postmenopausal women. *The Journal of Nutrition, Health & Aging*, **11**, 155–164.

Francis, S.T., Head, K., Morris, P.G. and Macdonald, I.A. (2006) The effect of flavanol-rich cocoa on the fMRI response to a cognitive task in healthy young people. *Journal of Cardiovascular Pharmacology*, **47** (Suppl. 2), S215–S220.

Frautschy, S.A. and Cole, G.M. (2010) Why pleiotropic interventions are needed for Alzheimer's disease. *Molecular Neurobiology*, **41**, 392–409.

Fuh, J.-L., Wang, S.-J., Lee, S.-J., Lu, S.-R. and Juang, K.-D. (2006) A longitudinal study of cognition change during early menopausal transition in a rural community. *Maturitas*, **53**, 447–453.

Gleason, C.E., Carlsson, C.M., Barnet, J.H. *et al.* (2009) A preliminary study of the safety, feasibility and cognitive efficacy of soy isoflavone supplements in older men and women. *Age and Ageing*, **38**, 86–93.

Henderson, V.W., St. John, J.A., Hodis, H.N. *et al.* (2012) Long-term soy isoflavone supplementation and cognition in women: a randomized, controlled trial. *Neurology*, **78**, 1841–1848.

Hendrickson, S.J. and Mattes, R.D. (2008) No acute effects of grape juice on appetite, implicit memory, and mood. *Food and Nutrition Research*, **52**. doi: 10.3402/fnr.v52i0.1891.

Ho, S.C., Chan, A.S., Ho, Y.P. *et al.* (2007) Effects of soy isoflavone supplementation on cognitive function in Chinese postmenopausal women: a double-blind, randomized, controlled trial. *Menopause*, **14**, 489–499.

Hurst, W.J., Krake, S.H., Bergmeier, S.C., Payne, M.J., Miller, K.B. and Stuart, D.A. (2011) Impact of fermentation, drying, roasting and Dutch processing on flavan-3-ol stereochemistry in cacao beans and cocoa ingredients. *Chemistry Central Journal*, **5**, 53.

Kennedy, D.O., Wightman, E.L., Reay, J.L. *et al.* (2010) Effects of resveratrol on cerebral blood flow variables and cognitive performance in humans: a double-blind, placebo-controlled, crossover investigation. *The American Journal of Clinical Nutrition*, **91**, 1590–1597.

Kreijkamp-Kaspers, S., Kok, L., Grobbee, D.E. *et al.* (2004) Effect of soy protein containing isoflavones on cognitive function, bone mineral density, and plasma lipids in postmenopausal women: a randomized controlled trial. *Journal of the American Medical Association*, **292**, 65–74.

Krikorian, R., Nash, T.A., Shidler, M.D., Shukitt-Hale, B. and Joseph, J.A. (2010) Concord grape juice supplementation improves memory function in older adults with mild cognitive impairment. *The British Journal of Nutrition*, **103**, 730–734.

Krikorian, R., Boespflug, E.L., Fleck, D.E. *et al.* (2012) Concord grape juice supplementation and neurocognitive function in human aging. *Journal of Agricultural and Food Chemistry*, **60**, 5736–5742.

Kritz-Silverstein, D., Von Muhlen, D., Barrett-Connor, E. and Bressel, M.A. (2003) Isoflavones and cognitive function in older women: the SOy and Postmenopausal Health In Aging (SOPHIA) study. *Menopause*, **10**, 196–202.

Kuriyama, S. (2008) The relation between green tea consumption and cardiovascular disease as evidenced by epidemiological studies. *The Journal of Nutrition*, **138**, 1548S–1553S.

Lacey, J.V., Jr, Mink, P.J., Lubin, J.H. *et al.* (2002) Menopausal hormone replacement therapy and risk of ovarian cancer. *Journal of the American Medical Association*, **288**, 334–341.

Letenneur, L., Proust-Lima, C., Le Gouge, A., Dartigues, J. and Barberger-Gateau, P. (2007) Flavonoid intake and cognitive decline over a 10-year period. *American Journal of Epidemiology*, **165**, 1364–1371.

Lin, Y.-S., Tsai, Y.-J., Tsay, J.-S. and Lin, J.-K. (2003) Factors affecting the levels of tea polyphenols and caffeine in tea leaves. *Journal of Agricultural and Food Chemistry*, **51**, 1864–1873.

Lindsay, J., Laurin, D., Verreault, R. *et al.* (2002) Risk factors for Alzheimer's disease: a prospective analysis from the Canadian Study of Health and Aging. *American Journal of Epidemiology*, **156**, 445–453.

Maggi, A., Ciana, P., Belcredito, S. and Vegeto, E. (2004) Estrogens in the nervous system: mechanisms and nonreproductive functions. *Annual Review of Physiology*, **66**, 291–313.

Martin, L.C. and Cooper, R. (2011) From herbs to medicines: a world history of tea—from legend to healthy obsession. *Alternative and Complementary Therapies*, **17**, 162–168.

Mastroiacovo, D., Kwik-Uribe, C., Grassi, D. *et al.* (2015) Cocoa flavanol consumption improves cognitive function, blood pressure control, and metabolic profile in elderly subjects: the Cocoa, Cognition, and Aging (CoCoA) study—a randomized controlled trial. *The American Journal of Clinical Nutrition*. doi: ajcn.092189.

Messina, M., Nagata, C. and Wu, A.H. (2006) Estimated Asian adult soy protein and isoflavone intakes. *Nutrition and Cancer*, **55**, 1–12.

Miles, C., Green, R., Sanders, G. and Hines, M. (1998) Estrogen and memory in a transsexual population. *Hormones and Behavior*, **34**, 199–208.

Miller, K.B., Stuart, D.A., Smith, N.L. *et al.* (2006) Antioxidant activity and polyphenol and procyanidin contents of selected commercially available cocoa-containing and chocolate products in the United States. *Journal of Agricultural and Food Chemistry*, **54**, 4062–4068.

Million Women Study Collaborators (2003) Breast cancer and hormone-replacement therapy in the Million Women Study. *The Lancet*, **362**, 419–427.

Natsume, M., Osakabe, N., Yamagishi, M. *et al.* (2000) Analyses of polyphenols in cacao liquor, cocoa, and chocolate by normal-phase and reversed-phase HPLC. *Bioscience, Biotechnology, and Biochemistry*, **64**, 2581–2587.

Nelson, H.D., Humphrey, L.L., Nygren, P., Teutsch, S.M. and Allan, J.D. (2002) Postmenopausal hormone replacement therapy: scientific review. *Journal of the American Medical Association*, **288**, 872–881.

Ng, T.P., Feng, L., Niti, M., Kua, E.H. and Yap, K.B. (2008) Tea consumption and cognitive impairment and decline in older Chinese adults. *The American Journal of Clinical Nutrition*, **88**, 224–231.

Nurk, E., Refsum, H., Drevon, C.A. *et al.* (2009) Intake of flavonoid-rich wine, tea, and chocolate by elderly men and women is associated with better cognitive test performance. *The Journal of Nutrition*, **139**, 120–127.

Pase, M.P., Scholey, A.B., Pipingas, A. *et al.* (2013) Cocoa polyphenols enhance positive mood states but not cognitive performance: a randomized, placebo-controlled trial. *Journal of Psychopharmacology*, **27**, 451–458.

Petersen, R.C., Smith, G.E., Waring, S.C., Ivnik, R.J., Tangalos, E.G. and Kokmen, E. (1999) Mild cognitive impairment: clinical characterization and outcome. *Archives of Neurology*, **56**, 303–308.

Pipingas, A., Silberstein, R.B., Vitetta, L. *et al.* (2008) Improved cognitive performance after dietary supplementation with a Pinus radiata bark extract formulation. *Phytotherapy Research*, **22**, 1168–1174.

Robert, C., Borella, E., Fagot, D., Lecerf, T. and De Ribaupierre, A. (2009) Working memory and inhibitory control across the life span: intrusion errors in the Reading Span Test. *Memory and Cognition*, **37**, 336–345.

Rohdewald, P.J. (2004) Pycnogenol®, French Maritime Pine Bark Extract, in *Encyclopedia of Dietary Supplements* (eds P.M. Coates, M.R. Blackman, G.M. Cragg, M. Levine, J. Moss and J.D. White), CRC Press, Boca Raton, FL.

Ryan, J., Croft, K., Mori, T. *et al.* (2008) An examination of the effects of the antioxidant Pycnogenol on cognitive performance, serum lipid profile, endocrinological and oxidative stress biomarkers in an elderly population. *Journal of Psychopharmacology*, **22**, 553–562.

Scholey, A. and Owen, L. (2013) Effects of chocolate on cognitive function and mood: a systematic review. *Nutrition Reviews*, **71**, 665–681.

Scholey, A.B., French, S.J., Morris, P.J., Kennedy, D.O., Milne, A.L. and Haskell, C.F. (2010) Consumption of cocoa flavanols results in acute improvements in mood and cognitive performance during sustained mental effort. *Journal of Psychopharmacology*, **24**, 1505–1514.

Scholey, A., Downey, L.A., Ciorciari, J. *et al.* (2012) Acute neurocognitive effects of epigallocatechin gallate (EGCG). *Appetite*, **58**, 767–770.

Scholey, A., Benson, S., Stough, C. and Stockley, C. (2014) Effects of resveratrol and alcohol on mood and cognitive function in older individuals. *Nutrition and Aging*, **2**, 133–138.

Setchell, K.D. (1998) Phytoestrogens: the biochemistry, physiology, and implications for human health of soy isoflavones. *The American Journal of Clinical Nutrition*, **68**, 1333S–1346S.

Shaywitz, S.E., Naftolin, F., Zelterman, D. *et al.* (2003) Better oral reading and short-term memory in midlife, postmenopausal women taking estrogen. *Menopause*, **10**, 420–426.

Singh-Manoux, A., Kivimaki, M., Glymour, M.M. *et al.* (2012) Timing of onset of cognitive decline: results from Whitehall II prospective cohort study. *British Medical Journal*, **344**, d7622.

Small, G.W. (2004) What does imaging add to the management of Alzheimer's disease? *CNS Spectrums*, **9**, 20–23.

Sofi, F., Cesari, F., Abbate, R., Gensini, G.F. and Casini, A. (2008) Adherence to Mediterranean diet and health status: meta-analysis. *British Medical Journal*, **337**, a1344.

Sorond, F.A., Lipsitz, L.A., Hollenberg, N.K. and Fisher, N.D.L. (2008) Cerebral blood flow response to flavanol-rich cocoa in healthy elderly humans. *Neuropsychiatric Disease and Treatment*, **4**, 433–440.

Tewari, S., Gupta, V. and Bhattacharya, S. (2000) Comparative study of antioxidant potential of tea with and without additives. *Indian Journal of Physiology and Pharmacology*, **44**, 215–219.

Theadom, A., Mahon, S., Barker-Collo, S. *et al.* (2013) Enzogenol for cognitive functioning in traumatic brain injury: a pilot placebo-controlled RCT. *European Journal of Neurology*, **20**, 1135–1144.

Thorp, A.A., Sinn, N., Buckley, J.D., Coates, A.M. and Howe, P.R. (2009) Soya isoflavone supplementation enhances spatial working memory in men. *British Journal of Nutrition*, **102**, 1348–1354.

Trebatická, J., Kopasová, S., Hradečná, Z. *et al.* (2006) Treatment of ADHD with French maritime pine bark extract, Pycnogenol®. *European Child and Adolescent Psychiatry*, **15**, 329–335.

Trock, B.J., Hilakivi-Clarke, L. and Clarke, R. (2006) Meta-analysis of soy intake and breast cancer risk. *Journal of the National Cancer Institute*, **98**, 459–471.

Truelsen, T., Thudium, D. and Grønbaek, M. (2002) Amount and type of alcohol and risk of dementia: the Copenhagen City Heart Study. *Neurology*, **59**, 1313.

Turner, R.S., Thomas, R.G., Craft, S. *et al.* (2015) A randomized, double-blind, placebo-controlled trial of resveratrol for Alzheimer disease. *Neurology*, **85**, 1383–1391.

Valls-Pedret, C., Lamuela-Raventos, R.M., Medina-Remon, A. *et al.* (2012) Polyphenol-rich foods in the Mediterranean diet are associated with better cognitive function in elderly subjects at high cardiovascular risk. *Journal of Alzheimer's Disease*, **29**, 773–782.

Van Erp-Baart, M., Brants, H., Kiely, M. *et al.* (2003) Isoflavone intake in four different European countries: the VENUS approach. *British Journal of Nutrition*, **89**, S25–S30.

Vanata, D.F. and Metzger, M.M. (2007) Acute soy isoflavone consumption does not impact visual-spatial or verbal memory among healthy young adults. *North American Journal of Psychology*, **9**, 379.

Vickland, V., Mcdonnell, G., Werner, J., Draper, B., Low, L.F. and Brodaty, H. (2010) A computer model of dementia prevalence in Australia: foreseeing outcomes of delaying dementia onset, slowing disease progression, and eradicating dementia types. *Dementia and Geriatric Cognitive Disorders*, **29**, 123–130.

Werner, C. and Engelhard, K. (2007) Pathophysiology of traumatic brain injury. *British Journal of Anaesthesia*, **99**, 4–9.

Wightman, E.L., Haskell, C.F., Forster, J.S., Veasey, R.C. and Kennedy, D.O. (2012) Epigallocatechin gallate, cerebral blood flow parameters, cognitive performance and mood in healthy humans: a double-blind, placebo-controlled, crossover investigation. *Human Psychopharmacology*, **27**, 177–186.

Wightman, E.L., Reay, J.L., Haskell, C.F., Williamson, G., Dew, T.P. and Kennedy, D.O. (2014) Effects of resveratrol alone or in combination with piperine on cerebral blood flow parameters and cognitive performance in human subjects: a randomised, double-blind, placebo-controlled, cross-over investigation. *British Journal of Nutrition*, **112**, 203–213.

Witte, A.V., Kerti, L., Margulies, D.S. and Floel, A. (2014) Effects of resveratrol on memory performance, hippocampal functional connectivity, and glucose metabolism in healthy older adults. *The Journal of Neuroscience*, **34**, 7862–7870.

Yang, H.-M., Liao, M.-F., Zhu, S.-Y., Liao, M.-N. and Rohdewald, P. (2007) A randomised, double-blind, placebo-controlled trial on the effect of Pycnogenol® on the climacteric syndrome in peri-menopausal women. *Acta Obstetricia et Gynecologica Scandinavica*, **86**, 978–985.

Zhu, N., Jacobs, D., Meyer, K. *et al.* (2015) Cognitive function in a middle aged cohort is related to higher quality dietary pattern 5 and 25 years earlier: the cardia study. *The Journal of Nutrition, Health & Aging*, **19**, 33–38.

Zuo, Y., Chen, H. and Deng, Y. (2002) Simultaneous determination of catechins, caffeine and gallic acids in green, Oolong, black and pu-erh teas using HPLC with a photodiode array detector. *Talanta*, **57**, 307–316.

Chapter 13

Curcumin and Cancer Metastasis

Ikuo Saiki

Abstract: Intrahepatic metastasis of hepatocellular carcinoma and mediastinal lymphatic metastasis of lung cancer are the most critical prognostic factors for cancer patients. Here, we analyze the antimetastatic potential of curcumin, a polyphenol isolated from *Curcuma Longa*, using murine hepatocellular carcinoma (CBO140C12) and Lewis lung carcinoma (LLC) cells. Daily oral administration of curcumin suppressed both the intrahepatic metastasis and the mediastinal lymph node metastasis in a dose-dependent manner, whereas the growth of implanted tumors was not affected. Moreover, curcumin inhibited the invasion of tumor cells through Matrigel-coated filters and also the production of matrix metalloproteinase-9 (MMP-9). Consistently, curcumin inhibited the adhesion and haptotactic migration to fibronectin and laminin without affecting the expression of integrin on the cell surface. As we showed by dual luciferase assay, curcumin downregulated the transcriptional activator protein-1 (AP-1), which in turn caused the inhibition of LLC cell invasion through the suppression of the urokinase-type plasminogen activator (u-PA) and its receptor (u-PAR). Finally, the formation of actin stress fibers was affected by treatment with curcumin. Collectively, these results suggested that curcumin suppressed the tumor metastasis mediated by the inhibition of several metastatic properties, in which the functional alteration of cytoskeletal organization, at least in part, could play an important role.

Keywords: curcumin, orthotopic implantation, intrahepatic metastasis, mediastinal lymph node metastasis, adhesion, migration, invasion, angiogenesis, activator protein-1 (AP-1), urokinase-type plasminogen activator (u-PA), u-PA receptor (u-PAR), matrix metalloproteinase-9 (MMP-9), stress fiber formation.

Recent Advances in Polyphenol Research, First Edition. Edited by Kumi Yoshida, Véronique Cheynier and Stéphane Quideau.

13.1 Introduction

Despite recent advances in diagnostic techniques for the early detection of hepatocellular carcinoma (HCC) and lung cancer and significant improvement of surgical procedures, the prognosis of patients with metastasizing HCC or lung cancer is generally poor, even in the early stages of cancer, as compared with other malignant neoplasms (Nagao *et al.*, 1987; Travis *et al.*, 1995; Ginsberg *et al.*, 1997; Terada, 1998). Intrahepatic metastasis and recurrence of HCC and mediastinal lymphatic metastasis of lung cancer are the most critical prognostic factors for cancer patients (Naruke *et al.*, 1988; Jefferson *et al.*, 1996). Therefore, it is necessary to investigate the mechanism responsible for intrahepatic metastasis and lymphatic metastasis, which should improve the future prognosis of patients with cancer.

To do this, we established a unique intrahepatic metastasis model in mice. Orthotopic implantation of a fragment of murine HCC, CBO140C12 cells, resulted in the formation of a solitary tumor nodule and intrahepatic metastasis in the liver (Sawada *et al.*, 2001). Previously, several models of intrahepatic metastasis produced by implantation of a tumor cell suspension into the mouse spleen and liver have been reported (Nakashima, 1976; Osada *et al.*, 1996; Genda *et al.*, 1999; Kuriyama *et al.*, 1999). However, these models do not include all the pathophysiological steps of cancer metastasis, such as release from the primary tumor, invasion into surrounding tissues, arrest in the capillary bed of distant organ, extravasation to organ parenchyma, and growth at apparently selective site, which are distant from the original tumor site (Fig. 13.1).

Orthotopic implantation of cell suspensions may induce changes in the nature and biological behavior of the original tumor (Fu *et al.*, 1991; Hoffman, 1991). In contrast, our

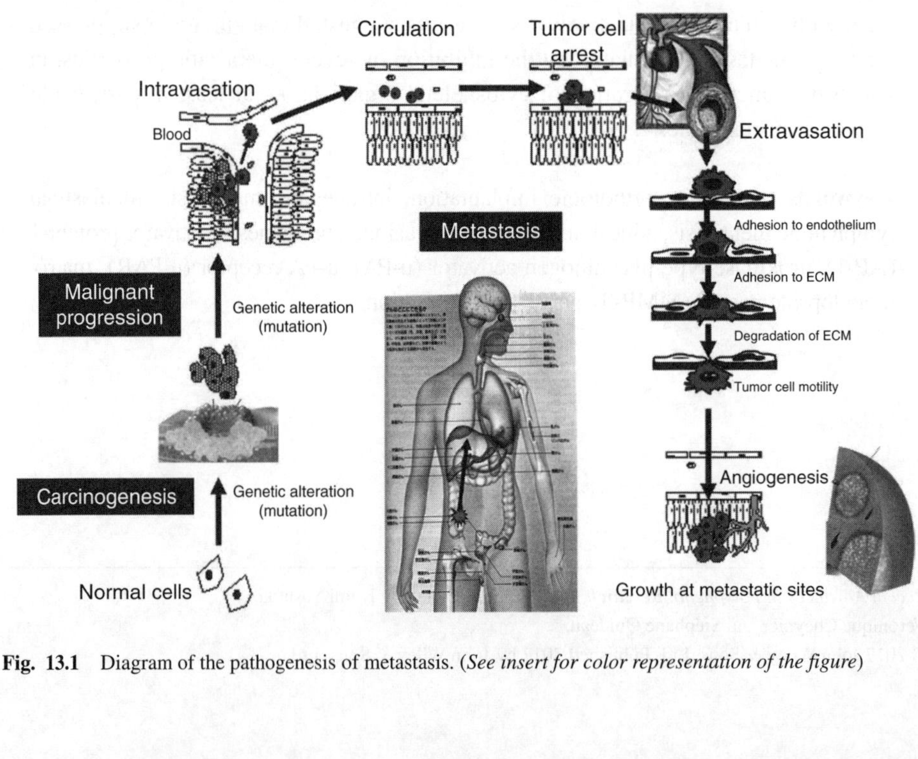

Fig. 13.1 Diagram of the pathogenesis of metastasis. (*See insert for color representation of the figure*)

model using an intact tissue fragment has several advantages, including the reflection of the cascades of metastasis, the high incidence of intrahepatic metastasis, and the ease of implantation (Sawada *et al.*, 2002). In addition, we have shown that the intrahepatic metastasis of murine HCC may be caused by enhancement of the invasive and adhesive activities of tumor cells through high expression of matrix metalloproteinase-9 (MMP-9), membrane type-1 matrix metalloproteinase (MT1-MMP), and β1 integrin (Sawada *et al.*, 2001, 2002; Tsuchiya *et al.*, 2002). Thus, our model provides new opportunities for the pathophysiological and pharmacological characterization of intrahepatic metastasis of HCC.

13.1.1 Antimetastatic mechanisms

The TNM Classification is an anatomically based system that records the primary (T) and regional nodal extent (N) of the tumor and the absence or presence of metastases (M). According to the new TNM revision (Mountain, 1997), the 5-year survival rate of patients with N2 lung cancer is only about 20%, even if the patients undergo curative resection (Vansteenkiste *et al.*, 1997). Therefore, more effective therapeutic approaches are needed to improve the prognosis of these patients. Although several investigators have reported experimental models for lung cancer, these models have had some important drawbacks, such as ectopic implantation and complicated procedures (Mclemore *et al.*, 1987; Howard *et al.*, 1991; Yano & Sone, 1997). Recently, we established a model of spontaneous lymphatic metastasis produced by orthotopic implantation of lung cancer cells (Doki *et al.*, 1999). Direct implantation of Lewis lung carcinoma (LLC) admixed with Matrigel into the left lobe of the lung caused the formation of a solitary tumor followed by metastasis to the mediastinal lymph node. This model should be useful for preclinical studies investigating therapeutic approaches for lung cancer treatment.

It has been reported that the overexpression of urokinase-type plasminogen activator (u-PA) and u-PA receptor (u-PAR) was correlated with lymphatic metastasis of lung cancer (Oka *et al.*, 1991; Nagayama *et al.*, 1994). Also, the prognosis of patients with higher expression of u-PA and u-PAR is significantly worse than that of patients with lower expression (Morita *et al.*, 1998). Both u-PA and u-PAR genes are regulated by the activator protein-1 (AP-1) (Murakami *et al.*, 1990). Interestingly, a variant of LLC cells with a high rate of mediastinal lymph node metastasis, obtained by the *in vivo* selection method of Fidler (1973), showed increased AP-1 activity compared with the original LLC cells. The AP-1 DNA-binding sites were identified in the promoter regions of mouse MMP-9 (Munaut *et al.*, 1999) and integrin (De Meirsman *et al.*, 1994) and play pivotal roles in their gene expression. Therefore, reducing the AP-1 activity may be a good strategy to prevent the mediastinal lymph node metastasis in lung cancer.

It was thought that lymphatic metastasis was a passive process in which detached tumor cells reached lymph nodes *via* drainage through preexisting local lymphatic vessels (Pepper, 2001). However, recent studies have identified lymphangiogenic growth factors and histochemical markers that discriminate between blood vessels and lymphatics, revealing that lymphangiogenesis may play a critical role in metastasis (Stacker *et al.*, 2002; Saharinen *et al.*, 2004). The expression of lymphangiogenic growth factors in a range of animal tumor models leads to the formation of lymphatic vessels either within or at the periphery of the tumors, and this is accompanied by enhanced lymphatic metastasis and, in some cases, by metastasis to distant organs (Stacker *et al.*, 2002). To evaluate lymphangiogenesis, we established a rat lymphatic endothelial line (TR-LE) cells from the thoracic duct of a transgenic rat harboring a temperature-sensitive simian virus 40 (SV40) large T-antigen and enhanced green fluorescent protein (EGFP) (Matsuo *et al.*, 2006a). TR-LE cells possess tube-forming ability on Matrigel and express the lymphatic endothelial markers VEGFR-3 (vascular endothelial growth factor receptor), LYVE-1 (a lymphatic endothelial receptor), Prox-1 (a homeobox gene product), and podoplanin (a glomerular podocyte membrane mucoprotein), together with the endothelial markers CD31, Tie-2, and VEGFR-2. This enabled us to conduct experiments on lymphangiogenesis by evaluating the ability of the formation of capillary-like tubes using lymphatic endothelial cell line.

13.1.2 Curcumin, a polyphenol from Curcuma longa

Turmeric powder from the rhizome of *C. longa* L. has been used for flavoring and as a colorant in many kinds of foods (Govindaraja, 1980) (Fig. 13.2). Traditionally, it has been utilized for the treatment of inflammatory disorders. Also, it has been used for the treatment of hepatitis, menstrual disorders, epilepsy, and skin disorders. The phenolic substances isolated from turmeric are known as curcuminoids, which exhibit antibacterial and anthelmintic activities and reduce cholesterol level. Many studies showed that curcumin possesses potent anti-inflammatory (Srivastava & Srimal, 1985), anticarcinogenic (Chauhan, 2002), and antimetastatic (Menon *et al.*, 1995, 1999; Ichiki *et al.*, 2001) activities.

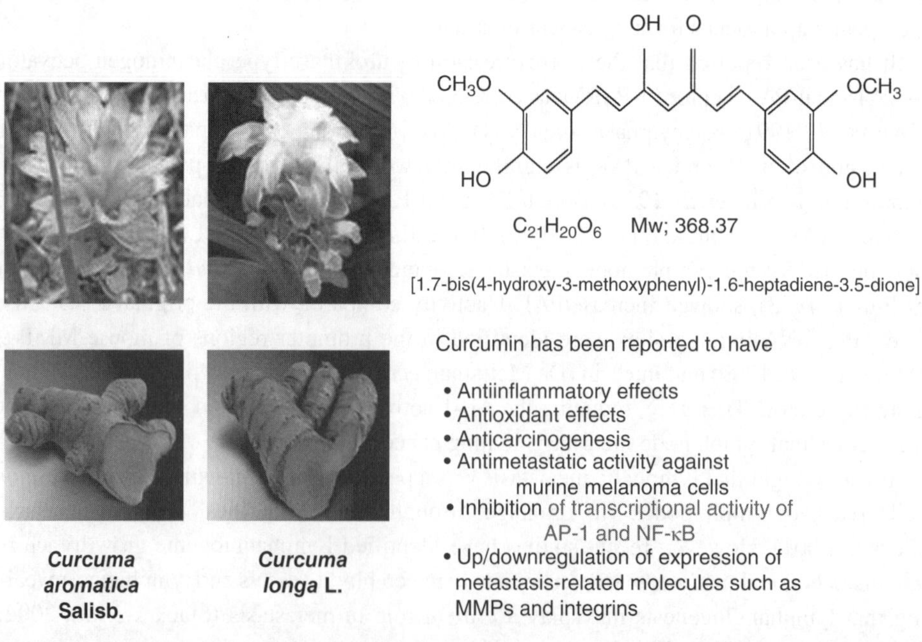

Fig. 13.2 Chemical structure of curcumin and its biological properties. (*See insert for color representation of the figure*)

Notably, it has been shown that the suppression of AP-1 transcriptional activity is a major part of the activities of curcumin (Bierhaus *et al.*, 1997).

Therefore, we attempted to examine whether mediastinal lymph node metastasis could be inhibited by treatment with curcumin, which inhibits AP-1 transcription, in our spontaneous mediastinal lymph node metastasis model for lung cancer (Doki *et al.*, 1999). We also investigated the effect of curcumin on the growth of primary tumors and the intrahepatic metastasis caused by orthotopic implantation of CBO140C12 tumor fragments and on its antimetastatic mechanism of action *in vitro*.

13.2 Effects of curcumin on intra-hepatic metastasis of liver cancer

13.2.1 Effect of curcumin on the growth of the implanted HCC and intrahepatic metastasis

We have previously demonstrated that intrahepatic metastasis might be established *via* the enhancement of invasion, adhesion, and migration through the marked expression of MMP-9, MT1-MMP, and b1 integrin in CBO140C12 cells (Sawada *et al.*, 2001; Tsuchiya *et al.*, 2002). The AP-1 binding site is located in the promoter region of some metastasis-related genes, including MMP-9 (Rosenberg, 1995; Munaut *et al.*, 1999), uPA (De Cesare *et al.*, 1996), uPA-R (Lee *et al.*, 1994).

VEGF (Garrido *et al.*, 1993) and some integrins (Rosmarin *et al.*, 1992; De Meirsman *et al.*, 1994). We first evaluated the effect of curcumin on the intrahepatic metastasis produced by inoculation of CBO140C12 tumor fragments into the liver. Curcumin (100 and 200 mg/kg) was administered p.o. daily for 20 days after the implantation. As shown in Fig. 13.3, curcumin markedly suppressed the intrahepatic metastasis in a dose-dependent manner, without causing loss of body weight or any other adverse effects. In contrast, the growth of CBO140C12 tumors at the implanted site was not affected. This result indicates that curcumin has a potent antimetastatic activity in an orthotopic implantation model of murine HCC.

In fact, the effect of anticancer drugs may be due to the suppression of tumor growth at implanted and metastatic sites without directly affecting the process of intrahepatic spread. These results further support the importance of targeting metastasis-specific processes to establish intrahepatic metastasis. A combination therapy with these two different types of drug may provide more information on the pharmacological potential of curcumin as a selective antimetastatic drug.

13.2.2 Effect of curcumin on tumor invasion and expression of invasion-related molecules

To examine the antimetastatic mechanisms of curcumin, an invasion assay was performed using Matrigel, a reconstituted basement membrane. Curcumin showed concentration-dependent inhibition of the invasion of CBO140C12 cells (Fig. 13.4). In contrast, curcumin

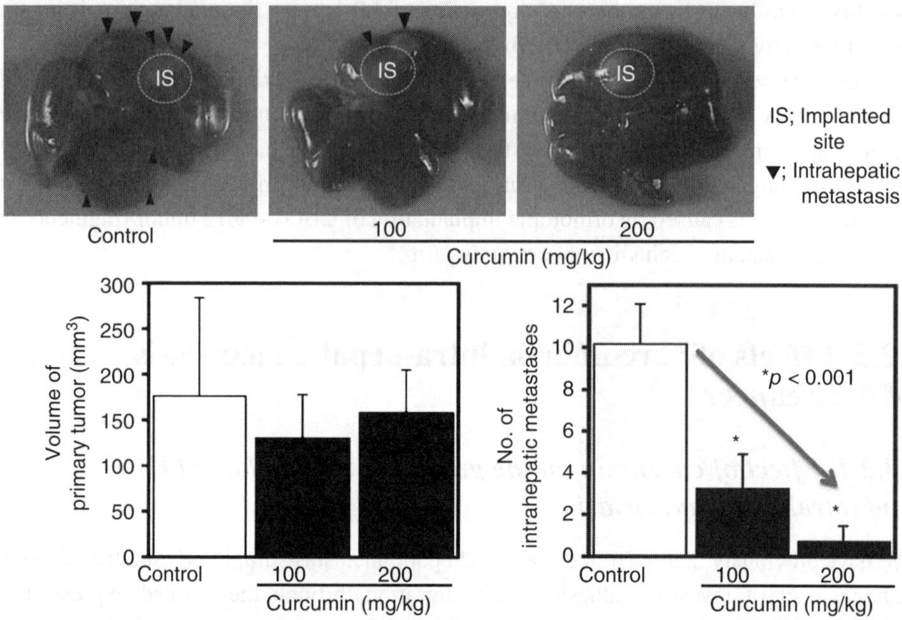

Fig. 13.3 Effect of curcumin on intrahepatic metastasis *in vivo*. (*See insert for color representation of the figure*)

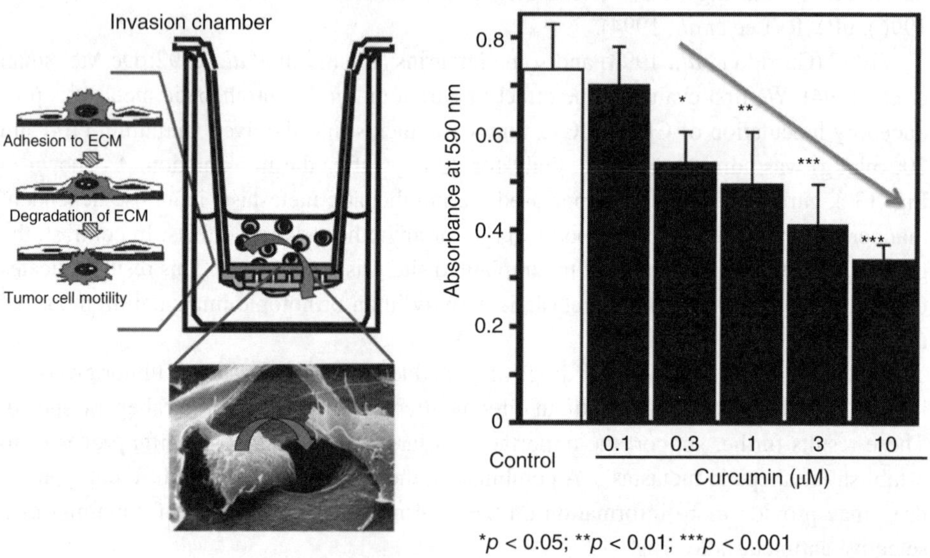

Fig. 13.4 Effect of curcumin on the invasion of CBO140C12 cells. (*See insert for color representation of the figure*)

at concentrations ranging from 0.1 to 10 µM showed no inhibitory effect on the proliferation of CBO140C12 cells *in vitro* (data not shown). The Matrigel invasion assay was subdivided into three essential steps, the tumor cell adhesion to ECM components such as fibronectin and laminin, the degradation of Matrigel by tumor-secreted proteases, and the haptotaxis on Matrigel-coated filters (Murata *et al.*, 1997). To clarify the inhibitory mechanism of tumor cell invasion through Matrigel, we next examined the effect of curcumin on the production of MMPs. Densitometric analysis of the zymogram showed that the degradation of gelatin substrate by MMP-9 produced by CBO140C12 cells was slightly inhibited by curcumin in a concentration-dependent manner.

13.2.3 Effect of curcumin on tumor cell adhesion to fibronectin, laminin, and poly-L-lysine substrates

Since tumor cell adhesion to extracellular matrices (ECM) is considered to be the first step in tumor cell invasion, we examined the effect of curcumin on the adhesion of CBO140C12 cells to laminin or fibronectin substrates. The tumor cells were treated with curcumin at concentrations ranging from 0.1 to 10 μM. Curcumin significantly inhibited the adhesion to laminin and fibronectin-coated substrates, but not to poly-L-lysine-coated plates (data not shown). In addition, curcumin significantly inhibited cell adhesion to Matrigel and type IV collagen (data not shown). These results suggest that curcumin inhibited the integrin-mediated cell adhesion.

13.2.4 Effect of curcumin on the expression of some integrin subunits

The ligation between ECM components and cell surface integrin is an essential step in cell adhesion. Many adhesive proteins such as fibronectin and vitronectin present in ECM and in the blood contain the arginine–glycine–aspartic acid (RGD) tripeptide as their cell recognition site (Ruoslahti & Pierschbacher, 1987). The RGD sequence is recognized by α5β1 and αvβ3 integrin. However, the binding modality between α3β1, α4β1, and α4β7 integrin and fibronectin was not mediated by RGD (Kunicki, 1989; Carter *et al.*, 1991; Mould *et al.*, 1991; Crowe *et al.*, 1994). The α6β1 integrin is widely known as a specific laminin receptor (Sonnenberg *et al.*, 1988).

To examine the inhibitory mechanism of cell adhesion to some ECM components by curcumin, we investigated the effect of curcumin on the expression in these integrin subunits. Flow cytometry analyses showed that all the integrin subunits indicated were expressed on cell surface and the severe quantitative changes were not detected when the cells were treated with 10 μM curcumin for 1 hour (Fig. 13.5). A similar result was obtained when the cells were treated with curcumin for 6 hours (data not shown). Cell surface expression of α5 subunit is slightly inhibited by curcumin; however, this could not fully explain the effect of curcumin on cell adhesion. In addition, mRNA expression of the integrins was confirmed by RT-PCR (data not shown). Altogether, these results suggested that the inhibition of adhesion by curcumin was not mediated by downregulation of the cell surface expression of integrin.

13.2.5 Effect of curcumin on the haptotactic migration

Haptotactic migration of tumor cells toward Matrigel-coated filters is another essential step in the Matrigel invasion assay. The filters in a Transwell chamber were precoated with laminin, one of the major components in Matrigel, on the lower surface. Curcumin strongly

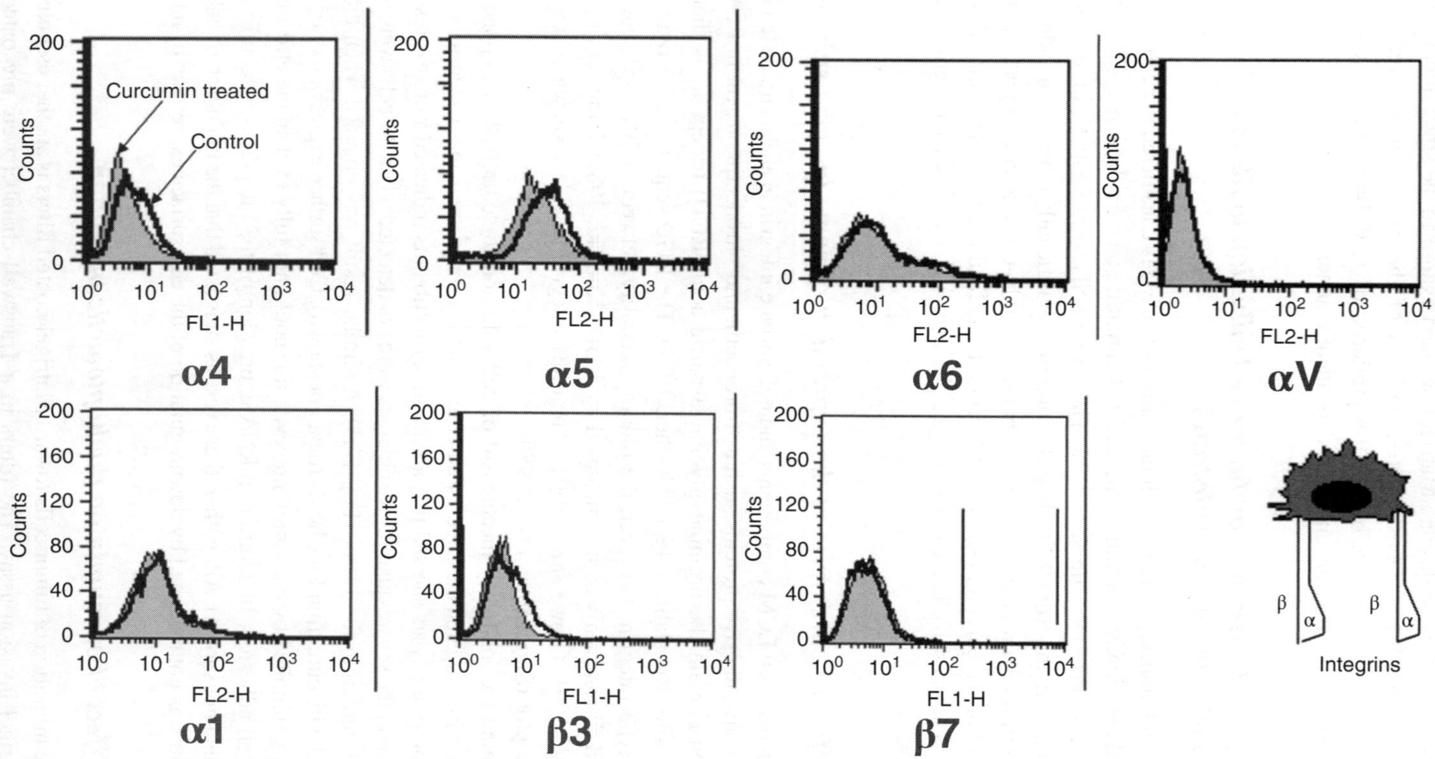

Fig. 13.5 Effect of curcumin on the expression of integrin α and β subunits on cell surface of tumor cells.

inhibited haptotactic migration of the tumor cells. A similar result was obtained in the haptotaxis to fibronectin-coated filters, but not to BSA-coated filters.

13.2.6 *Effect of curcumin on the formation of actin stress fibers*

The actin cytoskeleton and regulation of actin polymerization are widely accepted as a driving force for cell motility and adhesion. Integrin plays important roles in organizing the actin-containing cytoskeleton and in signal transduction from the ECM (Martin *et al.*, 2002). The Rho family of small GTP-binding proteins plays a central role in regulating actin polymerization (Mackay and Hall, 1998). Protein kinase C (PKC) regulates common pathways regulated by Rho family GTP-binding proteins such as RhoA, Cdc42, and Ras (Newton, 1995).

To further investigate the inhibitory mechanism of cell adhesion and migration by curcumin, we focused on the alteration of the stress fiber formation (Fig. 13.6). Serum-starved cells on fibronectin-coated slide were treated with or without curcumin for 3 hours. Fluorescent microscopic imaging showed a number of stress fibers in untreated cells and 0.5% DMSO-treated cells. In contrast, impaired stress fiber formation could be detected on noncoated culture slide (data not shown), suggesting that stress fiber formation is dependent on the interaction with extracellular matrix.

Treatment with curcumin at 10 μM resulted in the disappearance of stress fibers. A similar result was obtained in tumor cells exposed to 10 μM curcumin for 1 hour (data not shown). It was previously reported that curcumin suppressed TPA-induced PKC activation (Lin *et al.*, 1997). In addition, cell motility mediated through the Rho and Rho-associated coiled-coil forming protein kinase (p160ROCK) signaling pathway plays a crucial role in intrahepatic metastasis of human HCC (Genda *et al.*, 1999; Takamura *et al.*, 2001). These results raise the possibility that curcumin may inhibit the activation of one or some Rho family GTP-binding proteins through blocking PKC activation. However, further studies are necessary to clarify the precise inhibitory mechanism of stress fiber formation.

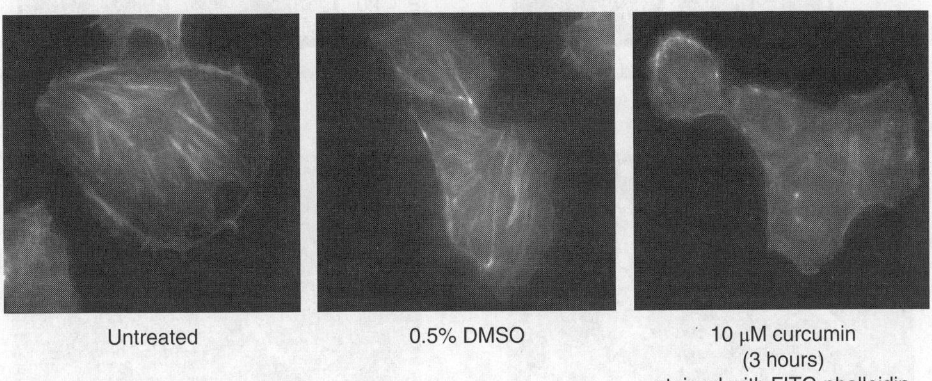

Fig. 13.6 Effect of curcumin on the formation of actin stress fiber on fibronectin-coated substrate. (*See insert for color representation of the figure*)

13.3 Effects of curcumin on lymp node metastasis of lung cancer

13.3.1 Comparison of metastatic properties of Lewis lung carcinoma (LLC) and its metastatic variant cell line

LLC-MLN is a highly metastatic line of LLC cells, which very frequently metastasizes to mediastinal lymph nodes after orthotopic implantation into the left lung. It was isolated by the *in vivo* selection method of Fidler (1973). When LLC or LLC-MLN cells were orthotopically implanted into the lung, LLC-MLN cells produced larger tumors at the implantation site and larger spontaneous metastases in the mediastinal lymph nodes, as compared with the original LLC cells (Fig. 13.7). The invasiveness of LLC or LLC-MLN cells was examined using Matrigel. LLC-MLN cells had significantly greater invasive ability. Since the overexpression of AP-1 was reported to be associated with lymphatic metastasis in lung cancer patients (Volm *et al.*, 1994), we also investigated AP-1 transcriptional activity in LLC or LLC-MLN cells using the dual luciferase assay system. Higher AP-1 transcriptional activity was observed in LLC-MLN cells than in LLC cells. This is in good agreement with the previous report that the overexpression of AP-1 is correlated with lymphatic metastasis in lung cancer patients (Volm *et al.*, 1994). Therefore, the downregulation of AP-1 is considered to be very important for reducing the lymph node metastasis of lung cancer.

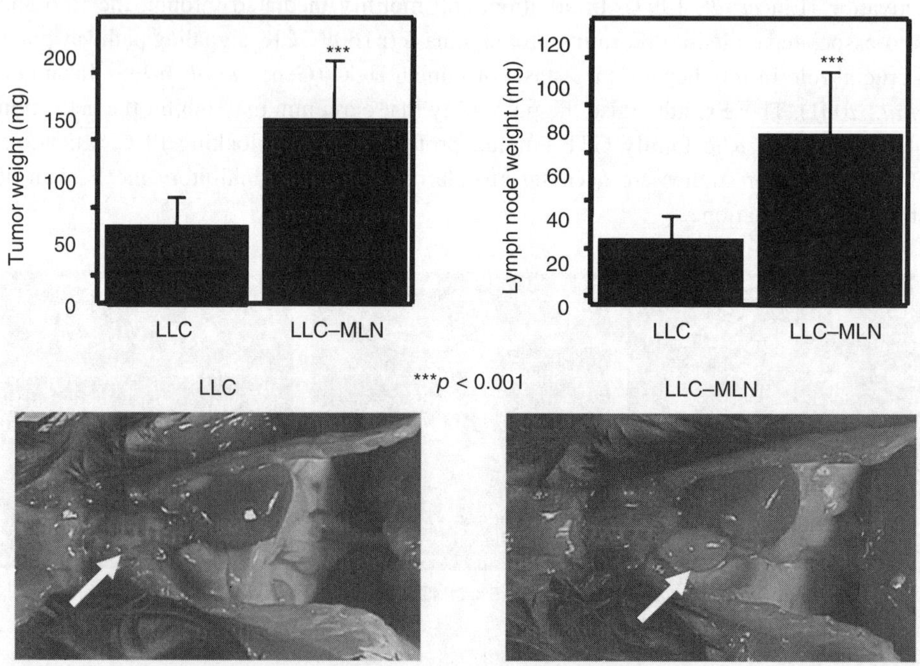

Fig. 13.7 Tumor growth and lymph node metastasis on day 15 after orthotopic implantation of LLC or LLC-MLN cells. (*See insert for color representation of the figure*)

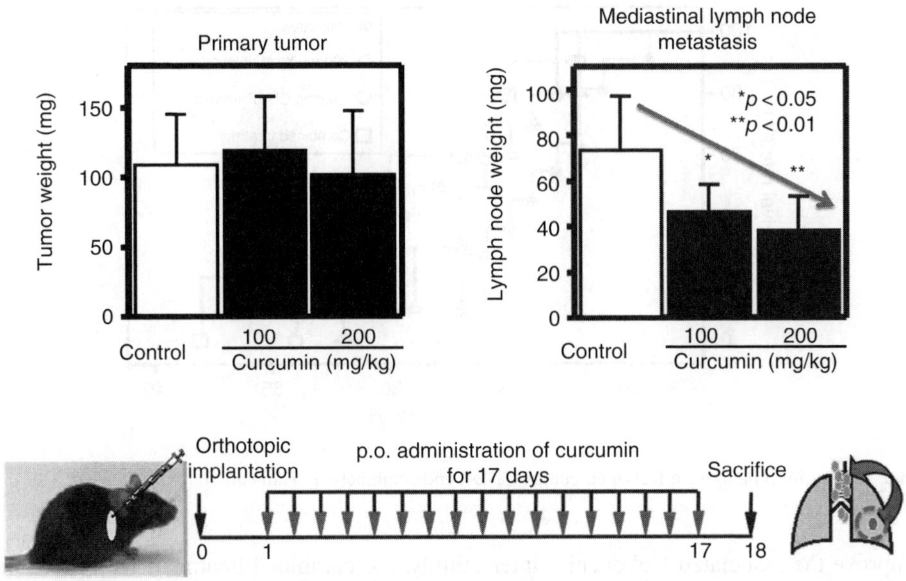

Fig. 13.8 Effect of curcumin on the growth of the inoculated tumor and mediastinal lymph node metastasis after the orthotopic implantation of LLC cells.

13.3.2 Effect of curcumin on the growth of the inoculated tumor and lymph node metastasis of orthotopically implanted LLC cells

Daily oral administration of curcumin failed to inhibit the growth of LLC tumors at the implantation site (Fig. 13.8). In contrast, metastasis of LLC cells to the mediastinal lymph nodes was significantly inhibited by curcumin in a dose-dependent manner (100 and 200 mg/kg: inhibition rate = 37 and 48%, respectively). Curcumin did not show any change in the body weight or any other adverse effect on the mice during the experiment (data not shown). This is highly consistent with our previous observation in the intrahepatic metastasis of orthotopically implanted HCC cells, in which curcumin inhibited the metastasis but not the primary tumor growth, as mentioned before.

13.3.3 Combined effect of curcumin and CDDP (cis-*diamine-dichloroplatinum) in the lung cancer model*

The prognosis of lung cancer patients who have recurrent metastatic disease is generally poor (Ginsberg *et al.*, 1997; Vansteenkiste *et al.*, 1997). Among the potent combined modalities, platinum-based chemotherapy is used as a standard modality used for the treatment of lung cancer patients (Livingston, 1997; Rigas, 1998). While curcumin (200 mg/kg) caused a significant inhibition of lymph node metastasis with no effect on the lung tumor growth, CDDP (7 mg/kg) significantly inhibited the growth of the tumor at the implantation site as well as the metastatic site (data not shown). However, dose-limiting side effects of CDDP, such as leukopenia, limit the clinical usage of CDDP even upon the cotreatment with recombinant human granulocyte-colony-stimulating factor (rhG-CSF), which can

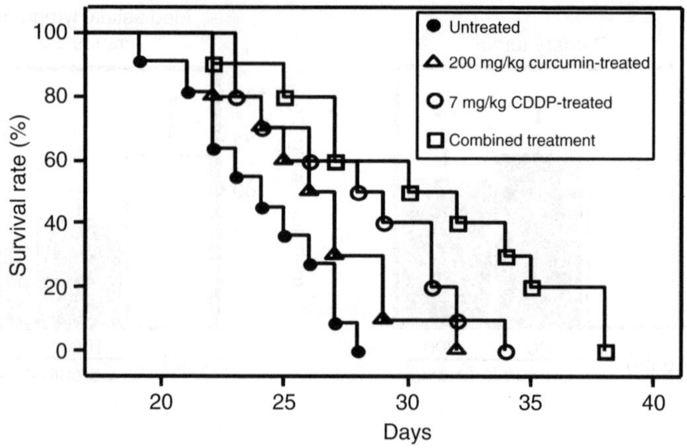

Fig. 13.9 Life-prolonging effect of curcumin, CDDP, and combination treatment.

improve the associated leukopenia. Interestingly, the combined treatment with curcumin and CDDP was effective to inhibit the tumor growth at the implantation site, as CDDP monotherapy, but resulted in a marked inhibition of lymph node metastasis as compared with CDDP monotherapy and/or no treatment. The weights of mediastinal lymph nodes in the combined treatment group were about the same as those of non-tumor-bearing mice. Combined treatment with curcumin did not improve the loss of body weight caused by CDDP treatment during the experiment. Therefore, the usage of noncytotoxic drugs as curcumin with anti-AP-1 transcriptional activity in combination with conventional anticancer drugs may be effective to improve the quality of life for patients with lung cancer.

We also examined the life-prolonging effect achieved with the combined treatment modality (Fig. 13.9). All the untreated control mice died of tumor burden within 28 days after the implantation. Similar enhancement of survival was observed in the groups of mice (10 mice per group), which had received either curcumin or CDDP alone. The combined treatment group showed a significantly prolonged survival time. These results indicate that the combination of curcumin and CDDP had an enhanced therapeutic effect on tumor metastasis and the survival rate ($p < 0.01$ vs. control). Since the mice died due to proliferation of the primary lung tumor rather than due to growth of the metastases in the mediastinal lymph nodes, surgical resection of the primary lung tumor might also be expected to produce a further life-prolonging effect in each treatment group in this model.

13.3.4 Effect of curcumin on the growth and invasion of LLC cells in vitro

For further examination of the antimetastatic properties of curcumin, we performed cell-invasion assay using Matrigel. Curcumin at noncytotoxic concentrations, less than 10 μM, caused dose-dependent inhibition of LLC cells invasion through the Matrigel (data not shown).

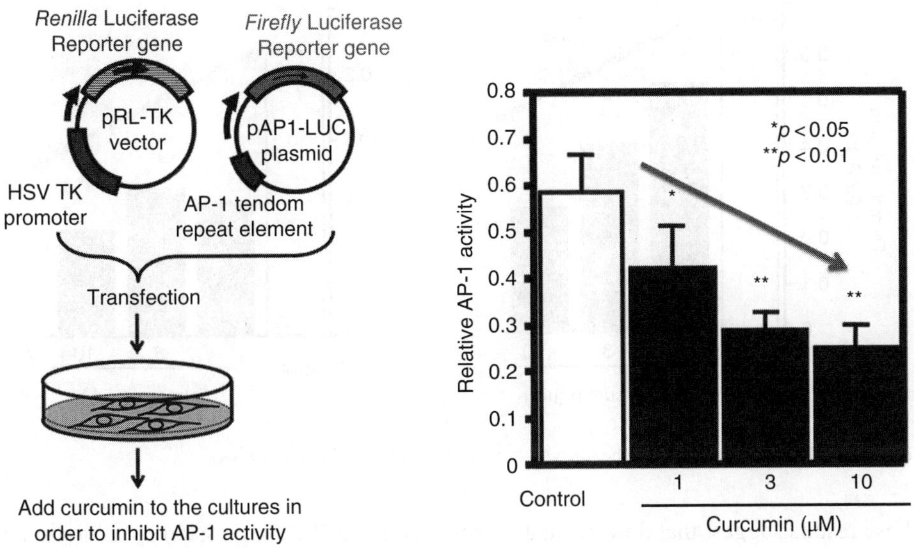

Fig. 13.10 Dual reporter assay for evaluation of anti-AP-1 activity of LLC cells treated with curcumin. (*See insert for color representation of the figure*)

13.3.5 Anti-AP-1 transcriptional activity of curcumin in LLC cells

Curcumin has been reported to inhibit the binding of AP-1 to its consensus DNA in some cell lines (Huang *et al.*, 1991; Park *et al.*, 1998). Therefore, we investigated the inhibitory effect of curcumin on the transcriptional activity of AP-1 using the dual luciferase assay system. Treatment of plasmid-transfected LLC cells with curcumin for 1 hour potently inhibited the AP-1 transcriptional activity in a concentration-dependent manner (Fig. 13.10). These results indicate that the anti-invasive activity of curcumin, resulting from anti-AP-1 transcriptional activity rather than direct cytotoxic activity, is mainly involved in the *in vivo* antimetastatic effect.

13.3.6 Effect of curcumin on the expression of mRNAs for u-PA and u-PAR in LLC

The gene expression of u-PA is considered to be transactivated by AP-1, because AP-1 binding motifs are located in the promoter regions of u-PA as well as u-PAR (De Cesare *et al.*, 1996; Lengyel *et al.*, 1996). Since reduction of u-PA activity in LLC cells by some inhibitors has been reported to inhibit the invasiveness (Tanaka *et al.*, 1981; Kobayashi *et al.*, 1994a, 1994b), we examined the effect of curcumin on the levels of the mRNAs for u-PA and u-PAR in LLC cells. RT-PCR showed that LLC cells expressed high mRNAs levels of u-PA and u-PAR. Treatment with curcumin for 3 hours resulted in a significant decrease in the mRNA for u-PA and u-PAR (Fig. 13.11). Thus, reduction of the mRNAs of u-PA and u-PAR by curcumin may be involved in the inhibition of the invasive ability of LLC cells. Although it has been reported that uPA significantly enhances tumor growth, curcumin may not affect the uPA-mediated enhancement of tumor cell proliferation.

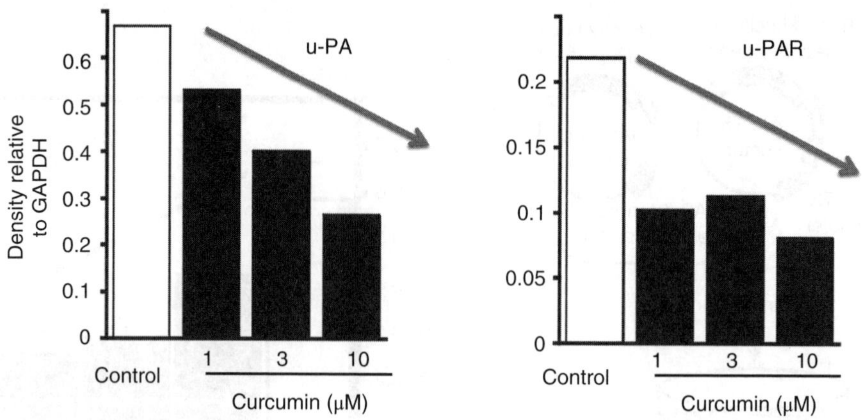

Fig. 13.11 Effect of curcumin on the mRNAs expression levels of u-PA and u-PAR.

These results suggest that downregulation of u-PA and u-PAR expression by inhibition of AP-1 transcriptional activity reduced the spontaneous lymphatic metastasis.

In fact, the c-Jun proto-oncogene product, which is a component of transcription factor AP-1, binds to cis-acting response elements (containing a $TGA^{G}/_{C}TCA$-like sequence) in the promoter region of some malignancy-related genes and regulates their transcriptional activation (Curran & Franza, 1988; Magrisso *et al.*, 1993; Volm *et al.*, 1993). Similarly, some metastasis-related genes, including the u-PA and u-PAR genes, have an AP-1 binding site in the promoter region (Lee *et al.*, 1994; De Cesare *et al.*, 1996). AP-1 is a major transcriptional enhancer for the transactivation of such genes. u-PA enhances the invasive ability of cancer cells through not only by converting plasminogen to plasmin but also by activating matrix metalloproteinases (MMPs) (Festuccia *et al.*, 1998) and epidermal growth factor (EGF) (Jarrad *et al.*, 1994). Activation of u-PA requires its binding with u-PAR. u-PAR is also regulated transcriptionally by AP-1. The u-PA-u-PAR complex may initiate a proteolytic cascade in which other proteases also influence cell migration and proliferation through multiple mechanisms. Moreover, it has been reported that exposure of cells to u-PA results in a significant increase in c-fos mRNA. The activation of the c-fos gene leads to an increase in AP-1 activity (Konakaova *et al.*, 1998).

On the other hand, AP-1 binding site exists in the promoter regions of some metastasis-related genes, such as some MMPs (Westermarck & Kahari, 1999), vascular endothelial growth factor (VEGF) (Garrido *et al.*, 1993), basic fibroblast growth factor (bFGF) (Shibata *et al.*, 1991), and cyclooxygenase-2 (COX-2) genes (Zhang *et al.*, 1999). There is a good correlation between VEGF or bFGF expression and angiogenesis of malignant tumors (Aonuma *et al.*, 1999). Moreover, recently, it has been reported that overexpression of COX-2 was well correlated with lymphatic metastasis of lung cancer (Hida *et al.*, 1998; Wolff *et al.*, 1998). Thus, the regulation of AP-1 is a candidate therapeutic modality for controlling the development of lymphatic metastasis of lung cancer.

In conclusion, a lymph node metastatic variant of LLC (LLC-MLN) cells had enhanced invasiveness and AP-1 transcriptional activity as compared with the original tumor. Curcumin inhibited the lymphatic metastasis of LLC cells *via* anti-invasive effects.

Repression of u-PA and u-PAR expression through downregulation of AP-1 activity was also associated with inhibition of tumor invasiveness. Combination of curcumin with the anticancer drug CDDP caused marked inhibition of tumor metastasis and prolonged the survival time of mice. Regulation of AP-1 activity may offer improved therapeutic efficacy for patients with lymphatic metastasis of lung cancer.

13.4 Effects of curcumin on tumor angiogenesis

13.4.1 Curcumin inhibits the formation of capillary-like tubes in rat lymphatic endothelial cells (TR-LE)

Berberine, an isoquinoline alkaloid and an active component of *Berberis aquifolium* (Oregon grape), *Berberis aristata* (tree turmeric), *Berberis vulgaris* (barberry), *Coptis chinensis* (coptis or goldthread), and *Hydrastis canadensis* (goldenseal), also has a wide range of pharmacological and biochemical effects including antimicrobial, antidiarrhea, anti-inflammatory, antiangiogenic, and anticancer activities. In our previous studies, the oral administration of curcumin and berberine inhibited spontaneous metastasis to mediastinal lymph nodes following the orthotopic implantation of LLC cells (Ichiki *et al.*, 2001; Mitani *et al.*, 2001). To evaluate the effect of curcumin and berberine on lymphangiogenesis, we performed an assay using TR-LE cells, in which the lengths of the capillary-like tubes formed on a murine tumor-derived matrix, Matrigel, were measured. Curcumin shortened the tubes in a dose-dependent manner with an IC50 of 3.0 µM, whereas 10 µM of berberine did not affect the formation of the tube-like network (Fig. 13.12). Curcumin had

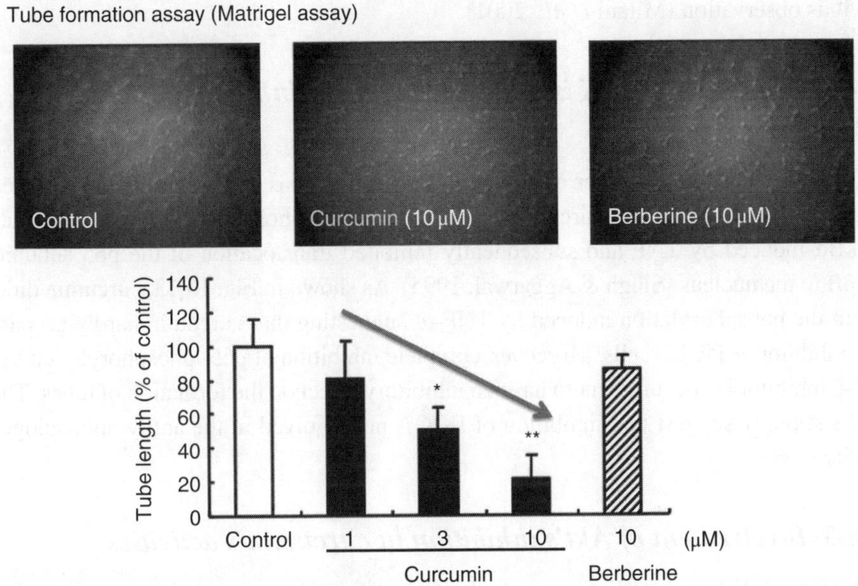

Fig. 13.12 Effect of curcumin on the tube-like formation lymphatic endothelial TR-LE cells. * $p<0.05$; ** $p<0.01$.

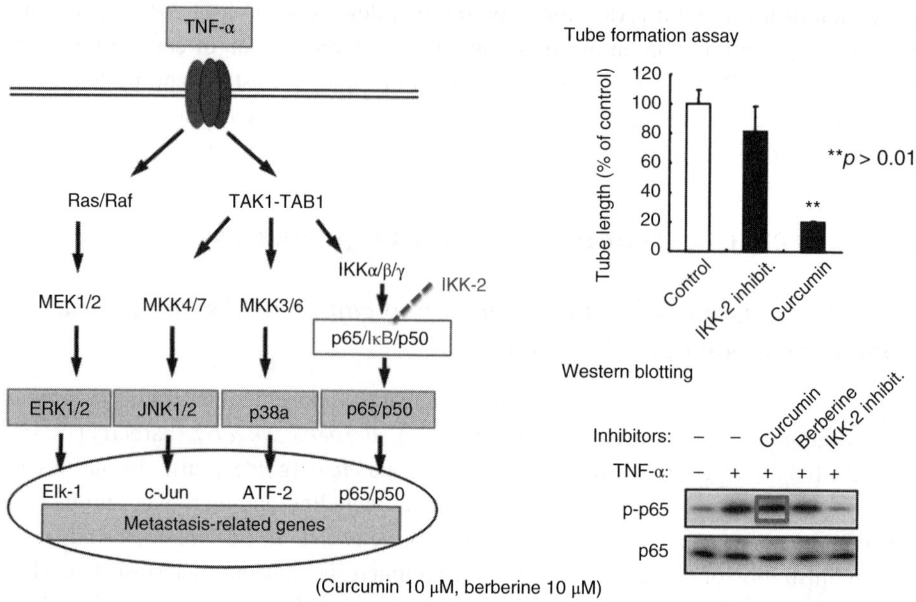

Fig. 13.13 Inhibition of IkB kinase is independent of antilymphangiogenic effect of curcumin.

no effect on cell adhesion to Matrigel, suggesting that it influences the postadhesion processes of migration, reorganization, and consolidation (Glesne *et al.*, 2006). In addition, berberine and curcumin did not act as cytotoxic agents. However, berberine, but not curcumin, exhibited a slight cytotoxic effect against LLC cells, which is consistent with our previous observation (Mitani *et al.*, 2001).

13.4.2 Inhibition of IKK is independent of the inhibitory effect of curcumin

NF-κB, which is a heterodimer of p65 and p50 subunits, has been identified as a target of curcumin. It is reported that curcumin abolished both the phosphorylation and degradation of IkBα induced by TNF and subsequently inhibited translocation of the p65 subunit of NF-βB to the nucleus (Singh & Aggarwal, 1995). As shown in Fig. 13.13, curcumin did not inhibit the phosphorylation induced by TNF-α, suggesting that curcumin hardly acts as an IKK inhibitor in TR-LE cells. Moreover, complete inhibition of p65- phosphorylation by an IKK-β inhibitor is not sufficient to have an inhibitory effect on the formation of tubes. These results strongly suggest that inhibition of IKK is not involved in the antilymphangiogenic activity of curcumin.

13.4.3 Involvement of Akt's inhibition in curcumin's activities

It is reported that curcumin suppresses the serine/threonine protein kinase Akt signaling pathway and subsequently exerts antiproliferative and apoptotic activities in leukemia cell lines (Hussain *et al.*, 2006; Tomita *et al.*, 2006). Akt is known to be an important component

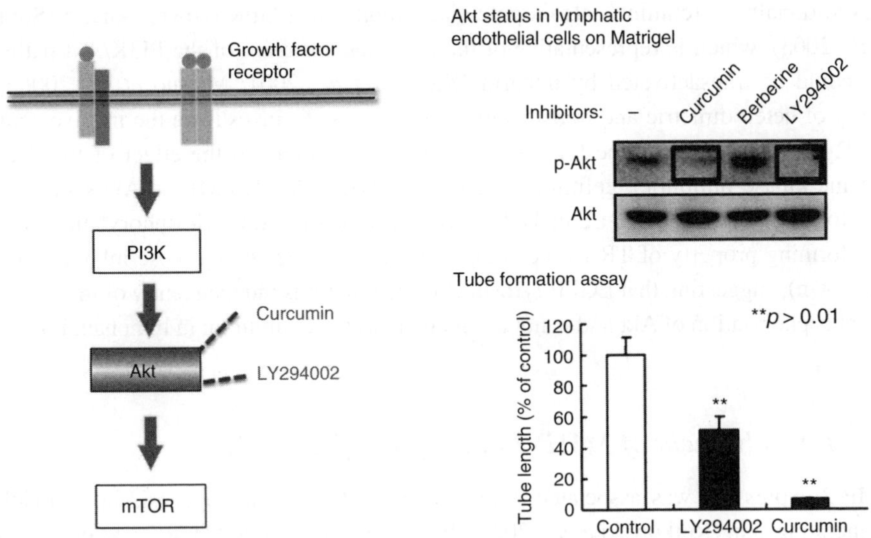

Fig. 13.14 Inhibitory effect of curcumin on Akt pathway of lymphatic endothelial cells.

of migratory and prosurvival signaling pathways and lies at a key junction in cellular signaling pathways leading to tumor angiogenesis (Shiojima & Walsh, 2002; Ma & Han, 2005). Recent investigations of different Akt isoforms have demonstrated that the amount of phosphorylated Akt is increased in tumor vessels and that Akt1, the predominant isoform in endothelial cells, regulates vascular maturation and permeability (Chen *et al.*, 2005). Akt phosphorylates endothelial NO synthase (eNOS) and thereby enhances endothelial NO synthesis and influences postnatal vessel growth (Ackah *et al.*, 2005; Ma & Han, 2005). However, little is known about the role of Akt in lymphangiogenesis.

Here, we investigated the effect of curcumin (10 μM) on Akt signaling by examining the tube-forming activity *in vitro* and the status of phosphorylated Akt in TR-LE cells 30 minutes after seeding onto Matrigel-coated dishes was assessed by Western blotting using anti-phospho-Akt (Ser473) antibody (Fig. 13.14). Phospho-Akt levels in TR-LE cells were reduced by treatment with curcumin (10 μM) but not berberine (10 μM). To determine whether inhibition of Akt plays a part in tube formation, we performed the tube-forming assay of TR-LE cells in the presence of LY294002 (20 μM). LY294002 significantly inhibited the tube formation (inhibitory rate = 48%), although to lesser extent than curcumin (Fig. 13.14). These results suggest that the Akt pathway is, at least in part, necessary for the tube-forming activity of TR-LE cells on Matrigel and might be a potential target for the antilymphangiogenic effect of curcumin. Here, we demonstrated, for the first time, that Akt plays an important role in the formation of capillary-like tubes by lymphatic endothelial cells. Thus, further dissection of the Akt pathway and elucidation of downstream effector molecules will lead to a better understanding of physiological and pathological lymphangiogenesis and may provide avenues for the development of a novel therapeutic strategy for cancer treatment.

Additionally, curcumin is also a potential inhibitor of EGFR tyrosine kinase (Squires *et al.*, 2003), which is representative of the enzymes upstream of the PI3K/Akt pathway and could be transactivated by integrin (Kalmes *et al.*, 2001; Matsuo *et al.*, 2006b), a family of heterodimetric and transmembrane receptors. To investigate the involvement of EGFR tyrosine kinase in the formation of tubes, we examined the effect of two EGFR tyrosine kinase inhibitors, gefitinib (1 μM) and PD153035 (10 μM), on Akt's status and tube-forming activity in TR-LE cells. Unexpectedly, neither Akt's phosphorylation nor the tube-forming property of TR-LE cells on Matrigel was affected by these inhibitors (data not shown), suggesting that EGFR tyrosine kinase functions independently of the constitutive phosphorylation of Akt and is not a principal target of curcumin in lymphangiogenesis *in vitro*.

13.4.4 Involvement of MMP-2 in lymphangiogenesis

MMP-2 expression was associated with the onset of lymphangiogenesis in a model of regenerating skin (Rutkowski *et al.*, 2006). In ovarian carcinoma, MMP-2 was closely correlated with VEGF-C expression, which was correlated with lymphangiogenesis and metastasis to the retroperitoneal lymph nodes (Ueda *et al.*, 2005). Moreover, elevated MMP-2 levels are associated with lymph node metastasis in gastric, oral, colorectal, and head and neck cancer (Langenskiold *et al.*, 2005; Patel *et al.*, 2005; Wiegand *et al.*, 2005; Wu *et al.*, 2006). It is also reported that the formation of tube-like structures by vascular endothelial cells *in vitro* depends on MMP-1, MMP-9, MT1-MMP, and MT3-MMP (Jeong *et al.*, 1999; Guenzi *et al.*, 2003; Jadhav *et al.*, 2004; Plaisier *et al.*, 2004). Given that curcumin exerts broad-spectrum inhibition of the gene expression of MMPs including MMP-1, -2, -3, -9, and -14, MT1-MMP (Shao *et al.*, 2002; Banerji *et al.*, 2004; Kim *et al.*, 2005), we cannot rule out the possibility of the involvement of these MMPs in the inhibitory effect of curcumin on the tube-forming activity of lymphatic endothelial cells and subsequent metastasis to lymph nodes. Thus, further investigations need to identify the subtypes of MMPs as pharmacological targets of curcumin.

As MMP-2/gelatinase A is vital during the formation of tubular networks of vascular endothelial cells (Nguyen *et al.*, 2001), we attempted to determine whether the inhibitory effect of curcumin could be mediated by MMP-2's activity. To investigate the influence of MMP, we examined the effect of an MMP inhibitor, GM6001, on the formation of tubes by TR-LE cells (Fig. 13.15). Treatment with GM6001 caused a 51.7% reduction in tube-forming activity. To assess the effect of curcumin on MMP-2's enzymatic activity, we performed gelatin zymography (Fig. 13.15). The degradation of a gelatin substrate by MMP-2 produced by TR-LE cells was completely inhibited by curcumin, but not by berberine.

Berberine did not affect the viability or tube-forming activity of lymphatic endothelial cells. Our previous study revealed that berberine reduces the invasiveness of LLC cells, accompanied by an inhibition of activator protein-1 (AP-1) and repression of urokinase-type plasminogen activator (u-PA) (Mitani *et al.*, 2001). It is likely that berberine is more selective to target cancer cells than noncancerous cells and that its inhibitory effect on lymph node metastasis greatly depends on inhibition of the metastatic properties of LLC cells, but not inhibition of lymphangiogenesis. Of interest, we found that berberine

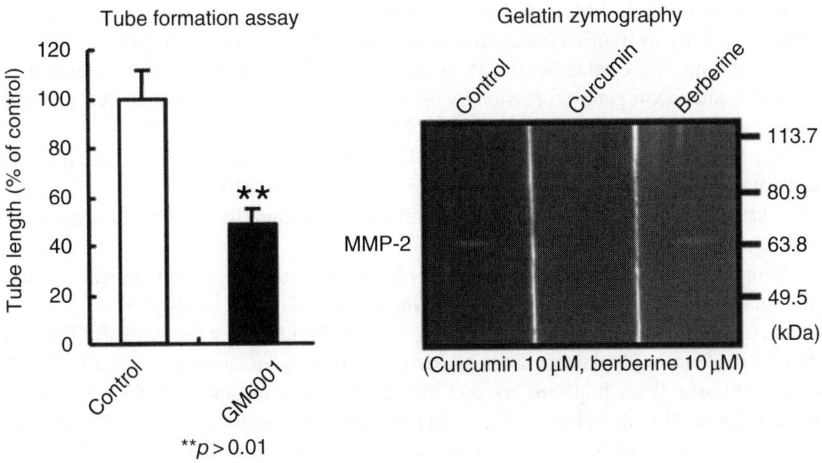

Fig. 13.15 Inhibitory effect of curcumin on MMP-2 activity of lymphatic endothelial cells.

significantly inhibits the formation of tubes by vascular endothelial TR-BE cells, derived from the inferior vena cava of a transgenic rat harboring a temperature-sensitive SV40 large T-antigen (data not shown). Future studies should focus on the difference in its activity and reveal a novel discrimination between vascular blood and lymphatic endothelial cells.

Altogether, here we showed curcumin as a potent antimetastatic agent, which can suppress tumor metastasis through the inhibition of several metastatic properties, including AP-1-dependent tumor cell invasion and migration and MMP-mediated inhibition of the lymphangiogenesis.

13.5 Conclusions

We demonstrated that curcumin suppressed intrahepatic metastasis of murine HCC and lymph node metastasis of lung cancer by inhibiting the invasion through preventing the production of some enzymes including MMP-9. Regression of u-PA and u-PAR expression through downregulation of AP-1 activity was also associated with inhibition of tumor invasiveness. The inhibitory effect of curcumin on the adhesion and migration of tumor cells might be mediated by functional alteration in cytoskeleton organization. Therefore, curcumin may provide an improved therapeutic benefit to tumor invasion and metastasis.

References

Ackah, E., Yu, J., Zoellner, S. *et al.* (2005) Akt1/protein kinase B alpha is critical for ischemic and VEGF-mediated angiogenesis. *Journal of Clinical Investigation*, **115**, 2119–2127.

Aonuma, M., Saeki, Y., Akimoto, T. *et al.* (1999) Vascular endothelial growth factor overproduced by tumor cells acts predominantly as a potent angiogenic factor contributing to malignant progression. *International Journal of Experimental Pathology*, **5**, 271–281.

Banerji, A., Chakrabarti, J., Mitra, A. and Chatterjee, A. (2004) Effect of curcumin on gelatinase A (MMP-2) activity in B16F10 melanoma cells. *Cancer Letters*, **211**, 235–242.

Bierhaus, A., Zhang, Y., Quehenberger, P. *et al.* (1997) The dietary pigment curcumin reduces endothelial tissue factor gene expression by inhibiting binding of AP-1 to the DNA and activation of NF-kappa B. *Thrombosis and Haemostasis*, **77**, 772–782.

Carter, W.G., Ryan, M.C. and Gahr, P.J. (1991) Epiligrin, a new cell adhesion ligand for integrin alpha 3 beta 1 in epithelial basement membranes. *Cell*, **17**, 599–610.

Chauhan, D.P. (2002) Chemotherapeutic potential of curcumin for colorectal cancer. *Current Pharmaceutical Design*, **8**, 1695–1706.

Chen, J., Somanath, P.R., Razorenova, O. *et al.* (2005) Akt1 regulates pathological angiogenesis, vascular maturation and permeability *in vivo*. *Nature Medicine*, **11**, 1188–1196.

Crowe, D.T., Chiu, H., Fong, S. and Weissman, I.L. (1994) Regulation of the avidity of integrin alpha 4 beta 7 by the beta 7 cytoplasmic domain. *Journal of Biological Chemistry*, **20**, 14411–14418.

Curran, T. and Franza, B.R., Jr (1988) Fos and Jun: the AP-1 connection. *Cell*, **55**, 395–397.

De Cesare, D., Palazzolo, M. and Blasi, F. (1996) Functional characterization of COM, a DNA region required for cooperation between AP-1 sites in urokinase gene transcription. *Oncogene*, **13**, 2551–2562.

De Meirsman, C., Schollen, E., Jaspers, M. *et al.* (1994) Cloning and characterization of the promoter region of the murine alpha-4 integrin subunit. *DNA and Cell Biology*, **13**, 743–754.

Doki, Y., Murakami, K., Yamaura, T., Sugiyama, S., Misaki, T. and Saiki, I. (1999) Mediastinal lymph node metastasis model by orthotopic intrapulmonary implantation of Lewis lung carcinoma cells in mice. *British Journal of Cancer*, **79**, 1121–1126.

Festuccia, C., Dolo, V., Guerra, F. *et al.* (1998) Plasminogen activator system modulates invasive capacity and proliferation in prostatic tumor cells. *Clinical and Experimental Metastasis*, **16**, 513–528.

Fidler, I.J. (1973) Selection of successive tumor lines for metastasis. *Nature*, **242**, 148–149.

Fu, X., Besterman, J.M., Monosov, A. and Hoffman, R.M. (1991) Models of human metastatic colon cancer in nude mice orthotopically constructed by using histologically intact patient specimens. *Proceedings of the National Academy of Sciences*, **88**, 9345–9349.

Garrido, C., Saule, S. and Gospodarowicz, D. (1993) Transcriptional regulation of vascular endothelial growth factor gene expression in ovarian bovine granulose cells. *Growth Factors*, **8**, 109–117.

Genda, T., Sakamoto, M., Ichida, T. *et al.* (1999) Cell motility mediated by Rho and Rho-associated protein kinase plays a critical role in intra-hepatic metastasis of human hepatocellular carcinoma. *Hepatology*, **30**, 1027–1036.

Ginsberg, R.J., Vokes, E.E. and Raben, A. (1997) Cancer of the lung, in *Cancer: Principles and Practice of Oncology*, vol. **1**, 5th edn (eds V.T. Devita, S. Helman, Jr and S.A. Rosenberg), Lippincott-Raven, New York, pp. 849–949.

Glesne, D.A., Zhang, W., Mandava, S. *et al.* (2006) Subtractive transcriptomics: establishing polarity drives *in vitro* human endothelial morphogenesis. *Cancer Research*, **66**, 4030–4040.

Govindaraja, V.S. (1980) Turmeric: chemistry, technology, and quality. *Critical Reviews in Food Science and Nutrition*, **12**, 199–301.

Guenzi, E., Topolt, K., Lubeseder-Martellato, C. *et al.* (2003) The guanylate binding protein-1 GTPase controls the invasive and angiogenic capability of endothelial cells through inhibition of MMP-1 expression. *EMBO Journal*, **22**, 3772–3782.

Hida, T., Yatabe, Y., Achikawa, H. *et al.* (1998) Increased expression of cyclooxygenase 2 occurs frequently in human lung cancers, specifically in adenocarcinomas. *Cancer Research*, **58**, 3761–3764.

Hoffman, R.M. (1991) Three-dimensional histoculture: origins and applications in cancer research. *Cancer Cells*, **3**, 86–92.

Howard, R.B., Chu, H., Zeligman, B.E. *et al.* (1991) Irradiated nude rat model for orthotopic human lung cancers. *Cancer Research*, **51**, 3274–3280.

Huang, T.S., Lee, S.C. and Lin, J.K. (1991) Suppression of c-jun/AP-1 activation by an inhibitor of tumor promotion in mouse fibroblast cells. *Proceedings of the National Academy of Sciences of the United States of America*, **88**, 5292–5296.

Hussain, A.R., Al-Rasheed, M., Manogaran, P.S. *et al.* (2006) Curcumin induces apoptosis via inhibition of PI3'-kinase/AKT pathway in acute T cell leukemias. *Apoptosis*, **11**, 245–254.

Ichiki, K., Mitani, N., Doki, Y., Hara, H., Misaki, T. and Saiki, I. (2001) Regulation of activator protein-1 activity in the mediastinal lymph node metastasis of lung cancer. *Clinical and Experimental Metastasis*, **18**, 539–545.

Jadhav, U., Chigurupati, S., Lakka, S.S. and Mohanam, S. (2004) Inhibition of matrix metalloproteinase-9 reduces *in vitro* invasion and angiogenesis in human microvascular endothelial cells. *International Journal of Oncology*, **25**, 1407–1414.

Jarrad, D.F., Blitz, B.F., Smith, R.C., Patai, B.L. and Rukstalis, D.B. (1994) Effect of EGF on prostate cancer cell line PC3 growth and invasion. *Prostate*, **24**, 46–53.

Jefferson, M.F., Pendleton, N., Faragher, E.B., Dixon, G.R., Myskow, M.W. and Horan, M.A. (1996) 'Tumor volume' as a predictor of survival after resection of non-small-cell lung cancer (NSCLC). *British Journal of Cancer*, **74**, 456–459.

Jeong, J.W., Cha, H.J., Yu, D.Y., Seiki, M. and Kim, K.W. (1999) Induction of membrane-type matrix metalloproteinase-1 stimulates angiogenic activities of bovine aortic endothelial cells. *Angiogenesis*, **3**, 167–174.

Kalmes, A., Daum, G. and Clowes, A.W. (2001) EGFR transactivation in the regulation of SMC function. *Annals of the New York Academy of Sciences*, **947**, 42–54.

Kim, S.Y., Jung, S.H. and Kim, H.S. (2005) Curcumin is a potent broad spectrum inhibitor of matrix metalloproteinase gene expression in human astroglioma cells. *Biochemical and Biophysical Research Communications*, **337**, 510–516.

Kobayashi, H., Gotoh, J., Fujie, M., Shinohara, H., Moniwa, N. and Terao, T. (1994a) Inhibition of metastasis of Lewis lung carcinoma by a synthetic peptide within growth factor-like like domain of urokinase in the experimental and spontaneous metastasis model. *International Journal of Cancer*, **57**, 727–733.

Kobayashi, H., Gotoh, J., Shinohara, H., Moniwa, N. and Terao, T. (1994b) Inhibition of the metastasis of Lewis lung carcinoma by antibody against urokinase-type plasminogen activator in the experimental and spontaneous metastasis model. *Thrombosis and Haemostasis*, **71**, 474–480.

Konakaova, M., Hucho, F. and Schleuning, W.D. (1998) Downstream targets of urokinase-type plasminogen-activator-mediated signal transduction. *European Journal of Biochemistry*, **253**, 421–429.

Kunicki, T.J. (1989) Platelet membrane glycoproteins and their function: an overview. *Blut*, **59**, 30–34.

Kuriyama, S., Yamazaki, M., Mitoro, A. *et al.* (1999) Hepatocellular carcinoma in an orthotopic mouse model metastasizes intrahepatically in cirrhotic but not in normal liver. *International Journal of Cancer*, **80**, 471–476.

Langenskiold, M., Holmdahl, L., Falk, P. and Ivarsson, M.L. (2005) Increased plasma MMP-2 protein expression in lymph node-positive patients with colorectal cancer. *International Journal of Colorectal Disease*, **20**, 245–252.

Lee, J.S., Favre, B., Hemming, B.A., Kiefer, B. and Nagamine, Y. (1994) Okadaic acid-dependent induction of the urokinase-type plasminogen activator gene associated with stabilization and autoregulation of c-jun. *Journal of Biological Chemistry*, **269**, 2887–2894.

Lengyel, E., Wang, H., Stepp, E. *et al.* (1996) Requirement of an upstream AP-1 motif for the constitutive and phorbol ester-inducible expression of the urokinase-type plasminogen activator receptor gene. *Journal of Biological Chemistry*, **271**, 23176–23184.

Lin, J.K., Chen, Y.C., Huang, Y.T. and Lin-Shiau, S.Y. (1997) Suppression of protein kinase C and nuclear oncogene expression as possible molecular mechanisms of cancer chemoprevention by apigenin and curcumin. *Journal of Cellular Biochemistry*, Suppl **28**, 39–48.

Livingston, R.B. (1997) Combined modality therapy of lung cancer. *Clinical Cancer Research*, **3**, 2638–2647.

Ma, F.X. and Han, Z.C. (2005) Akt signaling and its role in postnatal neovascularization. *Histology and Histopathology*, **20**, 275–281.

Mackay, D.J. and Hall, A. (1998) Rho GTPases. *Journal of Biological Chemistry*, **273**, 20685–20688.

Magrisso, I.J., Richmond, R.E., Carter, J.H., Pross, C.B., Gilfillen, R.A. and Carter, H.W. (1993) Immunohistochemical detection of RAS, JUN, FOS, and p53 oncoprotein expression in human colorectal adenomas and carcinoma. *Laboratory Investigation*, **69**, 674–681.

Martin, K.H., Slack, J.K., Boerner, S.A., Martin, C.C. and Parsons, J.T. (2002) Integrin connections map: to infinity and beyond. *Science*, **296**, 1652–1653.

Matsuo, M., Koizumi, K., Yamada, S. *et al.* (2006a) Establishment and characterization of conditionally immortalized endothelial cell lines from the thoracic duct and inferior vena cava of tsA58/EGFP doubletransgenic rats. *Cell and Tissue Research*, **326**, 749–758.

Matsuo, M., Sakurai, H., Ueno, Y., Ohtani, O. and Saiki, I. (2006b) Activation of MEK/ERK and PI3K/Akt pathways by fibronectin requires integrin alphav-mediated ADAM activity in hepatocellular carcinoma: a novel functional target for gefitinib. *Cancer Science*, **97**, 155–162.

Mclemore, T.L., Liu, M.C., Blacker, P.C. *et al.* (1987) Novel intrapulmonary model for orthotopic propagation of human lung cancers in athymic nude mice. *Cancer Research*, **47**, 5132–5140.

Menon, L.G., Kuttan, R. and Kuttan, G. (1995) Inhibition of lung metastasis in mice induced by B16F10 melanoma cells by polyphenolic compounds. *Cancer Letters*, **16**, 221–225.

Menon, L.G., Kuttan, R. and Kuttan, G. (1999) Anti-metastatic activity of curcumin and catechin. *Cancer Letters*, **1**, 159–165.

Mitani, N., Murakami, K., Yamaura, T., Ikeda, T. and Saiki, I. (2001) Inhibitory effect of berberine on the mediastinal lymph node metastasis produced by orthotopic implantation of Lewis lung carcinoma. *Cancer Letters*, **165**, 35–42.

Morita, S., Sato, A., Hayakawa, H. *et al.* (1998) Cancer cells overexpress mRNA of urokinase-type plasminogen activator, its receptor and inhibitors in human non-small-cell lung cancer tissue: analysis by Northern blotting and in situ hybridization. *International Journal of Cancer*, **78**, 286–292.

Mould, A.P., Komoriya, A., Yamada, K.M. and Humphries, M.J. (1991) The CS5 peptide is a second site in the IIICS region of fibronectin recognized by the integrin alpha 4 beta 1. Inhibition of alpha 4 beta 1 function by RGD peptide homologues. *Journal of Biological Chemistry*, **266**, 3579–3585.

Mountain, C.F. (1997) Revisions in the international system for staging lung cancer. *Chest*, **111**, 1710–1717.

Munaut, C., Salonurmi, T., Kontussari, S. *et al.* (1999) Murine matrix metalloproteinase 9 gene. 5'-upstream region contains cis-acting elements for expression in osteoclasts and migrating keratinocytes in transgenic mice. *Journal of Biological Chemistry*, **274**, 5588–5596.

Murakami, K., Yamaura, T., Suda, K. *et al.* (1990) TAC-101 (4-[3,5-bis (trimethylsilyl) benzamido] benzoic acid) inhibits spontaneous mediastinal lymph node metastasis produced by orthotopic implantation of Lewis lung carcinoma. *Japanese Journal of Cancer Research*, **90**, 1254–1261.

Murata, J., Ayukawa, K., Ogasawara, M., Fujii, H. and Saiki, I. (1997) α-Melanocytes-stimulating hormone blocks invasion of reconstituted basement membrane (Matrigel) by murine B16 melanoma cells. *Invasion & Metastasis*, **17**, 82–93.

Nagao, T., Inoue, S., Goto, S. *et al.* (1987) Hepatic resection for hepatocellular carcinoma. *Annals of Surgery*, **205**, 33–40.

Nagayama, M., Sato, A., Hayakawa, H., Urano, T., Takada, Y. and Takada, A. (1994) Plasminogen activator and their inhibitors in non-small cell lung cancer. *Cancer*, **73**, 1398–1405.

Nakashima, T. (1976) Vascular changes and hemodynamics in hepatocellular carcinoma, in *Hepatocellular Carcinoma* (eds K. Okuda and R.L. Peters), John Wiley & Sons, Inc., New York, pp. 169–203.

Naruke, T., Goya, T., Tsuchiya, R. and Suemasu, K. (1988) The importance of surgery to non-small-cell carcinoma of lung with mediastinal lymph node metastasis. *Annals of Thoracic Surgery*, **46**, 603–610.

Newton, A.C. (1995) Protein kinase C: structure, function, and regulation. *Journal of Biological Chemistry*, **270**, 28495–28498.

Nguyen, M., Arkell, J. and Jackson, C.J. (2001) Human endothelial gelatinases and angiogenesis. *International Journal of Biochemistry and Cell Biology*, **33**, 960–970.

Oka, T., Ishid, T., Nishino, T. and Sugimachi, K. (1991) Immunohistochemical evidence of urokinase-type plasminogen activator in primary and metastatic tumors of pulmonary adenocarcinoma. *Cancer Research*, **51**, 3522–3525.

Osada, T., Sakamoto, M., Ino, Y. *et al.* (1996) E-cadherin is involved in the intra-hepatic metastasis of hepatocellular carcinoma. *Hepatology*, **24**, 1460–1467.

Park, S., Lee, D.K. and Yang, C.H. (1998) Inhibition of fos-jun-DNA complex formation by dihydroguaiaretic acid and *in vitro* cytotoxic effects on cancer cells. *Cancer Letters*, **127**, 23–28.

Patel, B.P., Shah, P.M., Rawal, U.M. *et al.* (2005) Activation of MMP-2 and MMP-9 in patients with oral squamous cell carcinoma. *Journal of Surgical Oncology*, **90**, 81–88.

Pepper, M.S. (2001) Lymphangiogenesis and tumor metastasis: myth or reality? *Clinical Cancer Research*, **7**, 462–468.

Plaisier, M., Kapiteijn, K., Koolwijk, P. *et al.* (2004) Involvement of membrane-type matrix metalloproteinases (MT-MMPs) in capillary tube formation by human endometrial microvascular endothelial cells: role of MT3-MMP. *Journal of Clinical Endocrinology and Metabolism*, **89**, 5828–5836.

Rigas, J.R. (1998) Do newer chemotherapeutic agents improve survival in non-small cell lung cancer? *Seminars in Oncology*, **25**, 5–9.

Rosenberg, G.A. (1995) Matrix metalloproteases in brain injury. *Journal of Neurotrauma*, **12**, 833–842.

Rosmarin, A.G., Levy, R. and Tenen, D.G. (1992) Cloning and analysis of the CD18 promoter. *Blood*, **15**, 2598–2604.

Ruoslahti, E. and Pierschbacher, M.D. (1987) New perspectives in cell adhesion: RGD and integrins. *Science*, **23**, 491–497.

Rutkowski, J.M., Boardman, K.C. and Swartz, M.A. (2006) Characterization of lymphangiogenesis in a model of adult skin regeneration. *American Journal of Physiology—Heart and Circulatory Physiology*, **291**, H1402–H1410.

Saharinen, P., Tammela, T., Karkkainen, M.J. and Alitalo, K. (2004) Lymphatic vasculature: development, molecular regulation and role in tumor metastasis and inflammation. *Trends in Immunology*, **25**, 387–395.

Sawada, S., Murakami, K., Yamaura, T. *et al.* (2001) Intra-hepatic metastasis by orthotopic implantation of a fragment of murine hepatocellular carcinoma and its related molecules. *Tumor Biology*, **22**, 154–161.

Sawada, S., Murakami, K., Yamaura, T., Mitani, N., Tsukada, K. and Saiki, I. (2002) Therapeutic and analysis model of intra-hepatic metastasis reflects clinical behavior of hepatocellular carcinoma. *Japanese Journal of Cancer Research*, **93**, 190–197.

Shao, Z.M., Shen, Z.Z., Liu, C.C. *et al.* (2002) Curcumin exerts multiple suppressive effects on human breast carcinoma cells. *International Journal of Cancer*, **98**, 234–240.

Shibata, F., Baird, A. and Florkiewicz, R.Z. (1991) Functional characterization of the human basic fibroblast growth factor gene promoter. *Growth Factors*, **4**, 277–287.

Shiojima, I. and Walsh, K. (2002) Role of Akt signaling in vascular homeostasis and angiogenesis. *Circulation Research*, **90**, 1243–1250.

Singh, S. and Aggarwal, B.B. (1995) Activation of transcription factor NF-kappa B is suppressed by curcumin (diferuloylmethane). *Journal of Biological Chemistry*, **270**, 24995–25000.

Sonnenberg, A., Modderman, P.W. and Hogervorst, F. (1988) Laminin receptor on platelets is the integrin VLA-6. *Nature*, **1**, 487–489.

Squires, M.S., Hudson, E.A., Howells, L. *et al.* (2003) Relevance of mitogen activated protein kinase (MAPK) and phosphotidylinositol-3-kinase/protein kinase B (PI3K/PKB) pathways to induction of apoptosis by curcumin in breast cells. *Biochemical Pharmacology*, **65**, 361–376.

Srivastava, R. and Srimal, R.C. (1985) Modification of certain inflammation induced biochemical changes by curcumin. *The Indian Journal of Medical Research*, **81**, 215–223.

Stacker, S.A., Achen, M.G., Jussila, L., Baldwin, M.E. and Alitalo, K. (2002) Lymphangiogenesis and cancer metastasis. *Nature Reviews Cancer*, **2**, 573–583.

Takamura, M., Sakamoto, M., Genda, T., Ichida, T., Asakura, H. and Hirohashi, S. (2001) Inhibition of intra-hepatic metastasis of human hepatocellular carcinoma by Rho-associated protein kinase inhibitor Y-27632. *Hepatology*, **33**, 577–581.

Tanaka, N., Ogawa, H., Tanaka, K., Kinjo, M. and Kohga, S. (1981) Effect of tranexamic acid and urokinase on hematogenous metastasis of Lewis lung carcinoma in mice. *Invasion & Metastasis*, **1**, 149–157.

Terada, M. (1998) National Cancer Center Research Institute: history, current status, and priorities. *Nature Medicine*, **4**, 1221–1224.

Tomita, M., Matsuda, T., Kawakami, H. *et al.* (2006) Curcumin targets Akt cell survival signaling pathway in HTLV-I-infected T-cell lines. *Cancer Science*, **97**, 322–327.

Travis, W.D., Travis, L.B. and Devesa, S.S. (1995) Lung cancer. *Cancer*, **75**, 191–202.

Tsuchiya, Y., Sawada, S., Tsukada, K. and Saiki, I. (2002) A new pseudo-peptide of Arg-Gly-Asp (RGD) inhibits intra-hepatic metastasis of orthotopically implanted murine hepatocellular carcinoma. *International Journal of Oncology*, **20**, 319–324.

Ueda, M., Hung, Y.C., Terai, Y. *et al.* (2005) Vascular endothelial growth factor-C expression and invasive phenotype in ovarian carcinomas. *Clinical Cancer Research*, **11**, 3225–3232.

Vansteenkiste, J.F., De Leyn, P.R., Deneffe, G.J. *et al.* (1997) Survival and prognostic factors in resected N2 non-small cell lung cancer: a study of 140 cases. Leuven lung cancer group. *Annals of Thoracic Surgery*, **63**, 1441–1450.

Volm, M., Drings, P. and Wodrich, W. (1993) Prognostic significance of the expression of c-fos, c-jun and c-erbB-1 oncogene products in human squamous cell lung carcinomas. *Journal of Cancer Research and Clinical Oncology*, **119**, 507–510.

Volm, M., van Kaick, G. and Mattern, J. (1994) Analysis of c-fos, c-jun, c-erbB1, c-erbB2 and c-myc in primary lung carcinomas and their lymph node metastases. *Clinical and Experimental Metastasis*, **4**, 329–334.

Westermarck, J. and Kahari, V.M. (1999) Regulation of matrix metalloproteinase in tumor invasion. *FASEB Journal*, **13**, 781–792.

Wiegand, S., Dunne, A.A., Muller, H.H. *et al.* (2005) Metaanalysis of the significance of matrix metalloproteinases for lymph node disease in patients with head and neck squamous cell carcinoma. *Cancer*, **104**, 94–100.

Wolff, H., Saukkonen, K., Anttila, S. *et al.* (1998) Expression of cyclooxygenase-2 in human lung carcinoma. *Cancer Research*, **8**, 4997–5001.

Wu, Z.Y., Li, J.H., Zhan, W.H. and He, Y.L. (2006) Lymph node micrometastasis and its correlation with MMP-2 expression in gastric carcinoma. *World Journal of Gastroenterology*, **12**, 2941–2944.

Yano, S. and Sone, S. (1997) Novel metastasis model of human lung cancer cells representing different histological types in SCID mice depleted of NK cells. *Japan Journal of Cancer Chemistry*, **24**, 489–494.

Zhang, F., Altorki, N.K., Mestre, J.R., Subbaramaian, K. and Dannenberg, A.J. (1999) Curcumin inhibits cyclooxygenase-2 transcription in bile acid-and phorbol ester-treated human gastrointestinal epithelial cells. *Carcinogenesis*, **20**, 445–451.

Chapter 14

Phytochemical and Pharmacological Overview of *Cistanche* Species

Hai-Ning Lv, Ke-Wu Zeng, Yue-Lin Song, Yong Jiang and Peng-Fei Tu

Abstract: Cistanches Herba, the dried succulent stems of *Cistanche* species that are mainly distributed in the arid lands and warm deserts, has been widely used as a tonic for the treatments of kidney deficiency, impotence, female sterility, and constipation in traditional Chinese medical practices for centuries. *Cistanche* species are parasitic plants and their hosts are sand-fixing plants, such as *Haloxylon ammodendron* and *Tamarix* spp. In recent years, increasing attention has been paid to the *Cistanche* plants, since the wide cultivation of *Cistanche* plants and their hosts can not only supply the medicinal materials but also contribute greatly to the desert control. After decades of studies, diverse chemical classes, such as phenylethanoid glycosides, iridoids, lignans, and polysaccharides, have been disclosed to be the primary chemical constituents in *Cistanche* species, and antioxidation, neuroprotection, antiaging, vasorelaxation, hepatoprotection, and so on, act as their primary activities. Here, we made a review on the latest research progress regarding the chemical constituents and bioactivities of *Cistanche* plants.

Keywords: *Cistanche* species, review, chemical constituents, bioactivities, phenylethanoid glycosides

14.1 Introduction

The genus *Cistanche* Hoffmg. *et* Link (Orobanchaceae) consists of approximately 20 species all over the world, mainly distributed in arid lands and warm deserts in the Eurasia, for example, Iran, Afghanistan, Pakistan, Mongolia, and Northwestern China (Tan *et al.*, 2004). There are four species and one variety distributed in China, namely *Cistanche deserticola* Y.C. Ma, *C. tubulosa* (Schenk) R. Wight, *C. sinensis* G. Beck, *C. salsa* (C.A. Mey.) G. Beck, and *C. salsa* var. *albiflora* P.F. Tu *et* Z.C. Lou (Tu *et al.*, 1994a).

Recent Advances in Polyphenol Research, First Edition. Edited by Kumi Yoshida, Véronique Cheynier and Stéphane Quideau.

The dried succulent stems of *Cistanche* plants have been used as the traditional Chinese medicine (TCM) of Cistanches Herba (CH, "*Rou Cong Rong*" in Chinese), first recorded in Shen Nong's Chinese Materia Medica, for the treatments of kidney deficiency, impotence, female infertility, morbid leucorrhea, profuse metrorrhagia, and senile constipation (Jiangsu New Medical College, 1977; Chinese Academy of Medical Sciences Institute of Medicinal Plants, 1988). CH is a famous tonic and honored as "Desert Ginseng." It has been one of the most commonly used herbal drugs to treat kidney deficiency for centuries (Tu *et al.*, 1994a). Modern pharmacological investigations have revealed that CH could improve sexual function, prolong life span (antiaging), advance learning/memory ability (antidementia), and promote feces excretion (Chinese Pharmacopoeia Committee Editing, 2005).

Among the *Cistanche* plants, *C. deserticola* and *C. tubulosa* have been authenticated as the officinal plant sources of CH in Chinese Pharmacopoeia (since 2005 edition). However, the dried succulent stems of *C. salsa* and *C. sinensis* are also used as CH in Ningxia and Xinjiang autonomous regions of China due to the source shortage of CH (Tu *et al.*, 1997b; Chinese Pharmacopoeia Committee Editing, 2005). *Cistanche* species are perennial parasitic plants, attaching at the roots of sand-fixing plants, such as *Tamarix* species, *Haloxylon ammodendron*, and *Kalidium foliatum* (Luo *et al.*, 2002). In the past decades, the huge market demands have resulted in *Cistanche* plants on the edge of extinction. In particular, *C. deserticola* has been listed as Class II plants needing protection in China (Tu *et al.*, 1994a). In order to maintain sustainable development of CH, large-scale cultivations of *C. deserticola* and *C. tubulosa* have been carried out in the west of Inner Mongolia and the south of Xinjiang regions in China, which, to some extent, alleviates the source shortage and protects the wild resources of CH (Lei *et al.*, 2003a; Chen *et al.*, 2005).

Scientific research studies on *Cistanche* plants initiated in the 1980s (Karasawa & Komatsu, 2008), and the chemical investigations revealed that various compounds, including phenylethanoid glycosides (PhGs), iridoids, lignans, oligosaccharides, and polysaccharides, are the primary constituents of *Cistanche* plants (Lei *et al.*, 2003a; Chen *et al.*, 2005). In 2009, we reviewed the chemical analysis of this genus and the progress of the phytochemical investigations (Jiang & Tu, 2009). Recently, Wang and coworkers summarized the information on *C. deserticola* covering its botany, traditional usages, phytochemistry, and pharmacology (Wang *et al.*, 2012). Here, we update the latest progress concerning the phytochemical and bioactivity studies of *Cistanche* species.

14.2 Chemical constituents of *Cistanche* species

Cistanche species are rich sources of PhGs, iridoids, lignans, and polysaccharides, and PhGs are the characteristic principles of this genus. To date, 120 compounds have been isolated from *C. deserticola* (Cd), 75 compounds from Chinese *C. tubulosa* (Ct-C), 21 compounds from Pakistani *C. tubulosa* (Ct-P), 31 compounds from *C. salsa* (Csa), 11 compounds from *C. phelypaea* (Cp), and 20 compounds from *C. sinensis* (Csi). Table 14.1 shows their chemical types and distribution.

Table 14.1 Chemical types and their distribution in *Cistanche* species.

Species	Benzyl glycosides	PhGs	Iridoids	Lignans	Monoterpenoids	Others
Cd	4	41	12	14	1	48
Ct-C	—	33	21	3	5	13
Ct-P	—	11	2	5	1	2
Csa	3	10	3	—	3	12
Cp	—	5	4	—	—	2
Csi	—	12	6	1	—	1

Note: Cd, *C. deserticola*; Cp, *C. phelypaea*; Csa, *C. salsa*; Csi, *C. sinensis*; Ct-C, Chinese *C. tubulosa*; Ct-P, Pakistani *C. tubulosa*.

14.2.1 Phenylethanoid glycosides (PhGs)

PhGs have been regarded as the primary active components of *Cistanche* species. Until now, a total of 70 PhGs have been reported, including 4 monoglycosides, 41 diglycosides, and 25 triglycosides (Table 14.2). Those PhGs show typical structural properties: the monoglycosides only contain a glucosyl linked at the C-8 position of phenylethanol; the diglycosides generally contain a Rha $(1 \rightarrow 3)$ Glc linkage. Regarding the triglycosides, an additional glucosyl or rhamnosyl moiety usually links at C-6 position of the inside glucosyl. In general, there is a coumaroyl, caffeoyl, or feruloyl group linked at C-4 or C-6 position of the inside glucosyl.

14.2.2 Benzyl glycosides

Lei and coworkers isolated and identified three benzyl glycosides from *C. salsa* for the first time (Lei *et al.*, 2007). To date, six such compounds have been obtained from the genus *Cistanche*. In those benzyl glycosides, the glucosyl usually links directly with the benzyl alcohol aglycone, and a coumaroyl or caffeoyl group usually locates at C-4 or C-6 position of the glucosyl (Table 14.3).

14.2.3 Iridoids

Until now, 4 iridoid aglycones and 22 iridoid glycosides have been isolated from the *Cistanche* species (Tables 14.4 and 14.5). Their structural features can be summarized as follows: a glucosyl is generally linked at the C-1 position of aglycone, and the H-5 and H-9 are β-oriented; hydroxylation often occurs at C-8 and C-10 positions; sometimes, dehydration may occur between the hydroxyl of C-10 and that of C-1 or C-3 to form an epoxy substructure.

14.2.4 Monoterpenoids

Until now, there have been seven monoterpenoids purified from *Cistanche* plants, including two monoterpenes and five monoterpene glycosides (Table 14.6).

Table 14.2 PhGs from *Cistanche* species (**1–70**).

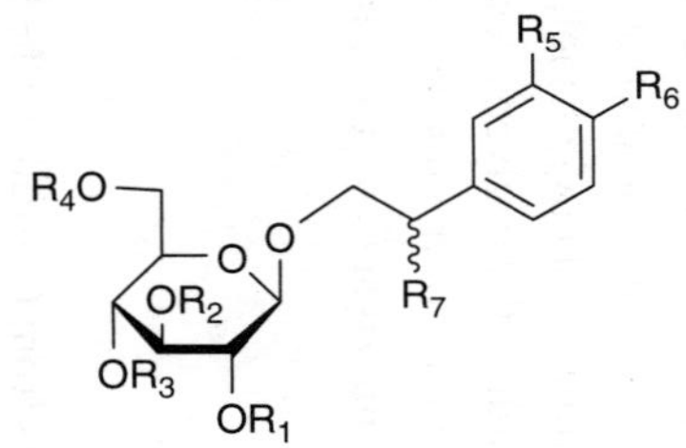

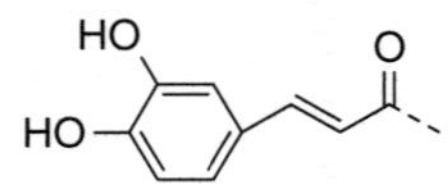

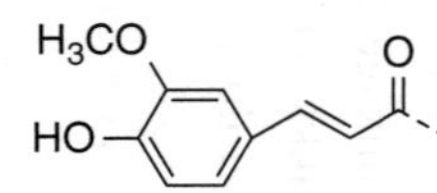

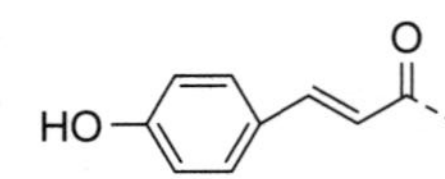

Cf: *trans*-caffeoyl | Fr: *trans*-feruloyl | Cm: *trans*-coumaroyl

c-Cf: *cis*-caffeoyl | *c*-Fr: *cis*-feruloyl | *c*-Cm: *cis*-coumaroyl

Glc: *β*-D-glucopyranosyl | Rha: *α*-L-rhamnopyranosyl | Ac: acetyl

TA-Rha: 2‴,3‴,4‴-triacetyl-α-L-rhamnopyranosyl

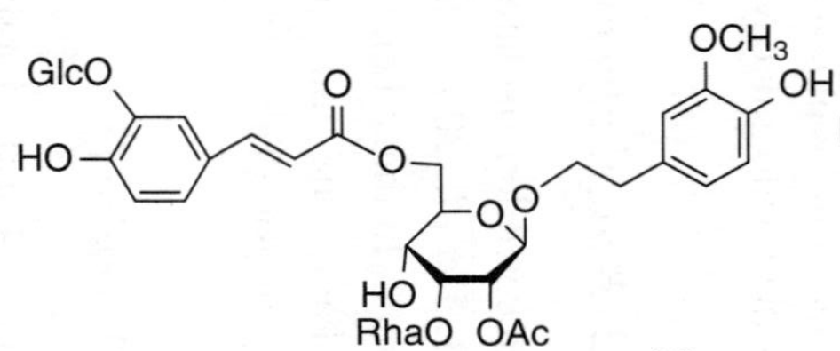

23

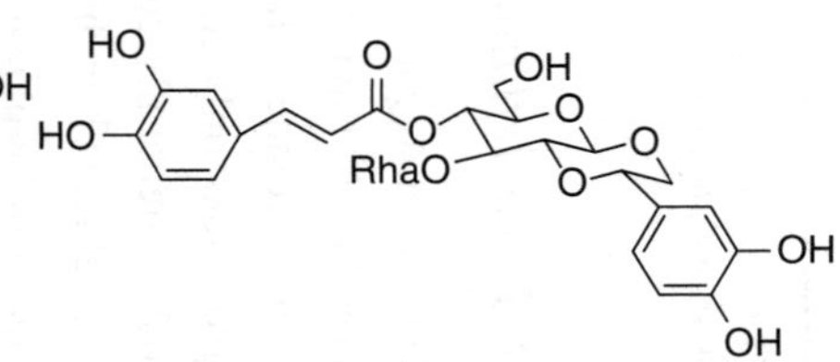

55

Compound name	R_1	R_2	R_3	R_4	R_5	R_6	R_7	Species	References
(**1**) 2′-Acetylacteoside	Ac	Rha	Cf	H	OH	OH	H	Cd, Cp, Ct-P, Ct-C, Csi, Csa	Kobayashi *et al.* (1984c, 1987), Du *et al.* (1993a), Deyama *et al.* (1995b), Tan *et al.* (2004), Luo *et al.* (2002), and Liu *et al.* (2013b)
(**2**) Tubuloside A	Ac	Rha	Cf	Glc	OH	OH	H	Cd, Cp, Ct-P, Ct-C	Kobayashi *et al.* (1987), Deyama *et al.* (1995a), Song *et al.* (2000), and Xiong *et al.* (1996)
(**3**) Tubuloside B	Ac	Rha	H	Cf	OH	OH	H	Cd, Ct-P, Csa, Ct-C, Csi	Kobayashi *et al.* (1987), Du *et al.* (1993a), Yang *et al.* (2005), and Morikawa *et al.* (2010b)
(**4**) Tubuloside C	Ac	Ta-Rha	Cf	Glc	OH	OH	H	Ct-P	Kobayashi *et al.* (1987)
(**5**) Tubuloside D	Ac	Ta-Rha	Cm	Glc	OH	OH	H	Ct-P	Kobayashi *et al.* (1987)
(**6**) Tubuloside E	Ac	Ta-Rha	Cm	H	OH	OH	H	Cd, Cp, Ct-P	Deyama *et al.* (1995a), Yoshizawa *et al.* (1990), and Liu *et al.* (2011)
(**7**) Cistansinenside A	Ac	Rha	Cf	H	OH	OMe	H	Csi	Tu *et al.* (2007)
(**8**) Salsaside D	Ac	Rha	Cf	H	H	OH	H	Cd, Csa	Liu *et al.* (2011) and Lei *et al.* (2007)
(**9**) Salsaside E	Ac	Rha	Cf	H	OMe	OH	H	Cd, Csa	Liu *et al.* (2011) and Lei *et al.* (2007)
(**10**) Salsaside F	Ac	Rha	H	Cm	OH	OH	H	Csa	Lei *et al.* (2007)
(**11**) Cistanoside H	Ac	Rha	H	H	OH	OH	H	Cd	Xu *et al.* (1994) and Karasawa *et al.* (1986a)
(**12**) Kankanoside H_1	Ac	Rha	Cm	Glc	OH	OH	H	Ct-C	Morikawa *et al.* (2010b)
(**13**) Kankanoside H_2	Ac	Rha	c-Cm	Glc	OH	OH	H	Ct-C	Morikawa *et al.* (2010b)
(**14,15**) Kankanosides J_1/J_2	Ac	Rha	Cf	H	OH	OH	OMe (*R/S*)	Ct-C	Pan *et al.* (2010)
(**16**) Cistansinenside B	Ac	Rha	Cf	Rha	OH	OMe	H	Csi	Liu *et al.* (2013b)
(**17**) 2′-*O*-Acetylpoliumoside	Ac	Rha	Cf	Rha	OH	OH	H	Csi	Liu *et al.* (2013b)
(**18**) *cis*-Tubuloside B	Ac	Rha	H	c-Cf	OH	OH	H	Cd	Nan *et al.* (2013a)
(**19**) Cistanoside J	Ac	Rha	H	Fr	OMe	OH	H	Cd	Nan *et al.* (2013b)
(**20**) *cis*-Cistanoside J	Ac	Rha	H	c-Fr	OMe	OH	H	Cd	Nan *et al.* (2013a)
(**21**) Cistanoside K	Ac	Rha	H	Cf	OMe	OH	H	Cd	Nan *et al.* (2013b)
(**22**) *cis*-Cistanoside K	Ac	Rha	H	c-Cf	OMe	OH	H	Cd	Nan *et al.* (2013a)
(**23**)Cistanoside N[a]								Cd	Nan *et al.* (2013b)
(**24**) Cistanoside L	H	Rha	H	Fr	OMe	OMe	H	Cd	Nan *et al.* (2013b)
(**25**) Cistanoside M	H	Rha	H	Cm	OMe	OH	H	Cd	Nan *et al.* (2013b)
(**26**) *cis*-Isocistanoside C	H	Rha	H	c-Cf	OMe	OH	H	Cd	Nan *et al.* (2013a)
(**27**) Acteoside	H	Rha	Cf	H	OH	OH	H	Cd, Cp, Ct-P, Ct-C, Csi, Csa	Kobayashi *et al.* (1984b, 1987), Du *et al.* (1993a), Deyama *et al.* (1995b), Song *et al.* (2000), Liu *et al.* (1993, 2013b), and Yang *et al.* (2009b)
(**28**) *cis*-Acteoside	H	Rha	c-Cf	H	OH	OH	H	Ct-C	Morikawa *et al.* (2010b)
(**29**) Cistanoside A	H	Rha	Cf	Glc	OMe	OH	H	Cd, Ct-C, Csa	Du *et al.* (1993a, 1993b), Yang *et al.* (2009b), and Kobayashi *et al.* (1984b)

(*Continued*)

Table 14.2 (Continued)

Compound name	R_1	R_2	R_3	R_4	R_5	R_6	R_7	Species	References
(**30**) Cistanoside B	H	Rha	Fr	Glc	OMe	OH	H	Cd	Xu *et al.* (1994) and Kobayashi *et al.* (1984b, 1986)
(**31**) Cistanoside C	H	Rha	Cf	H	OMe	OH	H	Cd	Kobayashi *et al.* (1984c, 1986) and Xu *et al.* (1994)
(**32**) Cistanoside D	H	Rha	Fr	H	OMe	OH	H	Cd	Kobayashi *et al.* (1984c, 1986)
(**33**) Cistanoside E	H	Rha	H	H	OMe	OH	H	Cd	Kobayashi *et al.* (1985a) and Sato *et al.* (1988)
(**34**) Cistanoside G	H	Rha	H	H	H	OH	H	Cd, Ct-C	Sato *et al.* (1988), Karasawa *et al.* (1986b), and Yoshikawa *et al.* (2006a)
(**35**) Decaffeoylacteoside	H	Rha	H	H	OH	OH	H	Cd, Ct-C	Song *et al.* (2000) and Karasawa *et al.* (1986a)
(**36**) Echinacoside	H	Rha	Cf	Glc	OH	OH	H	Cd, Cp, Ct-P, Ct-C, Csa, Csi	Kobayashi *et al.* (1987), Du *et al.* (1993a), Tan *et al.* (2004), Deyama *et al.* (1995b), Song *et al.* (2000), Liu *et al.* (2013b), and Yang *et al.* (2005)
(**37**) Isoacteoside	H	Rha	H	Cf	OH	OH	H	Cd, Ct-P, Ct-C, Csa, Csi	Kobayashi *et al.* (1987), Song *et al.* (2000), Liu *et al.* (2013b), Xiong *et al.* (1996), and Yang *et al.* (2005)
(**38**) Isosyringalide-3′-α-L-rhamnopyranoside	H	Rha	Cm	H	OH	OH	H	Cd, Ct-P	Yoshizawa *et al.* (1990) and Hayashi (2004)
(**39**) Osmanthuside B	H	Rha	Cm	H	H	OH	H	Cd	Kobayashi *et al.* (1984c) and Hayashi (2004)
(**40**) Salidroside	H	H	H	H	H	OH	H	Cd, Ct-C	Song *et al.* (2000), Morikawa *et al.* (2010b), Karasawa *et al.* (1986b), and Xu *et al.* (1999)
(**41**) Syringalide A-3′-α-L-rhamnopyranoside	H	Rha	Cf	H	H	OH	H	Cd, Ct-P, Ct-C	Song *et al.* (2000), Xiong *et al.* (1996), Morikawa *et al.* (2010b), and Yoshizawa *et al.* (1990)
(**42**) Cistantubuloside A	H	Rha	Cf	Glc	H	OH	H	Cd, Ct-C	Tu *et al.* (2002, 2006)
(**43**) Cistantubuloside B_1	H	Rha	Cm	Glc	OH	OH	H	Cd, Ct-C	Tu *et al.* (2002, 2006)
(**44**) Cistantubuloside B_2	H	Rha	c-Cm	Glc	OH	OH	H	Cd, Ct-C	Tu *et al.* (2002, 2006)
(**45**) Poliumoside	H	Rha	Cf	Rha	OH	OH	H	Csi	Liu *et al.* (2013b) and Tu *et al.* (2007)
(**46**) Jionoside D	H	Rha	Cf	H	OH	OMe	H	Csi	Tu *et al.* (2007)
(**47**) Kankanoside F	H	Rha	H	Glc	OH	OH	H	Ct-C	Yoshikawa *et al.* (2006a)
(**48**) Kankanoside G	H	Rha	H	Cf	H	OH	H	Ct-C	Yoshikawa *et al.* (2006a)
(**49**) Eutigoside A	H	H	H	Cm	H	OH	H	Csa	Yang *et al.* (2005)
(**50**) Kankanoside I	H	Rha	Cf	Glc	H	H	H	Ct-C	Morikawa *et al.* (2010b)
(**51,52**) Kankanosides K_1/K_2	H	Rha	Cf	Glc	OH	OH	OMe (*R/S*)	Ct-C	Pan *et al.* (2010)
(**53,54**) Cistantubuloside C_1/C_2	H	Rha	Cf	Glc	OH	OH	OH (*R/S*)	Ct-C	Tu *et al.* (2006)
(**55**) Crenatoside[a]								Ct-C	Song *et al.* (2000)
(**56**) Osmanthuside B6(*E*)	H	Rha	H	Cm	H	OH	H	Cd	Liu *et al.* (2011)
(**57**) Osmanthuside B6(*Z*)	H	Rha	H	c-Cm	H	OH	H	Cd	Liu *et al.* (2011)
(**58**) Plantainoside C	H	Rha	H	Fr	OH	OH	H	Cd	Liu *et al.* (2011)

(**59**) Isocampneoside I	H	Rha	H	Cf	OH	OH	OMe	Ct-C	Pan *et al.* (2010)
(**60,61**) Campneoside I	H	Rha	Cf	H	OH	OH	OMe (*R/S*)	Ct-C	Morikawa *et al.* (2010b) and Tu *et al.* (2006)
(**62,63**) Campneoside II	H	Rha	Cf	Rha	OH	OH	OH (*R/S*)	Csi, Ct-C	Liu *et al.* (2013b) and Morikawa *et al.* (2010b)
(**64**) Arenarioside	H	Rha	Cf	Xyl	OH	OH	H	Ct-C	Morikawa *et al.* (2010b)
(**65**) Wiedemanninoside C	H	Rha	Fr	Glc	OH	OH	H	Ct-C	Morikawa *et al.* (2010b)
(**66**) Isocistanoside C	H	Rha	H	Cf	OMe	OH	H	Cd	Nan *et al.* (2013b)
(**67**) Isomartynoside	H	Rha	H	Fr	OH	OMe	H	Cd	Nan *et al.* (2013a)
(**68**) Epimeridinoside A	H	Rha	H	Fr	OMe	OH	H	Cd	Nan *et al.* (2013b)
(**69**) 6′-Acetylsalidroside	H	H	H	Ac	H	OH	H	Cd	Nan *et al.* (2013b)
(**70**) Phenylethyl-glucopyranoside	H	H	H	H	H	H	H	Cd	Nan *et al.* (2013a)

[a] Structures are separately shown as earlier.

Table 14.3 Benzyl glycosides from *Cistanche* species (**71–76**).

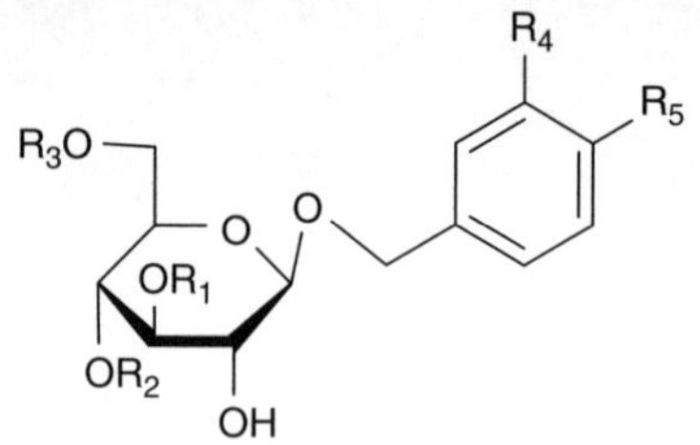

Compound name	R_1	R_2	R_3	R_4	R_5	Species	References
(**71**) Salsaside A	Rha	H	Cf	H	H	Csa	Lei *et al.* (2007)
(**72**) Salsaside B	Rha	Cf	H	H	H	Csa, Cd	Lei *et al.* (2007) and Nan *et al.* (2013a)
(**73**) Salsaside Ca/Cb	Rha	Cm	H	H	H	Csa	Lei *et al.* (2007)
(**74**) 3,4-Dimethoxybenzyl-β-D-glucoside	H	H	H	OMe	OMe	Cd	Nan *et al.* (2013a)
(**75**) 4-Hydroxybenzyl-β-D-glucoside	H	H	H	H	OH	Cd	Nan et al. (2013a)
(**76**) Benzyl-glucopyranoside	H	H	H	H	H	Cd	Nan et al. (2013a)

Table 14.4 Structures of iridoid aglycones (**77–87**).

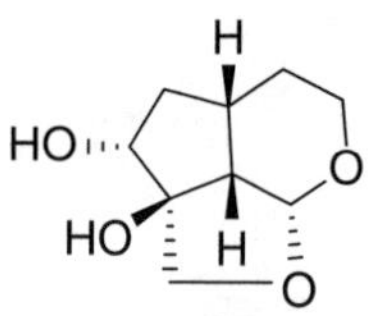

77

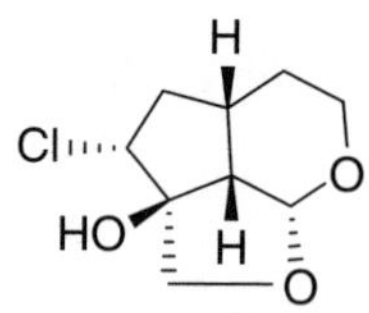

78

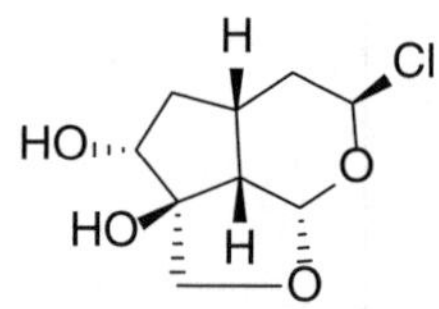

79

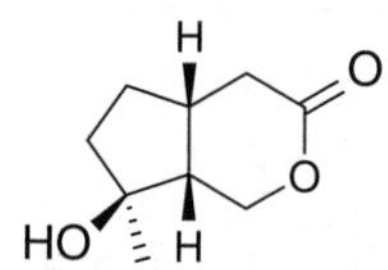

80

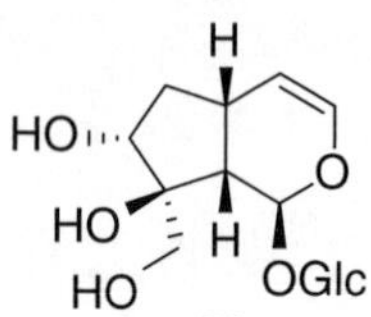

81

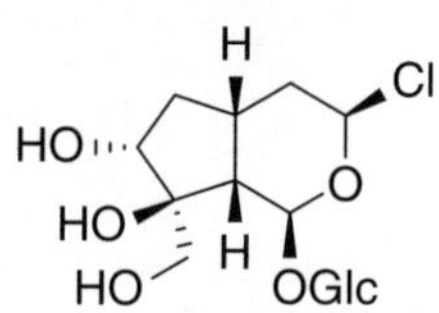

82

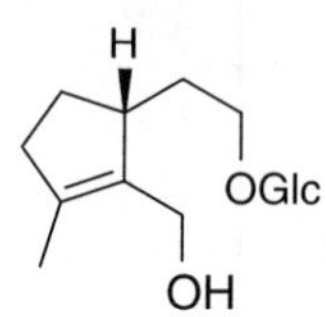

83

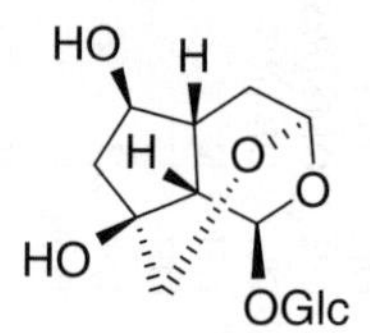

84

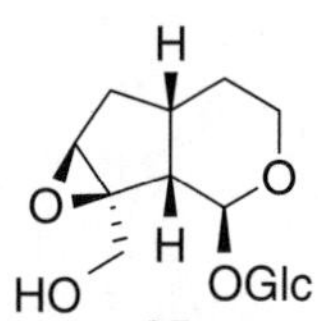

85

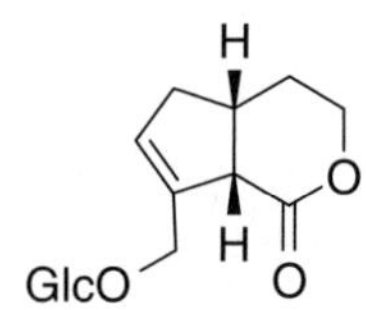

86

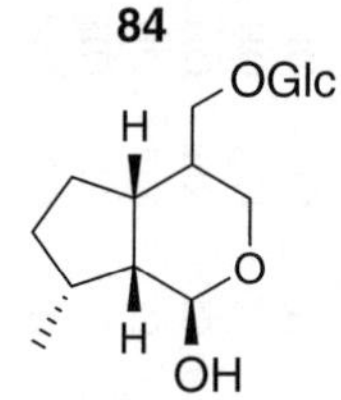

87

Compound name	Species	References
(**77**) Cistanin	Cd,Ct-C	Kobayashi *et al.* (1984a), Xie *et al.* (2006), and Xu *et al.* (1999)
(**78**) Cistachlorin	Cd,Ct-C	Kobayashi *et al.* (1984a) and Xie *et al.* (2006)
(**79**) Kankanol	Ct-C	Xie *et al.* (2006)
(**80**) Argyol	Ct-C	Xie *et al.* (2006)
(**81**) Kankanoside B	Ct-C	Xie *et al.* (2006)
(**82**) Kankanoside C	Ct-C	Xie *et al.* (2006)
(**83**) Kankanoside D	Ct-C	Xie *et al.* (2006)
(**84**) Phelypaeside	Cp	Deyama *et al.* (1995a)
(**85**) Kankanoside L	Ct-C	Morikawa *et al.* (2010a)
(**86**) Kankanoside M	Ct-C	Morikawa *et al.* (2010a)
(**87**) Kankanoside N	Ct-C	Morikawa *et al.* (2010a)

Table 14.5 Structures of iridoid glycosides (**88–102**).

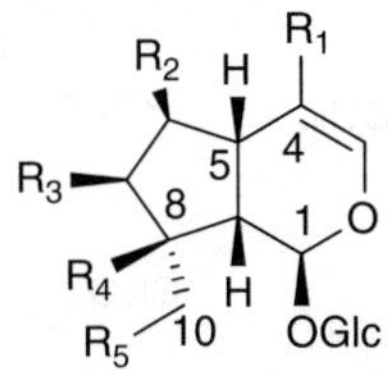

Compound name	R¹	R²	R³	R⁴	R⁵	Species	References
(**88**) Bartsioside	H	H	═══		OH	Cd, Ct-C	Xie *et al.* (2006) and Kobayashi *et al.* (1985b)
(**89**) 6-Deoxycatalpol	H	H	—O—		OH	Cd, Cp, Ct-C, Ct-P, Csa	Deyama *et al.* (1995a), Yoshizawa *et al.* (1990), Kobayashi *et al.* (1985b), and Tu *et al.* (1994b, 1997a)
(**90**) 8-Epideoxyloganic acid	COOH	H	H	H	H	Cd, Ct-C, Csi	Liu *et al.* (2013b), Morikawa *et al.* (2010a), and Kobayashi *et al.* (1985b)
(**91**) 8-Epiloganic acid	COOH	H	OH	H	H	Cd, Ct-C, Ct-P, Csa, Csi	Liu *et al.* (2013b), Yoshizawa *et al.* (1990), Du *et al.* (1993b), Tu *et al.* (1994b, 1997a), Kobayashi and Komatsu (1983), Luo *et al.* (1986), Song *et al.* (1999), and Yang *et al.* (2009a)
(**92**) Geniposidic acid	COOH	H	═══		OH	Cd, Ct-C	Xie *et al.* (2006), Morikawa *et al.* (2010a), Kobayashi *et al.* (1985b), and Song *et al.* (1999)
(**93**) Gluroside	H	H	H	OH	H	Cd, Cp, Ct-C	Deyama *et al.* (1995a), Xie *et al.* (2006), Morikawa *et al.* (2010a), and Kobayashi *et al.* (1985b)
(**94**) Leonuride (ajugol)	H	OH	H	OH	H	Cd, Cp, Ct-C, Csi, Csa	Deyama *et al.* (1995a), Liu *et al.* (2013b), Xie *et al.* (2006), Kobayashi *et al.* (1985b), and Yang *et al.* (2009a)
(**95**) Mussaenosidic acid	COOH	H	H	OH	H	Cd, Ct-C	Xie *et al.* (2006), Morikawa *et al.* (2010a), Kobayashi *et al.* (1985b), and Song *et al.* (1999)
(**96**) Adoxosidic acid	COOH	H	H	H	OH	Ct-C	Song *et al.* (1999)
(**97**) Kankanoside A	Me	H	H	OH	H	Cd, Ct-C	Nan *et al.* (2013a), Xie *et al.* (2006), and Morikawa *et al.* (2010a)
(**98**) Geniposide	COOMe	H	═══		OH	Cd, Csi	Liu *et al.* (2013b), Kobayashi *et al.* (1984a), and Tu *et al.* (2007)
(**99**) 8-Epiloganin	COOMe	H	OH	H	H	Csi	Liu *et al.* (2013b) and Tu *et al.* (2007)
(**100**) Mussaenoside	COOMe	H	H	OH	H	Csi	Liu *et al.* (2013b)
(**101**) Antirrhide	H	H	OH	═══		Ct-C	Xie *et al.* (2006)
(**102**) Catapol	H	OH	—O—		OH	Cd	Xu *et al.* (1999)

14.2.5 Lignans

To date, 2 lignan aglycones and 14 lignan glycosides have been isolated from the *Cistanche* plants (Table 14.7 and Fig. 14.1). Most of them are ditetrahydrofuran lignans.

Table 14.6 Structures of monoterpenoids (**103–109**).

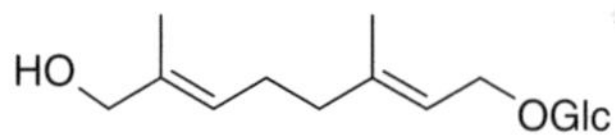

103

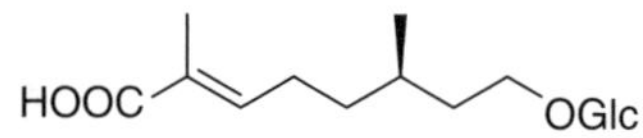

104

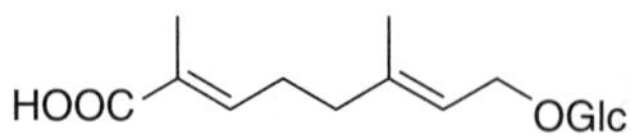

105

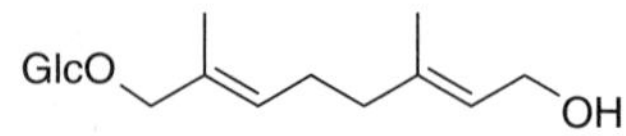

106

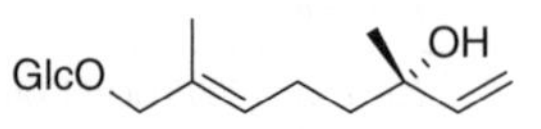

107

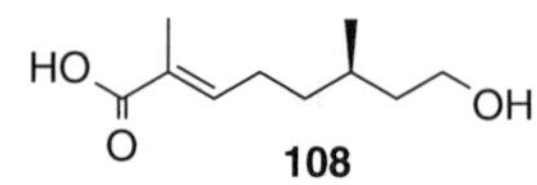

108

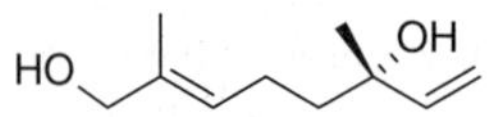

109

Compound name	Species	References
(**103**) 8-Hydroxygeraniol 1-β-D-glucopyranoside	Cd, Ct-P, Csa	Kobayashi and Komatsu (1983), Yang *et al.* (2008), and Yoshizawa *et al.* (1990)
(**104**) Kankanoside E	Ct-C	Morikawa *et al.* (2010a) and Xie *et al.* (2006)
(**105**) (2*E*,6*Z*)-8-β-D-Glucopyranosyloxy-2, 6-dimethyl-2,6-octadienoic acid	Ct-C	Morikawa *et al.* (2010a) and Xie *et al.* (2006)
(**106**) 8-Hydroxygeraniol 8-*O*-β-D-glucopyranoside	Ct-C, Csa	Morikawa *et al.* (2010a) and Yang *et al.* (2008)
(**107**) Betulalbuside A	Ct-C	Morikawa *et al.* (2010a)
(**108**) (2*E*,6*R*)-8-Hydroxy-2,6-dimethyl-2-octenoic acid	Csa	Yamaguchi *et al.* (1999)
(**109**) 8-Hydroxygeraniol	Ct-C	Song *et al.* (1999)

14.2.6 Polysaccharides

In the 1990s, the research works on polysaccharides mainly focused on the purification and characterization of the monosaccharide composition (Xue & Zhang, 1994; Naran *et al.*, 1995; Xue *et al.*, 1995; Kosikova *et al.*, 1999). Later, with the advancement of the modern isolation and spectroscopic techniques, the polysaccharides and their derivatives could be clarified for their complex structures. To date, there have been 13 polysaccharides identified from *C. deserticola* (Table 14.8).

14.2.7 Other types of compounds

There are some other compounds, such as phenolic glycosides, sterols, flavonoids, and alkaloids contained in the *Cistanche* plants (Table 14.9 and Fig. 14.2).

14.3 Bioactivities of the extracts and pure compounds from *Cistanche* species

Currently, there are three kinds of *Cistanche* plants for medicinal application, namely *C. tubulosa*, *C. deserticola*, and *C. salsa*. The primary bioactive ingredients of *Cistanche* are PhGs, which show remarkable antioxidation, antisenile dementia, anti-Parkinson's

Table 14.7 Structures of ditetrahydrofuran lignans from *Cistanche* species (**110–117**).

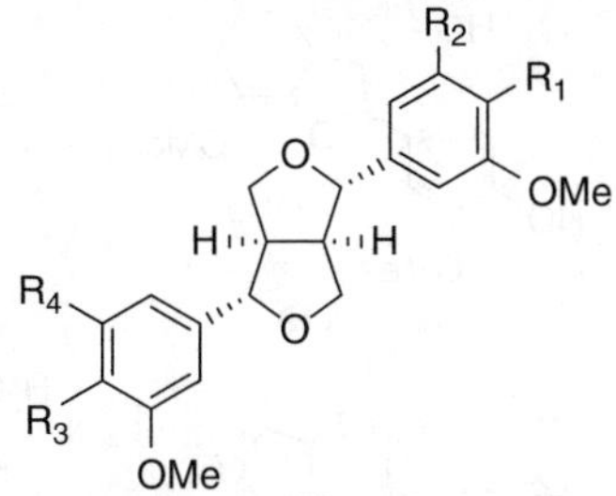

Compound name	R¹	R²	R³	R⁴	Species	References
(**110**) (+)-Pinoresinol	OH	H	OH	H	Cd	Karasawa *et al.* (1986b)
(**111**) (+)-Syringaresinol	OH	OMe	OH	OMe	Cd	Nan *et al.* (2013a)
(**112**) (+)-Pinoresinol-*O*-β-D-glucopyranoside	OH	H	OGlc	H	Ct-P	Yoshizawa *et al.* (1990)
(**113**) (+)-Syringaresinol-*O*-β-D-glucopyranoside	OH	OMe	OGlc	OMe	Cd, Ct-P, Ct-C	Yoshizawa *et al.* (1990), Nan *et al.* (2013a), Kobayashi *et al.* (1985a), and Song *et al.* (1999)
(**114**) Liriodendrin	OGlc	OMe	OGlc	OMe	Cd, Ct-P	Yoshizawa *et al.* (1990), Xu *et al.* (1994), Kobayashi *et al.* (1985a), and Tu *et al.* (1994b, 1997a)
(**115**) Eucommin A	OGlc	OMe	OH	H	Cd, Ct-C	Morikawa *et al.* (2010a) and Nan *et al.* (2013a)
(**116**) Isoeucommin A	OH	OMe	OGlc	H	Cd, Ct-C	Morikawa *et al.* (2010a) and Nan *et al.* (2013a)
(**117**) (+)-Pinoresinol monomethylether β-D-glucoside	OMe	H	OGlc	H	Cd	Nan *et al.* (2013a)
(**118**) Dehydrodiconifery alcohol 4-*O*-β-D-glucoside			—		Cd, Ct-P	Yoshizawa *et al.* (1990) and Nan *et al.* (2013a)
(**119**) Dehydrodiconifery alcohol γ′-*O*-β-D-glucoside			—		Cd, Ct-P	Yoshizawa *et al.* (1990) and Nan *et al.* (2013a)
(**120**) Isolariciresinol-9′-*O*-β-D-glucopyranoside			—		Csi	Liu *et al.* (2013b)
(**121**) Lariciresinol 4′-*O*-β-D-glucopyranoside			—		Cd	Nan *et al.* (2013a)
(**122**) Lariciresinol 4-*O*-β-D-glucopyranoside			—		Cd	Nan *et al.* (2013a)
(**123**) Conicaoside			—		Cd	Nan *et al.* (2013a)
(**124**) Citrusin A			—		Cd	Nan *et al.* (2013a)
(**125**) Alaschanioside A			—		Cd	Nan *et al.* (2013a)

disease (PD), immunoregulation, antiaging, hepatoprotection, and antimyocardial ischemia effects. The reports concerning the bioactive effects of *Cistanche* species and potential mechanisms are summarized as follows.

14.3.1 Antioxidation

The extract from *C. deserticola* could significantly decrease the lipid hydroperoxide (LPO) level in humans (Song *et al.*, 2003; He *et al.*, 2009b). Moreover, all three kinds of *Cistanche* plants showed inhibitory effects on LPO production in rabbit blood (Liang *et al.*, 2000). PhGs from *C. salsa* dramatically increased the glutathione peroxidase and superoxide

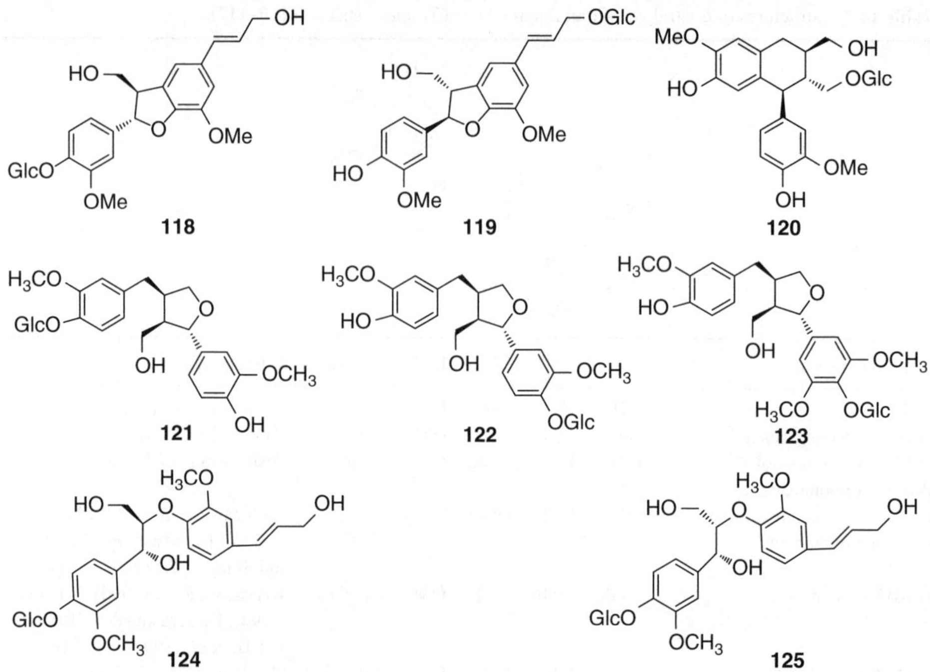

Fig. 14.1 Structures of other lignans (**118–125**).

dismutase (SOD) activities in mouse hearts, livers, and brains, nonetheless decreased the LPO and lipofuscin levels (Luo *et al.*, 2012). Therefore, PhGs were speculated as the primary antioxidation constituents of *Cistanche* species. Nine PhGs from *C. deserticola* showed the free-radical scavenging and antilipid peroxidation effects, and the structure–activity relationship assessments suggested that the number of phenolic hydroxy groups governed the antioxidation activity level (Gulinuer *et al.*, 2001; Liang *et al.*, 2015). Moreover, the methoxyl and acyl groups in PhGs could also contribute to the antioxidation activity.

Tubuloside B, an effective antioxidation compound from *Cistanche* plants, has been reported to significantly increase PC12 cell viability against H_2O_2-induced cytotoxicity and to decrease the apoptotic rate, and it could stabilize mitochondrial membrane protein and block caspase-3 activation, leading to the inhibitory effects on DNA degradation (Deng *et al.*, 2008b). Taken together, Echinacoside (ECH) and tubuloside B may be two prominent constituents contained in *Cistanche* plants being responsible for the antioxidant effects.

14.3.2 Neuroprotection

14.3.2.1 Anti-Parkinson's disease (PD)

ECH treatment significantly increased the levels of monoamine neurotransmitters including dopamine, 3,4-dihydroxyphenylacetic acid (DOPAC), and homovanillic acid (HVA) in 6-hydroxydopamine (6-OHDA) lesion rats (Chen *et al.*, 2007). ECH also protected PC12 cells against 6-OHDA-reduced cell toxicity through increasing the mitochondrial membrane

Table 14.8 The structural properties of polysaccharides from *C. deserticola*.

Type and name	Structural properties	Reference
(**126–128**) Pectic polysaccharides P_1–P_3	Containing homogalacturonan and rhamnogalacturonan RG-I sequences in different proportions, and neutral polysaccharides represented mainly by a highly branched (1 → 3,5)-α-arabinan. Mean Mw: P_1: 1.87×10^5 P_2: 1.25×10^5 P_3: 0.8×10^4	Ebringerova *et al.* (1997)
(**129**) Cistan A	Comprised of a complex of pectic arabino-3,6-galactan type II with slightly branched 3,5-α-L-arabinan residues.	Ebringerova *et al.* (2002)
(**130**) Mannoglucan	$[6)-\alpha-D-Glcp-(1]_n \rightarrow 6)-\beta-D-Manp-(1 \rightarrow$ 3 ↓ 1 $R \rightarrow 6)-\alpha-D-Manp-(1 \rightarrow [6)-\alpha-D-Glcp]_m$ 3 ↓ R $R = Glap(1 \rightarrow 6); Glap(1 \rightarrow X)$ Mean Mw: 6.8×10^4	Wu and Tu (2004a)
(**131**) Arabinogalactan (ACDP-2)	$\rightarrow [4)-\beta-D-Glap-(1]_n \rightarrow 4)-\beta-D-Glap-(1 \rightarrow 1)-\beta-D-Glap(\text{or } Glcp)$ 6 ↓ 1 Ara Mean Mw: 5.6×10^5	Wu *et al.* (2005)
(**132**) Linear glucan (CDP-4)	1,4-Linkage glc*p*: 1,6-linkage glc*p* = 3 : 1 Mean Mw: 1.4 × 104	Wu and Tu (2004b)
(**133–135**) Glucans 1–3	Having a backbone of α-(1 → 6)-glucan Mean Mw: 1: $>2 \times 10^6$ 2: 1.5×10^5 3: 3.3×10^4	Wu and Tu (2005)

(*Continued*)

Table 14.8 (Continued)

Type and name	Structural properties	Reference
(**136**) SPA	$\rightarrow[1)-\mathrm{Glc}-(6]_3\rightarrow 1)-\mathrm{Glc}-(6\rightarrow[1)-\mathrm{Glc}-(6]_7\rightarrow 1)-\mathrm{Glc}-(6\rightarrow[1)-\mathrm{Glc}-(6]_3\rightarrow$ (branches at O-3: 3 ↓ A on the first 1)-Glc-(6; 3 ↓ B on the second 1)-Glc-(6) A: $\rightarrow$[1)-Glc-$(4]_3\rightarrow$ 1)-Rha-(2 $\rightarrow$ 1)-Ara; B: $\rightarrow$1)-Glc-(3 $\rightarrow$ [1)-Man-$(4]_3\rightarrow$ 1)-Ara Mean Mw: 0.76×10^4	Zhao *et al.* (2005)
(**137**) CDA-1A	α-D-Glc*p*-(1 → 6)-α-D-Glc*p*-(1 → 6)-α-D-Glc*p* →4)-α-D-Glc*p*-(1 → [4)-α-D-Glc*p*-$(1]_5$ → 4)-α-D-Glc*p*-(1→ Mean Mw: 1.0×10^4	Dong *et al.* (2007)
(**138**) RG-I polysaccharide (CDA-3B)	Containing a typical rhamnogalacturonan backbone and arabinogalactan or arabinan branches. Mean Mw: 8.7×10^5	Dong *et al.* (2007)

Note: Gla*p*, galactopyranosyl; Glc*p*, glucopyranosyl; and Man*p*, mannopyranosyl.

Table 14.9 Other type compounds from *Cistanche* species.

Compound name	Species	References
(**139**) Syringin	Cd, Cp, Ct-P	Deyama *et al.* (1995a), Yoshizawa *et al.* (1990), Karasawa *et al.* (1986b), Xu *et al.* (1999), and Morikawa *et al.* (2010a)
(**140**) Demethyl syringing	Cd	Nan *et al.* (2013a)
(**141**) Coniferin	Cd, Ct-C	Nan *et al.* (2013a) and Morikawa *et al.* (2010a)
(**142**) Sinapic aldehyde 4-*O*-β-D-glucopyranoside	Ct-C	Morikawa *et al.* (2010a)
(**143**) Cistanoside F	Cd, Ct-C	Xiong *et al.* (1996), Morikawa *et al.* (2010b), Kobayashi *et al.* (1985a), Yoshikawa *et al.* (2006a), Hayashi (2004), Tu *et al.* (2006), and Jiao and Sun (1990)
(**144**) Cistanoside I	Cd	Karasawa *et al.* (1986a)
(**145**) Cistantubulose A1	Ct-C	Tu *et al.* (2006)
(**146**) Cistantubulose A2	Ct-C	Tu *et al.* (2006)
(**147**) Kankanose	Ct-C	Morikawa *et al.* (2010b) and Yoshikawa *et al.* (2006a)
(**148**) Daucosterol	Cd, Csa, Ct-C	Du *et al.* (1993b), Kobayashi *et al.* (1984a), Tu *et al.* (1997a), Luo *et al.* (1986), Yang *et al.* (2008), Qu *et al.* (1988), and Okura (1991)
(**149**) 20-Hydroxyecdysone	Ct-P	Yoshizawa *et al.* (1990)
(**150**) β-Sitosterol	Cd, Cp, Csa, Ct-C	Deyama *et al.* (1995a), Liu *et al.* (1993), Du *et al.* (1993b), Kobayashi *et al.* (1984a), Qu *et al.* (1986, 1988), and Okura (1991)
(**151**) β-Sitosterol glucoside 3-*O*-heptadecoicate	Cd, Csa	Tu *et al.* (1997a) and Yang *et al.* (2008)
(**152**) β-Sitosterol 3-*O*-β-D-xylopyranoside	Cd	Nan *et al.* (2013a)
(**153**) Panaxytriol	Cd	Nan *et al.* (2013a)
(**154**) 1-*O*-Hexadecanoyl-3-*O*-β-D-galactopyranosyl glycerol	Cd	Nan *et al.* (2013a)
(**155**) 2-Nonacosanone	Cd	Okura (1991)
(**156**) Triacontanic acid	Cd	Okura (1991)
(**157**) 1-Triacontanol	Cd	Liu *et al.* (1993) and Okura (1991)
(**158**) bis-2-Ehyl-hexyl-phthalate	Cd	Okura (1991)
(**159**) di-*n*-Butylphthalate	Cd	Nan *et al.* (2013a)
(**160**) Succinic acid	Cd, Csa, Ct-C	Liu *et al.* (1993), Kobayashi *et al.* (1984a), Qu *et al.* (1988), and Xue (1997)
(**161**) γ-Valerolactone	Cd	Tu *et al.* (1997a)
(**162**) Vanillic acid	Csa	Lei *et al.* (2003b)
(**163**) Androsin	Cd	Nan *et al.* (2013a)
(**164**) 3-Methyl-but-2-en-1-yl-β-D-glucopyranoside	Cd	Nan *et al.* (2013a)
(**165**) *n*-Butyl-α-D-fructofuranoside	Cd	Nan *et al.* (2013a)
(**166**) 4-Hydroxybenzeneethanol	Cd	Nan *et al.* (2013a)
(**167**) *p*-Hydroxybenzoic acid	Cd	Nan *et al.* (2013a)
(**168**) Ononin	Cd	Liu *et al.* (2011)
(**169**) D-Glucose	Cd, Csa, Ct-C	Xu *et al.* (1994), Tu *et al.* (1997a), Qu *et al.* (1988), and Xue (1997)
(**170**) D-Fructose	Ct-C	Xue (1997)
(**171**) Galactitol	Cd, Csi, Csa, Ct-C	Liu *et al.*, 2011 and Tu *et al.* (1994b, 1997a, 2007)
(**172**) D-Mannitol	Cd, Csa, Ct-C	Du *et al.* (1993b), Kobayashi *et al.* (1984a), Luo *et al.* (1986), and Qu *et al.* (1986)

(Continued)

Table 14.9 (Continued)

Compound name	Species	References
(**173**) Sucrose	Cd	Xu *et al.* (1994) and Tu *et al.* (1997a)
(**174**) 2*S*,3*S*,4*S*-Trihydroxypentanoic acid	Cd	Nan *et al.* (2013a)
(**175**) Adenosine	Cd	Nan *et al.* (2013a)
(**176**) 2,5-Dioxo-4-imidazolidinyl-carbamic acid	Cd, Csa	Xu *et al.* (1999) and Lei *et al.* (2003b)
(**177**) 2-Methanol-5-hydroxypyridine	Csa	Yang *et al.* (2008)
(**178**) Succinimide	Csa	Lei *et al.* (2003b)
(**179**) Nicotinamide	Cd	Nan *et al.* (2013a)
(**180**) Betaine	Cd, Csa	Liu *et al.* (1993), Yang *et al.* (2008), and Cao *et al.* (2004)
(**181**) *N*,*N*-Dimethyl glycine methyl ester	Cd	Jiao and Sun (1990)
(**182**) Allantoin	Cd	Liu *et al.* (2011)

potential as well as decreasing the mitochondria-mediated apoptosis rate, resulting in a significant neuroprotective effect (Wang *et al.*, 2015). Moreover, ECH exhibited neuroprotection on 1-methyl-4-phenyl-1,2,3,6-tetrahydropyridine (MPTP)-induced PD mice. ECH significantly reduced the MPTP-induced behavioral deficits and increased the striatal dopamine and dopamine metabolite levels (Geng *et al.*, 2004). Furthermore, ECH markedly reduced the caspase-3 activation in TNF-α-induced SHSH5Y cells and MPP^+-induced cerebellar granule neurons (CGNs) (Deng *et al.*, 2004b, 2005; Geng *et al.*, 2007). These findings indicated ECH as a promising agent for the treatment of PD.

Acteoside, another active compound from *Cistanche* species, could inhibit 1-methyl-4-phenylpyridinium ion (MPP^+)-induced SH-SY5Y cell apoptosis, showing anti-PD potential, and the mechanism is related to the protective effect on mitochondria function by increasing the Bcl-2 protein expression, decreasing the ROS production, and inactivating the caspase-3 apoptosis pathway (Deng *et al.*, 2008a). Additionally, acteoside also showed significantly protective effect on MPP^+-induced cerebellar granule neurons (CGNs), which may result from the inhibition of the expressions of the active caspase-3 fragments and the proteolytic poly (ADP-ribose) polymerase (PARP) fragments (Pu *et al.*, 2003). Tubuloside B has also been found to protect neuronal cells from MPP^+ and TNF-α-induced apoptosis. It significantly increased the cell viability, suppressed the intracellular accumulation of reactive oxygen species, and decreased the DNA fragmentation (Sheng *et al.*, 2002; Deng *et al.*, 2004a). In addition, campneoside II is a bioactive compound from *Cistanche* and can upregulate the CGNs viability against MPP^+ toxicity. It could inhibit the lactate dehydrogenase release and reverse DNA fragments formation, indicating an effective neuroprotective effect for PD (Pu *et al.*, 2001).

14.3.2.2 Cognitive improvement

Dementia is one of the major health risks facing elderly individuals. Fortunately, *C. tubulosa* is traditionally used to treat forgetfulness in China. *C. tubulosa* extract has been found to ameliorate the cognitive dysfunction caused by β-amyloid ($A\beta_{1-42}$) via blocking

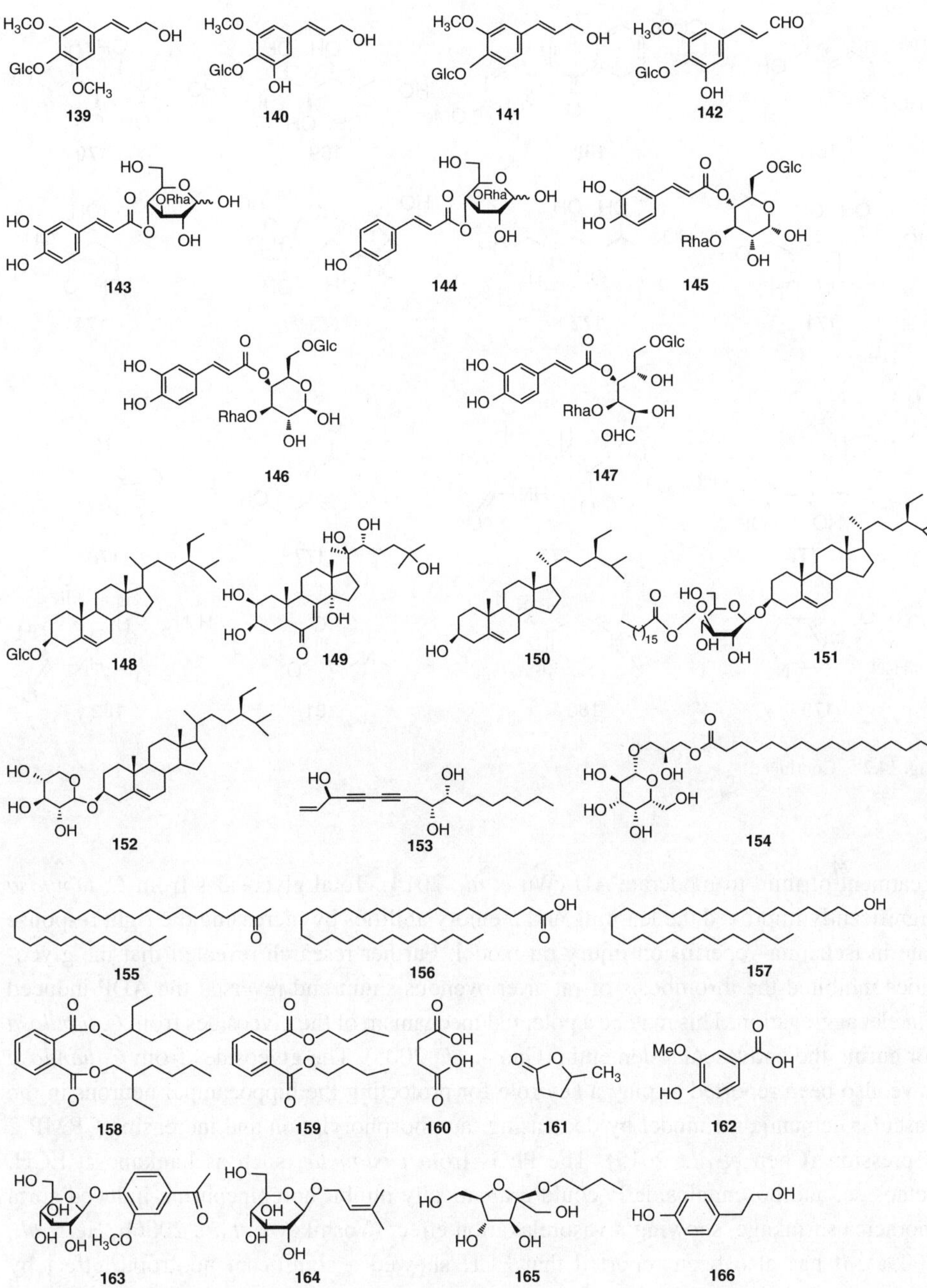

Fig. 14.2 Structures of other type compounds (**139–182**).

amyloid deposition and reversing cholinergic neuronal function (Guo *et al.*, 2013). *C. tubulosa* glycosides capsules treatment showed no obvious aggravation of the cognitive function in Alzheimer's disease (AD) patients, and the patients' conditions were stable throughout the study, suggesting that *C. tubulosa* might be a meaningful option for the

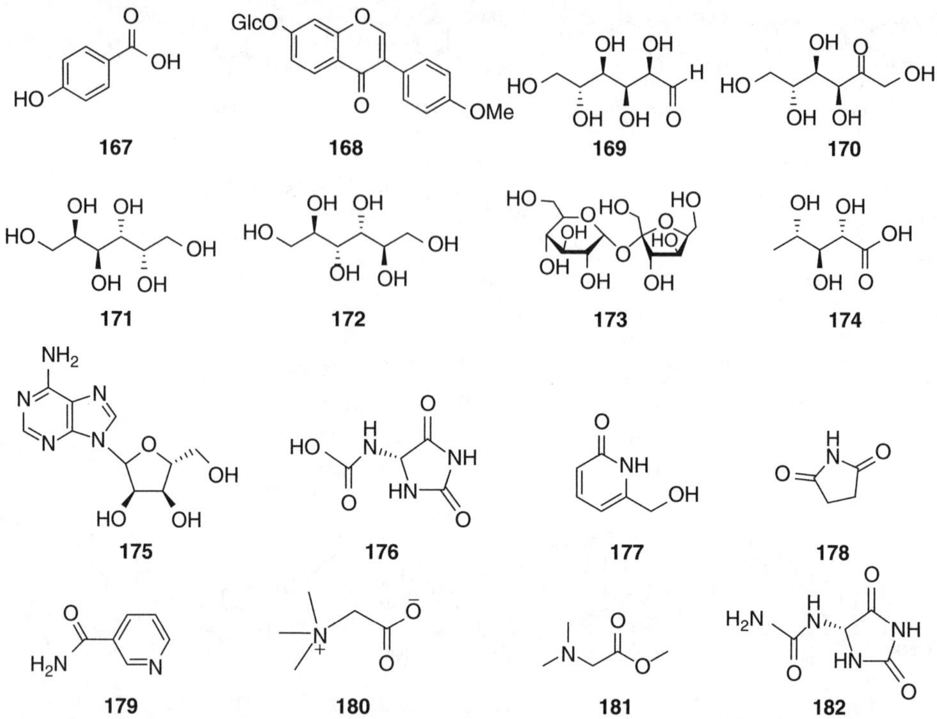

Fig. 14.2 (Continued)

treatment of mild to moderate AD (Wu *et al.*, 2014). Total glycosides from *C. tubulosa* significantly improved the learning and memory abilities by increasing the right response rate in ischemia–reperfusion injury rat model. Further research revealed that the glycosides inhibited the thrombosis of rat arteriovenous shunt and reversed the ADP-induced platelet aggregation. This may be a potential mechanism of the glycosides from *C. tubulosa* for curing the antivascular dementia (Gao *et al.*, 2005). The glycosides from *C. tubulosa* have also been reported to play a key role for protecting the hippocampal neurons in the vascular dementia rat model by decreasing tau phosphorylation and increasing CRMP-2 expression (Chen *et al.*, 2015). The PhGs from *Cistanche*, such as kankanose, ECH, acteoside, and boschnaloside F, could dramatically inhibit norepinephrine-induced aorta thoracica shrinkage, showing a vasorelaxation effect (Yoshikawa *et al.*, 2006b; He *et al.*, 2009a). It has also been reported that ECH showed a significant nootropic effect by increasing central cholinergic system function, facilitating protein and nucleic acid synthesis, and regulating the immune states on SAMS-P8 mice (Liu *et al.*, 2013a). Neurotrophic factors play vital roles in neuronal survival and the development of dementia. *Cistanche* extracts promoted the neuronal cell differentiation, neurite length, and synapse formation in the mouse hippocampus by increasing nerve growth factor (NGF) expression (Choi *et al.*, 2011). Moreover, acteoside also showed memory enhancement properties by significantly increasing the expressions of nerve growth factor and tropomyosin receptor kinase

A mRNA and protein in senescent mouse model induced by chronic intraperitoneal administration of D-gal and oral administration of $AlCl_3$ (Gao *et al.*, 2015b).

14.3.2.3 Sedation

C. deserticola extract decreased the rat autogenic movements including horizontal movement, moving time, and moving distance. In particular, its aqueous extract showed the highest effect. The potential mechanism may be resulted from the decrease of dopamine activity along with the increase of serotonin and γ-aminobutyric acid activities (Lu, 1998).

14.3.3 Vasorelaxation

Vascular endothelial dysfunction is an important risk factor that causes cardiovascular diseases, such as hypertension. Vascular endothelial cells play an important role in respond to various hormone stimuli by releasing endothelium-dependent vasodilators. Decrement of the endothelial NO has been observed in clinical patients and in hypertensive animal models. Therefore, the vasodilator drugs targeting vascular endothelial system should contribute to the relief of vascular pathology-associated diseases. Jin *et al.* explored the effects of PhGs from Cistanches Herba on high-altitude pulmonary hypertension (HAPH) rat models (Jin *et al.*, 2014). The data showed that PhGs treatment markedly decreased the pulmonary vascular thickening and inflammatory infiltration. PhGs also reduced the mean pulmonary artery pressure, right ventricular systolic pressure, and right ventricular hypertrophy index. Therefore, PhGs could be clinically used to improve the hemodynamics and right ventricular hypertrophy in high-altitude pulmonary hypertension (Jin *et al.*, 2014). Moreover, ECH plays a crucial role in vasorelaxant activity on the isolated rat thoracic aortic rings precontracted with phenylephrine and potassium chloride by concentration-dependently inducing an acute relaxation in endothelium-intact rings (30–300 µmol/l). Mechanism study revealed that the relaxation effect of ECH could be significantly inhibited in endothelium-denuded condition and in the presence of the endothelial nitric oxide synthase (eNOS) inhibitor (Gai *et al.*, 2015). Moreover, ECH significantly increased the cyclic guanosine monophosphate (cGMP) production in aortic rings contracted with phenylephrine. Therefore, ECH exerted the vasodilator action in rat thoracic aortic rings by mediating NO-cGMP pathway, indicating that ECH may be a promising candidate in the treatment of vascular diseases (Yoshikawa *et al.*, 2006b; He *et al.*, 2009a).

14.3.4 Antifatigue and longevity promotion

Tu and coworkers investigated the effects of ethanol extract of *C. deserticola* (ECD) on the life span of mice (Zhang *et al.*, 2014). In this report, the senescence-accelerated mouse prone 8 (SAM-P8) was involved as an animal model for aging study. Treatment SAM-P8 mice (8 months old) with ECD (150 and 450 mg/kg) by daily oral administrations for 4 weeks could effectively extend the life span in a dose-dependent manner. Moreover, ECD also showed regulatory effects on immunosenescence. Dietary supplementation of ECD

(100, 500, and 2500 mg/kg) on SAM-P8 mice for 4 weeks caused significant increases in both naive T and natural killer cells in blood and spleen cell populations. These findings revealed that *C. deserticola* could exhibit significant effects in prolonging the life span by antagonizing immunosenescence (Zhang *et al.*, 2014). Another study revealed that polysaccharides of *C. deserticola* could increase hydroxyproline level in mouse skins and delay the process of skin decrepitude (Cai *et al.*, 2010). Furthermore, *C. deserticola* also showed antifatigue effect by prolonging swimming time, decreasing the level of creatine kinase in blood, and improving the ultramicrostructure of skeletal muscle (Han *et al.*, 1993).

14.3.5 Anti-inflammation and immunoregulation

Colitis is a kind of acute or chronic digestive disease with the symptom of abdominal pain, loss of appetite, fatigue, and bloody diarrhea. Colitis is usually induced by autoimmune dysfunction (inflammatory bowel disease, IBD) and infection (infectious colitis). ECH has been reported to show anti-inflammatory activity in many studies. Du *et al.* investigated the effects of ECH-enriched extract of *C. tubulosa* on colitis. They established acute colitis B6 mouse model with dextran sulphate sodium (DSS) and treated the mouse with ECH (20 mg/kg/day) for 7 days. Results showed that the administration of ECH significantly inhibited the development of acute colitis. Pathological assays revealed that ECH suppressed the inflammatory injury on intestinal epithelium, nonetheless, showed mild effect on inflammatory cellular infiltration. Moreover, ECH could effectively increase the expression of transforming growth factor (TGF)-β1 and the number of $Ki67^+$ proliferating cells in colons. This study showed that ECH could exert protective effects on colitis, implying its clinical application values for IBD (Jia *et al.*, 2014).

Immunoregulation plays a crucial role in treating various physiological and pathological processes, including autoimmune diseases, cancer, diabetes, bone remodeling, and stem cell proliferation and differentiation. In the previous study, ACDP-2 that is a polysaccharide purified from the stems of *C. deserticola* showed obvious stimulating effects on the mouse lymphocyte proliferation (Wu *et al.*, 2005a). ACDP-2 treatment could obviously increase the volume and number of lymphocytes, suggesting that ACDP-2 can effectively activate the immune system. This observation was also confirmed by the DNA synthesis analysis using the incorporation of $[^3H]$-thymidine as a detecting probe. Results revealed that ACDP-2 (100 μg/ml) could significantly induce the lymphocyte proliferation to over 30-folds. Therefore, we speculate that ACDP-2 may, at least in part, to be the major immune regulatory ingredient of *Cistanche* (Wu *et al.*, 2005a). The aqueous extracts of *C. tubulosa* and *C. deserticola* significantly increased the phagocytosis of mouse peritoneal macrophages, and the polysaccharides from *C. deserticola* could effectively increase the spleen weight and cellular immune function (Dong *et al.*, 2007). In addition, D-galactosamine (GalN)/LPS treatment can induce severe hepatic injury by promoting immune overactivation, which is an inflammation-mediated hepatocellular injury process. Wen and coworkers reported that ECH provided an obvious protection against GalN/LPS-induced acute liver injury, leading the survival rate of mice increased (Li *et al.*, 2014).

14.3.6 Antitumor

It has been reported that both the ethyl acetate and aqueous extracts from three kinds of *Cistanche* plants (*C. tubulosa*, *C. deserticola*, and *C. salsa*) showed obvious peripheral blood lymphocyte activating effects, leading to the killing of human leukocythemia K562 cells. In particular, the extract of *C. tubulosa* showed the most significant effect (Zhang *et al.*, 1997). In addition, treatment with polysaccharides from *C. deserticola* for 10 days significantly inhibited the growth of mouse Lewis lung cancer and tumor S180 *in vivo*, and the tumor inhibitory rates were 42 and 46%, respectively (Shen *et al.*, 2005). Moreover, acteoside exhibited significant inhibitory effect on the proliferation of prostatic cancer cells (Mulani *et al.*, 2014).

14.3.7 Defecation promotion

C. deserticola and *C. salsa* extracts induced the ileum constriction in guinea pigs, and this effect could be inhibited by atropine, suggesting that the extracts exhibited cholinergic function associated with the defecation effect (Zhang *et al.*, 1993, 2003a, 2003b, 2009). Another research showed that all three kinds of *Cistanche* plants effectively promoted mouse intestine peristalsis, suppressed the intestine moisture absorption and shortened the defecation time (Duan *et al.*, 2012). After systematically pharmacodymatic studies, the galactitol and the oligosaccharides were deduced to be the active constituents for the defecation action of *Cistanche* species (Gao *et al.*, 2015a).

14.3.8 Hepatoprotection

Four PhGs from *C. deserticola* have been reported to show inhibitory effects on lipid peroxidation of NADPH-CCl_4-induced hepatomicrosomes and to suppress the release of aspartate aminotransferase *in vitro*. Studies showed that acteoside, isoacteoside, ECH, and 2′-acetylacteoside from *C. deserticola* exhibited significant protective effects on liver. These compounds inhibited the lipid peroxidation in liver cell microsomes, blocked the aspartate aminotransferase release, and reduced CCl_4 and D-aminogalactose-induced hepatotoxicity (Xiong *et al.*, 1998; Morikawa *et al.*, 2010b). In addition, ECH could effectively decrease the levels of ALT, AST, and MDA in blood, however, increase the SOD activity and downregulate the concentrations of caspase-3 and TNF-α, exhibiting an obvious hepatoprotection effect (Li *et al.*, 2014). Moreover, the polysaccharides from *C. deserticola* could effectively protect the liver by relieving anorexia and avoiding the weight loss on CCl_4-induced hepatic asthenia mouse model (Shimoda *et al.*, 2009; Xiong *et al.*, 2013).

14.3.9 Antimyocardial ischemia

The PhGs of *C. salsa* could increase the activity of superoxide dismutase and glutathione peroxidase in the mouse cardiac muscle after ischemia reperfusion (Siu & Ko, 2010; Wong & Ko, 2013). Moreover, β-sitosterol from *Cistanche* presented the protection activity on myocardial cells by upregulating mitochondrial glutathione redox cycling and inducing the mitochondrial ATP generation (Wong *et al.*, 2014). Hence, *C. salsa* is effective on protecting the myocardial ischemia.

14.3.10 Radiation resistance

Previous research showed that the PhGs of *C. salsa* could induce leukomonocyte enlargement and differentiation to T cells under ^{60}Co exposure, exhibiting a radiation-resistant action (Wu *et al.*, 2003). Moreover, PhGs could obviously increase the activity of superoxide dismutase in red blood cells, recover the content of nucleic acid in spleen, and inhibit the lipid peroxide production in liver. Histological analysis also revealed that *C. salsa* could repair the tissue structure and promote the nucleic acid synthesis in liver (Li *et al.*, 1997; Zhou *et al.*, 2008).

14.3.11 Tissue repairing

C. deserticola extract prevented the bone loss and promoted the bone mineralization, showing a novel effect for the treatment of osteoporosis (Li *et al.*, 2012). *C. deserticola* also promoted the mucosal tissue repaired by inducing the intestinal epithelial cell proliferation and increasing the TGF-β expression (Jia *et al.*, 2012).

14.4 Conclusions

Natural medicinal herbs are entity libraries containing a huge number of active compounds. Nowadays, natural medicines are still contributing to the lead compound discovery and novel drug development. In clinical trials, natural products account for more than 50% drug candidates. Therefore, in-depth exploration of the bioactive characters of natural herbs may be helpful for the precise elucidation of their active material basis and potential pharmacological mechanisms. The chemical and pharmaceutical investigations on *Cistanche* plants have provided enough evidences to support PhGs as the primary active ingredients. Although the pharmacological activities of *Cistanche* have been widely investigated for many years, most of the studies mainly focused on phenotype, such as morphocytology detection *in vitro* and immunohistochemistry/biochemical indicator analysis *in vivo*. In the future, more bioactivity researches should be applied using molecular biology and chemical biology technologies, which may further provide detailed information and solid evidences for clinical application and side effect prediction of *Cistanche*.

References

Cai, R.L., Yang, M.H., Shi, Y., Chen, J., Li, Y.C. and Qi, Y. (2010) Anti fatigue activity of phenylethanoid-rich extract from *Cistanche deserticola*. *Phytotherapy Research*, **24**, 313–315.

Cao, Z.J., Zhao, W.J. and Wu, X.P. (2004) Studies on the chemical constituents of the cultivated *Cistanche deserticola* Y.C. Ma. *Natural Product Research and Development*, **16**, 518–520.

Chen, S.X., He, S.H., He, X.Z. and Han, J.F. (2005) Advances in the chemistry and pharmacology of Cistanchis Herba. *Gansu Animal and Veterinary Sciences*, **34**, 41–44.

Chen, H., Jing, F.C., Li, C.L., Tu, P.F., Zheng, Q.S. and Wang, Z.H. (2007) Echinacoside prevents the striatal extracellular levels of monoamine neurotransmitters from diminution in 6-hydroxydopamine lesion rats. *Journal of Ethnopharmacology*, **114**, 285–289.

Chen, J., Zhou, S.N., Zhang, Y.M., Feng, Y.L. and Wang, S. (2015) Glycosides of *Cistanche* improve learning and memory in the rat model of vascular dementia. *European Review for Medical and Pharmacological Sciences*, **19**, 1234–1240.

Chinese Academy of Medical Sciences Institute of Medicinal Plants (1988) *Chinese Medicinal Herbal*, vol. **4**, 2nd edn, People's Medical Publishing House, Beijing, p. 347.

Chinese Pharmacopoeia Committee Editing (2005) *Chinese Pharmacopoeia (Part I, 2005 Edition)*, Chemical and Industrial Publisher, Beijing, p. 90.

Choi, J.G., Moon, M., Jeong, H.U., Kim, M.C., Kim, S.Y. and Oh, M.S. (2011) Cistanches Herba enhances learning and memory by inducing nerve growth factor. *Behavioural Brain Research*, **216**, 652–658.

Deng, M., Zhao, J.Y., Ju, X.D., Tu, P.F., Jiang, Y. and Li, Z.B. (2004a) Protective effect of tubuloside B on TNF alpha-induced apoptosis in neuronal cells. *Acta Pharmacologica Sinica*, **25**, 1276–1284.

Deng, M., Zhao, J.Y., Tu, P.F., Jiang, Y., Li, Z.B. and Wang, Y.H. (2004b) Echinacoside rescues the SHSY5Y neuronal cells from TNF-α-induced apoptosis. *European Journal of Clinical Pharmacology*, **505**, 11–18.

Deng, M., Zhao, J.Y., Tu, P.F., Jiang, Y. and Chen, J. (2005) Echinacoside rescues the SHSY5Y neuronal cells from TNF-α-induced apoptosis. *Chinese Pharmacological Bulletin*, **21**, 169–174.

Deng, M., Ju, X.D., Fan, D.S., Tu, P.F. and Zhang, J. (2008a) Verbascoside rescues the SHSY-5Y neuronal cells from MPP^+-induced apoptosis. *Chinese Pharmacological Bulletin*, **24**, 1297–1302.

Deng, M., Ju, X.D., Tu, P.F., Fan, D.S., Zhang, J. and Shen, Y. (2008b) Tubuloside B rescues the PC12 neuronal cells from H_2O_2-induced apoptosis. *Chinese Journal of Pathophysiology*, **24**, 1816–1821.

Deyama, T., Yahikozawa, K. and Al-Easa, H.S. (1995a) Constituents of plants growing in Qatar: part XXX. A new iridoid from *Cistanche phelypaea* (L) Cout. *International Journal of Chemistry (Calcutta, India)*, **6**, 107–112.

Deyama, T., Yahikozawa, K. and Al-Easa, H.S. (1995b) Constituents of plants growing in Qatar: part XXXVIII. Constituents of *Cistanche phelypaea*. *Qatar University Science Journal*, **15**, 51–55.

Dong, Q., Yao, J., Fang, J.N. and Ding, K. (2007) Structural characterization and immunological activity of two cold-water extractable polysaccharides from *Cistanche deserticola* Y. C. Ma. *Carbohydrate Research*, **342**, 1343–1349.

Du, N.S., Wang, H. and Yi, Y.H. (1993a) Isolation and identification of phenylethanoid glycosides from *Cistanche deserticola*. *Natural Product Research and Development*, **5**, 5–8.

Du, N.S., Zhou, P.W., Wang, J., Liu, C.X. and Li, W.J. (1993b) Studies on the constituents of *Cistanche tubulosa*. *Journal of China Pharmaceutical University*, **24**, 46–48.

Duan, Z., Jia, M., Zhang, Z.H., Wang, H., Zhou, H. and Yang, L.B. (2012) Variation in laxative action of *Cistanche deserticola* Y. C. Ma processed by different concentration of yellow wine. *Anhui Medical and Pharmaceutical Journal*, **16**, 438–439.

Ebringerova, A., Hromadkova, Z. and Machova, E. (1997) Isolation and characterization of mitogenic pectic polysaccharides from *Cistanche deserticola* Y. C. Ma. *Chemical Papers*, **51**, 289–294.

Ebringerova, A., Hromadkova, Z. and Hribalova, V. (2002) An immunomodulating pectic arabinogalactan from roots of *Cistanche deserticola*. *Chemical Papers*, **56**, 320–325.

Gai, X.Y., Wei, Y.H., Zhang, W. *et al.* (2015) Echinacoside induces rat pulmonary artery vasorelaxation by opening the NO-cGMP-PKG-BKCa channels and reducing intracellular Ca^{2+} levels. *Acta Pharmacologica Sinica*, **36**, 587–596.

Gao, C., Wang, C.Z., Wu, G.Z., Hu, M. and Tu, P.F. (2005) Effect of *Cistanchis* glycosides on learning and memory function of vascular dementia rats. *Chinese Traditional and Herbal Drugs*, **36**, 1852–1855.

Gao, Y.J., Jiang, Y., Dai, F. *et al.* (2015a) Study on the effect of *Cistanche deserticola* Y. C. Ma on moisturizing intestine to relaxing the bowels. *Modern Chinese Medicine*, **17**, 307–314.

Gao, L., Peng, X.M., Huo, S.X., Liu, X.M. and Yan, M. (2015b) Memory enhancement of acteoside (verbascoside) in a senescent mice model induced by a combination of d-gal and $AlCl_3$. *Phytotherapy Research*, **29**, 1131–1136.

Geng, X.C., Song, L.W., Pu, X.P. and Tu, P.F. (2004) Neuroprotective effects of phenylethanoid glycosides from *Cistanche salsa* against 1-methyl-4-phenyl-1,2,3,6-tetrahydropyridine (MPTP)-induced dopaminergic toxicity in C57 mice. *Biological and Pharmaceutical Bulletin*, **27**, 797–801.

Geng, X.C., Tian, X.F., Tu, P.F. and Pu, X.P. (2007) Neuroprotective effects of echinacoside in the mouse MPTP model of Parkinson's disease. *European Journal of Pharmacology*, **564**, 66–74.

Gulinuer, M., Liu, M.J. and Lu, J.F. (2001) Effect of Cistanoside compounds on oxidative stress and immunity. *Journal of Chinese Pharmaceutical Sciences*, **10**, 157–160.

Guo, Q., Zhou, Y., Wang, C.J. *et al.* (2013) An open-label, nonplacebo-controlled study on *Cistanche tubulosa* glycoside capsules (Memoregain(®)) for treating moderate Alzheimer's disease. *American Journal of Alzheimer's Disease and Other Dementias*, **28**, 363–370.

Han, L.C., Hou, J.F. and Ji, Z.H. (1993) Effects of *Cistanche deserticola* Y.C. Ma on serum creatine kinase and ultrastructures of skeletal muscles in mice. *Zhong guo Zhong Yao Za Zhi*, **18**, 743–745, 764.

Hayashi, K. (2004) Studies on the constituents of Cistanchis Herba. *Natural Medicines*, **58**, 307–310.

He, W.J., Fang, T.H., Ma, X., Zhang, K., Ma, Z.Z. and Tu, P.F. (2009a) Echinacoside elicits endothelium-dependent relaxation in rat aortic rings via an NO-cGMP pathway. *Planta Medica*, **75**, 1400–1404.

He, W.J., Fang, T.H. and Tu, P.F. (2009b) Research progress on pharmacological activities of echinacoside. *China Journal of Chinese Materia Medica*, **34**, 476–479.

Jia, Y.M., Guan, Q.N., Guo, Y.H. and Du, C.G. (2012) Echinacoside stimulates cell proliferation and prevents cell apoptosis in intestinal epithelial MODE-K cells by up-regulation of transforming growth factor-β1 expression. *Journal of Pharmacological Sciences*, **118**, 99–108.

Jia, Y.M., Guan, Q.N., Jiang, Y. *et al.* (2014) Amelioration of dextran sulphate sodium induced colitis in mice by echinacoside-enriched extract of *Cistanche tubulosa*. *Phytotherapy Research*, **28**, 110–119.

Jiang, Y. and Tu, P.F. (2009) Analysis of chemical constituents in *Cistanche* species. *Journal of Chromatography A*, **1216**, 1970–1979.

Jiangsu New Medical College (1977) *Dictionary of Traditional Chinese Drugs (upper volume)*, Shanghai Scientific & Technologic Publisher, Shanghai, p. 895.

Jiao, Y. and Sun, Y.J. (1990) Study on the chemical constituents of *Cistanche tubulosa* (Schenk) Wight of Xinjiang. *Chinese Traditional and Herbal Drugs*, **21**, 564–573.

Jin, C.L., Mao, X.M., Li, L.L. *et al.* (2014) Effect of phenylethanoid glycosides from Cistanches Herba hemodynamics on rats with high altitude pulmonary hypertension. *Chinese Journal of Experimental Traditional Medical Formulae*, **20**, 197–200.

Karasawa, H. and Komatsu, J. (2008) Studies on the constituents of Cistanchis Herba. I. *Yakugaku Zasshi*, **103**, 508–511.

Karasawa, H., Kobayashi, H. and Takizawa, N. (1986a) Studies on the constituents of Cistanchis Herba.VII. *Yakugaku Zasshi*, **106**, 562–566.

Karasawa, H., Kobayashi, H. and Takizawa, N. (1986b) Studies on the constituents of Cistanchis Herba. VIII. *Yakugaku Zasshi*, **106**, 721–724.

Kobayashi, H. and Komatsu, J. (1983) Constituents of Cistanchis Herba (1). *Yakugaku Zasshi*, **103**, 508–511.

Kobayashi, H., Karasawa, H. and Miyase, T. (1984a) Studies on the constituents of Cistanchis Herba. II. Isolation and structures of new iridoids, cistanin and cistachlorin. *Chemical & Pharmaceutical Bulletin*, **32**, 1729–1734.

Kobayashi, H., Karasawa, H. and Miyase, T. (1984b) Studies on the constituents of Cistanchis Herba. III. *Chemical & Pharmaceutical Bulletin*, **32**, 3009–3014.

Kobayashi, H., Karasawa, H. and Miyase, T. (1984c) Studies on the constituents of Cistanchis Herba IV. *Chemical & Pharmaceutical Bulletin*, **32**, 3880–3885.

Kobayashi, H., Karasawa, H. and Miyase, T. (1985a) Studies on the constituents of Cistanchis Herba. V. *Chemical & Pharmaceutical Bulletin*, **33**, 1452–1457.

Kobayashi, H., Karasawa, H. and Miyase, T. (1985b) Studies on the constituents of Cistanchis Herba. VI. Isolation and structure of a new iridoid glycoside, 6-deoxycatalpol. *Chemical & Pharmaceutical Bulletin*, **33**, 3645–3650.

Kobayashi, H., Karasawa, H. and Fukushima, S. (1986) Extraction of cistanosides from *Cistanche salsa*. Jpn. Kokai Tokkyo Koho. JP 61189290 A 19860822.

Kobayashi, H., Oguchi, H. and Takizawa, N. (1987) New pheylethanoid glycosides from *Cistanche tubulosa* (Schenk) Hook. f. I. *Chemical & Pharmaceutical Bulletin*, **35**, 3309–3314.

Kosikova, B., Ebringerova, A. and Naran, R. (1999) Characterization of lignin-carbohydrate fractions isolated from the wood parasite *Cistanche deserticola. Holzforschung*, **53**, 33–38.

Lei, L., Song, Z.H. and Tu, P.F. (2003a) Advances in research of chemical constituents in plants of *Cistanche* hoffing. et Link. *Chinese Traditional and Herbal Drugs*, **34**, 473–476.

Lei, L., Song, Z.H., Tu, P.F., Wu, L.J. and Chen, F.K. (2003b) Studies on chemical constituents of *Cistanche salsa. Chinese Traditional and Herbal Drugs*, **34**, 293–294.

Lei, L., Jiang, Y. and Liu, X.M. (2007) New glycosides from *Cistanche salsa. Helvetica Chimica Acta*, **90**, 79–85.

Li, L.L., Wang, X.W., Wang, X.F., Muhu, Y. and Du, N.S. (1997) Anti lipid Peroxidation and anti radiative action of glycosides in Herba Cistanches. *China Journal of Chinese Materia Medica*, **22**, 364–367.

Li, T.M., Huang, H.C., Su, C.M. *et al.* (2012) *Cistanche deserticola* extract increases bone formation in osteoblasts. *Journal of Pharmacy and Pharmacology*, **64**, 897–907.

Li, X.H., Gou, C.Y., Yang, H.S., Qiu, J.P., Gu, T. and Wen, T. (2014) Echinacoside ameliorates D-galactosamine plus lipopolysaccharide-induced acute liver injury in mice via inhibition of apoptosis and inflammation. *Scandinavian Journal of Gastroenterology*, **49**, 993–1000.

Liang, M.H., Sun, P.Y. and Cao, Y. (2000) Effect of glycosides of *Cistanche* on lipid peroxidation in hemorrhagic shock/reperfusion injury. *Journal of Chinese Microcirculation*, **4**, 230–231.

Liang, H.L., Jiang, X.J., Li, B. and Chen, D.L. (2015) Investigation of protective effect of phenethyl alcohol glycosides extracted from Herba Cistanchis on human sperm DNA with oxidative damage by Raman spectroscopy. *Journal of Guangzhou University of Traditional Chinese Medicine*, **32**, 121–125.

Liu, M.H., Liu, F.S. and Xu, J.P. (1993) Studies on the chemical constituents of *Cistanche deserticola* Y.C.Ma. *China Journal of Chinese Materia Medica*, **18**, 424–426.

Liu, X.M., Jiang, Y., Sun, Y.Q., Xu, X.W. and Tu, P.F. (2011) Study on chemical constituents of *Cistanche deserticola. Chinese Pharmaceutical Journal*, **46**, 1053–1058.

Liu, C.L., Chen, H., Jiang, Y. *et al.* (2013a) Effects of echinacoside on extracellular acetylcholine and choline levels of hippocampus and striatum of cerebral ischemia rats. *Yao Xue Xue Bao*, **48**, 790–793.

Liu, X.M., Li, J., Jiang, Y., Zhao, M.B. and Tu, P.F. (2013b) Chemical constituents from *Cistanche sinensis* (Orobanchaceae). *Biochemical Systematics and Ecology*, **47**, 21–24.

Lu, M.C. (1998) Studies on the sedative effect of *Cistanche deserticola. Journal of Ethnopharmacology*, **59**, 161–165.

Luo, S.F., Gu, Y. and Liu, Y.H. (1986) Studies on chemical constituents of *Cistanche salsa. China Journal of Chinese Materia Medica*, **11**, 41–42.

Luo, T.B., Chen, Y.N., Ren, W., Zhon, Y.M. and Zhong, W.Z. (2002) Interference control and chip waveform design in multirate DS-CDMA communication systems. *Arid Zone Research*, **1**, 56–66.

Luo, Q.J., Wang, Y.S. and Huang, K.Q. (2012) Protective effect of *Cistanche deserticola* on skeletal muscle oxidative injury in high intensity training rats. *Journal of Zhan Jiang Normal College*, **33**, 132–135.

Morikawa, T., Pan, Y. and Ninomiya, K. (2010a) Iridoid and acyclic monoterpene glycosides, kankanosides L, M, N, O, and P from *Cistanche tubulosa*. *Chemical & Pharmaceutical Bulletin*, **58**, 1403–1407.

Morikawa, T., Pan, Y., Ninomiya, K. *et al.* (2010b) Acylated phenylethanoid oligoglycosides with hepatoprotective activity from the desert plant *Cistanche tubulosa*. *Bioorganic & Medicinal Chemistry*, **18**, 1882–1890.

Mulani, S.K., Guh, J.H. and Mong, K.K. (2014) A general synthetic strategy and the anti-proliferation properties on prostate cancer cell lines for natural phenylethanoid glycosides. *Organic & Biomolecular Chemistry*, **12**, 2926–2937.

Nan, Z.D., Jiang, Y. and Tu, P.F. (2013a) Studies on the chemical constituents and biological activities of *Cistanche deserticola* cultured in Tarim desert. Doctoral Dissertation of Peking University.

Nan, Z.D., Zeng, K.W., Shi, S.P., Zhao, M.B., Jiang, Y. and Tu, P.F. (2013b) Phenylethanoid glycosides with anti-inflammatory activities from the stems of *Cistanche deserticola* cultured in Tarim desert. *Fitoterapia*, **89**, 167–174.

Naran, R., Ebringerova, A. and Badgaa, D. (1995) Carbohydrate components of the holoparasite *Cistanche deserticola* general characteristics of the underground part. *Chemical Papers*, **49**, 35–38.

Okura, T. (1991) Study on *Cistanche tubulosa* in Inner Mongolia. *Foreign Medical Sciences*, **13**, 52.

Pan, Y., Morikawa, T. and Ninomiya, K. (2010) Bioactive constituents from Chinese natural medicines. XXXVI. Four new acylated phenylethanoid oligoglycosides, kankanosides J1, J2, K1, and K2, from stems of *Cistanche tubulosa*. *Chemical & Pharmaceutical Bulletin*, **58**, 575–578.

Pu, X.P., Li, X.R., Li, H.N., Tu, P.F., Song, Z.H. and Li, C.L. (2001) Campneoside II of *Cistanche tubulosa* (Schenk) R. Wight protects neurons from apoptosis induced by neurotoxin 1-methyl-4-phenylpyridinium (MPP^+). *Journal of Peking University (Health Sciences)*, **33**, 217–220.

Pu, X.P., Song, Z.H., Li, Y.Y., Tu, P.F. and Li, H.N. (2003) Acteoside from *Cistanche salsa* inhibits apoptosis by 1-methyl-4-phenylpyridinium ion in cerebellar granule neurons. *Planta Medica*, **69**, 65–66.

Qu, S.H., Du, N.S., Jin, Y.X. and Lian, Z. (1986) Preliminary study on the chemical constituents of *Cistanche tubulosa* (Schenk) Wight of Xinjiang. *Journal of Xinjiang Medical University*, **9**, 161–162.

Qu, S.H., Du, N.S. and Wu, J.L. (1988) The chemical constituents of *Cistanche tubulosa* (Schenk) Wight of Xinjiang. *Chinese Traditional and Herbal Drugs*, **19**, 24.

Sato, T., Kobayashi, H. and Kojima, A. (1988) Antistress drugs containing cistanosides. Jpn. Kokai Tokkyo Koho. JP 63198627 A 19880817.

Shen, J.H., Yang, L.M., Zhang, L.N., Baoin, K. and Guo, J.F. (2005) Studies on the anti-tumor effect of five kinds of Chinese Herb. *Acta Academiae Medicinae Neimongol*, **27**, 300–302.

Sheng, G.Q., Pu, X.P., Lei, L., Tu, P.F. and Li, C.L. (2002) Tubuloside B from *Cistanche salsa* rescues the PC12 neuronal cells from 1-methyl-4-phenylpyridinium ion-induced apoptosis and oxidative stress. *Planta Medica*, **68**, 966–970.

Shimoda, H., Tanaka, J., Takahara, Y., Takemoto, K., Shan, S.J. and Su, M.H. (2009) The hypocholesterolemic effects of *Cistanche tubulosa* extract, a Chinese traditional crude medicine, in mice. *The American Journal of Chinese Medicine*, **37**, 1125–1138.

Siu, A.H. and Ko, K.M. (2010) Herba Cistanche extract enhances mitochondrial glutathione status and respiration in rat hearts, with possible induction of uncoupling proteins. *Pharmaceutical Biology*, **48**, 512–517.

Song, Z.H., Mo, S.H., Chen, Y., Tu, P.F., Zhao, Y.Y. and Zheng, J.H. (1999) Studies on chemical constituents of *Cistanche tubulosa* (Schenk) R. Wight. *China Journal of Chinese Materia Medica*, **25**, 728–730.

Song, Z.H., Tu, P.F., Zhao, Y.Y. and Zheng, J.H. (2000) Phenylethanoid glycosides from *C. tubulosa*. *Chinese Traditional and Herbal Drugs*, **31**, 808–810.

Song, Z.H., Lei, L. and Tu, P.F. (2003) Advances in research of pharmacological activity in plants of *Cistanche* hoffing. et Link. *Chinese Traditional and Herbal Drugs*, **34**, 113–115.

Tan, D.Y., Guo, Q.S. and Wang, C.L. (2004) Study on the status quo of *Cistanche deserticola* and its exploitation and utilization in China. *Forest Resource Management*, **33**, 29–32.

Tu, P.F., He, Y.P. and Lou, Z.C. (1994a) A survey on the origin and resource protection of *Cistanche*. *Chinese Traditional and Herbal Drugs*, **25**, 205–208.

Tu, P.F., He, Y.P. and Lou, Z.C. (1994b) Chemical constituents of the fresh inflorescences of desert living *Cistanche* (*C. deserticola*). *Chinese Traditional and Herbal Drugs*, **25**, 451–452.

Tu, P.F., He, Y.P. and Lou, Z.C. (1997a) Studies on the chemical constituents of the cultivated desert living *Cistanche* (*C. deserticola*). *Natural Product Research and Development*, **9**, 7–10.

Tu, P.F., Wang, B., Deyama, T., Zhang, Z.G. and Lou, Z.C. (1997b) Analysis of phenylethanoid glycosides of Herba Cistanchis by RP-HPLC. *Acta Pharmaceutica Sinica*, **32**, 294–300.

Tu, P.F., Lei, L. and Song, Z.H. (2002) Phenethyl alcohol glycosides of *Cistanche deserticola*. Faming Zhuanli Shenqing. CN 1379036 A 20021113.

Tu, P.F., Song, Z.H. and Shi, H.M. (2006) Aryethyl glycosides and oligosaccharide from the stem of *Cistanche tubulosa*. *Helvetica Chimica Acta*, **89**, 927–935.

Tu, P.F., Song, Z.H. and Jiang, Y. (2007) Chemical constituents of *Cistanche sinensis*. *Journal of Asian Natural Products Research*, **9**, 79–84.

Wang, T., Zhang, X.Y. and Xie, W.Y. (2012) *Cistanche deserticola* Y. C. Ma, "Desert Ginseng": a review. *The American Journal of Chinese Medicine*, **40**, 1123–1141.

Wang, Y.H., Xuan, Z.H., Tian, S. and Du, G.H. (2015) Echinacoside protects against 6-hydroxydopamine-induced mitochondrial dysfunction and inflammatory responses in PC12 cells via reducing ROS production. *Evidence-based Complementary and Alternative Medicine*, **2015**, 189239.

Wong, H.S. and Ko, K.M. (2013) Herba Cistanches stimulates cellular glutathione redox cycling by reactive oxygen species generated from mitochondrial respiration in H_9C_2 cardiomyocytes. *Pharmaceutical Biology*, **51**, 64–73.

Wong, H.S., Chen, N., Leong, P.K. and Ko, K.M. (2014) β-Sitosterol enhances cellular glutathione redox cycling by reactive oxygen species generated from mitochondrial respiration: protection against oxidant injury in H_9C_2 cells and rat hearts. *Phytotherapy Research*, **28**, 999–1006.

Wu, X.M. and Tu, P.F. (2004a) Chemical properties of a mannoglucan from *Cistanche deserticola*. *Pharmazie*, **59**, 815–816.

Wu, X.M. and Tu, P.F. (2004b) Studies on the chemical structure of polysaccharide CDP-4 isolated from *Cistanche deserticola*. *Journal of Peking University (Health Sciences)*, **36**, 24–26.

Wu, X.M. and Tu, P.F. (2005) Isolation and characterization of α-(1→6)-glucans from *Cistanche deserticola*. *Journal of Asian Natural Products Research*, **7**, 823–828.

Wu, L.Y., Wang, X.W., Asi, Y. and Jiang, X.Y. (2003) Protective effect of glycosides of *Cistanche* on T lymphocyte function of ^{60}Co γ irradiated mice. *Journal of Xinjiang Medical University*, **26**, 558–560.

Wu, X.M., Gao, X.M., Tsim, K.W. and Tu, P.F. (2005) An arabinogalactan isolated from the stems of *Cistanche deserticola* induces the proliferation of cultured lymphocytes. *International Journal of Biological Macromolecules*, **37**, 278–282.

Wu, C.R., Lin, H.C. and Su, M.H. (2014) Reversal by aqueous extracts of *Cistanche tubulosa* from behavioral deficits in Alzheimer's disease-like rat model: relevance for amyloid deposition and central neurotransmitter function. *BMC Complementary and Alternative Medicine*, **26**, 202.

Xie, H.H., Morikawa, T. and Matsuda, H. (2006) Monoterpene constituents from *Cistanche tubulosa*-chemical structures of kankanosides A–E and kankanol. *Chemical & Pharmaceutical Bulletin*, **54**, 669–675.

Xiong, Q.B., Kadota, S., Tani, T. and Namba, T. (1996) Antioxidative effects of phenylethanoids from *Cistanche deserticola*. *Biological & Pharmaceutical Bulletin*, **19**, 1580–1585.

Xiong, Q., Hase, K., Tezuka, Y., Tani, T., Namba, T. and Kadota, S. (1998) Hepatoprotective activity of phenylethanoids from *Cistanche deserticola*. *Planta Medica*, **64**, 120–125.

Xiong, W.T., Gu, L., Wang, C., Sun, H.X. and Liu, X. (2013) Anti-hyperglycemic and hypolipidemic effects of *Cistanche tubulosa* in type 2 diabetic db/db mice. *Journal of Ethnopharmacology*, **150**, 935–945.

Xu, W.H., Qiu, S.X., Zhao, J.H. *et al.* (1994) Studies on the chemical constituents of desert living *Cistanche* (*Cistanche deserticola*). *Chinese Traditional and Herbal Drugs*, **25**, 509–513.

Xu, C.H., Yang, J.S., Lv, R.M. and Yang, S.S. (1999) Studies on the chemical constituents of *Cistanche tubulosa*. *Chinese Traditional and Herbal Drugs*, **30**, 244–246.

Xue, D.J. (1997) Studies on the chemical constituents of *Cistanche tubulosa* (Schenk) Wight. *China Journal of Chinese Materia Medica*, **22**, 170–171.

Xue, D.J. and Zhang, M. (1994) The analysis of polysaccharides from three kinds of *Cistanche deserticola*. *Journal of Chinese Medicinal Materials*, **17**, 36–37.

Xue, D.J., Zhang, M. and Wu, X.H. (1995) Study on anti-aging active ingredients of *Cistanche deserticola*. *China Journal of Chinese Materia Medica*, **20**, 687–689.

Yamaguchi, K., Shinohara, C. and Kojima, S. (1999) (2E,6R)-8-Hydroxy-2,6-dimethyl-2-octenoic acid, a novel anti-osteoporotic monoterpene, isolated from *Cistanche salsa*. *Bioscience, Biotechnology, and Biochemistry*, **63**, 731–735.

Yang, J.H., Du, N.S. and Kasimu, R. (2005) Isolation and elucidation of phenylethanoid glycosides from cultivated *Cistanche salsa*. *Journal of Chinese Pharmaceutical Sciences*, **14**, 242–245.

Yang, J.H., Hu, J.P. and Rena, K. (2008) Studies on chemical constituents of cultivated *Cistanche salsa*. *Journal of Chinese Medicinal Materials*, **31**, 1663–1665.

Yang, J.H., Hu, J.P. and Rena, K. (2009a) Studies on the iridoid glycosides of cultivated *Cistanche salsa*. *Lishizhen Medicine and Materia Medica Research*, **20**, 522–523.

Yang, J.H., Hu, J.P., Rena, K. and Du, N.S. (2009b) Structure-activity relationships of phenylethanoid glycosides in plants of *Cistanche salsa* on antioxidative activity. *Journal of Chinese Medicinal Material*, **32**, 1067–1069.

Yoshikawa, M., Matsuda, H. and Morikawa, T. (2006a) Phenylethanoid oligoglycosides and acylated oligosugars with vasorelaxant activity from *Cistanche tubulosa*. *Bioorganic & Medicinal Chemistry*, **14**, 7468–7475.

Yoshikawa, M., Matsuda, H., Morikawa, T., Xie, H., Nakamura, S. and Muraoka, O. (2006b) Phenylethanoid oligoglycosides and acylated oligosugars with vasorelaxant activity from *Cistanche tubulosa*. *Bioorganic & Medicinal Chemistry Letters*, **14**, 7468–7475.

Yoshizawa, F., Deyama, T., Takizawa, N., Usmanghani, K. and Ahmad, M. (1990) The constituents of *Cistanche tubulosa* (Schenk) Hook. f. II. *Chemical & Pharmaceutical Bulletin*, **38**, 1927–1930.

Zhang, Y., Wu, H., Wang, S.N., Zheng, H.C. and Xu, S.K. (1993) Study on laxative effects of Herba Cistanches and its processing products. *Chinese Traditional Patent Medicine*, **15**, 20–21.

Zhang, H.Q., Xu, A.H. and Li, D.D. (1997) Pharmacological action of *Cistanches deserticola* polysaccharides on tumor bearing mice. *Chinese Journal of Integrative Medicine*, **17**, 155–156.

Zhang, B.S., Chen, S.H., Zhao, X.W. and Liu, S.H. (2003a) Study on the dosage-effect relationship of laxative action of galactitol from *Cistanche deserticola*. *Chinese Journal of Information on Traditional Chinese Medicine*, **10**, 28–29.

Zhang, B.S., Zhao, X.W., Chen, S.H., Liu, R.H. and Cheng, H.P. (2003b) Laxative action of separated parts from Herba Cistanches. *Chinese Journal of Information on Traditional Chinese Medicine*, **10**, 31–32.

Zhang, T., Wei, T., Gao, Z.L., Zhang, J. and Qin, F. (2009) Study on comparing the Runchang purge and anti-fatigue effects of *Cistanche deserticola* and the tissue culture of *Cistanche deserticola*. *Science and Technology of Food Industry*, **30**, 155–156.

Zhang, K., Ma, X., He, W.J. *et al.* (2014) Extracts of *Cistanche deserticola* can antagonize immunosenescence and extend life span in senescence-accelerated mouse prone 8 (SAM-P8) mice. *Evidence-Based Complementary and Alternative Medicine*, **2014**, 1–14.

Zhao, W., Yan, H., Liang, Z.Y., Zhang, Y.J. and Jiao, X.Q. (2005) Structural analysis of water-soluble polysaccharide SPA isolated from the stem of the *Cistanche deserticola* Ma. *Chemical Journal of Chinese Universities*, **26**, 461–463.

Zhou, Y.C., Yi, C.Z., Hu, Y.X., Hu, Y.M., Chen, W.L. and Deng, B.B. (2008) Experimental study on the antiradiation and antifatigue effects of soft capsule made of *Cistanche* and acanthopanax senticosus and jujube. *China Tropical Medicine*, **8**, 35–37.

Index

Page numbers in *italics* refer to illustrations; those in **bold** refer to tables

Recent Advances in Polyphenol Research, First Edition. Edited by Kumi Yoshida, Véronique Cheynier and Stéphane Quideau.